Ecological Studies

Analysis and Synthesis

Edited by

W. D. Billings, Durham (USA)   F. Golley, Athens (USA)
O. L. Lange, Würzburg (FRG)   J. S. Olson, Oak Ridge (USA)

Volume 19

Water and Plant Life

Problems and Modern Approaches

Edited by
O. L. Lange   L. Kappen   E.-D. Schulze

With 178 Figures

Springer-Verlag   Berlin Heidelberg New York   1976

The picture on the cover is a modified version of Fig. 5 on p. 81.

ISBN 3-540-07838-X Springer-Verlag Berlin · Heidelberg · New York
ISBN 0-387-07838-X Springer-Verlag New York · Heidelberg · Berlin

Library of Congress Cataloging in Publication Data. Main entry under title: Water and plant life. (Ecological studies; v. 19) Includes bibliographies and index. 1. Plant-water relationships. I. Lange, Otto Ludwig. II. Kappen, L., 1935–. III. Schulze, Ernst-Detlef, 1941–. IV. Series. QK870.W37. 581.1'9212. 76-42191.

This work is subject to copyright. All rights are reserved, whether the whole or part of the material is concerned, specifically those of translation, reprinting, re-use of illustrations, broadcasting, reproducing by photocopying machine of similar means, and storage in data banks.
Under § 54 of the German Copyright Law where copies are made for other than private use, a fee is payable to the publisher, the amount of the fee to be determined by agreement with the publisher.

© by Springer-Verlag Berlin · Heidelberg 1976.
Printed in Germany.

The use of registered names, trademarks, etc. in this publication does not imply, even in the absence of a specific statement, that such names are exempt from the relevant protective laws and regulations and therefore free for general use.

Typesetting and printing: Zechnersche Buchdruckerei, Speyer.
Bookbinding: Brühlsche Universitätsdruckerei, Gießen.

Introduction

Water is essential for life and without water no life exists. The liquid surrounding of an aqueous solution is the conditio sine qua non for most of the physiological responses and as such, water is as decisive for the occurrence of a single enzymatic reaction as it is for the global zonation of world vegetation. It is no wonder that scientists since early times have made every effort to describe and understand the functional interrelationships between water and the phenomenon of life. During the past half of this century, these endeavours in the field of botany have been marked by steps which might be symbolized by a series of books such as "The plant in relation to water" (J. Maximov, 1929), and "Die Hydratur der Pflanze in ihrer physiologisch-ökologischen Bedeutung" (H. Walter, 1931), then "Pflanze und Wasser" (Vol. III of the Encyclopedia of Plant Physiology, edited by O. Stocker, 1956), and "Plant-water relations" (R. O. Slatyer, 1967), or the treatment of "Displacement of water and its control of biochemical reactions" (S. Levin, 1974).

Recently, general interest in the different aspects of water relations in plants has increased greatly all over the world. It is realistic to assume that now at least three new publications in this field are appearing every day. There are two reasons for this currently very high scientific activity. On the one hand our growing biochemical and biophysical knowledge and capability sets a scientific challenge to gain closer insights into the fundamental processes of plant-water interrelationships. On the other hand, mankind has more and more to face and to solve practical problems which are connected with water physiology and water ecology of plants. We are forced to enlarge the food production of a world which to a great extent is limited by lack of water. In addition it has become obvious that man's obligation is to manage the water resources within the ecosystems of the globe. This is no longer simply a matter of having "nice clean" water but a requirement for continuing human existence. All this has also led to the increasing interest in problems concerning plant existence in relation to water in the subdisciplines of applied botany, such as agronomy, forestry, and land management.

It is nearly impossible to cover the immense bulk of information which has been presently accumulated in these fields of biology. The task is rendered more difficult because experimental scientific work today is essentially the work of specialists. The separation of their limited areas of investigation handicaps the understanding of the function of whole plants and systems. However, a synthesis is necessary, because the ultimate aim of botany must remain the explanation of plant life in all its complexity.

It is perhaps ecology which increasingly stimulates the cooperation and integration of the different disciplines in biology. In order to explain the existence and behavior of the different plant types and plant communities in their environment, the various aspects of plant structure and plant function must be analyzed. This knowledge needs to be integrated in order to draw ecological conclusions, therefore, analysis as well as synthesis characterizes the task of ecological biology. We try to accomplish this in the present volume by presenting, from an ecological point of view, an analytical as well as a synoptic survey of the large field of "Water and plant life". We have not tried to summarize all known facts in this book. However, we will describe the level that our knowledge has reached, point out the problems which have recently arisen, and provide recommendations for future research. We hope that the obvious gaps in our understanding will stimulate and facilitate further work.

In September 1974 two scientific conferences were held at Würzburg where biophysicists, biochemists, physiologists, ecologists and geobotanists from all over the world came together to discuss amongst other subjects the problems of plant-water relations. Within its first meeting, the International Association of Plant Physiologists organized a symposium on Plant Productivity and Water Relations. At the same time the Deutsche Botanische Gesellschaft devoted a section of its biennial congress to the subject Pflanze und Wasser. We took the opportunity of these meetings to invite scientists representing the different fields of water-relations research to collaborate in producing this book. The authors were asked to integrate their results into a broad discussion covering the entire field of their chapters. In this way, the reader should obtain the most comprehensive picture possible of actual problems in plant-water research.

The scope of the volume, consequently, is the importance of water for plant life throughout the different levels of plant existence, from the molecular level of the cytoplasm to cells and tissues, and from the entire plant organism to ecosystems and vegetation zones. The text is divided into seven parts. Short prefaces for each part provide a brief summary of the chapters included and attempt to show the interrelationships between them as well as links between the different parts. In the first Part of the book the fundamentals of water relations are discussed, and special attention is paid to the definition and description of water status within the plant tissues and within the soil-plant-atmosphere continuum. Parts 2 and 3 consider the processes of water uptake and water loss and their regulation. In Part 4 the interrelationships between water stress and the metabolic activity of the plants are discussed especially in respect to ultrastructural, biochemical and hormonal aspects. The subjects of Part 5 are the different types of the photosynthetic CO_2 pathways which are known to be adaptive biochemical responses of plants to water supply and temperature under different environmental conditions. Part 6 is devoted to water status as a determinant in the productivity of plants and plant communities including aspects of artificial irrigation of plant cultures. The importance of the water factor for plant distribution and vegetation pattern is demonstrated exemplarily in Part 7 of this volume.

The editors are indebted to their author colleagues for their understanding cooperation in preparing the chapters of this book. Thanks are due also to Mr. M. Englert for his valuable assistance in preparing the English edition of many of the manuscripts and Mrs. I. Urlaub for her careful help during the editorial work. Last, but not least, we gratefully acknowledge the agreement of Dr. K. F. Springer to publish this book as a volume of Ecological Studies. His continuing support has been invaluable. Within this series specially selected aspects of water relations of plants have already been considered. The physics of soil and salts in ecosystems is treated in Vol. 4 (ed. Hadas et al.). The special ecological and physiological problems of plants in saline environments are discussed in Vol. 15 (ed. Poljakoff-Mayber and Gale). Volume 5 (ed. Yaron et al.) deals with arid-zone irrigation and its physiological implications. Methods of studying plant-water relations are critically reviewed by Slavík in Vol. 9 of Ecological Studies. Certain aspects of these foregoing volumes have been summarized and included in the present book, which aims to outline an overall view of plant-water relations. This book is primarily intended to serve as a source of information for scientists such as biologists and agronomists. However, we hope that it will be found suitable also for use as a textbook for advanced students in biology, and will thus stimulate young people to discuss and consider the various ecological aspects of water in biological systems.

Würzburg/Bayreuth, November 1976

O. L. LANGE
L. KAPPEN
E.-D. SCHULZE

Contents

Part 1 **Fundamentals of Plant Water Relations** 1
        Preface 2

A. The Structure of Water in the Biological Cell
      G. Peschel 6
  I.   Introduction 6
  II. Evidence for Structured Aqueous Boundary Layers 7
  III. Thermal Anomalies in Biological Tissues 10
  IV. Properties of Aqueous Electrolyte Layers 13
  V. Conclusions 16
      References 16

B. The States of Water in the Plant—Theoretical Consideration
      J. J. Oertli 19
  I.   Introduction 19
  II. Physiological Importance of Processes and Properties
      Involving Water 20
  III. Metabolism and Water Relations 28
  IV. Conclusions 30
      References 31

C. The Soil–Plant–Atmosphere Continuum
      J. J. Oertli 32
  I.   Introduction 32
  II. Description of the Turgor Pressure as a Function of
      Environmental Variables 33
  III. Water Flow in the SPAC as a Link Between Plant and
      Environment 34
  IV. The Solute-free Transport System 36
  V. Effects of Solutes in the SPAC 37
  VI. Changes in Resistances or Potential Differences 39
  VII. Conclusions 40
      References 40

D. The Water Status in the Plant—Experimental Evidence
      H. Richter 42
  I.   Introduction 42

	II.	Current Methods for the Determination of Total Water Potential and Its Components	43
	III.	The Range of Water Potentials Hitherto Determined and the Continuum Conditions Favoring Extreme Values	46
	IV.	The Component Potentials Adjusting Total Water Potential in the Plant Body: Ranges and Changes	51
	V.	Why does Water Potential in a Plant Change?	53
	VI.	Conclusions	54
		References	55

Part 2 Water Uptake and Soil Water Relations 59
Preface 60

A. Root Extension and Water Absorption
M. M. CALDWELL 63

I.	Introduction	63
II.	Water Movement Through the Soil–Plant–Atmosphere Continuum: Limitations in the Liquid Phase	63
III.	Root Extension and Facilitation of Water Uptake in Unexplored Soil Regions	64
IV.	Root Extension Within the Rooted Zone: A Case for Avoidance of Localized Rhizospheric Resistances	67
V.	Conclusions	82
	References	84

B. Resistance to Water Flow in the Roots of Cereals
E. L. GREACEN, P. PONSANA, and K. P. BARLEY 86

I.	Introduction	86
II.	Anatomy of Cereal Roots	86
III.	Zone of Water Absorption	87
IV.	Forces Causing Flow of Water	89
V.	Resistance to Flow	91
VI.	Effect of Root Resistance on Withdrawal of Water from the Soil	95
VII.	Conclusions	98
	References	99

C. Soil Water Relations and Water Exchange of Forest Ecosystems
P. BENECKE 101

I.	Introduction	101
II.	Water Balance	104
III.	Fundamental Equations and Principles	120
IV.	Simulation of Evapotranspiration and Percolation	123
V.	Conclusions	128
	References	129

Contents XI

Part 3 Transpiration and Its Regulation 133
Preface 134

A. **Energy Exchange and Transpiration**
   D. M. GATES 137
   I. Introduction 137
   II. Gas Diffusion 137
   III. Energy Balance 138
   IV. Transpiration 139
   V. Wind Speed Influence 139
   VI. Leaf Temperature Affected by Transpiration 141
   VII. Conclusions 147
   References 147

B. **Water Permeability of Cuticular Membranes**
   J. SCHÖNHERR 148
   I. Introduction 148
   II. Cuticular Transpiration—Early Observations and Hypotheses 149
   III. The Concept of the Polar Pathway Through Lipid Membranes 150
   IV. Conclusions 157
   References 158

C. **Physiological Basis of Stomatal Response**
   J. LEVITT 160
   I. Introduction 160
   II. Biochemical Processes Leading to Movement 161
   III. Conclusions: Ability of the Mechanism to Explain the Known Facts 167
   References 167

D. **Current Perspectives of Steady-state Stomatal Responses to Environment**
   A. E. HALL, E.-D. SCHULZE, and O. L. LANGE 169
   I. Introduction 169
   II. Measurement of Stomatal Responses to Environment . . . 170
   III. Steady-state Stomatal Responses to Environment 171
   IV. Stomatal Responses to Diurnal Changes in Environment . . 183
   V. Conclusions and Future Research Directions 184
   References 185

E. **Water Uptake, Storage and Transpiration by Conifers: A Physiological Model**
   R. H. WARING and S. W. RUNNING 189
   I. Introduction 189
   II. Description of the Model 189
   III. Applications 198
   IV. Conclusions 200
   References 200

Part 4	Direct and Indirect Water Stress	203
	Preface	204

A. Water Stress, Ultrastructure and Enzymatic Activity
J. Vieira da Silva 207
- I. Introduction 207
- II. Effects of Water Stress on Hydrolytic Enzymatic Activity . . 208
- III. Effects of Water Stress on the Ultrastructure of the Cell . . 212
- IV. Relationships of Ultrastructural Alteration and Hydrolytic Enzyme Decompartmentation and Activation, with Alteration of Chloroplasts and Mitochondria Metabolism 217
- V. Conclusions 220
- References 221

B. Water Stress and Hormonal Response
C. Itai and A. Benzioni 225
- I. Introduction 225
- II. Endogenous Hormonal Changes Due to Water Stress . . . 226
- III. The Physiological Significance of Hormonal Effects 232
- IV. A Hypothetical Model for the Role of Hormones in Plant Adaptation to Water Stress 238
- V. Conclusions 238
- References 240

C. Carbon and Nitrogen Metabolism Under Water Stress
M. Kluge 243
- I. Introduction 243
- II. Carbon Metabolism Under Water Stress 244
- III. Nitrogen Metabolism Under Water Stress 247
- IV. Biochemical Aspects of Desiccation Resistance 249
- V. Conclusions 250
- References 250

D. Water Stress During Freezing
U. Heber and K. A. Santarius 253
- I. Introduction 253
- II. Frost Injury 253
- III. Frost Resistance 257
- IV. Conclusions 265
- References 266

E. Cell Permeability and Water Stress
O. Y. Lee-Stadelmann and E. J. Stadelmann 268
- I. Introduction 268
- II. Principles of Cell Permeability 269
- III. Quantitative Determination of Permeability 270
- IV. Alterations of Cell Permeability by the Plant Water Deficit . 272

V.	Possible Mechanisms for Changes in Cell Permeability by Plant Water Stress	277
VI.	Conclusions	278
	References	279

F. Water Stress and Dynamics of Growth and Yield of Crop Plants
T. C. Hsiao, E. Fereres, E. Acevedo, and D. W. Henderson 281

I.	Introduction	281
II.	Overview of Growth and Yield as Affected by Water	282
III.	Some Behavior Observed in the Field	289
IV.	Concluding Remarks	301
	References	303

Part 5 Water Relations and CO_2 Fixation Types ... 307
Preface ... 308

A. Crassulacean Acid Metabolism (CAM): CO_2 and Water Economy
M. Kluge ... 313

I.	Introduction	313
II.	Carbon Metabolism of CAM Plants	313
III.	Gas Exchange of CAM Plants	316
IV.	Ecological Aspects of CAM	318
V.	Conclusions	320
	References	320

B. Balance Between C_3 and CAM Pathway of Photosynthesis
K. Winter and U. Lüttge ... 323

I.	Introduction	323
II.	Adaptation to Salinity	324
III.	Environmental Control of Photosynthetic Pathways	325
IV.	Regulation of the Balance between C_3 and CAM	328
V.	Ecological Aspects	332
	References	332

C. C_4 Pathway and Regulation of the Balance Between C_4 and C_3 Metabolism
W. Huber and N. Sankhla ... 335

I.	Introduction	335
II.	Carbon Metabolism of C_4 Plants	335
III.	General Characteristics of C_4 Plants	341
IV.	Factors Affecting Shift	346
V.	Natural C_3–C_4 Intermediates	352
VI.	Ecological Implications	354
VII.	Conclusions	356
	References	357

D. Ecophysiology of C$_4$ Grasses
M. M. Ludlow 364
I. Introduction 364
II. Environmental Conditions 365
III. Physiological Responses to Environmental Conditions . . . 366
IV. Ecological Implications 377
V. Conclusions: Future Research 380
References 380

Part 6 Water Relations and Productivity 387
Preface 388

A. The Use of Correlation Models to Predict Primary Productivity from Precipitation or Evapotranspiration
H. Lieth 392
I. Introduction 392
II. Construction of Correlation Models and Geographical Patterns (Surfaces) 396
III. Some Examples of Correlation Models of Net Primary Productivity versus Water Factor 399
IV. Accuracy of Correlation Models 403
V. Conclusions 406
References 406

B. The Use of Simulation Models for Productivity Studies in Arid Regions
H. van Keulen, C. T. de Wit, and H. Lof 408
I. Introduction 408
II. The Structure of the Model 409
III. Description of the Model ARID CROP 410
IV. Validation of the Model 413
V. Application of the Model 418
VI. Conclusions 419
References 420

C. Irrigation and Water Use Efficiency
J. F. Bierhuizen 421
I. Introduction 421
II. Efficiency of Water Supply 421
III. Transpiration/Photosynthesis Relationships 423
IV. Some Agronomic Aspects 428
V. Conclusions 430
References 430

D. Estimating Water Status and Biomass of Plant Communities by Remote Sensing
B. G. Drake 432
I. Introduction 432

	II.	Water Stress, Reflectance, and Temperature of Single Leaves.	432
	III.	Reflectance and Biomass of Communities.	434
	IV.	Conclusions	435
		References	437

E. Plant Production in Arid and Semi-Arid Areas
M. EVENARI, E.-D. SCHULZE, O. L. LANGE, L. KAPPEN, and U. BUSCHBOM 439

    I. Introduction 439
    II. Survey of Phytomass, Net Annual and Relative Annual Production of Some Main Vegetation Units of the Globe . . 440
    III. Phytomass and Production of Some Arid and Semi-Arid Vegetation Units and their Annual Fluctuations 441
    IV. Permanent Phytomass 445
    V. Potential Production 448
    VI. Recovery 448
    VII. Conclusions 449
    References 450

F. Water Content and Productivity of Lichens
G. P. HARRIS 452

    I. Introduction 452
    II. Productivity of Lichens 452
    III. Water Relations of Lichens 454
    IV. Thallus Water Content and Physiological Response 457
    V. Conclusions: Water Relations and Productivity—a Synthesis 463
    References 466

Part 7 Water and Vegetation Patterns 469
Preface 470

A. Water Relations and Alpine Timberline
W. TRANQUILLINI 473

    I. Introduction 473
    II. Water Relations of Trees at the Timberline 476
    III. Causes of Winter Desiccation of Trees at Timberline . . . 478
    IV. Conclusions: Ecophysiological Analysis of the Alpine Timberline and its Dynamics 487
    References 489

B. The Water Factor and Convergent Evolution in Mediterranean-type Vegetation
E. L. DUNN, F. M. SHROPSHIRE, L. C. SONG, and H. A. MOONEY 492

    I. Introduction 492
    II. Environmental Stresses in Mediterranean-type Climates . . 493
    III. Ecological Significance of Leaf Structure 496
    IV. Seasonal Patterns of Photosynthesis, Water Relations and Productivity 498

	V.	Evolutionary Consequences of Mediterranean-type Environmental Stresses 501
	VI.	Conclusions 503
		References 503
	C.	**The Water-Photosynthesis Syndrome and the Geographical Plant Distribution in the Saharan Deserts**
		O. STOCKER 506
	I.	Introduction 506
	II.	The Floristic and Physiognomic Aspects of the Sahara . . . 506
	III.	The Water-Photosynthesis Syndrome in the Northern and in the Southern Sahara 509
	IV.	Holarctic and Palaeotropic Constitution Types 516
	V.	Conclusions 519
		References 520

Index of Plant Species 523
Subject Index 527

Contributors

ACEVEDO, E.	Facultat de Agronomia, Universidad de Chile, Casilla 1004, Santiago, Chile
BARLEY, K. P. †	Formerly: Waite Agricultural Research Institute, University of Adelaide, Glen Osmond, S.A. 5064/Australia
BENECKE, P.	Institut für Bodenkunde und Waldernährung, Universität Göttingen, Büsgenweg 2, 3400 Göttingen/Fed. Rep. Germany
BENZIONI, A.	Research and Development Authority, Ben-Gurion University of the Negev, Beer Sheva/Israel
BIERHUIZEN, J. F.	Department of Horticulture, Agricultural University, Haagsteeg 3, Wageningen/The Netherlands
BUSCHBOM, U.	Botanisches Institut der Universität Würzburg, Mittlerer Dallenbergweg 64, 8700 Würzburg/Fed. Rep. Germany
CALDWELL, M. M.	Department of Range Science and the Ecology Center, Utah State University, Logan, UT 84322/USA
DRAKE, B. G.	Radiation Biology Laboratory, Smithsonian Institution, 12441 Parklawn Dr., Rockville, MD 20852/USA
DUNN, E. L.	Department of Botany, University of Georgia, Athens, GA 30602/USA
EVENARI, M.	Department of Botany, Hebrew University, Jerusalem/Israel
FERERES, E.	Laboratory of Plant-Water Relations, Department of Land, Air and Water Resources, Veihmeyer Hall, University of California, Davis, CA 95616/USA
GATES, D. M.	Biological Station, University of Michigan, Ann Arbor, MI 48109/USA
GREACEN, E. L.	CSIRO, Division of Soils, Private Bag No. 1 P.O., Glen Osmond, S.A. 5064/Australia
HALL, A. E.	Department of Plant Sciences, University of California, Riverside, CA 92502/USA

HARRIS, G. P.	Department of Botany, McMaster University, 1280 Main Street West, Hamilton, Ontario L8S 4Kl/Canada
HEBER, U.	Botanisches Institut der Universität Düsseldorf, Universitätsstr. 1, 4000 Düsseldorf/Fed. Rep. Germany
HENDERSON, D. W.	Laboratory of Plant-Water Relations, Department of Land, Air and Water Resources, Water Science and Engineering Section, University of California, Davis, CA95616/USA
HSIAO, T. C.	Laboratory of Plant-Water Relations, Department of Land, Air and Water Resources, Water Science and Engineering Section, University of California, Davis, CA95616/USA
HUBER, W.	Institut für Botanik und Mikrobiologie, Technische Universität, Arcisstr. 21, 8000 München/Fed. Rep. Germany
ITAI, C.	Department of Biology, Ben-Gurion University of the Negev, Beer Sheva/Israel
KAPPEN, L.	Botanisches Institut der Universität Würzburg, Mittlerer Dallenbergweg 64, 8700 Würzburg/Fed. Rep. Germany
KEULEN, H. VAN	Department of Theoretical Production Ecology, State Agricultural University, De Dreijen 2, Wageningen/The Netherlands
KLUGE, M.	Botanisches Institut der Technischen Hochschule Darmstadt, Schnittspahnstr. 3–5, 6100 Darmstadt/Fed. Rep. Germany
LANGE, O. L.	Botanisches Institut der Universität Würzburg, Mittlerer Dallenbergweg 64, 8700 Würzburg/Fed. Rep. Germany
LEE-STADELMANN, O. Y.	Department of Horticultural Science and Landscape Architecture, University of Minnesota, St. Paul, MN 55108/USA
LEVITT, J.	Department of Horticultural Science and Landscape Architecture, University of Minnesota, St. Paul, MN 55101/USA
LIETH, H.	Department of Botany, University of North Carolina, Chapel Hill, NC 27514/USA
LOF, H.	Department of Theoretical Production Ecology, State Agricultural University, De Dreijen 2, Wageningen/The Netherlands

LUDLOW, M. M.	Division of Tropical Crops and Pastures, CSIRO, Cunningham Laboratory, Mill Road, St. Lucia, QLD, 4067/Australia
LÜTTGE, U.	Botanisches Institut der Technischen Hochschule Darmstadt, Schnittspahnstr. 3–5, 6100 Darmstadt/Fed. Rep. Germany
MOONEY, H. A.	Department of Biological Sciences, Stanford University, Stanford, CA 94305/USA
OERTLI, J. J.	Botanisches Institut der Universität Basel, Schönbeinstr. 6, 4056 Basel/Switzerland
PESCHEL, G.	Institut für Physikalische Chemie, Universität Würzburg, Markusstr. 9–11, 8700 Würzburg/Fed. Rep. Germany
PONSANA, P.	Central Region Agricultural Centre, Chainat/Thailand
RICHTER, H.	Botanisches Institut, Universität für Bodenkultur, Gregor Mendel-Str. 33, 1180 Wien/Austria
RUNNING, S. W.	Department of Forest and Wood Sciences, Colorado State University, Fort Collins, CO 80521/USA
SANKHLA, N.	Botany Department, University Jodhpur, Jodhpur/India
SANTARIUS, K. A.	Botanisches Institut der Universität Düsseldorf, Universitätsstr. 1, 4000 Düsseldorf/Fed. Rep. Germany
SCHÖNHERR, J.	Institut für Botanik und Mikrobiologie der Technischen Universität München, Arcisstr. 21, 8000 München 2/Fed. Rep. Germany
SCHULZE, E.-D.	Lehrstuhl für Pflanzenökologie, Universität Bayreuth, Am Birkengut, 8580 Bayreuth/Fed. Rep. Germany
SHROPSHIRE, F. M.	Department of Plant Sciences, University of California, Riverside, CA 92502/USA
SONG, L. C.	Department of Biology, California State University, Fullerton, CA/USA
STADELMANN, E. J.	Department of Horticultural Science and Landscape Architecture, University of Minnesota, St. Paul, MN 55108/USA
STOCKER, O.	Botanisches Institut der Technischen Hochschule Darmstadt, Schnittspahnstr. 3–5, 6100 Darmstadt/Fed. Rep. Germany

Tranquillini, W.	Außenstelle für subalpine Waldforschung, Forstliche Bundesversuchsanstalt, Rennweg 1, 6020 Innsbruck/ Austria
Vieira da Silva, J.	Laboratoire d'Ecologie Générale et Appliquée, Université Paris VII, 2 Place Jussieu, 75221 Paris Cedex/ France
Waring, R. H.	School of Forestry, Oregon State University, Corvallis, OR 97331/USA
Winter, K.	Botanisches Institut der Technischen Hochschule Darmstadt, Schnittspahnstr. 3–5, 6100 Darmstadt/ Fed. Rep. Germany
de Wit, C. T.	Department of Theoretical Production Ecology, State Agricultural University, De Dreijen 2, Wageningen/ The Netherlands

Part 1
Fundamentals of Plant Water Relations

Preface

In the first Part of this book the reader is introduced to the fundamental properties of water within the plant body. He will learn the function of water as a solvent medium, and the importance of its status at the molecular level of the living system, in cells and tissues, as well as in whole plants as links in the soil–plant–atmosphere continuum.

Most of the decisive biochemical processes of plant life take place in the aqueous phase of cells. Biologists usually consider liquid water to be a more or less uniform medium. However, experimental evidence has demonstrated that this is not correct. Even liquid water has a certain "structure" according to a local and always transient ordering of its molecules via hydrogen bonds. There is still much discussion about the nature of this solvent water structure, and several different models are under consideration at the present time (see Hübner et al., 1970). The organization of bulk water in the cell, the interactions between water molecules, their associations with solutes and interfaces, and the displacement of water in various ways control biochemical reactions (see Lewin, 1974). Probably the most important of these structural phenomena within biological systems is the order of water and aqueous solutions near interfaces. Such aqueous boundary layers are described and discussed by Peschel in the first chapter (Part 1:A) of this volume. He points out the thermal anomalies of water as an apparent result of this structure, and he interprets possible biological consequences of these "kinks" in the properties of aqueous systems. Not many physical chemists are engaged in treating this kind of problem, and a lack of experimental evidence with living material is obvious. Hopefully, the chapter by Peschel will help to stimulate further interdisciplinary work among physicists, chemists and biologists in this interesting and important field.

The overall water relations of an intact plant cell are determined by the semi-permeability of the protoplast. This controls the flux of water and allows the generation of turgor pressure within the cell according to differences in the solute concentrations between the vacuole and its surroundings. In 1877 Pfeffer published his fundamental article "Osmotische Untersuchungen" in which for the first time, the decisive importance of osmosis for these processes was detected. He formulated the rules governing water uptake by the osmotic cell in relation to hydrostatic pressure conditions, and thus prepared the foundations for our present understanding of plant water relationships. Just one hundred years ago (in 1876) Wilhelm Pfeffer wrote in the preface of his famous publication: "Hoffentlich wird dieses den Anstoss geben, daß auch Andere thätig auf einem Gebiet eingreifen, welchem die Arbeitskraft eines Einzelnen nicht entfernt gewachsen ist" ("hopefully this will give the impulse to other scientists to undertake research in this field for which the resources of one single individual are not nearly sufficient"). His hopes have been fulfilled by the efforts of innumerable biologists to define, measure, and explain water conditions of plant cells and plant tissues since that time. Few subjects in the field of botany present such a large diversity of discussions and proposals, the terminology of which have sometimes led to serious misunderstandings (for further information see the table: Terminology and symbols of values of water state, in Slavík, 1974). Only within the last ten years have the new concepts, based on thermodynamic ideas, found increasing general acceptance and thereby unified the expressions characterizing water properties.

According to these concepts, the state of water is described by its chemical potential within the system under study in relation to that of pure free water. This term is a measure of the capacity of the water at a point under consideration to do work compared with the work capacity of pure free water. It is defined by the "water potential", the relative chemical potential of the system concerned per partial molal volume of water. (For detailed and critical explanation of the concept of water potential and its theoretical background see Slatyer and Taylor, 1960, and the monographs by Slatyer, 1967; Kramer, 1969; Walter and Kreeb, 1970; Oertli, 1971.) The dimension of water potential as a measure of specific free energy is energy per unit of volume. However, this is dimensionally equivalent to pressure, so that it can be expressed also in units of bars or atmospheres.

The total water potential (Ψ) is determined by different features of a system, and consists of a series of mutually independent components[1]. The specific free energy of water, for instance, is decreased due to compounds dissolved (osmotic partial component), or due to interrelationships with interfacial borders (matric potential component), and increased with increasing hydrostatic pressure (pressure potential component). The total water potential in biological systems is usually negative, that of pure free water being taken to be zero. Consequently, Ψ_π

---

[1] For the convenience of the reader and in order to avoid misunderstandings, the symbols to describe water potential and its components are unified throughout this book. If a special definition is necessary, the component potential is indicated by a subscript (a letter without subscript means the total water potential), such as

Ψ_π, Ψ_s: osmotic, solute potential component
Ψ_p: pressure potential component
Ψ_m: matric potential component.

In a similar way, the system to which specification of water potential is related is indicated by a superscript such as

Ψ^{leaf}, Ψ^l: leaf water potential
Ψ^{soil}, Ψ^s: soil water potential.

Consequently, Ψ_p^{xylem} means the pressure potential component of the xylem sap; Ψ_m^{soil} means the soil matric potential component; and $\Psi^{mesophyll}$ means the total water potential in the mesophyll.

and Ψ_m are always negative in contrast to Ψ_p, the turgor pressure which is usually positive:

$$(-)\Psi = (-)\Psi_\pi + (+)\Psi_p + (-)\Psi_m.$$

After a physical and chemical description of the water status in the plant from biological points of view, the decisive question is, which property of water controls the different functions of the organism? This important problem is discussed in Part 1:B by Oertli, who describes a theoretical analysis of properties and processes as causes of the first signs of injury in plant water relations. Without doubt, water transport within the cell, from cell to cell, and through the plant as a whole, takes place along gradients of decreasing total water potential as the driving force. However, it becomes obvious from Oertli's discussion that values of total water potential, on the other hand, do not play the key role in the water dependencies of metabolic processes. He points out the physiological importance of the pressure component and of pressure differentials for many plant functions, and he concludes that a primary injury results from insufficient turgor pressure relations by affecting mechanical equilibria. This idea will be taken up again in Part 4:F of this volume, where Hsiao et al. discuss growth and yield of plants as affected by water stress.

The ecological significance of water conditions for plant existence and vitality cannot be studied without direct reference to the environmental conditions under which the plant grows in the field. Plant water status is determined by the pattern of water exchange with its surroundings, being linked into the water fluxes which occur within the soil–plant–atmosphere continuum. This system is described in the next chapter (Part 1:C) by Oertli. He investigates the ecophysiology of plant water by establishing a causal link between its critical functions and the environment. Again, emphasis is placed on turgor pressure, which certainly plays an important physiological role in cellular water relations and which is described as a function of environmental variables.

Following these theoretical considerations, the last chapter of this part on fundamentals of plant water relations is devoted to our experimental information on the water conditions of plants. Methods of determining the status of water in living plants have been greatly improved over the last decades. Thermocouple psychrometers, hygrometers recording dew point depression, and the pressure chamber are used world-wide for measuring total water potential; osmometers have been developed that allow the determination of the solute potential within minute probes of plant sap (see Slavík, 1974; Walter and Kreeb, 1970). Successful attempts are even being made to adapt thermocouple psychrometers for measuring leaf water potential in situ while only slightly disturbing the leaf environment (Brown and van Haveren, 1972). These improvements of the methodological requirements have led to rapidly increasing accumulation of quantitative information on the water conditions of wild and cultivated plants in their different habitats. Of special interest are the absolute ranges of extreme values of water potential and its components, as well as diurnal and annual fluctuations in relation to water balance and metabolic activity under the various ecological situations. In Part 1:D, Richter gives an extensive review of our present knowledge of experimental evidence of water status in plants. After a critical summary of currently available methods in this field of research, there follows a survey of the hitherto observed ranges of total water potential and different potential components in different plant types. The continuum conditions which favor extreme values are discussed. The chapter is completed with a consideration of why prominent and characteristic changes in water potential exist within different plant compartments.

References

Brown, R. W., van Haveren, B. P. (eds.): Psychrometry in water relations research. Utah Agr. Exp. Station, Logan: Utah State Univ. 1972.
Hübner, G., Jung, K., Winkler, E.: Die Rolle des Wassers in biologischen Systemen. Berlin: Akademie-Verlag 1970.
Kramer, P. J.: Plant and soil water relationships: A modern synthesis. New York: McGraw-Hill 1969.
Lewin, S.: Displacement of water and its control of biochemical reactions. London and New York: Academic Press 1974.
Oertli, J. J.: A whole system approach to water physiology in plants. Advan. Frontiers Plant Sci. **27**, 1–200; **28**, 1–73 (1971).
Pfeffer, W.: Osmotische Untersuchungen. Leipzig: Verlag Wilhelm Engelmann 1877.
Slatyer, R. O.: Plant-water relationships. London and New York: Academic Press 1967.
Slatyer, R. O., Taylor, S. A.: Terminology in plant-soil-water relations. Nature **187**, 922–924 (1960).
Slavík, B.: Methods of studying plant water relations. Ecological Studies, Vol. 9. Berlin-Heidelberg-New York: Springer 1974.
Walter, H., Kreeb, K.: Die Hydratation und Hydratur des Protoplasmas der Pflanzen und ihre öko-physiologische Bedeutung. Protoplasmatologia, Vol. II/C/6. Wien-New York: Springer 1970.

A. The Structure of Water in the Biological Cell

G. PESCHEL

I. Introduction

The question of possible structures of bulk water is still unsolved and the theoretical concepts of structure based on experimental evidence are to a large degree fragmentary and sometimes too crude and contradictory. Reviews about bulk water structure are given by a number of authors (Franks, 1972; Eisenberg and Kauzmann, 1969; Frank, 1965).

The intricate problem of water structure near interfaces, which is commonly encountered in biological tissue, is still a subject of controversy since a number of experimental results which point to differently structured water in boundary layers are rejected by some scientists, who regard them as artificial effects. The majority of authors, however, are quite sure that the existence of a rather thick and structurally changed aqueous boundary layer near a solid wall is not a figment of imagination but a highly realistic concept (Henniker, 1949; Drost-Hansen, 1965, 1969, 1971; Peschel and Adlfinger, 1971). Solid–liquid interfaces in biological tissue occur, e.g., at cell surfaces and membranes. These are thought to act as momentum sinks which might stabilize structures intrinsically latent in bulk water by damping thermal fluctuations and thus inducing a long-range oriented sheet of water molecules which are subject to restrictions in their mobility. This immobilization effect was found to extend over some hundred Ångstroms deep into the bulk aqueous phase. Widely differing views concerning bulk water structure are still held by various authors. A number of models for water structure are found in literature, which are compatible with much of the experimental evidence. At the present stage most of the models are classified as continuum, specific structure and cluster models.

The common feature of all models is the local ordering of water molecules via hydrogen bonds in a tetrahedral four-coordinated arrangement. In the case of a continuum model all local domains in the bulk phase should be equal in structure, with hydrogen bonds suffering continuous distortions and bending (Pople, 1951). The specific structure models involve a framework which is apt to take up further water molecules on interstitial lattice positions. The most outstanding interstitial model is that proposed by Pauling (1959), who worked out some clathrate frameworks similar to those built up by gases solved in water. These structural moieties are considered to be labile pentagonal dodecahedrons surrounding an unbounded water molecule.

The cluster models suggest the existence of ice-like clusters with a short lifetime of only about 10^{-11} sec which exchange with surrounding monomers. Such models are described by Nemethy and Scheraga (1962) and by Frank and Wen (1957). Following more recent results evidence is accumulating that water structure can be treated best by a mixture model which allows for the existence of so-called shortlived flickering clusters. Possibly the structured entities are Ice-Ih, the high pressure ice polymorphs, clathrate cage clusters or other types of clusters.

The problem of the cause of long-range ordering of water molecules near interfaces and its manner of arrangement is still unsolved. In the following, different results are presented which might help to understand the structuring of water near interfaces and in biological tissue.

II. Evidence for Structured Aqueous Boundary Layers

1. Water Near Interfaces

Evidence for long-range structural order of water near interfaces is particularly based on measurements of the disjoining pressure and the anomalous viscosity which are typical for such layers.

A first study of the disjoining pressure in water layers bounded by two or one solid surfaces was reported by Deryagin and Obuchov (1936) and Deryagin and Kusakov (1939) respectively. In this procedure a solid plate or a gas bubble, both within the aqueous phase, were brought near another solid plate. The repulsive forces in the intervening layer due to some sort of rigidity of the oriented phase give rise to the so-called 'disjoining pressure'. In order to avoid the problem of plane-parallelity the present author, in determining the disjoining pressure in thin water layers between highly polished fused silica plates, used a plane/spherical plate system (Peschel and Adlfinger, 1971). The results are depicted in Fig. 1 for a plate distance $H = 500$ Å. The distance dependence of the disjoining pressure P_D is assumed to be

$$P_D = C \exp(-nh)$$

with C and n being parameters which are to be determined by experimental data. In view of the present findings it should be stressed that water structuring near interfaces extends to a considerable distance on the molecular scale.

In view of these results it is tempting to anticipate a shear-dependent 'boundary' viscosity. In fact, Churaev et al. (1971) succeeded in detecting anomalous viscosity of water in thin capillaries with diameters of only about 800 Å. Peschel and Adlfinger (1969), in moving a spherically formed fused silica plate immersed in water towards a planar one, obtained similar results; the 'boundary' viscosity of water turned out to be enhanced by an order of magnitude (Fig. 2).

Hori (1956) studied the freezing point of water between glass surfaces and found that no freezing can be observed in thin layers even at temperatures

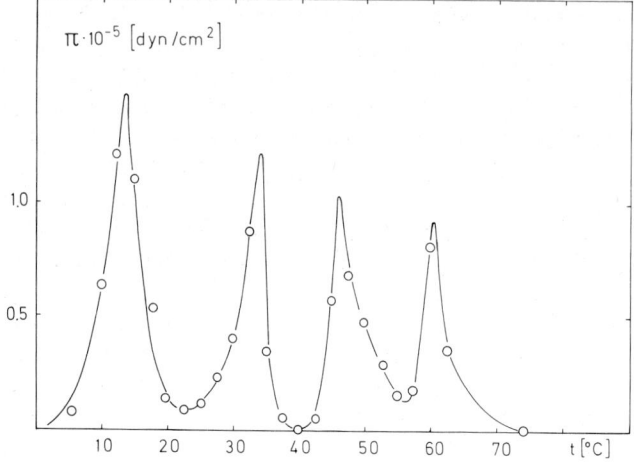

Fig. 1. Temperature dependence of the disjoining pressure in thin layers of water between hydroxylated fused silica plates. Plate distance 500 Å. (After Peschel and Adlfinger, 1971)

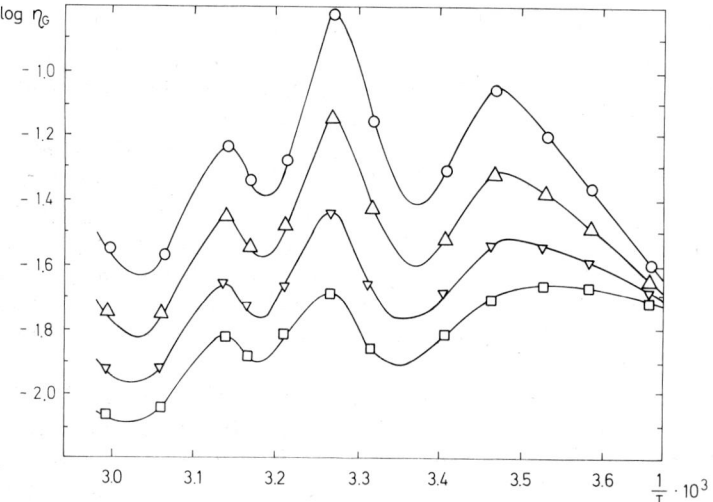

Fig. 2. Logarithm of boundary layer viscosity η_G vs. $1/T$ for 20°C. Plate distances ○ 300 Å; △ 500 Å; ▽ 700 Å; □ 900 Å. Thin water film bounded by two hydroxylated fused silica plates. (After Peschel and Adlfinger, 1969)

as low as −90°C. A most interesting effect was detected by Zorin and Churaev (1968). Investigating the polymolecular adsorption of water vapor on quartz surfaces they found sections of two structural modifications of adsorbed layers of different thicknesses corresponding to the same vapor pressure. The reason for this effect is still a matter of dispute. Further unequivocal proof of the formation of structured water layers near solid surface was given by Metsik and Timoshchenko (1971), who found an anomalously high thermal conductivity in thin water layers between stacks of mica sheets. These selected observations warrant the conclusion that water near solid surfaces forms a fluid phase distinguished by a number of uncommon physical properties.

To be exact, the special arrangement of vicinal water should depend on the proximity of nonpolar, polar or ionic surfaces. This point is stressed by Drost-Hansen (1971), who discusses this problem in detail. Nevertheless, systematic experiments in this field are still lacking.

According to the common view the ordering of structured water decreases with increasing distance from the solid wall. However, the incompatibility of the particular boundary layer structure with that of bulk water might lead to the existence of a zone of enhanced disorder far from the surface; but no progress has been made hitherto in substantiating this view.

The purpose of this chapter is to deal with the structure of water in biological tissue. Water in such systems acts as a solvent for a large number of electrolytes and nonelectrolytes. Examination of the state of aqueous solutions near solid walls has yielded results of outstanding importance, particularly for living organisms.

2. Aqueous Solutions Near Interfaces

It is well established that in thin layers of aqueous electrolyte solutions additional electrostatic forces play an important role.

The electrolyte double layers forming at each of the two charged bounding surfaces will overlap at sufficiently small layer thicknesses and give rise to electrical repulsive forces. Following current theories of this electrostatic problem (Verwey and Overbeek, 1948) a rather high electrostatically generated disjoining pressure should be anticipated at low electrolyte concentration; with rising concentration the electrostatic disjoining pressure drops constantly to very low values.

Experimental confirmation of the theory proved to be a very difficult task. Roberts (1972) used a glass–rubber contact immersed in an electrolyte solution to determine the repulsive forces predicted by theory. Barclay et al. (1972) investigated the overlap of electrical double layers in disperse systems. Both studies revealed that the disjoining pressure exerted in such layers has a structural component due to long-range ordering of water molecules near surfaces. These findings clearly show that the surface zone of an electrolyte solution, too, exhibits a three-dimensional matrix of polarized water molecules. Evidence in agreement with these results has come from the present author and will be dealt with in Section IV.

As aqueous solutions represent a large part of biological tissue, appropriate investigations should provide information about anomalous structuring of surface water. Most work in this field was done using nuclear magnetic resonance (NMR). Water in cells shows broader NMR signals than in the bulk, which supports the idea of organized water structure (Odeblad et al., 1956; Sussman and Chin, 1967; Hazlewood et al., 1969; Cope, 1969). It should be mentioned that the NMR results can only yield a mean value over a continuous distribution of structures or motional restrictions of water, according to the properties of the structure inducing interfaces.

Further evidence for anomalous water properties in biological systems comes from intracellular freezing patterns (Rapatz and Luyet, 1959) indicating that water inside a cell can be appreciably supercooled. More details can be found in a paper by Ling (1970). It is, e.g., highly probable that the extended form of desoxyribonucleic acid has a structure on its surface that fits into the water lattice. This is a typical case in which much of the surrounding water gets organized in an ice-like structure. The term 'ice-like', however, which is frequently used in literature to characterize surface water, is certainly too extreme, since the molecular mobility of organized water might be restricted on the time-scale by at most a factor of 10.

III. Thermal Anomalies in Biological Tissues

Recalling the results in Fig. 1 it is obvious that the uncommon temperature dependence of the disjoining pressure or the long-range water organization is an important factor in describing surface phenomena in aqueous systems.

Disjoining pressure maxima are encountered close to 15°, 30°, 45°, and 61°C. Following the idea of Drost-Hansen (1965, 1969, 1971) the extrema separate temperature ranges in which specific arrangements of organized water exist adjoining a surface. At the "characteristic" temperatures structural transitions of a higher order occur within the surface phase. Around these prominent temperatures there is enhanced molecular randomness, and thus, via a cooperative and epitaxial mechanism, the formation of a long-range oriented water sheet is entropically favored.

The present author showed (Peschel and Belouschek, 1975) that this phenomenon also occurs in thin electrolyte layers where the effect is superimposed on the electrostatic repulsion effect, which exhibits only a weak and by no means so dramatic temperature response.

The thermal anomalies which are carefully listed by Drost-Hansen (1965, 1969, 1971) have become known as "kinks" in a number of physical properties of aqueous systems rich in interfaces or of biological tissues. Particularly, a large number of examples is available concerning complex physiological processes in diverse organisms. This is consistent with the view that kinetic and transport phenomena in biology are closely related to water structure. There is no room here to detail all findings referring to 'kinks' in biological systems; some typical examples may be given.

Figure 3 shows the growth of a sulfate-reducing bacterium (Oppenheimer and Drost-Hansen, 1960). This plot strongly resembles that of the disjoining pressure results obtained by the present author (Fig. 1). Growth optima are found near 12°, 26°, and 39°C. The temperatures of minimum activity coincide with those of the disjoining pressure maxima.

Figure 4 depicts the growth of *Streptococcus faecalis* which is clearly depressed at about the characteristic temperatures (Davey et al., 1966). The intrinsic reason

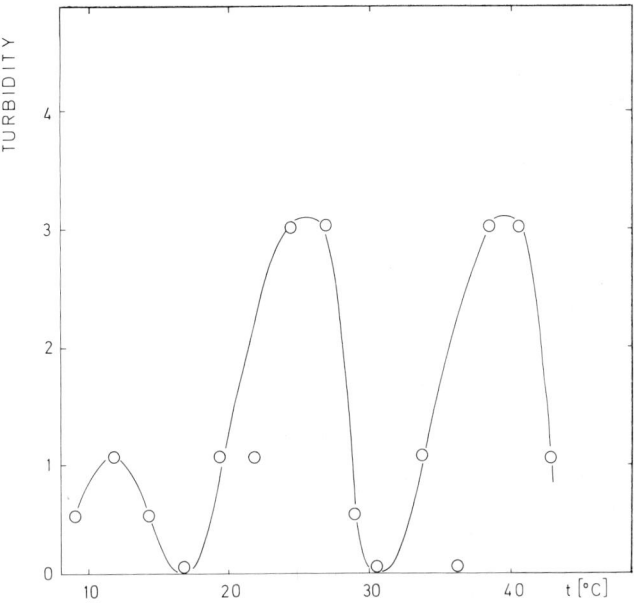

Fig. 3. Growth of a sulfate-reducing bacterium, probably a *Clostridium sp.*, determined by turbidity in relation to temperature. (After Oppenheimer and Drost-Hansen, 1960)

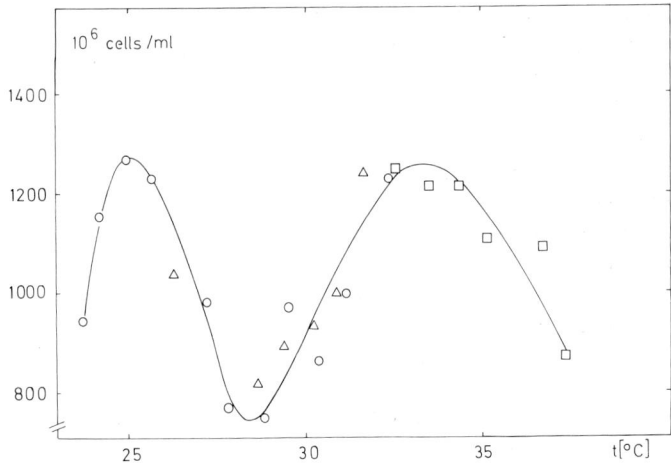

Fig. 4. Number of cells of *Streptococcus faecalis* incubated for 16 h at different temperatures; measurements on first (○), second (△), and third (□) day. (After Davey et al., 1966)

for this dramatic effect is still not clear. Diffusional processes and the movements of fluids across the membranes might be restricted. Moreover it appears likely that even the DNA molecule may be affected by the structural transition of organized water.

Heinzelmann et al. (1972) investigated the surface tension of a cell-free leakage medium of *Saccharomyces cerevisiae*. The temperature dependence of the surface tension of the medium, which is depicted in Fig. 5, exhibits sharp breaks at the prominent temperatures, where a higher local order in the surface zone under the participation of surface-active substances might be assumed.

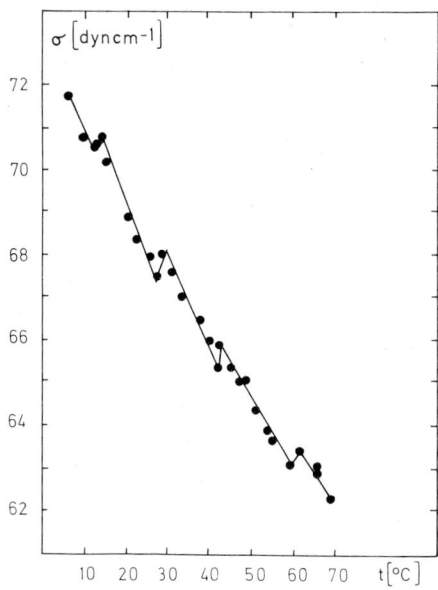

Fig. 5. Surface tension of cell-free leakage medium of *Saccharomyces cerevisiae* vs. temperature. (After Heinzelmann et al., 1972)

Water transport in plants seems to be considerably influenced by temperature. The problem of water movement across a maize root was taken up by Brouwer (1965). For three different mechanisms of water transport in a thermostated root he plotted the water conductivity against the root temperature (Fig. 6). The author showed that the water conductivity per atmosphere suction tension or pressure difference was minor with an osmotic suction tension, and largest with pressure operative in the outer medium. The data for transpiration lie within these limits.

The striking moment in Fig. 6 is that two curves exhibit sharp breaks at about 15°C and 32°C. This effect points to the occurrence of structured water in the root tissue which suffers a structural transition at about this temperature. The pressure flow, for which such a break is detectable, only at about 32°C, is, as Brouwer argues, less controlled by protoplasm which, on the other hand, is known to be a system rich in interfaces and thus capable of producing relatively high amounts of structured water. In this context it should be stressed that the long-range order of water flowing near interfaces is strongly shear-dependent (Peschel and Adlfinger, 1970).

Even highly developed organisms are subjected to the anomalous properties of vicinal water. As an example a very instructive graph by Drost-Hansen (1965) is shown (Fig. 7). The distribution of body temperatures of about 160 mammals is centred around a mean temperature of about 38°C. A look at Fig. 1 reveals that this temperature corresponds approximately to a disjoining pressure minimum.

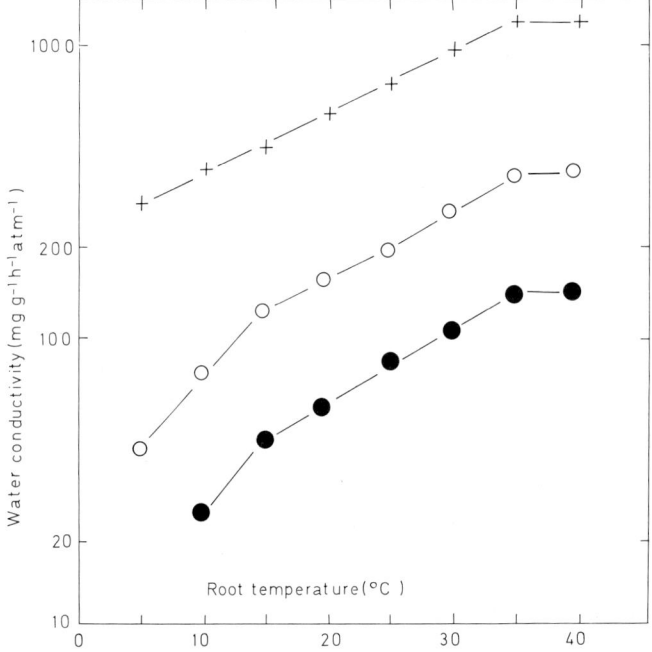

Fig. 6. Relation between root temperature and water conductivity of maize roots for different processes; logarithmic scale. Flow induced by hydrostatic pressure difference (+—+), transpiration (o—o), bleeding (●—●). (After Brouwer, 1965)

In summary it can be stated that the cooperative order-disorder phenomena in thin aqueous layers within biological tissue account for dramatic changes in many manifestations of life and seem to play a major role in biological processes.

IV. Properties of Aqueous Electrolyte Layers

The assumption that solid surfaces at very small separation distances are kept apart by the sum of an electrostatic and a structural component of the disjoining pressure when brought into an aqueous electrolyte solution seems to be justified according to the results of Roberts (1972) and Barclay et al. (1972). Investigations by the present author clearly confirmed this concept and

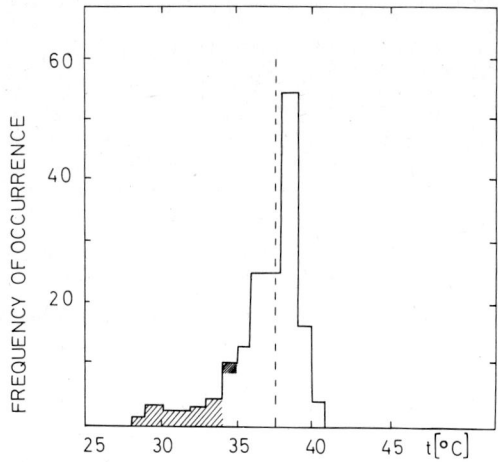

Fig. 7. Distribution of body temperatures of mammals. (After Drost-Hansen, 1965)

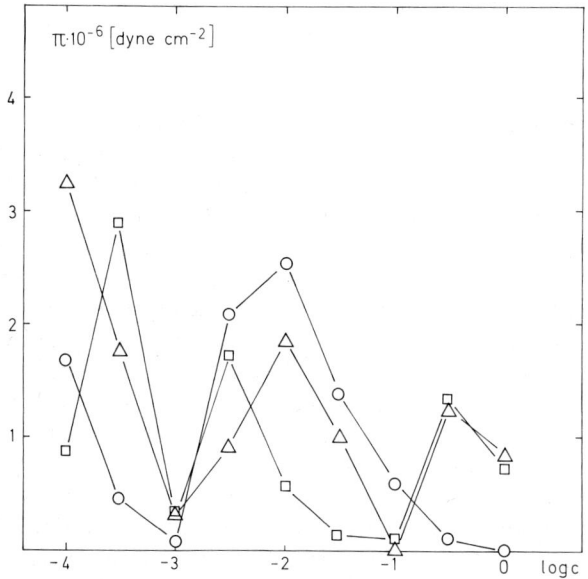

Fig. 8. Concentration dependence of disjoining pressure in thin layers of aqueous electrolyte solutions between hydroxylated fused silica plates. ○ LiCl; △ NaCl; □ KCl; plate distance 100 Å; temperature 20°C. (After Peschel and Belouschek, 1975)

provided implications for the state of water in the biological cell. In Fig. 8 the concentration dependence of the experimentally found disjoining pressure is shown for a plate distance of 100 Å for LiCl, NaCl, and KCl as solutes. For very low concentrations the disjoining pressure follows roughly the electrostatic theory (Verwey and Overbeek, 1948). But the extrema at about the concentrations of 10^{-2} M and 0.5–0.8 M (M expressed in mole/l) cannot be explained as electrostatic effects.

It now seems well established that cooperative changes take place in solution structure when the electrolyte concentration is as high as about 1 M (Vaslow, 1966). This degree of concentration might be critical since here all solvent molecules participate somehow in the formation of voluminous hydration clusters around the solute molecules. An increase in the solute concentration then leads to a cooperative transformation or breakdown of the hydration clusters. Since in this concentration range the transformation produces enhanced disorder, the solid surface, as in the case described above, might epitaxially influence the vicinal solution and favor a long-range orientation of water. The respective disjoining pressure maximum is discernible in every respect in Fig. 8.

At about neutral pH fused silica surfaces are negatively charged (as proteins are in biological tissue) and exert attractive forces on the cations in their proximity. Thus, the cations will become concentrated near the charged surface in an adsorption plane about 3 Å distant from the surface. This is the closest approach of cations to the surface because of intercalated hydration water.

By use of the simple Boltzmann distribution law

$$n_+ = n_{0+} \cdot \exp \frac{-ze\psi}{kT}$$

the cation concentration n_+ (number of ions per cm^3) close to the surface can be calculated. n_{0+} is the number of cations per cm^3 in the bulk solution, ψ is the potential in the adsorption plane, k the Boltzmann constant, T the temperature, e the elementary charge, and z the valency of the cation. In this way it can be shown that the 'critical' solute concentration of about 1 M can be achieved near such a charged solid surface when the solution far away from the surface still has a concentration of about 10^{-2} M. In other words, at this concentration cations can be accumulated near the wall by an adsorptive process so that all surrounding water molecules are elements of hydration clusters. Such immobilized water sheets may serve as a matrix for a long-range orientation of more distant water molecules deep into the solution (Fig. 8). This evidence suggests the presence of cage-like clusters near the wall.

These hydration cages are thought to be rather unstable in bulk solution but apparently more stabilized near a wall. In fact, the Li$^+$-ion which is known to have a marked immobilization effect on its hydration shell produces the strongest disjoining pressure maximum at about 10^{-2} M.

As outlined above, organisms seem to prefer conditions of low water structuring at interfaces. The plots in Fig. 8 exhibit two minimum values, one about 10^{-3} M and the other at about 10^{-1} M. It is undoubtedly worth noting that living cells are isotonic with a 0.1 M NaCl solution. In view of this new evidence it is difficult to reject the idea that in the course of evolution the development of organisms—particularly that of mammals—took the pathway of minimum structural organization of water in cells.

Ling (1969) in his extensive work tries to introduce a new physical model for the living cell on the basis of intracellular water having a higher organization than water outside the cell. To account for the different intra- and extracellular ionic levels Ling attributes to the intracellular water a lower activity and argues against the common concept of ion pumps. Similar considerations are found in Wiggins (1971, 1973).

Our results do not exclude water structuring in cells, but considering the findings discussed above they allow only for moderate orientational order in the cell, otherwise biological processes might be seriously hampered.

V. Conclusions

Any prediction about the arrangement of water molecules near a solid surface is still somewhat speculative. The occurrence of distinct structural entities in the hydration sheets on solid surfaces provides evidence for a mixture model for water structure. The epitaxial orientation effect seems to be dependent on whether or not the first molecular water layer directly adsorbed on the surface fits into the bulk water structure. In the first case it seems likely that the vicinal water structure becomes enhanced.

Considering the results discussed above it appears that knowledge about the state of water in the biological cell is important in understanding the mechanism of many biological processes. It has been argued that biological tissue is far too complex to represent a system appropriate for water structure research. As Tait and Franks (1971) point out in their excellent paper, there is a need for interdisciplinary research in this field. Particularly physicists and physical chemists prefer to treat model systems which might provide more detailed knowledge about water near interfaces.

References

Barclay, L., Harrington, A., Ottewill, R. H.: The measurement of forces between particles in disperse systems. Kolloid Z. Z. Polymere **250**, 655–666 (1972).
Brouwer, R.: Water movement across the root. In: The state and movement of water in living organisms (ed. G. E. Fogg), pp. 131–149. Symp. Soc. Exp. Biol. No. XIX, Cambridge: University Press 1965.
Churaev, N. V., Sobolev, V. D., Zorin, Z. M.: Measurement of viscosity of liquids in quartz capillaries. Spec. Discuss. Faraday Soc. **1**, 213–220 (1971).
Cope, F. W.: Nuclear magnetic resonance evidence using deuterated water for structured water in muscle and brain. Biophys. J. **9**, 303–319 (1969).
Davey, C. B., Miller, R. J., Nelson, L. A.: Temperature-dependent anomalies in the growth of microorganisms. J. Bacteriol. **91**, 1827 (1966).
Deryagin, B. V., Kusakov, M.: Anomalous properties of thin polymolecular films. V. An experimental investigation of polymolecular solvate (adsorbed) films as applied to the development of a mathematical theory of the stability of colloids. Acta Physicochim. (URSS) **10**, 25–44 (1939).
Deryagin, B. V., Obuchov, E.: Anomalien dünner Flüssigkeitsschichten. III. Ultramakrometrische Untersuchungen der Solvathüllen und des "elementaren" Quellungsaktes. Acta Physicochim. (URSS) **5**, 1–22 (1936).
Drost-Hansen, W.: The effects of biologic systems of higher-order phase transitions in water. Ann. N. Y. Acad. Sci. **125**, 471–501 (1965).
Drost-Hansen, W.: Structure of water near interfaces. Ind. Eng. Chem. **61**, 10–47 (1969).

Drost-Hansen, W.: Structure and properties of water at biological interfaces. In: Chemistry of the cell interface, Part B (ed. H. D. Brown), pp. 1–184. New York: Academic Press 1971.
Eisenberg, D., Kauzmann, W.: The structure and properties of water. Oxford: Clarendon Press 1969.
Frank, H. S.: The structure of water. Federation Proc., Suppl., 1–11 (1965).
Frank, H. S., Wen, W.-Y.: Ion-solvent interaction. Structural aspects of ion-solvent interaction in aqueous solutions: A suggested picture of water structure. Discuss. Faraday Soc. **24**, 133–140 (1957).
Franks, F.: Structural models. In: Water. A comprehensive treatise. Vol. 1. The physics and physical chemistry of water (ed. H. S. Frank), pp. 515–543. New York-London: Plenum Press 1972.
Hazlewood, C. F., Nichols, B. L., Chamberlain, N. F.: Evidence for the existence of a minimum of two phases of ordered water in skeletal muscle. Nature (London) **222**, 747 (1969).
Heinzelmann, H., Kraepelin, G., Bogen, H. J.: Leakage-Produkte von Hefen. I. Diskontinuitäten in der Oberflächenspannung des Mediums und im Wachstum der Zellen. Arch. Mikrobiol. **82**, 300–310 (1972).
Henniker, J. C.: The depth of the surface zone of a liquid. Rev. Mod. Phys. **21**, 322–341 (1949).
Hori, T.: On the supercooling and evaporation of thin water films. Low Temp. Sci., A **15**, 34–42 (1956).
Ling, G. N.: A new model for the living cell. A summary of the theory and recent experimental evidence in its support. Intern. Rev. Cytol. **26**, 1–61 (1969).
Ling, G. N.: The physical state of water in living cells and its physiological significance. Intern. J. Neurosci. **1**, 129–152 (1970).
Metsik, M. S., Timoshchenko, G. T.: New data on the thermal conductivity of thin films of water. In: Research in surface forces, Vol. 3 (ed. B. V. Deryagin), pp. 34–35. New York-London: Consultants Bureau 1971.
Nemethy, G., Scheraga, H. A.: Structure of water and hydrophobic bonding in proteins. I. A model for the thermodynamic properties of liquid water. J. Chem. Phys. **36**, 3382–3400 (1962).
Odeblad, E., Bhar, B. N., Lindstrom, G.: Proton magnetic resonance of human red blood cells in heavy-water exchange experiments. Arch. Biochem. Biophys. **63**, 221–225 (1956).
Oppenheimer, C. H., Drost-Hansen, W.: A relationship between multiple temperature optima for biological systems and the properties of water. J. Bacteriol. **80**, 21–24 (1960).
Pauling, L.: The structure of water. In: Hydrogen bonding (ed. L. Hadzi), pp. 1–6. London: Pergamon Press 1959.
Peschel, G., Adlfinger, K. H.: Temperaturabhängigkeit der Viskosität sehr dünner Wasserschichten zwischen Quarzglasoberflächen. Naturwissenschaften **56**, 558 (1969).
Peschel, G., Adlfinger, K. H.: Viscosity anomalies in liquid surface zones. IV. The apparent viscosity of water in thin layers adjacent to hydroxylated fused silica surfaces. J. Colloid. Interf. Sci. **34**, 505–510 (1970).
Peschel, G., Adlfinger, K. H.: Thermodynamic investigations of thin liquid layers between solid surfaces. II. Water between entirely hydroxylated fused silica surfaces. Z. Naturforsch. **26a**, 707–715 (1971).
Peschel, G., Belouschek, P.: A new method for investigating the structure of thin electrolyte layers between solid surfaces. Fortschrittsberichte über Kolloide und Polymere (in press, 1975).
Pople, J. A.: Molecular association in liquids. II. A theory of the structure of water. Proc. Roy. Soc. Ser. A **205**, 163–178 (1951).
Rapatz, G., Luyet, B.: Recrystallisation at high sub-zero temperatures in gelatin gels subjected to various cooling treatments. Biodynamica **8**, 85–105 (1959).
Roberts, A. D.: Direct measurement of electrical double-layer forces between solid surfaces. J. Colloid Interf. Sci. **41**, 23–34 (1972).
Sussman, M. V., Chin, L.: Liquid water in frozen tissue. Study by nuclear magnetic resonance. Science **151**, 324–325 (1967).

Tait, M. J., Franks, F.: Water in biological systems. Nature (London) **230**, 91–94 (1971).
Vaslow, F.: The apparent molal volumes of the alkali metal chlorides in aqueous solution and evidence for salt-induced structure transitions. J. Phys. Chem. **70**, 2286–2294 (1966).
Verwey, E. J., Overbeek, J. T. G.: Theory of the stability of lyophobic colloids. Amsterdam: Elsevier Publ. Co. 1948.
Wiggins, P. M.: Water structure as a determinant of ion distribution in living tissue. J. Theor. Biol. **32**, 131–146 (1971).
Wiggins, P. M.: Ionic partition between surface and bulk water in a silica gel. A biological model. Biophys. J. **13**, 385–398 (1973).
Zorin, Z. M., Churaev, N. V.: Observation of stepped films of water on the surface of quartz in polarized light. Kolloidn. Zh. **30**, 371–374 (1968).

B. The States of Water in the Plant Theoretical Consideration

J. J. OERTLI

I. Introduction

In spite of a great deal of research in the field of plant water relations, opinions differ about the nature and site of primary lesions through which suboptimal water relations affect plant metabolism. Knowledge of this property is necessary not only for an understanding of the cause-and-effect interaction between water and plant metabolism, but also for any meaningful connection of plant water relations with the soil–plant–atmosphere continuum (SPAC). The major task of ecophysiology is to describe, formulate, and analyze such links of plant properties and processes with the environment, a task that obviously requires the critical function of water to be known.

Currently the opinion is widely held that the water potential (= the quantitative measure for moisture stress) is most critical for plant behavior (Kramer and Brix, 1965; Slatyer, 1966; Hoffman and Splinter, 1968; Oechel et al., 1972; Yang and Dejong, 1968), but critical theoretical and experimental evidence is lacking and the contention must be considered unproven. In earlier periods the primary interest of water physiology focused on the importance of turgor pressure, because of the observation that a lack of turgidity consistently had a deleterious effect on plant behavior. Since an osmotic adjustment to saline moisture stress is often observed, Eaton in the USA and Walter in Germany started a search more than forty years ago for other explanations for this salinity-induced growth reduction of plants. Whereas Eaton today believes that the main injury of salinity is biochemical in nature (Eaton et al., 1971), Walter and Kreeb (1970) describe a property of water, the hydrature, which varies in plants even when the turgor pressure is maintained through an osmotic adjustment and which is thus correlated with the salt-induced growth reduction. Weatherly (1950) proposes a relative turgidity or relative water content as a suitable expression for the status of water in plants. The reference state for this property is the water content of cells when in equilibrium with pure free water at 1 bar pressure. Occasionally, other aspects of plant water have been considered, including water content, rate of transpiration, and degree of hydration of cytoplasm or of specific chemical compounds.

This chapter represents a theoretical analysis of properties and processes with regard to their relevance in causing the primary lesion in plant water relations. A number of problems arise with such an analysis. Firstly, it is often difficult to identify the critical aspect of water relations because, e. g., a favorable plant response is usually correlated with an increase in water content, transpiration rate, water potential, relative turgidity, hydrature, and turgor pressure; only through experiments in which some of these properties are made to change in the opposite direction is it possible to find the critical aspect. Secondly, other difficulties in the search for the primary lesion arise when the type of injury varies with the nature and intensity of suboptimal moisture conditions. For example, the mechanism that retards metabolism in a dry seed is different from the interaction that causes the primary lesion in a mesophytic plant leaf under ordinary drought. A third type of difficulty arises from the interdependence of many properties, e. g. of water potential and turgor pressure. In this case the more direct relation between water and metabolism should be chosen. Fourthly, it is desirable to include more aspects of plant water relations in this discussion than might have been done with a narrow, though perhaps more precise interpretation of the term "state of water".

II. Physiological Importance of Processes and Properties Involving Water

1. Water Content

The term "water deficit" refers to a lowered water content, although some scientists have identified it with low water potentials. These two definitions of water deficit require different descriptions of plant water relations because there is no unique relation between water content and water potential.

It is difficult to see why the water content over its usual range should be of critical importance. Of course, many plant properties, such as the heat capacity, change with the water content, but these effects are physiologically insignificant.

Relative Turgidity. Closely related to the water content, but also to pressure relations, is the relative turgidity. It is the ratio between the water content of a cell and the water content at full turgidity, whereby full turgidity is obtained at equilibrium with water at its standard state, i. e. pure free water at 1 bar pressure (Weatherly, 1950). The relative turgidity is indicative of the deviation of the actual situation from the possible maximum. But conditions at this maximum will vary from cell to cell depending on the vacuolar osmolalities, so that maximum relative turgidity may still represent a condition that is suboptimal for plant metabolism. For example in the extreme case of $\Psi=0$ ($=100\%$ relative turgidity) and $\Psi_s \sim 0$ one obtains $\Psi_p \sim 0$, i. e. a wilting plant. As a consequence of the variable reference state it is possible that a wilting plant has a higher relative turgidity than another turgid one. Such facts restrict the values of relative turgidity as a universal expression of the status of water in plants. It is of some usefulness in the study of changes if applied to different stages of an individual plant, but it is very problematic for comparisons between plants.

2. Rate of Transpiration

It is well established that plants of vigorous growth also transpire heavily. However, a causal correlation between the two phenomena is difficult to find; most likely both are related to open stomata. It appears that possible direct mechanisms of interaction are not sufficiently effective to explain the principal phases of plant water relations.

a) Transport of Solutes. The velocity of solute transport increases and falls with the rate of transpiration. Since there is a partial compensation of solute concentration and velocity in the xylem (low velocities—high concentrations and vice versa), the case where a low transpiration rate induces nutrient deficiencies, is probably rare.

b) Evaporative Cooling of the Leaf. There is little doubt that occasionally leaf tissues overheat because of insufficient cooling, but again this is hardly the critical site of interference, if only for the one reason that stomata often tend to close during noon hours, at the very time when cooling is most needed for survival.

3. Energy Relations

a) The Water Potential. Much attention has been paid recently to the water potential as an important property in plant water relations. This trend has been furthered by an increasing awareness by biologists of the possibilities offered by physical chemistry to describe biological systems. Interest in this field has also increased because of recent improvements in measuring techniques (psychrometers, pressure bomb). It is, however, unproven that the water potential directly plays a key role in plant water relations. As a chemical potential it has the following important properties: (1) It is a criterion for determining whether a process is at equilibrium or not; (2) at non-equilibrium it is usually a criterion for determining the direction of a spontaneous reaction; (3) potential differences are driving forces and as such can sometimes be used in rate laws describing processes.

At this point two specific cases will be analyzed that have been of considerable interest in the literature.

The Availability of Water. Soil water with a potential lower than -1500 Joules kg^{-1} (lower than -15 bars) is sometimes considered unavailable in the sense that the water potential gradient necessary for the flow of water across the SPAC has disappeared. The reason for this is that plants usually wilt under such conditions. Since the water potential difference between soil and atmosphere can easily amount to -10^5 Joules kg^{-1} (-1000 bars) the above-mentioned change in soil water potential has only a negligible effect on the driving force. It is obvious that in each case water must flow spontaneously from the soil to the atmosphere and that the driving forces in the two cases are essentially equal. Thus if very little water moves from a dry soil through the plant to the atmosphere it must be for a different reason than reduced availability expressed by the soil water potential.

Chemical Equilibrium of Reactions Involving Water. As an example of a reaction of the type

$$aA + nH_2O = bB$$

the hydration of a protein is studied:

$$\text{Protein} + nH_2O = \text{Protein}(H_2O)_n.$$

Although it is not usually done, this process of hydration can be treated thermodynamically like a chemical reaction. A reduction in water potential by 750 Joules kg^{-1} (7.5 bars)—and experience shows that this is a rather significant decrease—corresponds to a change in equilibrium relative humidity of about 0.5%. Thus, if the activity of water at the standard state is 1, a decrease in water potential by 7.5 bars will reduce the activity to 0.995. Consequently, the position of the equilibrium described by

$$K = \frac{(\text{Protein}(H_2O)_n)}{(\text{Protein})(H_2O)^n}$$

will hardly be changed as long as n is small. For $n=10$, the factor $(H_2O)^n$ is 0.95, which is still a negligible change; for $n=100$, the factor becomes 0.61. Thus it is clear that, unless n is very large, the water potential will not have a critical effect on the position of chemical equilibria. This conclusion is supported by experimental evidence. Walter (1963) published data showing a change of cytoplasmic water content of about 60% upon a decrease of relative humidity in an equilibrium atmosphere from 100 to 96%; hence, the water content changed by about 1% for each reduction of the water potential by 1 bar or 100 Joules kg^{-1}. This is a rather small change of hydration.

This analysis of the water potential's role in chemical reactions is only valid if the activities of other reactants do not change simultaneously. For reactions in the plant system such situations are rare because the agents that change the water potential, principally pressure and solutes, also affect the chemical potential of other reactants.

b) Components of the Water Potential. One can speak of a pressure and a solute content, but there is no real, unambiguous property that could be called a pressure potential component or an osmotic potential component. This statement becomes clear when a change from pure water at 1 bar pressure to water in a solution at a different pressure (similar to cell sap) is considered. If in a first pathway I, the pressure is changed first, followed by a change in solute content, then the associated potential changes $\Delta\Psi_p$ and $\Delta\Psi$, might reasonably be called the pressure and the osmotic components of the water potential, e.g. in a vacuolar sap. If in a pathway II solutes are added first, succeeded by a change in pressure, then of course, the final water potentials will be identical for the two pathways but the pressure and osmotic components will differ. Usually, the differences will be negligible, for which reason the terminology of "potential components" can still be used. However, there is an important consequence: since these components refer to specific pathways that exist only in the mind of an experimenter and usually not in nature, they cannot generally be of physiological importance. Furthermore, it is difficult to expect a component to play a critical role if the total overall changes of the water potential in the biological range are of a negligible magnitude.

The Pressure-free Component (often equivalent to the osmotic component). The water potential is affected by variables such as pressure and solutes. In a chemical reaction, as for example the above hydration of proteins, the pressure not only affects the activity of water, a collegative property to the water potential, but it also changes the activities of all other participants. If the partial molal volumes of reactants and products are equal, i.e.

$$\bar{V}_{Protein} + n\bar{V}_{H_2O} = \bar{V}_{Protein(H_2O)_n}$$

then the pressure has no effect on the position of the chemical equilibrium and the degree of hydration is related to the difference water potential minus pressure effect on water potential:

$$\Psi - \Psi_p.$$

This difference, if expressed as a relative humidity, is essentially the same as the hydrature introduced by Walter (Walter and Kreeb, 1970) for the cytoplasm as a whole. It is now necessary to analyze the physiological significance of this water potential difference: (1) The pressure-free component of the water potential should be an indicator of the hydration of the cytoplasm, hence it remains to be shown that this degree of hydration critically affects metabolic processes. In the range of moisture conditions that is usually of interest in mesophytic plants it is not known whether and in what way hydration is of importance. Moreover, excepting special cases (dry seeds, xerophytes), changes in protein hydration are probably much too small to account for drastic injuries caused by suboptimal water relations. (2) A unique relation between pressure-free component of the water potential and hydration exists only when other factors, e.g. solutes, influence only the water potential. Unfortunately, a change in osmolality will also affect activities of all components of a system. (3) The concept developed in this section is restricted to the case of no volume change in the process considered. Substantial deviations from these ideal conditions are occasionally to be expected. (4) Neglect of pressure effects in water relations leads to an erroneous assessment of physiological consequences of suboptimal water relations because, at least sometimes, pressures play a dominant role. (5) As stated above, the hydrature is closely related to the pressure-free component of the water potential. By definition (Walter and Kreeb, 1970) it must be applied to the cytoplasm as a whole. A major problem with this approach is that there is no unique relation between hydration of the entire cytoplasm and the hydrature. The hydrature, like the water potential and the relative turgidity, also presents the difficulty that its numerical value in a wilting plant can be higher or lower than in a turgid one. Finally it does not seem justified to ignore the structural cytoplasmic organization. For this reason plant water relations must be discussed in terms of fundamental processes and states such as chemical reactions or the degree of swelling of organelles rather than of the cytoplasm as a whole.

The Pressure Component. The pressure component of the water potential relates to a change in water potential and its effect must not be confused with that of the pressure itself. Usually, a good correlation between the vacuolar pressure potential component and plant behavior is observed, but the cause-and-effect interaction should perhaps be sought between pressures and metabolism. In the first case the concern is with free energy, in the second, with forces.

4. Pressure Relations

With regard to physiological relevance one has to distinguish between effects of the absolute pressure and pressure differences across membranes.

a) Effects of the Absolute Pressure. It has already been mentioned that the pressure not only affects the water potential but also the chemical potential of each other component in a system. In this section pressure relations will be analyzed in more detail.

Suppose the reaction

$$A + B = C$$

is an equilibrium so that

$$\frac{a_C}{a_A \, a_B} = K$$

(K = equilibrium constant, a_A, a_B, a_C are the activities). Upon an increase in pressure, all the activities are changed according to

$$d \ln a = \frac{\bar{V}}{RT} dP$$

and the new product of activities is

$$\frac{a'_C}{a'_A \, a'_B} = \frac{a_C}{a_A \, a_B} \exp\left(\frac{\Delta P}{RT} \Delta \bar{V}\right).$$

Obviously, the reaction is no longer at equilibrium unless $\Delta \bar{V} = 0$, i.e. unless there is no volume change with the chemical reaction.

It is more practical to calculate a new equilibrium constant K' based on a new standard state at the new pressure. One obtains

$$\ln \frac{K'}{K} = -\frac{P \Delta \bar{V}}{RT}.$$

A volume increase during the reaction ($\Delta \bar{V}$ positive) will shift the equilibrium toward the reactants, a volume decrease toward the products. These statements are also obvious from the principle of Le Chatelier.

At a given temperature T, the ratio K'/K depends on the product $P\Delta\bar{V}$. Fig. 1 shows the effect of P on this ratio of equilibrium constants for three

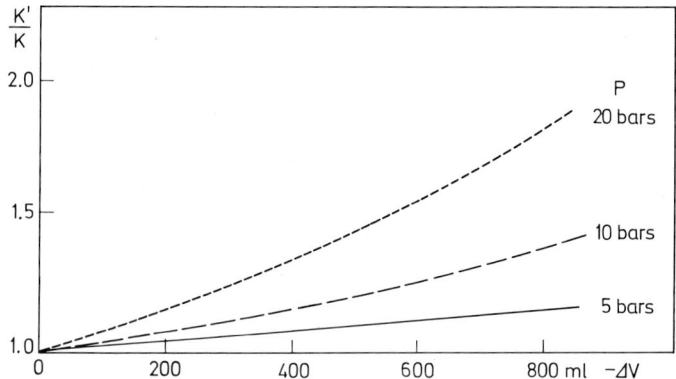

Fig. 1. Effect of pressure on the equilibrium constant of chemical reactions as a function of the volume change that occurs during the reaction

turgor pressures, 5, 10, and 20 bars, all within the range of possible internal cellular pressures. It follows from this figure that turgor pressure fluctuations have only negligible effects on the position of chemical equilibria of reactions that involve few and low molecular weight compounds, because in this case molar volume changes will be small and hardly exceed a few ml. In contrast, molar volumes of macromolecules are very large so that volume changes of several 100 ml are easily possible. In these situations, the turgor pressure can affect chemical reaction equilibria measurably. One such reaction is the hydration of proteins which would become a pressure-sensitive process. Whether the resulting degree of protein hydration has a metabolic effect and whether the effect is stimulatory or inhibitory are still open questions.

So far the position of the chemical equilibrium has been studied. Pressure effects on reaction rates can be handled in a similar way if the pressure effect on the activation reaction is considered (Hochochka and Somero, 1973).

There can be little doubt today that the fine structure of the cytoplasm plays an important role in regulating metabolism. In this connection the question arises whether the absolute pressure has an effect on this structure, in particular on the degree of swelling of organelles. Since swelling and shrinking are related to pressure differences it is preferable to discuss these aspects first and to postpone treatment of the absolute pressure.

b) Pressure Differences. The swelling of the cell as a whole as well as that of organelles including the vacuole, mitochondria, and chloroplasts is governed by pressure differences. Of these, the *turgor pressure* is probably best known. Much evidence is available that such pressure differences play an important role in the water physiology of plants.

The turgor pressure is critical for the mechanical support of most leaves. If cell walls are not stretched by a positive turgor pressure, most plant leaves will wilt, and it is common knowledge that this is an unhealthy situation which, if prolonged, will lead to the death of many tissues.

It has been shown frequently that the function of organelles such as mitochondria (Christiansen, 1968) depends on the degree of swelling, but swelling and shrinking are related to changed pressure differences across surrounding membranes. The fact that the magnitude of pressure differences across some organelle membranes might be considerably smaller than in the case of turgor pressure does not diminish a possible physiological importance.

A crucial role is commonly attributed to the turgor pressure in the process of cell elongation and in a few cases the effect has been demonstrated experimentally.

The function of guard cells surrounding the stomatal openings is almost certainly regulated by turgor pressure, although closure and opening of stomata also respond to such environmental factors as CO_2 of the air, temperature, and light, besides water relations directly.

c) Injury Mechanisms. The mechanism by which a reduced rate of cell elongation or a closure of stomata will cause a lower plant production is readily apparent. In other situations mechanisms remain more speculative. For example the degree of swelling could affect the activity of mitochondria in several ways. The permeability of the membrane surrounding the organelle could respond to swelling so that the injury would be one of changed flow characteristics of substrates and products. Alternatively, the steric arrangement of enzymes attached to membrane surfaces might change with swelling and shrinking and thus affect metabolic activities.

Two different cases must be distinguished with respect to the cell as a whole when moisture relations become suboptimal:

Cytorrhysis. Normally the osmotic component of cell wall water will be negligible and primarily variations of the non-osmotic component must occur. Here, it is irrelevant whether reference is made to a matric or a pressure component; mechanically the hydrostatic pressure of the water is changed. The cellular water potential always tends to equilibrate with that of cell wall water. Since, in contrast to the cell wall, vacuoles contain a substantial negative osmotic component, it follows that the vacuolar hydrostatic pressure is higher than that of the cell wall water, i.e. the pressure difference between internal and external water is always positive even though both pressures might be negative. As a consequence of this positive pressure difference, plasmolysis is impossible although the turgor pressure itself could be negative.

The outside pressure on a leaf remains atmospheric and is transmitted to the interior via the cell wall system. The leaf structure is better suited to withstanding an excess internal vacuolar pressure through elastic extension than to balancing this excess external pressure through compression of cell walls. As a result, cell walls in mesophytic tissues tend to collapse and cells will shrivel. This process, called cytorrhysis, which was described long ago (Steinbrink, 1910), is thus a major mechanism of injury under severe drought. Thanks to sclerotic cells many desert plants are partly protected against this type of injury.

Whether and how cytorrhysis affects plant metabolism remains to be shown. Water under tension (turgor pressure < -1 bar) is in a metastable state with respect to the gaseous phase at the respective vapor pressure of water. Should bubble formation occur then the cell water would rapidly be transferred to neighboring cells. The transfer might even involve some cytoplasma along plasmadesmata. The resulting injury could be permanent even if, as has been shown, the cell volume is refilled upon improvement of moisture relations.

Plasmolysis. In rare cases the osmolality of cell wall water can be expected to be higher than that in vacuoles. Only in these cases will plasmolysis occur. Such conditions could occasionally be fulfilled under saline moisture stress or when guttation drops are drying out and are gradually being reabsorbed by the leaf.

The immediate interaction of plasmolysis with metabolism is again a matter of speculation. Perhaps a rupture of plasmodesmata causing a disruption of intercellular connections can lead to a permanent injury (Karzel, 1926).

d) Effect of the Absolute Pressure on the Fine Structure of the Cytoplasma. It is now possible to analyze the effect of the absolute pressure on the degree of swelling of organelles. At thermodynamic equilibrium the excess internal pressure potential component is balanced by a correspondingly lower osmotic component. It follows that as long as the difference in osmolalities across a membrane remains constant, the difference in pressure potential components is also maintained irrespective of the level of actual pressures. Hence, with good approximation, the fine structure of the cytoplasm, inasmuch as it is affected by swelling, is independent of the absolute pressure.

e) The Question of Osmotic Adjustment. In saline media, plants often perform osmotic adjustments, which means that the vacuolar osmolality of tissues increases parallel to the salinity of the root medium. Indeed, within the relatively short time of one day after a transfer to a saline solution, plants often seem to have regained turgidity. In spite of this adjustment growth is reduced. This dilemma of growth reduction while turgidity is maintained has led to considerable controversies about the mechanism of injury resulting from salinity.

An osmotic adjustment is not inconsistent with a growth reduction that is related to turgor pressure. However, one has to observe expanding cells rather than mature cells in which the adjustment is completed in a matter of a day. The solute requirement for an osmotic adjustment of a cell increases linearly with the decrease of the external water potential. Since the rate of fulfilment of this requirement, namely the rate of solute transport, obeys saturation kinetics it is evident that with increasing salinity the rate of tissue extension must progressively become smaller. The actual situation, however, is more complicated: theoretical and experimental findings point toward an optimum concentration at which the rate of cell elongation is at a maximum (Oertli, 1975).

III. Metabolism and Water Relations

At this point, as an interim summary, the question can be posed whether general statements are now possible on how metabolism is affected by the state of water. Instead of analyzing systematically various properties of water for their physiological relevance the emphasis is now reversed by first looking at the living system and then investigating its relation to water. This living system is characterized by processes and by equilibrium states the water relations of which will now be described.

a) Chemical Equilibrium; *Effect of Internal Pressure.* If no volume change is involved in the reaction, the equilibrium position remains unchanged upon an increase in pressure. If the volume increases, then the higher pressure changes the equilibrium toward the reactants; if the volume decreases, the corresponding change is toward the products. Effects on reactions with compounds of low molecular weight are negligible in the normal range of turgor pressure; with macromolecules or with series of reactions a significant metabolic interference by pressure variations is not impossible.

Effect of Internal Solutes. Similar to the pressure which not only affects the water potential but also the chemical potential of other components, the addition of any solute also changes the chemical potential of other molecular species in a solution besides water. A reaction is at chemical equilibrium when

$$\sum v\mu = 0$$

(v = stoichiometric coefficient, μ = chemical potential). This sum remains constant if changes in chemical potentials, weighted by the stoichiometric coefficients, cancel. Otherwise the chemical equilibrium is disturbed. Suppose one reactant is water. Its potential is lowered by the addition of a solute (assumed to be non-participating in the reaction) but the overall effect depends on the changes in activity of all participants.

A more definite statement is possible if the mole fraction

$$N_x = \frac{n_x}{\sum n}$$

is substituted for the activities. The equilibrium constant for a reaction of the type

$$A + H_2O = B$$

is

$$K_1 = \frac{n_B \sum n}{n_A n_{H_2O}}$$

and for one of a reaction type

$$A + H_2O = B + C + D$$

the equilibrium constant is

$$K_2 = \frac{n_B n_C n_D}{n_A n_{H_2O} \sum n}.$$

In each case the water potential is lowered on addition of a non-reacting solute, but in the former case the equilibrium is shifted to the left, and in the latter to the right.

Many solutes that are of interest because of their osmotic effect also interact with participants of a reaction, changing their chemical nature. This may be an important aspect of salt injury, but it has little connection with water relations.

Effect of the External Water Potential. A change of the water potential immediately outside the cell, i.e. in the wall, causes an influx or efflux of water until equilibrium is reestablished and the internal water potential has adjusted to the new conditions. At this state, both pressure and osmolality within the cell will have changed and their effects on all reactants of a chemical reaction must be considered.

b) Chemical Reaction Rates. What has been said of the equilibrium holds equally for reaction rates if effects of pressure and osmolality are applied to the formation of activated compounds.

It has often been postulated that the degree of hydration of many compounds is critical for reaction rates. One must thus distinguish between a main reaction and the many hydration reactions. Pressure and solutes will interfere with either type of reaction. An analogous distinction must apply to the hydration of an enzyme protein and the reaction which it catalyzes. For the hydration process the equilibrium is important; for the main reactions the interest is usually in the rate. Unfortunately little or nothing is known about these aspects of water relations. It is not even possible to state whether, for instance, an increase in hydration stimulates or inhibits enzymatic activities.

c) Membrane Equilibria. Membrane equilibria are characterized by unequal solute distributions, by electrical potential differences across membranes and by pressure differences. Thus, it is hardly possible to isolate water relations from other aspects of metabolism. The cytoplasm is compartmentalized through a system of semipermeable membranes, and solute and pressure distributions are not homogeneous. Pressure differences are related to swelling of organelles. As long as differences in osmolalities are maintained, compensating pressure differences more or less independent of the absolute pressure and absolute osmolality must occur.

The function of at least some organelles is known to change with swelling and shrinking, but the mechanism of interaction remains obscure. Effects on membrane permeabilities of some substrates or on the steric arrangement of enzymes attached to membrane surfaces are definitely possibilities for the site of injuries. Consider now a chemical reaction involving water that occurs inside an organelle. An increase in external osmolality will cause shrinkage of the organelle, the internal osmolality increases, the pressure decreases. The response of the chemical reaction to these changes cannot be predicted without further knowledge; it can be either stimulatory or inhibitory. Such examples demonstrate how meaningless it is to discuss a hydration of the cytoplasm as a whole and to expect a unique relationship between hydration and metabolism.

d) Transport Processes Across Membranes. Changes in mechanical equilibrium through stretching of membranes could possibly affect permeabilities. Regarding water, the physiological consequences are probably minor because permeabilities, at least for the exchange between vacuoles and the immediate surroundings, are always sufficiently high to allow for a rapid adjustment of equilibrium.

e) Mechanical Equilibria. Possible effects of mechanical equilibria at the subcellular level have been briefly considered in connection with membrane equilibria. Membrane equilibria also affect mechanical equilibria at the cellular and at the organ level, where they can be of great physiological importance. The mechanical support of most leaves depends on the turgidity of the tissue. Wilting, caused by a loss of turgidity, is well known to be an unhealthy condition, although the mechanism through which plant metabolism is injured is not completely clear. Short-term wilting from which plants appear to recover fully may only cause a temporary interruption of photosynthesis, whereas during a prolonged wilting period a permanent injury may result from cytorrhysis.

The stomatal aperture, and through it the stomatal resistance, are also determined by a mechanical equilibrium in which the turgor pressure of the guard cells plays a key regulatory role.

f) Movement of Cells and Organs. As with chemical and membrane equilibria, the mechanical equilibrium at the cellular and organ level also has its dynamic counterpart. Motions related to turgor pressure are well known in the plant kingdom. The ejection of fern spores or sporangia, or of seeds of some higher plants are water-related processes that play a role in the propagation of species. In other cases like that of the nastic movements of *Mimosa pudica* a physiological significance is less clear.

IV. Conclusions

In the course of evaluating various aspects of plant water relations, the conclusion has been reached that a primary lesion results from insufficient turgor pressure relations through affecting mechanical equilibria. The exact mechanism through which inadequate turgidity injures plants is not known in all cases. In severely wilting leaves, cytorrhysis with folded and collapsed cell walls and with emptied cell volumina resulting from cavitation of water under tension is a definite possibility for causing permanent damage.

Over the past decades the main interest in plant-water interactions has focused on biochemical effects. However, a correlation of the biochemical pattern with moisture stress does not prove that the primary lesion consists of direct interferences. On theoretical grounds one would expect chemical reaction equilibria and rates as well as swelling of organelles to be relatively independent of the water potential and of the turgor pressure. The biochemical system of a plant cell as well as the fine structure are thus buffered to some degree against fluctuations or turgor pressure and water potential. However, it is not possible to predict generally the direction and the magnitude of a chemical reaction's response to a change in water potential, in pressure, or in osmolality; each individual situation must be investigated separately.

References

Christiansen, K.: Integrity and respiration of red beet mitochondria isolated and examined in different sucrose concentrations. Physiol. Plantarum **21**, 510–518 (1968).
Eaton, E. M., Olmstead, W. R., Taylor, O. C.: Salt injury to plants with special reference to cations versus anions and ion activities. Plant Soil **35**, 533–547 (1971).
Hochochka, P. W., Somero, G. N.: Strategies of biochemical adaptation. Philadelphia, London, Toronto: W. B. Saunders Co. 1973.
Hoffman, G. H., Splinter, W. E.: Water potential measurements of an intact plant-soil system. Agron. J. **60**, 408–413 (1968).
Karzel, R.: Über die Nachwirkung der Plasmolyse. Jb. Wiss. Bot. **65**, 551–591 (1926).
Kramer, P. J., Brix, H.: Measurement of water stress in plants. UNESCO Arid Zone Res. **25**, 343–351 (1965).
Oechel, W. C., Strain, B. R., Odening, W. R.: Tissue water potential, photosynthesis, ^{14}C-labeled photosynthate utilization, and growth in the desert shrub *Larrea divaricata* Cav. Ecol. Monographs **42**, 127–141 (1972).
Oertli, J. J.: Effects of external solute supply on cell elongation in barley coleoptiles. Z. Pflanzenphysiol. **74**, 440–450 (1975).
Slatyer, R. O.: An underlying cause of measurement discrepancies in determination of osmotic characteristics in plant cells and tissues. Protoplasma **62**, 34–43 (1966).
Steinbrink, C.: Weiteres über den Kohäsionsmechanismus von Laubmoosblättern. Ber. Deut. Botan. Ges. **28**, 19–30 (1910).
Walter, H.: Zur Klärung des spezifischen Wasserzustandes im Plasma und in der Zellwand bei der höheren Pflanze und seine Bestimmung, Part I. Ber. Deut. Botan. Ges. **76**, 40–53 (1963).
Walter, H., Kreeb, K.: Hydratation und Hydratur des Protoplasmas der Pflanzen und ihre ökophysiologische Bedeutung. Wien: Springer 1970.
Weatherly, P. E.: Studies in the water relations of the cotton plant I. New Phytologist **49**, 81–87 (1950).
Yang, S. J., Dejong, E.: Measurement of internal water stress in wheat plants. Can. J. Plant. Sci. **48**, 89–95 (1968).

C. The Soil–Plant–Atmosphere Continuum

J. J. OERTLI

I. Introduction

Since the beginning of plant physiology the human mind has been intrigued with the flow of water in the soil-plant-atmosphere continuum (SPAC)[1]. In the last century a prime interest focused on the mechanism of flow from roots to plant tops. This problem seems to have been solved by the cohesion theory, proposed and elaborated by Böhm (1893), Askenasy (1895), and Dixon and Joly (1895). The boldness that was originally necessary to propose such a theory in apparent contradiction to the accepted teaching of physics is evident not only from the language in which scientific debates were conducted in those days, but also from doubts that are expressed even in these days in particular by thinking students.

Farmers, must have observed from the early days of agriculture that water can be extracted from soils by plants only to a limited extent and that failure of soils to supply sufficient quantities of water will reduce crop production. The fact that computer programs are now written with the purpose of optimizing water-use efficiency is a clear sign that the problem is not outdated.

In this chapter the ecophysiology of plant water is investigated by establishing a causal link between critical functions of plant water and the environment. Through a functional connection via the SPAC it will become possible to explain many plant responses to environmental variables. An actual system involves exchanges of matter, energy, and momentum and it can on occasions be far from isothermal. Here only a relatively simple version restricted to isothermal conditions can be discussed.

It is self-evident that a better understanding of water ecophysiology is necessary to solve the important practical problem in agriculture of optimizing the water-use efficiency.

---

[1] Permeabilities of plant membranes for water are probably sufficiently high normally to prevent establishment of large water potential differences across plant membranes, i.e. a thermodynamic continuum of the water potential exists across the soil–plant–atmosphere system. However, the continuum is absent for pressures and osmolalities: plant membranes are locations of discontinuities. Pressures and osmolalities represent important aspects of plant water relations; for this reason the term SPAC is used here only as a concession to uniformity.

II. Description of the Turgor Pressure as a Function of Environmental Variables

The primary metabolic lesion from suboptimal water relations is not well understood, but the indication is that injury often occurs because of inadequate pressure relations (see Oertli, this volume Part 1:B). For this reason an attempt is made here to express pressure relations in terms of environmental variables. As an example the turgor pressure is chosen. It is the difference between the pressure acting inside the cell on the wall and 1 bar. Other pressure differences could have been chosen as well. Then the treatment would have been somewhat more complicated but not fundamentally different. To a good approximation the turgor pressure P can be expressed by (Oertli, 1971)

$$\bar{V}P \approx \Psi_p = \Psi - \Psi_s$$

where $\bar{V}$ is the partial molal volume of water, P the turgor pressure, Ψ, Ψ_p, and Ψ_s total, pressure, and vacuolar osmotic water potentials. The osmotic component is related to the cellular solute content. Van't Hoff's law is an example of such a relation:

$$\Psi_s = -\bar{V}\frac{nRT}{V}.$$

If elastic properties of cell walls are known, the volume V can be expressed as a function of the fixed Volume V_0 at incipient plasmolysis, i.e. at the stress-free state of cell walls. Hooke's law can now be applied to describe the expansion of a cell wall. In order to simplify the formulation, it is assumed that a prismatic cell expands only in a longitudinal direction. One obtains

$$V = V_0 \frac{A}{E A_w} P$$

where A is the cross-section of the cell perpendicular to the direction of expansion, E the linear modulus of elasticity and A_w the cross-section of the expanding cell walls. Combining these expressions one obtains the following quadratic equation for the turgor pressure:

$$\bar{V}P = \Psi - \frac{\bar{V}RTn}{V_0\left(\frac{A}{E A_w}P + 1\right)}.$$

It follows that the turgor pressure depends on the water potential, the solute concentration at incipient plasmolysis n/V_0, and on certain cell wall properties. Each of these variables must now be expressed as a function of the environment.

At present, this is not possible for cell wall properties and also for the osmolality the task is rather too difficult. This chapter, therefore, is restricted to a description of water potential relations. Solutes will be included inasmuch as they play a role in the link of water potentials between environment and plant cell.

III. Water Flow in the SPAC as a Link Between Plant and Environment

1. Potential Changes in the Flow System

Following proposals of Huber (1924), Gradmann (1928), and van den Honert (1948) the SPAC is described analogous to the flow of electrical current in an Ohm's system. Accordingly, the flow in any section is given by

$$J = \frac{-\Delta \Psi}{R}$$

where J is the flux in mass/time, $-\Delta \Psi$ the potential drop in energy units, and R the resistance, defined by the above equation. In the plant, the transport of water occurs in an intricately complicated system of parallel and series flows. The leaf water potential Ψ^l is related to the soil water potential Ψ^s through

$$\Psi^l = \Psi^s + \Delta \Psi.$$

2. Initial and Final States

For a description of flows under steady-state conditions, all fluxes should start at the same equipotential surface and terminate at a different equipotential surface. The question is where to locate such surfaces. They should remain practically invariable during steady-state flow. For this reason they must be put in a place where the water potential remains more or less constant over a considerable period. A free water table, for example, could fulfil these conditions for the initial state.

3. Resistances

Since the resistance to flow is defined by the quotient $R = -\Delta \Psi / J$, attention must be paid to the fact that the so-called diffusive resistance $1/D$ in the gaseous phase has a different physical dimension and is not directly comparable to R. For example, in a series flow the two resistances are not additive. A relation between them is obtained from Fick's first law of diffusion

$$J = D A\, dc/dx = D A\, \Delta\Psi/\Delta\Psi\, dc/dx = \frac{D A \Delta \Psi}{R_g T \Delta \ln c} dc/dx = -\Delta \Psi / R$$

(R_g = univ. gas constant, A = area, c = concentration, x = distance). It follows that

$$R = \frac{R_g T \Delta x \Delta \ln c}{D A \Delta c}.$$

For a branched flow system, as is normally the case in plants, the resistances of parallel flows can be combined by taking the harmonic mean

$$\frac{1}{R} = \frac{1}{R_1} + \frac{1}{R_2} + \cdots$$

where the various resistances must be taken between equipotential surfaces.

4. The Turgor Pressure

From the above equations an expression for the turgor pressure at a certain location is obtained:

$$P = \frac{\Psi^s - \sum J R_i - \Delta \Psi_g^l - \Psi_s^l}{\bar{V}}$$

where Ψ_s must be substituted as shown above. R_i is the resistance in a section of the series flow, $J R_i$ the corresponding potential drop and $\sum J R_i$ the potential drop over the entire length. For a branched conduit, R_i is the harmonic mean of resistances in parallel. This is the equation that connects the turgor pressure to environmental factors. The turgor pressure is lower, the more negative the soil water potential Ψ^s is; the higher the transpiration rate J, the higher the resistance R, the taller the plant, and the less negative the osmotic water potential component is.

This relatively simple approach has been chosen for pedagogic reasons. The flux J through equipotential surfaces is constant along the flow system. Therefore effects of principal resistances are immediately apparent and amenable for qualitative discussion. But it can be considered an approximation only and, depending on the morphology of plants, a quantitative treatment can seriously be in error for two main reasons:

1. The position of the equipotential surface is not horizontal, but this is a requirement if the harmonic mean of resistances in a branched circuit is taken at the same time as the potential loss due to elevation is treated as a separate term.

2. The position of an equipotential surface is not invariable. A change in resistance, e.g. at a different side branch from the one concerned, will change the shape and the position of an equipotential surface.

Such difficulties could be solved by treating each conduit separately and formulating the flow for the whole system analogous to Kirchhoff's first and second laws, a task that in detail would require an extremely large number of simultaneous equations. Separate equations may even have to be applied to individual vessels or bundles since it is not impossible that the water potential varies from one side of a tree trunk to the other at the same elevation.

Richter et al. (1972) only consider an isolated branch circuit and again in analogy to electricity give the frictional loss in water potential as

$$\Delta \Psi = \sum J_n R_n.$$

If we recall that the aim is to describe the water potential at a particular location in terms of environmental variables (and of resistances and position above ground) it becomes clear that Richter's formulation requires the same amount of work, namely an analysis of the whole system analogous to an application of the two laws by Kirchhoff (see also Richter, this volume Part 1:D).

IV. The Solute-free Transport System

1. Steady State

Suppose water flows from a soil through plants to the atmosphere. With given initial and final water potentials as well as a definite plant system a certain distribution of water potentials, decreasing in the direction of flow, is established. If now a particular cell is considered, its water potential and through it the turgor pressure is higher the higher the initial and the lower the final water potential is; it is also the higher the lower the resistance before and the higher the resistance after the cell and vice versa.

The soil resistance decreases with the water content because of a decrease of the cross-section filled with water and also because of a decrease of the effective diameter of pores that are still filled with water (see Part 2 of this volume). Consequently a dry soil means lower turgor pressure for two reasons, namely because of a reduced soil water potential at the source and also because of a larger potential loss during flow in the soil.

Increased xylar resistances can result from wilting diseases during which the vessels become plugged or from cavitation of water columns under tension after which conduits are air-filled. In either case the potential loss is increased and the turgor pressure reduced.

An increased resistance after the point of interest, viz. the leaf cell, will raise the water potential and the turgor pressure. This happens with stomatal closure. Thus, an important function of stomata is to preserve foliar water potential and turgor pressure. Since this turgor-preserving action also applies to guard cells, stomata may open again after the turgor pressure has been restored. This calls for a new steady state at which the turgor pressure is again reduced and stomata will close anew. The resulting rhythmic behavior has indeed been observed (Hopmans, 1969).

Under good growing conditions, potential losses in soils due to frictional flow are negligible, but as the permanent wilting point is approached these losses gain more importance. Considerable uncertainty exists about potential drops across roots. The values reported in the literature vary greatly. A few hundred Joules kg^{-1} (a few bars) may perhaps be a reasonable figure for normally transpiring plants. Within the xylem under average flow conditions the potential loss from frictional flow is about equal in magnitude to the loss from a change in elevation. Differing observations by Scholander et al. (1965) have recently been proved to be erroneous (Richter et al., 1972). Quantitative data on this subject will be presented in the following chapter (Part 1:D) of this volume by Richter.

2. Transient States

Transient state flow of water in the SPAC can be described by combining flow equations with the law of mass conservation. For a linear rate law like the Darcy equation one obtains for a one-dimensional situation

$$\frac{\partial \Theta}{\partial T} = \frac{\partial}{\partial x}\left(D_\Theta \frac{\partial \Theta}{\partial x}\right) = \frac{\partial}{\partial x}\left(\frac{K_{on}}{C_\Theta} \frac{\partial \Theta}{\partial x}\right)$$

where Θ is the water content, $D_\Theta = K_{on} \, d\Psi/d\Theta$ is the diffusivity, $C_\Theta = d\Theta/d\Psi$ is the water capacity and K_{on} the conductivity. Capacities exist in soils and in plants. In elastic systems such as cells the capacities depend on the elastic properties of the walls.

Any change in environmental conditions or in resistances defines a new steady state with a definite distribution of water potentials. During the adjustment to the new steady state water is lost or gained from tissues, thus transpiration may exceed water absorption or vice versa.

V. Effects of Solutes in the SPAC

A reduction of the osmotic water potential component in the conductive system by solutes will—if not compensated by a higher pressure—lower the turgidity. Whether a compensation occurs or not depends on the nature of the conductive system, i.e. on the characteristics of the "membranes". Depending on their permeabilities a change in water potential due to solutes may or may not be effective as a driving force. Of particular interest is a situation in which such a change occurs without an effect on the water transport, so that the foliar water potential changes but the transpiration rate which is determined largely by flow conditions in the gaseous phase stays constant.

1. The Flux Equation

Thermodynamics of irreversible processes furnishes an equation for membrane transport as

$$J_{H_2O} \approx J_v = -L_P(\Delta \Psi_p + \sigma \Delta \Psi_s)$$

where J_{H_2O} is the flux of water, J_v the volume flux, L_P, L_{PD} are linear flux coefficients, $\sigma = L_{PD}/L_P$ is the reflection coefficient.

For a perfectly semipermeable membrane which allows only the passage of water but not of solutes, the reflection coefficient $\sigma = 1$ and the osmotic potential component is also as effective as the pressure component. The flux is thus proportional to the total potential drop across the membrane.

For the other extreme, a completely leaky membrane, the reflection coefficient is $\sigma = 0$, the osmotic component is ineffective and the volume flux is proportional to the pressure component. One might object to referring to a membrane in this situation because it applies to the flow of any open body of water. It is justifiable, however, to treat it as a limiting case of a membrane.

2. Effects of Osmotic Potential Differences in Leaky Membranes

The term $\Delta \Psi_s$ can vary widely. As a consequence the total potential difference

$$\Delta \Psi = \Delta \Psi_p + \Delta \Psi_s$$

can become larger or smaller without affecting the flow rate through a leaky membrane. The potential difference could even reverse its sign without changing the direction of flow. In this case water will flow against its chemical potential gradient. If such an uphill flow should occur under steady state conditions it is further necessary that the gradient of osmolalities in the conductive system is maintained. These conditions are fulfilled in plants through continuous influxes of solutes at one location and effluxes at another place. Such reversals of water potential gradients under steady state flow conditions in plants have been observed repeatedly (Oertli, 1966; Taerum, 1973). Whether an uphill transport of this nature is called active is merely a question of definition.

3. "Resistance" in the Presence of Solute Gradients

Applying the previous definition for resistance

$$R = \frac{-\Delta \Psi}{J}$$

to a partially semipermeable membrane in the presence of solute gradients, and setting $K = 1/R$ for the conductivity, one obtains for the flow the following equation:

$$J_v = -K \Delta \Psi = -\frac{\Delta \Psi}{R} = -L_P(\Delta \Psi_p + \sigma \Delta \Psi_s).$$

It follows for the conductivity that

$$K = 1/R = \frac{L_P \Delta \Psi_p + L_{DP} \Delta \Psi_s}{\Delta \Psi_p + \Delta \Psi_s} = \frac{L_P \Delta \Psi_p + L_{DP} \Delta \Psi_s}{\Delta \Psi}.$$

The conductivity is thus a weighted mean of the coefficients L_P and L_{PD}. A flow against the decreasing potential gradient now becomes equivalent to a flow with a negative resistance.

4. Physiological Consequences

These extra potential losses or gains originating from gradients in osmolality of leaky membrane systems can have physiological significance. The water potential is either increased or decreases relative to the flow in a perfectly semipermeable membrane. The foliar turgor pressure increases or decreases in relation to the water potential. Under saline conditions when the supply of salts to leaf cells could exceed their capacity to remove them from the extracellular space, the continuous transpirational loss of water could lead to a residual extracellular accumulation of solutes. Situations could develop which lead to plasmolysis of leaf cells. The evidence for such a residual accumulation of solutes has been discussed by Oertli (1971).

VI. Changes in Resistances or Potential Differences?

The inability of an osmolality gradient to induce a flow of water in a leaky membrane system has other interesting consequences besides making a flow of water possible against a decreasing potential gradient. Suppose an experimenter applies a constant total potential difference $\Delta\Psi$ with variable potential components to a perfectly semipermeable membrane system. Suppose he observes that with increasing external salinity the flux of water decreases. The investigator will probably conclude that the resistance has increased since, after all, the applied potential gradient remained constant. This conclusion is not necessarily correct, as is evident from an analysis of a simple model:

outside	boundary layer B	"membrane" M	inside.

By hypothesis the inside consists of pure water at a constant pressure, the outside consists of a solution in which osmolalities and pressures are varied in such a way that the water potential remains constant. Flow is from left to right and the "membrane" is perfectly semipermeable.

On the outside, a convective flow of solution occurs toward the membrane. Only water will pass across the semipermeable barrier M and solutes will accumulate residually in the boundary layer. A back diffusion will develop as a consequence of this concentration gradient and steady state conditions are reached when back diffusion and rate of accumulation are equal. At this stage an osmotic potential difference $\Delta\Psi_{sB}$ across the boundary layer can be observed. For the flux of H_2O in sequence through boundary layer B and membrane M one finds

$$J_{H_2O} = -L_{P_B}(\Delta\Psi_{pB} + \sigma_B \Delta\Psi_{sB}) = -L_{P_M}(\Delta\Psi_{pM} + \sigma_M \Delta\Psi_{sM}).$$

The boundary layer is equivalent to a leaky membrane, hence $\sigma_B = 0$, the membrane M is semipermeable so that $\sigma_M = 1$. The total potential drop across membrane plus boundary layer is

$$\Delta \Psi = \Delta \Psi_{pB} + \Delta \Psi_{sB} + \Delta \Psi_{pM} + \Delta \Psi_{sM}.$$

Combining these equations one obtains for the flux

$$J = -\frac{L_B L_M}{L_B + L_M} (\Delta \Psi - \Delta \Psi_{sB}).$$

The term $L_B L_M/(L_B + L_M)$ representing the overall conductivity is unchanged and independent of salinity, but the effectiveness of the constant potential difference $\Delta \Psi$ has been reduced by $\Delta \Psi_{sB}$. Even though none of the flux coefficients (conductivities) have changed and, the total potential drop has not altered, the flux rate is reduced with increasing salinity because of the development of potential gradients in the boundary layer, an osmotic potential gradient that is inefficient as a driving force. It follows that salt-induced "resistance changes" reported in the literature may not necessarily be true resistance changes but rather potential gradients that are inefficient for flow. Evidence for such a situation in a plant system has been reported by Oertli and Richardson (1968).

VII. Conclusions

A primary task of ecophysiology is to find functional links between plant metabolism and the environment. Before such links can be investigated in detail, the key processes through which the environment interacts with the plant must be known, i.e. information on the site of the primary lesion must be available. Turgor pressure, which certainly plays an important physiological role in cellular water relations, depends on the vacuolar osmolality and on elastic properties of the cell wall besides the water potential.

The tissue water potential itself can be expressed as a function of the external (soil) water potential and of potential changes along the flow system of water from soil through the plant to the atmosphere. In this flow system, potential changes occur because of frictional losses, because of changes in elevation, as well as because of osmolality gradients in imperfectly semipermeable membranes. In this last case, potential losses or gains can be observed without any concomitant change of the water flux rate. Such ineffective solute-induced potential gradients lead to changed apparent resistances which can even change their algebraic sign in the extreme case when the potential gain exceeds potential losses.

References

Askenasy, E.: Über das Saftsteigen. Botan. Zentralbl. **62**, 237–238 (1895).
Böhm, J.: Capillarität und Saftsteigen. Ber. Deut. Botan. Ges. **11**, 203–212 (1893).

Dixon, H. H., Joly, J.: On the ascent of sap. Roy. Soc. (London), Phil. Trans. B **186**, 563–576 (1895).
Gradmann, H.: Untersuchungen über die Wasserverhältnisse des Bodens als Grundlage des Pflanzenwachstums. Jb. Wiss. Botan. **69**, 1–100 (1928).
Honert, T. H. van den: Water transport as a catenary process. Faraday Soc. Discuss. **3**, 146–153 (1948).
Hopmans, P. A. M.: Types of stomatal cycling and their water relations in bean leaves. Z. Pflanzenphysiol. **60**, 242–254 (1969).
Huber, B.: Die Beurteilung des Wasserhaushaltes der Pflanze. Ein Beitrag zur vergleichenden Physiologie. Jb. Wiss. Botan. **64**, 1–120 (1924).
Oertli, J. J.: Active water transport in plants. Physiol. Plantarum **19**, 809–817 (1966).
Oertli, J. J.: A whole-system approach to water physiology in plants. Advan. Frontiers Plant Sci. **27**, 1–200; **28**, 1–73 (1971).
Oertli, J. J., Richardson, W. F.: Effects of external salt concentrations on water relations in plants IV. Soil Sci. **105**, 177–183 (1968).
Richter, H., Halbwachs, G., Holzner, W.: Saugspannungsmessungen in der Krone eines Mammutbaumes. Flora **61**, 401–420 (1972).
Scholander, P. F., Hammel, H. F., Bradstreet, F. D., Hemmingsen, E. A.: Sap pressure in vascular plants. Science **148**, 339–346 (1965).
Taerum, R.: Occurrence of inverted water potential gradients between soil and bean roots. Physiol. Plantarum **28**, 471–475 (1973).

D. The Water Status in the Plant
Experimental Evidence

H. RICHTER

I. Introduction

In Part 1:B and 1:C of this volume Oertli gives a detailed theoretical treatment of water in the plant body and in the soil–plant–atmosphere continuum. This leaves the task of providing quantitative data for at least some of the parameters discussed. The choice of these parameters and of the papers dealing with them has to be rather subjective, since even the most recent literature is far too voluminous to be covered completely.

A thorough description of water status in the plant should contain:

1. Information on the numeric value of total water potential at each point in the plant body,
2. Information on the state of the soil–plant–atmosphere continuum requiring these potentials, and
3. Information on the forces adjusting total potential in different compartments of the plant body.

A set of two equations may be used to formalize such a description. The first gives total water potential as a function of demands from the continuum:

$$\Psi = \Psi_{\text{stat}}^{\text{soil}} + \Psi_g + \Psi_f \qquad (1)$$

where Ψ = total water potential, $\Psi_{\text{stat}}^{\text{soil}}$ = static soil water potential, Ψ_g = gravitational potential, and Ψ_f = frictional potential.

The second equation describes total water potential as the result of independent agents lowering the free energy of water in body fluids compared with the reference state:

$$\Psi = \Psi_s + \Psi_m + \Psi_p \qquad (2)$$

with s, m, and p indicating osmotic, matric, and pressure (hydrostatic) terms. This is a generalized description for an isothermal system. Since not all three factors are of importance everywhere in the plant body, one may derive from Eq. (2) specialized equations applicable to single compartments only (e. g. vacuole, protoplasm, cell wall, xylem; cf. Brown, 1972).

Thus, Eq. (1) describes demands and Eq. (2), responses. It will be the task of this chapter to report on the currently available methods for determining the components of these two equations, to assess variability and ranges of total and component potentials, and to mention endogenous and exogenous causes for changes.

II. Current Methods for the Determination of Total Water Potential and Its Components

This section gives a short survey of current approaches to the experimental investigation of water status. A recent monograph (Slavìk, 1974) is recommended as a source of more detailed information. It seems appropriate to follow here the course suggested by Eqs. (1) and (2) and discuss their terms in the above order.

a) Total Water Potential. Presently there are two experimental methods in widespread use for the determination of Ψ, namely thermocouple psychrometry and the pressure chamber technique. Psychrometry is claimed to provide a straightforward measurement of total water potential, typically in leaf parenchyma or other soft tissues. The pressure chamber, on the other hand, can measure only the pressure component of xylem potentials. However, since concentrations of osmotically active substances in the xylem sap are very small in most cases and matric components are completely absent, the difference between total water potential and xylem pressure potential is usually negligible. Both chamber and psychrometer determinations are restricted by a number of problems which make the various comparative studies reported in the literature less reliable than one would desire. Thus preference remains, to a certain extent, a matter of faith, since an independent standard is not available. In-depth information on these two modern methods may be found in recent monographs (Brown and van Haveren, 1972; Ritchie and Hinckley, 1975). Older methods are evaluated by Barrs (1968) as well as by Slavík (1974). Sampling for all these methods usually results in the destruction of the plant tissue. Only dewpoint hygrometry (Neumann and Thurtell, 1972; Neumann et al., 1974; Campbell and Campbell, 1974), an advanced thermocouple technique, seems to provide a means for nondestructive, repeated measurements of leaf water potentials in a limited range of objects.

b) Static Soil Water Potential. Methods for the determination of this rather complex term are reviewed by Hillel (1971) and Slavík (1974). Again, thermocouple techniques are of predominant importance. In an indirect approach, plant water potentials, under conditions preventing water flow through the plant body, are used as an indicator for Ψ_{stat}^{soil} in the root zone. The validity of this method depends on a state of complete equilibrium between soil and plant.

c) Gravitational Potential. Work must be done to transport water reversibly and isothermally to a specified elevation h above the reference point in the system (i.e. the absorbing zone of the roots). The exact value for $\Psi_g = -\varrho g h$

depends on the density of water ϱ (which changes with temperature) and on the gravity constant g (which is somewhat variable with geographic location). However, the total is always very close to $0.1\,\mathrm{bars\,m^{-1}}$. Since one needs no complicated instruments when determining Ψ_g, there were correct numerical values at hand at a time when not even an order of magnitude could be given for any other term in Eq. (1). Consequently, the relative importance of gravitational potential was vastly overestimated especially by tree physiologists.

d) Frictional Potential. Work must be done to overcome friction of water in the pathway. Theoretically, Ψ_f may be given as the sum, along the branched conduit from the bulk of soil S to point P in the plant body, of products from partial fluxes f_i and partial resistances r_i. Thus,

$$\Psi_f = -\sum_S^P f_i \cdot r_i. \tag{3}$$

In practice, Ψ_f must be determined from Eq. (1). Calculations following Eq. (3) are blocked by the sheer difficulty of direct determinations of fluxes and resistances. In the soil, resistances vary with the saturation state, and pathways for fluxes toward each root are extremely complex. There are only very few reliable experimental data, and not even basic parameters for a theoretical treatment are out of discussion (Williams, 1974). Conduits in the plant body are always branched. Thus, potentials in a given twig or leaf will be influenced by the amount of water flowing toward parallel twigs or leaves and sharing some sections of the conduit (Richter, 1973). Transpirational water flow out of single branches or leaves, although easy to measure, does not therefore permit calculation of a total resistance in the pathway from Eq. (3), unless fluxes through all the parallel transpiring organs are also known. A recent development (Sheriff, 1973, 1974) provides for the direct measurement of fluxes through narrow stems. This approach as well as the heat balance techniques for woody tissue could perhaps yield information on the magnitude of all the series-linked volume fluxes in the above-ground part of the pathway. Experimental determination of partial resistances, on the other hand, is a lengthy task not taken up by many dedicated experimenters thus far. Heine (1971) evaluates data for wood and xylem resistances determined by pushing water through stem sections under the action of a known pressure head. Calculations of the relative magnitude of resistances in root, stem and leaves for different herbaceous species and for some trees are discussed by Jarvis (1975). Certainly more experimental work would be desirable.

In Eq. (2), the terms on the right side may assume widely different values according to the precise location of the point in the plant body where the water status is under consideration. Experimental methods for the determination of these terms are often applicable to a single compartment only.

e) Osmotic Potential. Methods for its determination may be divided into cell methods and methods for expressed fluids. The most important "cell method" is limiting plasmolysis which gives osmotic potential at the point of zero turgor potential of living cells in an excised tissue. Most routine determinations are, however, carried out on expressed cell sap or xylem fluid. Cryoscopy, psychrometry and vapor pressure osmometry may be used for both. Correct preparation of the sap sample is of crucial importance. For vacuolar sap, this involves the killing of the tissue and the application of pressure (Walter and Kreeb, 1970), whereas uncontaminated xylem sap may be collected by centrifugation or with a pressure chamber. The pressure chamber may also be used for the determination of an average osmotic potential of whole organs (Scholander et al., 1965; Tyree et al., 1973). Values for the protoplasm and its organelles, although of great interest far beyond water relations research, are not available without recourse to biochemical separation techniques which greatly interfere with solute distribution in the cell compartments. Detailed studies of osmotic potentials in the interfibrillar solution of cell walls seem to be completely lacking as well. Evidently, solute content in the cell wall interstices is closely linked to the solute content of xylem sap. In both cases, the general assumption is one of a minor, almost negligible concentration of osmotically active substances. However, this may not be justified even for xylem sap, especially with plants from salt sites (Kaplan and Gale, 1974). In the cell walls of transpiring organs, salts may be further concentrated when water is removed faster by transpiration, than salts are taken up into the protoplasm.

f) Matric Potential. Adsorption on colloids and capillarity effects lower the potential of water. This "matric potential" is certainly unimportant in the lumen of xylem elements, and measurable effects in the vacuole are probably restricted to cases of high slime content. The influence of this term on the water status in protoplasm and cell wall, on the other hand, must be important. Unfortunately, dependable experimental assays are largely restricted to artificial substrates or isolated cell walls (Wiebe, 1966), whereas measurements on organized plant tissue (review by Miller, 1972) have repeatedly come under heavy criticism.

g) Pressure (Hydrostatic) Potential. Approaches and results are very different depending on the compartment. Pressures in the xylem are usually negative, thus lowering total water potential in the xylem sap. Since pressure potential is generally by far the most important single factor of total xylem potential, pressure chamber determinations are considered, and often safely so, to yield good estimates not only of pressure potential but also of total water potential. Some authors assume hydrostatic potential in the cell wall to be always equal to zero, since intercellular spaces bordering on the walls contain air under a pressure of 1 bar (e.g. Brown, 1972). This view seems not to be justified, since intermicellar and interfibrillar spaces of the cell wall are direct extensions of the xylem lumina, and they must of necessity be at the same potential. In the immediate vicinity of macromolecules, water potential is of course reduced by adsorption, but the forces involved have no long-range effect. Potentials in the center of microscopic and submicroscopic spaces are probably lowered by a state of tension which extends from the xylem ducts right into the walls. Experimental proof for this negative pressure is, however, lacking.

Positive hydrostatic pressure, on the other hand, acts on protoplasm and vacuole. Direct methods for the determination of turgor pressure are unreliable, except where pressure transducers are brought into contact with suitable single cells (e.g. Zimmermann et al., 1969). Thus, determinations are generally carried out by calculating differences between total water potential and osmotic potential of the vacuole. The value thus obtained is also the correct one for the cytoplasm.

In retrospect, one sees that all the elements of Eq. (1) are open to direct determination, except frictional potential. The equation may therefore be solved. Equation (2) can be solved with good reliability for two compartments in the plant body, namely xylem and vacuole. For protoplasm and cell wall, at least two terms cannot be determined directly, and a solution is therefore impossible.

It has been mentioned that the above equations are applicable to isothermal systems only. Movement of water over a nonisothermal region requires application of the thermodynamics of irreversible processes. However, Spanner (1972) provides theoretical evidence that the traditional concept may be safely used under most ecological conditions, since temperature gradients are small over distances where diffusion plays a major role in transport processes.

III. The Range of Water Potentials Hitherto Determined and the Continuum Conditions Favoring Extreme Values

Total water potential in plant tissues has to balance the demands imposed by the soil–plant–atmosphere continuum as formalized in Eq. (1). It is sound, therefore, to look for the highest potentials in well-watered plants prevented from transpiration, and thus from frictional potential losses. Under such conditions, some plants will even develop positive pressure in the xylem and demonstrate guttation. However, this does not indicate positive values for total water potential, the occurrence of which is rather unlikely (O'Leary, 1970). In field-grown plants, friction is minimal during the night, especially just before dawn, when stomata are closed and resaturation of the cells has already been completed. To their dismay, some authors found even such "pre-dawn" potentials to be considerably lower than those of the soil. This is not too surprising, however, since species differ widely in the night-time opening of their stomata (Hübl, 1963), and appreciable xylem or soil resistances may preclude complete resaturation before dawn (Richter, 1974; Hinckley and Bruckerhoff, 1975).

According to Eq. (1), very low potentials could be due to low soil water potential, great height of the sampling point above root level, and friction. The classic paper of the pressure chamber technique (Scholander et al., 1965) has already given a good survey of the ranges of potentials encountered in widely contrasting environments. Tables 1 to 5 present some reliable data for low potentials in plants adapted to different habitats. Only the lowest value has been entered for species dealt with in more than one paper. Most of the results come from field studies using different sampling intensities. Thus, there is no guarantee that the reported minima always approximate the lowest level attainable. In general, there are more data available for woody plants than for herbs, and more for plants from dry sites than for those from mesic sites. Among herbs, cultivated species have received predominant attention thus far.

The lowest potentials were found on sites of very low soil potential (Table 1). Some desert plants reach minima of less than -100 bars. Relatively high

Table 1. Minimum values for total water potential in desert plants

Species	$\Psi_{(min)}$ (-bars)	Habitat	Reference
Artemisia herba-alba	163	Negev desert	Kappen et al., 1972
Reaumurea negevensis	108	Negev desert	Schulze et al., 1973
Zygophyllum dumosum	88	Negev desert	Schulze et al., 1973
Franseria deltoidea	85	Sonora desert	Halvorson and Patten, 1974
Larrea divaricata	82	"California"	Scholander et al., 1965
Hammada scoparia	80	Negev desert	Schulze et al., 1973
Eriogonum fasciculatum	76	Sonora desert	Halvorson and Patten, 1974
Krameria grayi	74	Sonora desert	Halvorson and Patten, 1974
Simmondsia chinensis	62	Sonora desert	Halvorson and Patten, 1974
Juniperus californica	59	"California"	Scholander et al., 1965
Atriplex polycarpa	55	"California"	Scholander et al., 1965
Prosopis juliflora	45[a]	Colorado desert	Strain, 1970
Cercidium microphyllum	36	Sonora desert	Halvorson and Patten, 1974
Fouquieria splendens	19	"California"	Scholander et al., 1965
Opuntia basilaris	18[b]	Colorado desert	Szarek, 1974

All determinations were made with the pressure chamber, except (a) where a solution exchange technique, and (b) where a thermocouple psychrometer were used.

Table 2. Minimum values for total water potential in plants from sites with pronounced drought periods

Species	$\Psi_{(min)}$ (-bars)	Country (Habitat)	Reference
Juniperus phoenicea	70	France (Garrigue)	Duhme, 1974
Acacia harpophylla	60[a]	Australia	van de Driessche et al., 1971
Hippocrepis comosa	56	Germany (Triassic Limestone)	Lösch and Franz, 1974
Amelanchier ovalis	55	France (Garrigue)	Duhme, 1974
Cornus mas	52	France (Garrigue)	Duhme, 1974
Rosmarinus officinalis	48	France (Garrigue)	Duhme, 1974
Astrebla lappacea	48[a]	Queensland (dry grassland)	Doley and Trivett, 1974
Pyrus amygdaliformis	45	France (Garrigue)	Duhme, 1974
Juniperus oxycedrus	44	France (Garrigue)	Duhme, 1974
Quercus coccifera	44	France (Garrigue)	Duhme, 1974
Acer monspessulanum	41	France (Garrigue)	Duhme, 1974
Quercus douglasii	41[b]	California (Coast Range)	Griffin, 1973
Quercus agrifolia	37[b]	California (Coast Range)	Griffin, 1973
Lavandula latifolia	37	France (Garrigue)	Duhme, 1974
Buxus sempervirens	34	France (Garrigue)	Duhme, 1974
Prosopis glandulosa	32	Texas (grassland)	Haas and Dodd, 1972

All values from pressure chamber determinations, (a) from laboratory experiments under controlled conditions; (b) "pre-dawn" values.

values characterize drought-avoiding specialists with a capability either for tapping deep-lying water resources *(Cercidium microphyllum)*, for shedding leaves and stopping transpiration *(Fouquieria splendens)*, or for filling large reservoirs

Table 3. Minimum values for xylem pressure potential ($\approx$ total water potential) in mangrove plants. (From Scholander, 1968)

Species	$\Psi_{(min)}$ (-bars)
Sonneratia alba	57
Avicennia marina	54
Aegialitis annulata	52
Aegiceras corniculatum	52
Ceriops tagal	50
Osbornia octodonta	50
Lumnizera littorea	48
Excoecaria agallocha	37

Table 4. Minimum values for total water potential in woody plants at mesic sites

Species	$\Psi_{(min)}$ (-bars)	Country	Height of sampling point (m)	Reference
Malus domestica	26	England	5?	Goode and Higgs, 1973
Sequoiadendron giganteum	25	Austria	25	Richter et al., 1972
Picea abies	22	Austria	19	Halbwachs, 1970
Pyrus communis	21	Australia	4?	Klepper, 1968
Taxus baccata	20	Austria	3	Richter, 1974
Prunus serotina	20	West Virginia	6?	Kochenderfer and Lee, 1973
Pseudotsuga menziesii	20	Oregon	25	Waring and Cleary, 1967
Pinus resinosa	19	Minnesota	7	Sucoff, 1972
Pinus contorta	19	Colorado	6?	Mark and Reid, 1971
Betula pendula	19	Austria	12	Halbwachs, 1970
Vitis vinifera	19	Australia	2?	Klepper, 1968
Quercus prinus	18	West Virginia	6?	Kochenderfer and Lee, 1973
Picea sitchensis	18	Scotland	6	Hellkvist et al., 1974
Pinus sylvestris	17	Sweden	1.5	Hellkvist, 1973
Quercus rubra	16	West Virginia	6?	Kochenderfer and Lee, 1973
Tsuga canadensis	16	Connecticut	0.7	Turner and de Roo, 1974
Camellia sinensis	15	Kenia	2?	Carr, 1971
Phoenix sylvestris	15	India	6	Milburn and Davis, 1973

All values from periods during the growing season without pronounced soil drought. Determination by pressure chamber except for the last entry, where a solution exchange technique was used.
? = height estimated from general descriptions in the paper.

in the body with water at a high potential during rainy periods (succulents, e.g. *Opuntia* species).

Data from sites with pronounced drought periods (Table 2) show a broad overlap with those from deserts, as do the values for mangrove trees (Table 3) rooted in sea water ($\Psi = -24$ to -26 bars). Woody (Table 4) and herbaceous species (Table 5) from mesic sites cover a comparable range of potentials. Their minima during an ordinary growing season do not reach the lowest values in the first three tables. However, there are examples for rather low potentials

Table 5. Minimum values for total water potential in more or less mesophilic herbs

Species	$\Psi_{(min)}$ (-bars)	Substrate	Reference
Phragmites communis	43[a]	lake water	Tuschl, 1970
Triticum aestivum	31[b]	dry soil	Campbell and Campbell, 1974
Hordeum vulgare	29[b]	PEG solution, -20 bars	Singh et al., 1972
Ambrosia trifida	27[b]	dry soil	Zabadal, 1974
Phaseolus vulgaris	26[b]	saline solution, -4 bars	Kirkham et al., 1974
Triticum durum	25[a]	nutrient solution, -0.2 bars	Richter and Ruckenbauer (unp.)
Zea mays	22[a]	moderately dry soil	Brevedan and Hodges, 1973
Sorghum bicolor	21[a]	moderately dry soil	Turner, 1974
Solanum tuberosum	20[b]	moderately dry soil	Shepherd, 1973
Nicotiana tabacum	20[a]	dry soil	de Roo, 1969
Lolium perenne	19[a]	dry soil	Jackson, 1974
Vicia faba	16[a]	dry soil	Kassam, 1973
Cardaria draba	15[a]	well-watered soil	Türk, 1970
Beta vulgaris	14[b]	dry soil	Biscoe, 1972

[a] Pressure chamber.
[b] Thermocouple techniques.

in evergreen conifers during the winter and in early spring, a period of pronounced water stress (Larcher, 1972).

Obviously, minimum values of total water potential are somehow adapted to the habitat, especially its soil. Differences in plants at the same site and similarities at contrasting sites indicate, however, considerable influence of plant factors in the continuum.

One of these plant factors could be height. Investigators of plant water relations have long been fascinated by the question of how trees get their water into transpiring crowns 60 or 80 m above ground. Looking at the minima for trees, it is evident that the gravitational height effect is unimportant in relation to the total. However, the height of a plant also has an indirect effect: it increases the conduit length leading up to a certain point, and in a conduit of given structure, friction must increase with length. Thus trees should presumably develop markedly lower potentials due to friction than herbs. Another look at the data for herbaceous plants shows, however, some surprisingly low values even for species rooted in lake water or nutrient solution. Ψ_f is obviously about equal in herbs and in trees, including the hygrohalophytic mangrove species (Table 3) and tree seedlings (Waring and Cleary, 1967; cf. Table 4).

Very negative values for Ψ_f may be caused by either large fluxes or large resistances [Eq. (3)]. Hellkvist et al. (1974) compile data for resistances from various sources. They show that overall resistances (roughly estimated from the expression $\Delta\Psi \cdot F^{-1}$, potential difference between substrate and transpiring leaves divided by total volume transpired) are as high in small herbs as in high trees. Thus, there seems to be an environmental pressure toward an appreciable frictional potential loss in plants regardless of their size.

Soil water relations at periodically dry sites might impose such a pressure. Soil-moisture characteristic curves show that by far the greater part of the water stored in soils with a low salt content is released in the low-suction range (Hillel, 1971). Favorable light conditions would tend to keep stomata wide open just during periods of highest evaporative demand, namely the noon hours of sunny days. A too-efficient conduit which transports large amounts of water through the plant body without lowering total leaf water potential might then lead to a rapid depletion of water resources in the soil. Frictional potential losses could provide feedback which triggers a preservation mechanism: a lowering of total water potential toward the vicinity of values for osmotic potentials whenever there is an excess flow of water might induce turgor loss and hormonal responses leading to stomatal closure.

This hypothesis does not rule out alternative or additional mechanisms for the restriction of flow during periods of high evaporative demand: a recently discovered direct response of stomata to changes in atmospheric humidity (Lange et al., 1971) seems to play an important role in the field (Lösch and Franz, 1974), even providing for midday recovery of potentials in certain plants at arid sites.

The second factor contributing to Ψ_f is the fluxes in the system. It is the most important prediction of the Huber-Gradmann-van den Honert theory that fluxes are inversely proportional to the largest resistance in the soil–plant–atmosphere continuum, which lies in the gas-phase diffusion pathway between leaf interior and surrounding free air. The opening state of stomata is therefore crucial for fluxes, and closing responses indicate reactions of the plant preventing maximum water flow through the system.

Duhme (1974) derived an estimate for the state of stomata from infiltration pictures and used this together with total water potential for an evaluation of 26 shrubs and trees in a Mediterranean environment. Low water potentials in a species were considered a sign of drought tolerance, whereas pronounced closing reactions of stomata indicated drought avoidance. Duhme could thus arrange his material in different groups characterized by their tolerance-avoidance behavior in the course of the day and of the year. His data show that guard cell reactions have significant influence on differences in minimum potentials at a given site.

However, the opening state of the stomata is not the only factor determining the magnitude of fluxes. Flow is driven by the potential difference between soil and atmosphere; the vapor pressure deficit of the air is a good yardstick for this difference, since potentials in the soil are very close to zero in comparison with those in the free atmosphere. Elfving et al. (1972) therefore used the quotient from vapor pressure deficit and stomatal resistance, determined with a porometer, as an indicator for flux changes in citrus trees under different evaporative demand. This approach could be valuable for the interpretation of responses in plant communities along the lines shown by Duhme (1974). However, caution must be used in the evaluation of such data: such fluxes per unit leaf area cannot give a quantitative basis for the comparison of total fluxes in different plant species or even specimens. Partial fluxes in Eq. (3) have the dimensions of volume per unit time and depend on the total area of the transpiring surface, which is, by necessity, different in different plants (Richter, 1973).

IV. The Component Potentials Adjusting Total Water Potential in the Plant Body: Ranges and Changes

The sum of continuum factors [Eq. (1)] is species-specific: this requires different total potentials in plants at the same site and during the same day-course. The counterbalance in the body is provided by the right side of Eq. (2), with vastly different values for the three terms. It will be worthwhile to survey ranges and changes of osmotic, matric and hydrostatic potentials, and to ask whether there is any single compartment responsible for the possible minimum of total water potential in a given plant. In such a "limiting compartment", the lowest physiologically tolerable sum total of component potentials would have to be less negative than in other compartments. One would expect potential-controlling responses, e. g. regulation of stomatal opening, to depend on the water status in this compartment.

The cell wall may easily be ruled out for this important role. The matrix substances must behave like a gel, coming into vapor pressure equilibrium even with dry air and thus developing several hundred bars of matric potential. The microcapillaries in the wall will perhaps become emptied at somewhat higher potentials when hydrostatic tensions overcome cohesive forces. This process must, however, begin at far higher potentials in the wide xylem elements. Thus wall potentials could presumably easily adjust to the lowest values hitherto found in desert plants.

This possibility has been in dispute concerning the xylem. Here, the pressure potential is the only important term on the right side of Eq. (2); direct determinations (cf. Greenidge, 1957; Oertli, 1971) yield vastly different values for the tensions a capillary water column can withstand without breaking. Generally it is assumed that tensions encountered in xylem conduits do not cause column breaking in artificial capillaries, but the situation may be different in vessels and tracheids. Various papers by Milburn (cf. Milburn and McLaughlin, 1974) show cavitation to occur in leaves of herbaceous plants quite frequently, with the rate increased by rapid transpiration and, therefore, higher tension in the xylem. It seems that root pressure during the night is required to refill the empty spaces. There is no evidence, however, that cavitation could limit potentials in the plant body and thus determine the environmental demand to which a species is adapted.

Attention must be concentrated on cytoplasm and vacuole. These two compartments are characterized by the same hydrostatic potential (turgor) and by the same sum of osmotic and matric potentials.

Turgor potential is usually positive, ranging from the absolute value for vacuolar Ψ_s down to zero. Zero turgor potential defines the "wilting point" of cells and tissues. Ψ is then numerically equal to Ψ_s. There has been a long-standing conviction that, at least for malacophyllous plants, this point of zero turgor pressure must be about the lowest value the total potential can reach. There have been occasional reports on negative turgor in sclerophylls, but Kreeb (in Walter and Kreeb, 1970) summarizes that, according to his experience, negative turgor is more or less without ecophysiological relevance. However, recent observations on *Artemisia herba-alba* gave extremely negative values for the turgor potential (−31 bars; Kappen et al., 1972). Such states of tension in the cell lumen would clearly be very important. Walter (1973) has doubts about these values, since the reliability of the pressure chamber technique for measuring very low xylem pressure potentials has not been sufficiently established. Other recent papers give evidence for some 4 or 5 bars of negative turgor even above −20 bars of total potential (Turner, 1974; Lösch and Franz, 1974); thus, occurrence, frequency and magnitude of negative turgor can only be clarified by more experimental work. Theory does not seem to rule out moderate negative turgor, even in malacophylls.

On the whole, however, the range for minima of total water potential is set by osmotic and matric potentials. Minimum total potential and minimum osmotic potential of the vacuole were correlated wherever comparative determinations were made. Some $\Psi_s^{vacuole}$-values for desert species from Walter (1931, 1973) are shown in Table 6. Arrangement according to osmotic potentials gives

Table 6. Minimum values for osmotic potential in desert plants. (From Walter, 1931, 1973)

Species	$\Psi_{s(min)}$ (-bars)
Artemisia herba-alba	92
Zygophyllum dumosum	73
Larrea divaricata	>55
Franseria deltoidea	>53
Atriplex polycarpa	52
Simmondsia chinensis	49
Prosopis juliflora	41
Cercidium microphyllum	37
Fouquieria splendens	21
Opuntia spp.	16–19

approximately the order already known from Table 1. Values for Tables 1 and 6 were determined independently with most of them on different sites and even in different years. One cannot be sure that both entries for a species are close to the absolute minimum, but the vast differences for some of the plants suggest that a scrutiny for the occurrence of negative turgor would be useful.

It is not possible here to review the literature on osmotic potential which is by far the best documented component of Ψ. Walter (cf. 1960) and his school compiled data from all over the world and convincingly demonstrated the relevance of Ψ_s of individual plants and "osmotic spectra" of communities for the interpretation of plant distribution and adaptation to different habitats. This is in itself proof of the intimate connection between Ψ_s and minimum total Ψ. Walter (cf. Walter and Kreeb, 1970) stresses the fact that the hydration state of the protoplasm is only dependent on the osmotic potential of the vacuole, not on pressure potential or total water potential. The minimum value for $\Psi_s^{vacuole}$ must then be in tune with the requirements of functioning protoplasm in a given species. Thus the protoplasm would have to be acknowledged as the "limiting compartment" which, by determining the osmotic potential of the vacuole, regulates the minimum total potential available for counterbalancing the demands from the continuum.

Ψ_s^{vacuole} shows diurnal and seasonal variations reflecting both active and passive changes in the concentration of vacuolar substances. Active changes are caused by secretion into or removal from the vacuole of substances other than water, whereas passive changes are due to fluctuations in water content. Walter differentiates between "hydrostable" and "hydrolabile" species; this reflects different adaptations to periods of water stress. It would certainly be worthwhile to study changes of total potential in plants already assigned to these groups because of their regulation of Ψ_s.

V. Why does Water Potential in a Plant Change?

The range of water potentials, continuum states that favor extreme values, and the ranges of component potentials adjusting total Ψ in different compartments have now been outlined. This final section will briefly describe some important factors leading to conspicuous changes in total water potential.

The most obvious role is played by environmental events which change the conditions of the continuum. On clear days, plant potentials follow a pronounced day-course from pre-dawn maxima to minima in the early afternoon and back again during the evening and in the night. This coincides with the day-course of transpiration, where morning minima change to maxima in the middle of the day. The pattern is strongly modified by transient climatic events. Stansell et al. (1973) showed rapid reactions of total potentials resulting from changes in radiation intensity and, presumably, in transpiration by clouds. Similar results are obtained when vapor pressure deficits vary suddenly, e. g. just before a shower (Richter, 1974).

Waring and Cleary (1967) and Duhme (1974) describe the effects of soil drought on day-courses of Ψ. Moderately reduced pre-dawn values due to falling soil water potentials are compensated by an additional decline in midday potentials, so that the range for frictional fluctuations during the day remains the same. A further sharp decline in soil potentials reduces the differences in plant potentials until pre-dawn and noon values become nearly equal. One can assume that active changes in osmotic potential are induced under stress, but that this adaptation to soil drought is limited by the necessity to keep up a certain hydration state of the protoplasm.

Static soil water potential shows diurnal fluctuations, especially at sites with a pronounced day-course of soil temperatures. This must have influence on total plant potential, but detailed comparative studies have not been reported thus far.

Fluxes depend on the opening state of stomata. Since the stomata are part of a feedback system, they can show cyclic opening and closing movements under constant environmental conditions (Barrs, 1971; Brogårdh and Johnsson, 1974). Water potentials follow these oscillations closely. There is some evidence that a measurable cycling of potentials of an endogenous nature may occur even in the field (Duhme, 1974). Hormones have a strong influence on the stomatal mechanism (see Itai, this volume Part 4:B). It is thus not surprising that age and pretreatment of a plant or an organ modify onset and extent of an opening or closing response and, thus, the variability of total potentials with time (e.g. Unterscheutz et al., 1974). Air pollutants interfere with stomatal mechanisms (Poovaiah and Wiebe, 1973); this fact can help with the explanation of changes in water potentials observed in polluted areas (e.g. Halbwachs, 1970).

Variations in resistances may be reversible or irreversible. Reversibly variable resistances have been described for some herbaceous species, where they lead to a plateau value of leaf water potential despite drastic changes in transpiration rate (Weatherley, 1970; Stoker and Weatherley, 1971). The mechanism for this potentially important phenomenon still remains in dispute. Irreversible changes in resistance occur regularly during the normal course of development. Longitudinal growth of roots and shoots increases the resistances, whereas root branching and widening of the stem diameter due to cambial activity lower them by supplying conduits in parallel. As has been shown, these processes are kept in step, with the result being best described as morphological homeostasis reducing the resistance to flow in proportion to the increase in volume flowing (Jarvis, 1975). The variability of resistances in high trees was first analyzed by Huber (1928). It leads, in combination with differences in stomatal opening and irradiation, to a complex pattern of total water potentials in the crown; one cannot explain this pattern as a result of the height gradient, Ψ_g, alone (Richter, 1972, 1974; Hellkvist et al., 1974).

Infections and injuries may interfere strongly with plant resistances; for example, *Ceratocystis ulmi* blocks water conductance in elm trees (MacHardy and Beckman, 1973), due to occlusion of the vessels by a glycopeptide, the action of which is reflected by increased plant resistances and lowered total potential (van Alfen and Turner, 1975). Nemec (1975) shows similar processes in diseased citrus plants. A good survey of other pertinent data is given in a symposium report edited by Durbin (1973). Mechanical injuries to the roots (Havranek and Tranquillini, 1972), or fire scars on the stem base (Rundel, 1973) increase the resistances, which leads to measurable effects on water relations of the crown.

Diurnal, seasonal and developmental changes in the state of the continuum may lead to adaptive responses in the protoplasm and ultimately, a lower value for $\Psi_s^{vacuole}$ may result. As has been mentioned, much information exists on this osmotic adjustment and the reader is referred to the reviews already cited.

VI. Conclusions

Total water potential Ψ, which is the best measure for the momentary, transient state of water in the plant, depends on complex interactions between demands from the soil–plant–atmosphere continuum and responses in different compartments of the plant body modifying the plant part of the continuum by a feedback mechanism. Recent development in experimental methods has shifted the general interest toward direct determination of total Ψ. As research is not immune to fashion, this could detract due attention from older methods and results. It should, however, by now be clear that isolated values for total water potential do not reveal too much about the significance of this property for the plant. Recent papers show that additional use of other techniques like soil water studies, cryoscopy or porometry, resulting in a careful partitioning of both demands and responses, can greatly increase the biological information. Such combined studies of total water potential and its components are apparently becoming rapidly accepted as one of the most useful approaches in modern water relations research.

Acknowledgements. It is a pleasure to thank Prof. K. Burian for valuable literature and Dr. A. V. S. Pedeliski for reading the paper and improving the style. Thanks are likewise due for continuous support of my work by Fonds zur Förderung der Wissenschaftlichen Forschung (Projects 1 465 and 1 686).

References

Barrs, H. D.: Determination of water deficits in plant tissues. In: Water deficits and plant growth (ed. T. T. Kozlowski), Vol. 1, pp. 235–368. New York and London: Academic Press 1968.

Barrs, H. D.: Cyclic variations in stomatal aperture, transpiration, and leaf water potential under constant environmental conditions. Ann. Rev. Plant Physiol. **22**, 223–236 (1971).

Biscoe, P. V.: The diffusion resistance and water status of leaves of *Beta vulgaris*. J. Exp. Botany **23**, 930–940 (1972).

Brevedan, E. R., Hodges, H. F.: Effects of moisture deficit on ^{14}C translocation in corn (*Zea mays* L.). Plant Physiol. **52**, 436–439 (1973).

Brogårdh, T., Johnsson, A.: Oscillatory transpiration and water uptake of *Avena* plants. IV. Transpiratory response to sine shaped light cycles. Physiol. Plantarum **31**, 311–322 (1974).

Brown, R. W.: Determination of leaf osmotic potential using thermocouple psychrometers. In: Psychrometry in water relations research (ed. R. W. Brown, B. P. van Haveren), pp. 198–209. Logan, Utah: Utah State Univ. 1972.

Brown, R. W., van Haveren, B. P. (ed.): Psychrometry in Water Relations Research. Logan, Utah: Utah State Univ. 1972.

Campbell, G. S., Campbell, M. D.: Evaluation of a thermocouple hygrometer for measuring leaf water potential in situ. Agron. J. **66**, 24–27 (1974).

Carr, M. K. V.: The internal water status of the tea plant *(Camellia sinensis)*: Some results illustrating the use of the pressure chamber technique. Agr. Meteorol. **9**, 447–460 (1971).

de Roo, H. C.: Water stress gradients in plants and soil-root systems. Agron. J. **61**, 511–515 (1969).

Doley, D., Trivett, N. B. A.: Effects of low water potentials on transpiration and photosynthesis in Mitchell grass *(Astrebla lappacea)*. Australian J. Plant. Physiol. **1**, 539–550 (1974).

Driessche, R. van den, Connor, D. J., Turnstall, B. R.: Photosynthetic response of brigalow to irradiance, temperature and water potential. Photosynthetica **5**, 210–217 (1971).

Duhme, F.: Die Kennzeichnung der ökologischen Konstitution von Gehölzen im Hinblick auf den Wasserhaushalt. Dissertationes Botanicae Vol. 28. Lehre: J. Cramer 1974.

Durbin, R. D. (ed.): Symposium on water stress: pathogenic induction and influence on metabolism and disease development. Phytopathology **63**, 451–472 (1973).

Elfving, D. C., Kaufmann, M. R., Hall, A. E.: Interpreting leaf water potential measurements with a model of the soil–plant–atmosphere continuum. Physiol. Plantarum **27**, 161–168 (1972).

Goode, J. E., Higgs, K. H.: Water, osmotic and pressure potential relationships in apple leaves. J. Hort. Sci. **48**, 203–215 (1973).

Greenidge, K. N. H.: Ascent of sap. Ann. Rev. Plant Physiol. **8**, 237–256 (1957).

Griffin, J. R.: Xylem sap tension in three woodland oaks of central California. Ecology **54**, 152–159 (1973).

Haas, R. H., Dodd, J. D.: Water-stress patterns in honey mesquite. Ecology **53**, 674–680 (1972).

Halbwachs, G.: Vergleichende Untersuchungen über die Wasserbewegung in gesunden und fluorgeschädigten Holzgewächsen. Zentralbl. Ges. Forstwesen **87**, 1–22 (1970).

Halvorson, W. L., Patten, D. T.: Seasonal water potential changes in Sonoran desert shrubs in relation to topography. Ecology **55**, 173–177 (1974).

Havranek, W.: Tranquillini, W.: Untersuchungen über den Versetzschock bei der Lärche. Wachstum und Wasserhaushalt nach dem Versetzen. Mitt. Forstl. Bundesversuchsanstalt Wien **96**, 11–135 (1972).

Heine, R. W.: Hydraulic conductivity in trees. J. Exp. Botany **22**, 503–511 (1971).

Hellkvist, J.: The water relations of *Pinus sylvestris*. II. Comparative field studies of water potential and relative water content. Physiol. Plantarum **29**, 371–379 (1973).

Hellkvist, J., Richards, G. P., Jarvis, P. G.: Vertical gradients of water potential and tissue water relations in Sitka spruce trees measured with the pressure chamber. J. Appl. Ecol. **11**, 637–668 (1974).

Hillel, D.: Soil and water. Physical principles and processes. New York and London: Academic Press 1971.

Hinckley, T. M., Bruckerhoff, D. N.: The effects of drought on water relations and stem shrinkage of *Quercus alba*. Can. J. Botany **53**, 62–72 (1975).

Huber, B.: Weitere quantitative Untersuchungen über das Wasserleitungssystem der Pflanzen. Jb. Wiss. Bot. **67**, 877–959 (1928).

Hübl, E.: Über das stomatäre Verhalten von Pflanzen verschiedener Standorte im Alpengebiet und auf Sumpfwiesen der Ebene. Sitz.-Ber. Österr. Akad. Wiss., Mathem.-Naturw. Klasse **172**, 1–84 (1963).

Jackson, D. K.: The course and magnitude of water stress in *Lolium perenne* and *Dactylis glomerata*. J. Agr. Sci. **82**, 19–27 (1974).

Jarvis, P. G.: Water transfer in plants. In: Heat and mass transfer in the biosphere (ed. D. A. de Vries, N. H. Afgan), Vol. I, pp. 369–394. Washington, D.C.: Scripta Book Co. 1975.

Kaplan, A., Gale, J.: Modification of the pressure bomb technique for measurement of osmotic potential in halophytes. J. Exp. Botany **25**, 663–668 (1974).

Kappen, L., Lange, O. L., Schulze, E.-D., Evenari, M., Buschbom, U.: Extreme water stress and photosynthetic activity of the desert plant *Artemisia herba-alba* Asso. Oecologia (Berl.) **10**, 177–182 (1972).

Kassam, A. H.: The influence of light and water deficit upon diffusive resistance of leaves of *Vicia faba* L. New Phytologist **72**, 557–570 (1973).

Kirkham, M. B., Gardner, W. R., Gerloff, G. C.: Internal water status of kinetin treated, salt-stressed plants. Plant Physiol. **53**, 241–243 (1974).

Klepper, B.: Diurnal pattern of water potential in woody plants. Plant Physiol. **43**, 1931–1934 (1968).

Kochenderfer, J., Lee, R.: Indexes to transpiration by forest trees. Oecol. Plant. **8**, 175–184 (1973).

Lange, O. L., Lösch, R., Schulze, E.-D., Kappen, L.: Responses of stomata to changes in humidity. Planta (Berl.) **100**, 76–86 (1971).

Larcher, W.: Der Wasserhaushalt immergrüner Pflanzen im Winter. Ber. Deut. Botan. Ges. **85**, 315–327 (1972).

Lösch, R., Franz, N.: Tagesverlauf von Wasserpotential und Wasserbilanz bei Pflanzen verschiedener Standorte des fränkischen Wellenkalkes. Flora **163**, 466–479 (1974).

MacHardy, W. E., Beckman, C. H.: Water relations in American elm infected with *Ceratocystis ulmi*. Phytopathology **63**, 98–103 (1973).

Mark, W. R., Reid, C. P. P.: Lodgepole pine—dwarf mistletoe xylem water potentials. Forest Sci. **17**, 470–471 (1971).

Milburn, J. A., Davis, T. A.: Role of pressure in xylem transport of coconut and other palms. Physiol. Plantarum **29**, 415–420 (1973).

Milburn, J. A., McLaughlin, M. E.: Studies of cavitation in isolated vascular bundles and whole leaves of *Plantago major* L. New Phytologist **73**, 861–871 (1974).

Miller, L. N.: Matric potentials in plants: means of estimation and eco-physiological significance. In: Psychrometry in water relations research (ed. R. W. Brown, B. P. van Haveren), pp. 211–217. Logan, Utah: Utah State Univ. 1972.

Nemec, S.: Vessel blockage by myelin forms in Citrus with and without rough-lemon decline symptoms. Can. J. Botany **53**, 102–108 (1975).

Neumann, H. H., Thurtell, G. W.: A Peltier cooled thermocouple dewpoint hygrometer for in situ measurement of water potential. In: Psychrometry in water relations research (ed. R. W. Brown, B. P. van Haveren), pp. 103–112. Logan, Utah: Utah State Univ. 1972.

Neumann, H. H., Thurtell, G. W., Stevenson, K. R.: In situ measurements of leaf water potential and resistance to water flow in corn, soybean and sunflower at several transpiration rates. Can. J. Plant. Sci. **54**, 175–184 (1974).

Oertli, J. J.: The stability of water under tension in the xylem. Z. Pflanzenphysiol. **65**, 195–209 (1971).

O'Leary, J. W.: Can there be a positive water potential in plants? Bioscience **20**, 858–859 (1970).

Poovaiah, B. W., Wiebe, H. H.: Influence of hydrogen fluoride fumigation on the water economy of soybean plants. Plant Physiol. **51**, 396–399 (1973).

Richter, H.: Wie entstehen Saugspannungsgradienten in Bäumen? Ber. Deut. Botan. Ges. **85**, 341–351 (1972).

Richter, H.: Frictional potential losses and total water potential in plants: a re-evaluation. J. Exp. Botany **24**, 983–994 (1973).

Richter, H.: Erhöhte Saugspannungswerte und morphologische Veränderungen durch transversale Einschnitte in einem Taxus-Stamm. Flora **163**, 291–309 (1974).

Richter, H., Halbwachs, G., Holzner, W.: Saugspannungsmessungen in der Krone eines Mammutbaumes *(Sequoiadendron giganteum)*. Flora **161**, 401–420 (1972).

Ritchie, G. A., Hinckley, T. M.: The pressure chamber as an instrument for ecological research. Advan. Ecol. Res. **9**, 165–254 (1975).

Rundel, P. W.: The relationship between basal fire scars and crown damage in giant sequoia. Ecology **54**, 210–213 (1973).

Scholander, P. F.: How mangroves desalinate seawater. Physiol. Plantarum **21**, 251–261 (1968).

Scholander, P. F., Hammel, H. T., Bradstreet, E. D., Hemmingsen, E. A.: Sap pressure in vascular plants. Science **148**, 339–346 (1965).

Schulze, E.-D., Lange, O. L., Kappen, L., Buschbom, U., Evenari, M.: Stomatal responses to changes in temperature at increasing water stress. Planta (Berl.) **110**, 29–42 (1973).

Shepherd, W.: A simple thermocouple psychrometer for determining tissue water potential and some observed leaf-maturity effects. J. Exp. Botany **24**, 1003–1013 (1973).

Sheriff, D. W.: A new apparatus for the measurement of sap flux in small shoots with the magnetohydrodynamic technique. J. Exp. Botany **23**, 1086–1095 (1973).

Sheriff, D. W.: Magnetohydrodynamic sap flux meters: an instrument for laboratory use and the theory of calibration. J. Exp. Botany **25**, 675–683 (1974).

Singh, T. N., Aspinall, D., Paleg, L. G.: Proline accumulation and varietal adaptability to drought in barley: a potential metabolic measure of drought resistance. Nature New Biol. **236**, 188–190 (1972).

Slavík, B.: Methods of studying plant water relations. Ecological Studies, Vol. 9. Berlin-Heidelberg-New York: Springer 1974.

Spanner, D. C.: Plants, water and some other topics. In: Psychrometry in water relations research (ed. R. W. Brown, B. P. van Haveren), pp. 29–39, Logan, Utah: Utah State Univ. 1972.

Stansell, J. R., Klepper, B., Browning, V. D., Taylor, H. M.: Plant water status in relation to clouds. Agron. J. **65**, 677–678 (1973).

Stoker, R., Weatherley, P. E.: The influence of the root system on the relationship between the rate of transpiration and depression of leaf water potential. New Phytologist **70**, 547–554 (1971).

Strain, B. R.: Field measurements of tissue water potential and carbon dioxide exchange in the desert shrubs *Prosopis juliflora* and *Larrea divaricata*. Photosynthetica **4**, 118–122 (1970).

Sucoff, E.: Water potential in red pine: soil moisture, evapotranspiration, crown position. Ecology **53**, 681–686 (1972).

Szarek, S. R.: Physiological mechanisms of drought adaptation in *Opuntia basilaris* Engelm. et Bigel. Dissertation, Riverside, California 1974.

Türk, R.: Der Einfluß der Umweltfaktoren auf die Saugspannung höherer Pflanzen. Dissertation Wien 1970.

Turner, N. C.: Stomatal behavior and water status of maize, sorghum, and tobacco under field conditions. II. At low soil water potential. Plant Physiol. **53**, 360–365 (1974).

Turner, N. C., de Roo, H. C.: Hydration of eastern hemlock as influenced by waxing and weather. Forest Sci. **20**, 19–24 (1974).

Tuschl, P.: Die Transpiration von *Phragmites communis* Trin. im geschlossenen Bestand des Neusiedler Sees. Wiss. Arb. Bgld. **44**, 126–186 (1970).

Tyree, M. T., Dainty, J., Benis, M.: The water relations of hemlock (*Tsuga canadensis*). I. Some equilibrium water relations as measured by the pressure bomb technique. Can. J. Botany **51**, 1471–1480 (1973).

Unterscheutz, P., Ruetz, W. F., Geppert, R. R., Ferrell, W. K.: The effect of age, pre-conditioning, and water stress on the transpiration rates of Douglas fir (*Pseudotsuga menziesii*) seedlings of several ecotypes. Physiol. Plantarum **32**, 214–221 (1974).

van Alfen, N. K., Turner, N. C.: Influence of a *Ceratocystis ulmi* toxin on water relations of elm (*Ulmus americana*). Plant Physiol. **55**, 312–316 (1975).

Walter, H.: Die Hydratur der Pflanze und ihre physiologisch-ökologische Bedeutung. Jena: Fischer 1931.

Walter, H.: Einführung in die Phytologie. Vol. 3, Pt. 1. Stuttgart: Ulmer 1960.

Walter, H.: Die Vegetation der Erde in öko-physiologischer Betrachtung. Vol. 1: Die tropischen und subtropischen Zonen. 3rd Ed. Stuttgart: Fischer 1973.

Walter, H., Kreeb, K.: Die Hyratation und Hydratur des Protoplasmas der Pflanze und ihre öko-physiologische Bedeutung. Wien and New York: Springer 1970.

Waring, R. H., Cleary, B. D.: Plant moisture stress: evaluation by pressure bomb. Science **155**, 1248–1254 (1967).

Weatherley, P. E.: Some aspects of water relations. Advan. Bot. Res. **3**, 171–206 (1970).

Wiebe, H. H.: Matric potential of several plant tissues and biocolloids. Plant Physiol. **41**, 1439–1442 (1966).

Williams, J.: Root density and water potential gradients near the plant root. J. Exp. Botany **25**, 669–674 (1974).

Zabadal, T. J.: A water potential threshold for the increase of abscisic acid in leaves. Plant Physiol. **53**, 125–127 (1974).

Zimmermann, U., Räde, H., Steudle, E.: Kontinuierliche Druckmessung in Pflanzenzellen. Naturwissenschaften **56**, 634 (1969).

Part 2
Water Uptake and Soil Water Relations

Preface

In poikilohydric plants such as algae, fungi, lichens and most of the mosses, the water status of the thalli always tends to match the ambient water conditions. These plants are unable to maintain a long-lasting imbalance in their water potential in relation to the environment because they have no protection against excessive water loss. Thus, their metabolic activity, their productivity, and their growth under natural conditions are restricted solely to moist periods. This is the reason why the success of poikilohydric thallophytes in colonizing terrestrial habitats is limited and their biomass always remains small. During the evolution of life on earth, the final occupation of the terrestrial area of the biosphere was only possible after the homoiohydric structure of plants had been developed. The decisive advantage of the homoiohydric higher plants is the fact that here the sites of water uptake and water loss are separated from each other. Special organs have been generated in order to control water supply, water transport within the plant, and transpiration. These characteristics enable higher plants provided with roots, stems, and foliage to maintain favorable water conditions and hence metabolic activity within their tissues even under dry atmospheric conditions. However, within the soil–plant–atmosphere system the plant is linked into the water flux which leads from a usually restricted source in the soil to the sink of unlimited capacity in the ambient air. Thus, the water status of the plant body must be regulated by a sensitive control of water uptake and water loss. These two important processes which govern the water balance of the plant are the subject of Parts 2 and 3 of this volume.

The appearance of a plant, a plant culture, or a plant community in nature gives almost no information about the below-ground structure of the vegetation. However, it must be realized that the plants have to invest a considerable portion of their productivity into their root systems, primarily in order to ensure the necessary water supply. In coniferous forests, for example, the share of roots in total phytomass amounts to 21–25%; in tropical arid open woodlands the below-ground phytomass constitutes up to 30–40%; and in desert communities it can reach even 90% of the total phytomass (Bazilevich and Rodin, 1968). The roots of desert plants can penetrate deeper than 20 m into the ground; and the total length of all roots of a single rye plant amounts to 79 km (Walter, 1960; Baumeister and Reichart, 1969). In spite of the great importance of roots for the existence, growth and competition of plants, our knowledge of root-system dynamics, its productivity and its functioning under field conditions is still meager. The key reasons for this, undoubtedly, are the technical difficulties in analyzing the plant processes which take place invisibly below ground and which in many instances are complicated by the interactions of lower plants, mainly the mycorrhiza. In recent years, methods and theories have been improved that facilitate the study of root behavior in relation to moisture conditions of the soil and allow a closer understanding of water uptake and water supply of intact plants and plant communities (see Carson, 1974; Hadas et al., 1973; Hillel, 1971). This is clarified in the present Part 2 of this volume.

There are some few plant habitats where soil water is not restricted and an excess of water even limits root growth and functioning. This is the case when water in the soil displaces most of the air from the pore space and causes an oxygen deficiency. Considerable differences in resistance against deficient aeration occur among plant types (Kramer, 1969). Many special morphological and anatomical adaptations to permanent or periodical flooding are known, such as the pneumatophores of mangroves, the root knees of certain trees living on wet sites, and the aerenchyma in stems and roots of swamp and marsh plants apparently providing an air supply. Roots of swamp plants also show metabolic adaptations to the oxygen-poor conditions of their environments (Crawford, 1972). Intolerant species are injured by flooding due to the production of increased amounts of ethanol resulting from an increased rate of glycolysis induced by the anaerobic conditions. Flood-tolerant species, on the other hand, are able to avoid this excess ethanol accumulation by a metabolic switching over to malic acid production under anaerobic conditions. Thus they can exist in high-water-table habitats.

Situations where terrestrial plants are always sufficiently supplied with water or even suffer from an excess are not the rule. Usually soil water supply is permanently or at least temporarily limited. When studying uptake of water under conditions of water deficiency, it must be taken into account that not all stored soil water is at the plants' disposal. The classical concept of "soil water availability" (this and the following according to Hillel, 1971) states that water is only available throughout a definable range of soil-moisture content from an upper limit ("field capacity") to a lower limit ("permanent wilting point"). It was claimed that plant functions remain unaffected by decrease in soil water content until the permanent wilting point is reached. Another theory postulated that water availability continuously decreases with decreasing moisture content of the soil. However, in recent years, water uptake by plants has been proved to be a much more complicated dynamic process (see reviews by Gardner, 1968; Newman, 1974). It depends (1) on properties of the soil which determine the status of water and its flow to the roots, e.g. the hydraulic soil conductivity; (2) on properties of the plant, such as the water potential which can be achieved by the roots, the structure of the root system, the ability of the root system to extend, and the anatomical features of the roots in respect to water absorption; and (3) on the momentary meteorological conditions which determine the needs of the plant for water and thus control the water potential gradients within the soil–plant–atmosphere continuum.

These dependencies are analyzed and elucidated for different plant types in the following two chapters. Caldwell (Part 2:A) discusses water absorption and root extension of established perennial plants. He stresses the significance of continual re-exploration of the same soil mass by growth of the root system, and he develops a model to illustrate quantitatively the resistance to water uptake in a root system where root permeability is not uniform and where localized rhizospheric resistances develop. Resistances to water flow in the roots of annual cereals are described and formulated by Greacen et al. (Part 2:B). In the last chapter of this Part (Part 2:C), the problems of soil water relations, water uptake and water fluxes are considered from the point of view of a whole ecosystem. Benecke treats the water exchange of forest communities and compares the water balance of a beech stand with that of a spruce stand under identical meteorological conditions. The results, which are a contribution to the International Biological Program (see Ellenberg, 1971), make it obvious that utilization of the progress that has been achieved in theoretical soil physics in the past years allows a new and promising approach to understanding the functioning of the water dynamics of ecosystems, and might help to manage their water economy in a beneficial way.

References

Baumeister, W., Reichart, G.: Lehrbuch der Angewandten Botanik. Stuttgart: Gustav Fischer 1969.
Bazilevich, N. I., Rodin, L. E.: Reserves of organic matter in underground sphere of terrestrial phytocoenoses. In: Methods of productivity studies in root systems and rhizosphere organisms (eds. M. S. Ghilarov, V. A. Kovda, L. N. Novichkova-Ivanova, L. E. Rodin, V. M. Sveshnikova), pp. 4–8. Leningrad: NAUKA 1968.
Carson, W. E. (ed.): The plant root and its environment. Charlottesville: Univ. Press Virginia 1974.
Crawford, R. M. M.: Physiologische Ökologie: Ein Vergleich der Anpassung von Pflanzen und Tieren an sauerstoffarme Umgebung. Flora **161**, 209–223 (1972).
Ellenberg, H. (ed.): Integrated experimental ecology. Ecological Studies, Vol. 2. Berlin-Heidelberg-New York: Springer 1971.
Gardner, W. R.: Availability and measurement of soil water. In: Water deficits and plant growth, Vol. I, (ed. T. T. Kozlowski), pp. 107–135. New York and London: Academic Press 1968.
Hadas, A., Swartzendruber, D., Rijtema, P. E., Fuchs, M., Yaron, B. (ed.): Physical aspects of soil water and salts in ecosystems. Ecological Studies, Vol. 4. Berlin-Heidelberg-New York: Springer 1973.
Hillel, D.: Soil and Water. New York and London: Academic Press 1971.
Kramer, P. J.: Plant and soil water relationships: A modern synthesis. New York: McGraw-Hill 1969.
Newman, E. I.: Root and soil water relations. In: The plant root and its environment (ed. E. W. Carson), pp. 363–440. Charlottesville: Univ. Press Virginia 1974.
Walter, H.: Grundlagen der Pflanzenverbreitung I. Standortslehre. Stuttgart: Eugen Ulmer 1960.

A. Root Extension and Water Absorption

M. M. CALDWELL

I. Introduction

This chapter is not intended to be an extensive review of root–soil water relationships. Balanced coverage of this field has been most adequately undertaken by texts such as Slatyer (1967) and Kramer (1969). Furthermore, various facets of the subject have been well covered in recent reviews by Weatherley (1970), Gardner (1968), Cowan and Milthorpe (1968), and particularly by Newman (1974). Characteristics of root growth, development, and senescence have been reviewed by Head (1973). Instead, it is intended to address the specific topic of the role of root extension in water absorption.

Extension of roots into previously unexplored areas of the soil by annual plants or incompletely established perennial species is readily envisaged as an often necessary venture in order to obtain moisture, as will be discussed. However, the more intriguing question of the significance of continual re-exploration of the same soil mass by established perennial plant root systems in relation to water absorption will receive particular emphasis. The energetic cost of new root extension in established perennial plant root systems becomes apparent when belowground productivity is assessed. For example, reports recently emanating from the International Biological Programme attribute over half of the annual primary production of an established deciduous forest and a mesic fescue meadow and about 75% of the annual production of shortgrass prairie and shrub steppe communities to be in the form of new root production (Harris and Coleman, personal communication; Speidel and Weiss, 1972; Caldwell and Camp, 1974). Yet is this annual investment of energy in re-exploration of this same soil mass particularly necessary for water uptake?

II. Water Movement Through the Soil–Plant–Atmosphere Continuum: Limitations in the Liquid Phase

Oertli has discussed the soil–plant–atmosphere continuum and the catenary theory of water movement in Part 1:C of this volume. When van den Honert (1948) described the catenary theory of water movement in response to water potential gradients, his prime intention was to show that the vapor phase resistance is the largest and hence dominant resistance in the catena as depicted in

$$E = \frac{\Psi^a - \Psi^e}{R_a} = \frac{\Psi^e - \Psi^x}{R_{sh}} = \frac{\Psi^x - \Psi^r}{R_r} = \frac{\Psi^r - \Psi^s}{R_s} \quad (1)$$

where E is the flux of water through the catena; Ψ^s, Ψ^r, Ψ^x, Ψ^e, and Ψ^a are the water potentials in the bulk soil, at the root-soil interface, in the xylem of the root, at the evaporating surfaces of the leaves, and in the atmosphere, respectively; and R_s, R_r, R_{sh}, and R_a constitute the resistances to water movement in the soil, from the root-soil interface to the root xylem (usually considered a direct radial pathway), from the xylem of the root to the evaporating surfaces in the leaves, and from the evaporating surfaces in the leaves to the open atmosphere, respectively.

Although there seems to be general concurrence with van den Honert's hypothesis of the dominant resistance being in the vapor phase, at the same time it is well recognized that stomatal closure and hence accentuation of the vapor phase resistance, R_a, is often the result of decreased plant tissue water potential which is a consequence of what is often termed as an absorption lag. Since the water potential of the plant shoot normally increases over the nocturnal period in the absence of renewal of the soil moisture reservoir, an appreciable resistance somewhere in the liquid phase of the catena is clearly implicated (Elfving et al., 1972). The ecological significance of the liquid phase resistance is also indicated in studies of plant photosynthesis and transpiration that depict curtailment of gas exchange due to increased stomatal diffusion resistance that is in turn often linked with decreased plant water potential in the mid- to latter part of the day (e.g., Schulze et al., 1972; De Puit and Caldwell, 1973).

Where in the catena does the most significant resistance in the liquid phase exist? Since most of the resistances of the catena are in series and it is usually not possible to measure Ψ^r or Ψ^x directly, it is difficult to partition the component resistances experimentally and they must therefore be inferred. Detailed discussions of the component liquid phase resistances are contained in Cowan and Milthorpe (1968), Gardner (1968), Weatherley (1970), and Newman (1974). No attempt will be made to retrace and review the transport of water in the liquid phase through the soil and plant components since this has been so adequately done by these recent reviews. Nevertheless, there are a number of facets of this subject which do relate to the question at hand—namely, is the continual extension of roots necessary for water uptake? Much of the following discussion will therefore involve components of the liquid phase resistance.

III. Root Extension and Facilitation of Water Uptake in Unexplored Soil Regions

The movement of soil water in response to water potential gradients is specifically represented by an equation of the form

$$\frac{\delta W}{\delta t} = -k \frac{\delta \Psi}{\delta z} \quad (2)$$

where $\delta W/\delta t$ is the rate of water movement, $\delta \Psi/\delta z$ is the water potential gradient, and k is capillary conductivity. This conductivity term is usually represented by an expression of the form

$$k = \frac{a}{|\Psi|^n} \tag{3}$$

where n is typically between 2 and 3 for fine-textured soils but can range up to 5 for coarse-textured soils, and a is a constant (Lang and Gardner, 1970). In the discussion of this chapter, Ψ^s will be considered to be composed entirely of matric potential with the osmotic potential being negligible. Isothermal conditions are also assumed. Soil moisture diffusivity, D, is an analogous term when the water gradient is expressed as volumetric water content, Θ. Expression of these relationships is interchangeable and primarily a matter of convenience.

Equations (1) and (2) are basically Darcy equations which assume a Newtonian behavior of flow (see also Benecke, this volume Part 2:C). Water movement in soils can exhibit non-Darcy behavior as Newman (1974) has discussed. However, for the purpose of this discussion it is assumed that water movement in soil follows Eq. (2), or its analog involving diffusivity and water content, without regard for hysteresis.

When soils are only partially wetted, the capillary conductivity is usually too small to support significant water movement to the root over great distances and extension of roots into moist regions of the soil profile is necessary. Gardner (1968) has discussed the relationship between the extension of a front of roots towards deeper regions of the soil profile and upward water movement to the roots.

Gardner (1968) has presented a flow equation assuming a constant weighted mean diffusivity, $\bar{D}$(cm$^2 \cdot$d^{-1}), and constant upward flux or transpiration rate, E, (cm$\cdot$d^{-1})

$$W = Et = \pi \bar{D}(\Theta_i - \Theta_w)^2/4E = \sqrt{\pi \bar{D} t}(\Theta_i - \Theta_w)/2 \tag{4}$$

where Θ_i (cm$^3 \cdot$cm^{-3}) is the initial water content below the rooted zone, and Θ_w is a lower limit of water content which might, for example, correspond to the permanent wilting point or any other prescribed limit, and W (cm) is the total quantity of water in this one-dimensional scheme which is moved into the rooted zone over the time period in question, t (days). This equation is for an indefinitely deep soil or would also apply for a finite soil system where the transpiration rate is rather high and $\bar{D}$ is small. Gardner further depicted the total amount of water made available to the root system, by a combination of upward movement of water toward the roots and the additional amount of water provided by extension of the root system to greater depths, when a steady state has been achieved

$$W = V(\Theta_i - \Theta) + (\Theta_i - \Theta)\bar{D}/U. \tag{5}$$

The first component of this equation reflects the amount of water made available, due to root extension where V (cm) is the total increase in length of the root system in this one-dimensional system. The second component of this equation is the additional amount of water made available by upward diffusion through the soil. The rate of root extension is designated as U (cm·d^{-1}), and the amount of water made available per unit time by root extension is $U(\Theta_i - \Theta_w)$. Equation (5) assumes that root growth is continuing over a long enough period of time so that root growth predominates over water movement, $U > \sqrt{\pi \bar{D}/t}$; transpiration rate E is also assumed to remain constant and a steady state is achieved where

$$E/U = (\Theta_i - \Theta). \tag{6}$$

The value Θ is then the water content to which water is removed under these steady state conditions. This one-dimensional scheme which assumes a constant $\bar{D}$ and E presents a simple means of evaluating the approximate importance of soil moisture movement toward the rooted zone in relation to root extension into areas of the soil profile not already permeated by functional roots. It is apparent from Eq. (4) that with each succeeding time increment there will be a diminishing supply of water to the rooted zone due to water movement. If $\bar{D}$ were not held constant and was allowed to decrease with decreasing soil moisture content, this curtailment of water supply through time would, of course, be much more pronounced. Under the steady-state conditions assumed by Eq. (5), the contribution of root extension to the total water supply can be easily recognized. For example, for a daily elongation rate of 2 cm·d^{-1}, $\bar{D}$ of 10 cm^2·d^{-1} (a reasonable value for moist soil), and a water content depletion $(\Theta_i - \Theta)$ of 0.04 cm^3·cm^{-3}, the contribution from root extension to W over a 30-day period is 92% of the total, whereas upward diffusion of water through the soil accounts for only 8%. Under the same conditions but with a root elongation rate of 1 cm·d^{-1}, root extension would still account for 75% of W. These, too, would seem to be liberal estimates because of the assumption of constant $\bar{D}$. Such depictions, though confined by a number of assumptions, quantitatively bear out that which has long been purported; namely, that root growth into new soil regions is of decided advantage for the acquisition of water by plants. This thesis is, however, underpinned by the assumption that water movement upward through the root xylem encounters a negligible resistance as compared to movement through the soil. This assumption is generally valid and will be discussed later. The importance of root extension for annual plants or establishing perennial species would seem to encounter little argument.

A decided competitive advantage is exhibited by annual grasses such as *Bromus tectorum* and *Taeniatherum asperum*, which are able to extend their roots swiftly in the early spring at low soil temperatures and hence avail themselves of the soil moisture much more efficiently than the establishing perennial bunchgrass (*Agropyron spicatum*), which is apparently not able to undergo rapid root extension in cold soils (Harris and Wilson, 1970). Such examples of competitive efficiency for water being linked to root extension of annual species or establishing perennial species invading new soil regions further support the general thesis.

IV. Root Extension Within the Rooted Zone: A Case for Avoidance of Localized Rhizospheric Resistances

1. Rhizospheric Resistances and an Optimized Perennial Root System

Although root extension into previously unexplored regions of the soil profile is generally recognized to reap benefit for the plant, root extension within the region of the soil profile where the plant already has a well-established root system is less immediately satisfying from a teleological point of view. This is particularly the case when the nature and magnitude of resistances to soil moisture movement within the rooted zone are considered. In this situation, rather than a front of water moving toward the rooted zone, moisture movement from a cylinder surrounding each root element is normally taken as the basis of a model for resistance to soil moisture movement. These are often termed as rhizospheric resistances.

Radial flow toward the root located in the central axis of this soil cylinder is represented by the differential equation

$$\frac{\delta \Theta}{\delta t} = \frac{1}{r} \frac{\delta}{\delta r} \left(r D \frac{\delta \Theta}{\delta r} \right) \tag{7}$$

where Θ is the water content on a volume basis and r is the radial distance from the axis of the root (Gardner, 1960). As recognized earlier, D is itself quite dependent on water content. For a more tractable solution of Eq. (7), Gardner derived a steady state approximation by setting $\delta \Theta / \delta t$ equal to zero and Cowan (1965) used a steady rate equation by setting $\delta \Theta / \delta t$ as a constant, but not zero. The two approximations give very similar results, particularly for reasonably moist soils (Lawlor, 1972; Williams, 1974). Passioura and Cowan (1968), however, showed that the steady rate approximation provides a somewhat closer fit when compared with a numerical analysis of Eq. (7) when soil moisture contents are lower. The steady rate equation of Cowan (1965) is

$$\Theta_r = \bar{\Theta}_s - \frac{Q}{2D} \left[\frac{r_2^4}{(r_2^2 - r_1^2)} \ln \left(\frac{r_2}{r_1} \right) - \frac{3 r_2^2 - r_1^2}{4} \right] \tag{8}$$

where r_1 (cm) is the radius of the root and r_2 (cm) is the radius of the soil cylinder from which water is being extracted, Θ_r (cm$^3 \cdot$cm^{-3}) is the water content at the root-soil interface and $\bar{\Theta}_s$ is water content of the bulk soil, and hence at r_2, and Q is the rate of extraction of water per volume of soil (cm$^3 \cdot$cm$^{-3} \cdot$d^{-1}). Assuming a uniform root distribution in a given volume of soil, r_2 may be related to the density of rooting, L_v (cm root$\cdot$cm^{-3} soil) by $r_2 = 1/(\pi L_v)^{1/2}$. Equations (7) and (8) assume that the rate of uptake of water per unit length of root is uniform, that the root has sole access to the water within the prescribed cylinder, and Eq. (8) assumes that all water taken up by the root comes from this cylinder and no moisture moves into the cylinder from other parts of the soil. This latter assumption is quite reasonable if a uniform rooting density and uniform uptake along the entire root system length is assumed. The steady state equation of Gardner (1960), based on slightly different assumptions, takes the form

$$\Psi^r = \Psi^s - \frac{Q}{4\pi k L_v} \ln\left[\frac{r_2^2}{r_1^2}\right] \tag{9}$$

where Ψ^r and Ψ^s are soil water potentials at the root-soil interface and in the bulk soil.

Gardner (1960) and Cowan (1965) applied their approximations of Eq. (7) to illustrate that even over very short distances on the order of several mm appreciable rhizosphere resistances could develop. This is in contrast to the one-dimensional situation discussed earlier, where appreciable water movement over distances one to two orders of magnitude greater is expected (Gardner, 1968).

This cylindrical model, however, involves convergence of water toward the center of the cylinder and the consequent development of a gradient of decreasing soil moisture content toward the root and therefore increasing resistance as is depicted in Fig. 1.

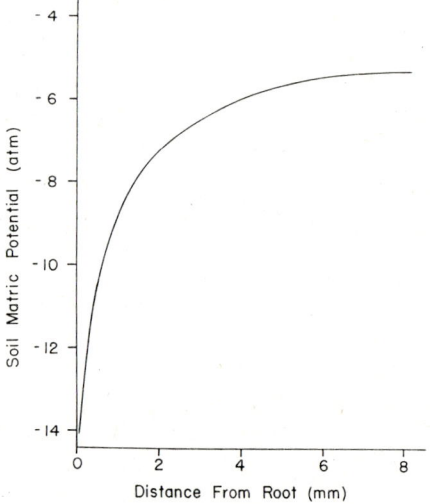

Fig. 1. Soil matric potential gradient from the root-soil interface to the bulk soil at 8 mm from the root under conditions of steady state flow at 2.5 mm day^{-1} with a soil moisture content in the bulk soil of 0.155 cm^3 cm^{-3}. (After Cowan, 1965)

Although other papers (Newman, 1969a; Lawlor, 1972; Williams, 1974; Newman, 1974) have not taken issue with the basic equations of Gardner (1960) and Cowan (1965), they have questioned the magnitude of the input parameters used in these models. An inspection of the Eqs. (8) and (9) reveals that the reduction in Θ_r or Ψ^r which, of course, reflects the rhizospheric resistance, is particularly dependent on Q, D or k, and L_v, which determines r^2 as well as being directly involved in Eq. (9). Root diameter is, however, much less influential in these equations. Newman (1969a) reviewed available data on rooting densities and water uptake rates from the literature and concluded that the values of Q and L_v used by Gardner and Cowan were simply unrealistic for field conditions, and that appreciable depressions of Ψ^r and hence significant rhizospheric resistances, R_s, would probably not occur until Ψ^s approached -15 atm. In a companion paper (Newman, 1969b) experimental evidence was reviewed where appreciable R_s were indicated and it was concluded that in most cases there was no conclusive empirical evidence to suggest that R_s was important as long as Ψ^s was above -15 atm. Lawlor (1972) followed suit with an analysis of a stand of *Lolium perenne* and reached a similar conclusion; namely, that R_s would not be very important until the soil dried to the wilting point, taken as -15 atm. In a similar vein, Hansen (1974a, b) inferred R_s to be unimportant for young wheat plants growing in small volumes of soil until the bulk soil had dried to -12 to -15 atm, and Williams (1974) reviewed data on rooting density and also concurred with these authors.

To return to the basic question of this chapter regarding the importance of continual extension of root elements within the rooted zone for water absorption, a cursory synopsis at this point might quickly lead to a null hypothesis. In fact, from the standpoint of water absorption, the physical models of water flow discussed thus far would suggest that the optimal root system for an established perennial plant, which does not enjoy the advantage of a free water table, would be a uniformly diffuse yet static root system that is uniformly permeable over the entire length of the root system. As long as the rooting density is sufficient in the zones of the profile where soil moisture is available, formation and extension of new roots would hardly seem necessary except to replace nonfunctional roots. Furthermore, a selection for long-lived root elements with low maintenance costs would also seem to be of advantage.

Since many field observations indicate that there does appear to be an annual renewal of an appreciable fraction of the fine root elements and that root extension takes place throughout the established root system of most observed perennial plant root systems during the growing season (Head, 1973; Speidel and Weiss, 1974; Bhar et al., 1970), apparently plants are not following the optimal strategy outlined by the physical models. Since this new root element growth can constitute a substantial energy cost for these plants as outlined in the Introduction, an explanation of this behavior would at least be satisfying for our concept of teleology, if not for selective value in the evolution of these species.

Circumstantial observations derived from subterranean root-soil observation chambers also suggest that this root extension may be of significance for water uptake. Observations in the field of the root systems of arid steppe shrubs indicate a progression of root element growth from the upper to lower regions of the soil profile during the course of the spring-summer growing season (Fig. 2) and (Fernandez and Caldwell, 1975). Even though these were established shrubs with root systems thoroughly pervading the zone of annual soil moisture recharge, there was an apparent correspondence between soil moisture depletion and the zones of new root element growth during this period. Furthermore, the ability of one species, *Atriplex confertifolia*, to maintain a certain amount of photosynthetic activity and concomitant transpiratory water loss during the driest portion of the year, as compared to another shrub species, was correlated with the higher rate of root element extension in this species at this time of year. Since the actual quantity of new root elements was usually less than 5% of the existing root mass at any particular soil horizon, this association of new root growth and water extraction cannot be explained simply by short-term

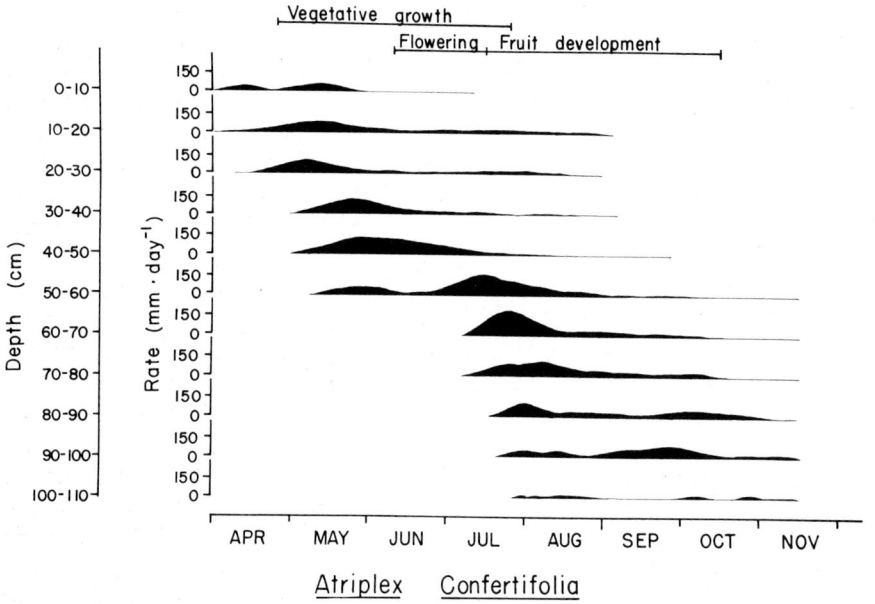

Fig. 2. Progression of new root growth during the course of the growing season for an established stand of *A. confertifolia* expressed as mean daily rate of root elongation at 10-cm intervals in the soil. The period of active shoot elongation denoted as vegetative growth and the principal periods of flowering and fruit development are also indicated. (From Fernandez and Caldwell, 1975)

changes in bulk rooting density at each horizon. This is in contrast to the reports of Taylor and Klepper (1971) who reported substantial changes in the distribution of the total root system with depth in the profile due to new root growth during the comparatively short period of a few weeks for young cotton plants. In this case, new root growth was proportionately large enough for these annual plants to effect substantial changes in total rooting density with depth in the profile, and hence could influence water extraction capacity. These changes corresponded to the irrigation cycle for this crop.

2. Variation in the Root Resistance Term

Equation (7) and its steady-rate and steady-state approximations assume that the uptake of water is uniform along the entire length of the root system. Naturally, if for any reason a fraction of the root elements or portions of individual roots are not equally effective in water absorption, then the effective localized value of Q, which is normally derived from a knowledge of transpiration per unit leaf or ground area, could change dramatically with a resultant increase in localized rhizospheric resistances. Because of the sensitivity of these relationships to the localized variation in effective root resistance, R_r, and the consequent implications for R_s and root extension, some attention must be directed to the nature of the root component of resistance to water transport in the catena.

The classical depiction of root growth and morphologic development requires only a brief statement (see also Greacen et al., this volume Part 2:B). The meristematic region of the root is normally conceived to be shielded by a root cap protecting the root as it forces its way through the soil. Behind the meristematic region is the zone of cell elongation. A zone of root hair development is behind the elongation zone, where most of the water and nutrient absorption is normally conceived to take place. Behind the zone of root hair development, roots are usually noted to undergo a darkening of color, often to a dark brown. This is normally attributed to suberization. If this suberization were to curtail water uptake greatly by the older portions of the root, then the model depicted by Eqs. (7), (8) and (9), although still theoretically applicable, would certainly lead to very different results. In this case, since the percentage of unsuberized root tips would only constitute a very small fraction, usually less than 5%, of the total root length, the effective Q might increase some 20-fold with the result that appreciable rhizospheric resistances could quickly develop. Water extraction from the soil would also be very uneven. In this situation, the continual formation and extension of new root tips would seem to be of decided advantage for the plant. However, as Kramer (1969) and Newman (1974) point out, suberized roots are not necessarily disfunctional in terms of water uptake. As Newman (1974) discussed, although suberin is, indeed, a water-impermeable compound, the patterns of suberin deposition in the root are not always so arranged as to seriously impede water uptake. In the case of periderm formation, the active root system is surrounded by a reasonably continuous layer of suberin. Nevertheless, there are still cracks, lenticels, wounds, etc., where water might be taken up, as will be discussed later.

The oft-cited study of Brouwer (1953), in which measurements of permeability were made along various sections of a broad bean *(Vicia faba)* root, indicated that at low flux rates the greatest permeability was near the root tip. Hansen (1974a) showed a similar phenomenon for wheat roots. At greater flux rates, however, the permeability of the broad bean root increased to a much greater extent several cm from the tip than it did in the region where permeability had been greatest closer to the tip at low flux rates. At the higher flux rates, the greatest permeability was, therefore, several cm from the tip (Brouwer, 1953). As Newman (1974) suggested, acceptable generalizations about differential permeability along the first 10 cm or so of young roots cannot be safely made. Of course, these broad bean and wheat roots did not have a periderm.

Satisfactory comparisons between unsuberized roots and roots suberized with a periderm are difficult to find. Kramer and Bullock (1966) attempted such comparisons and found: (1) that suberized roots did have a significant permeability as had been reported in earlier literature (e.g., Hayward et al., 1942), (2) that permeability of the suberized roots increased with increasing diameter which was thought to be attributed to a greater number of disruptions in the periderm, and (3) that for both suberized and unsuberized root segments there was a high degree of variability among samples. Although the average permeability of the suberized root segments was only a fraction of that of the unsuberized segments, they concluded that since the unsuberized portion of the root system usually constituted less than 5% of the total root surface area, less than 25% of the total seasonal water absorption should be attributed to the unsuberized portion of the root system. However, these conclusions were based merely on the proportion of unsuberized and suberized components of the root system and their relative permeabilities. Numerous reports of water absorption by suberized roots have been discussed and reviewed by Kramer (1969), which corroborate the findings of Kramer and Bullock (1966). It might seem then that the entire root system of perennial plants might be active in water absorption and that the small portion of the root system composed of unsuberized root tips may play a relatively minor role, in which case the model of the hydrologically optimized perennial plant root system might be approached. However, a closer inspection of the exact pattern of water uptake by the root system does not necessarily justify this assumption.

Although the concept of water absorption being restricted to the region immediately behind the growing root tips is sufficiently well countered by experimental evidence so as to reject it as a realistic depiction, the concept of the root system as being uniformly permeable also seems untenable. Because nonuniform water uptake may lead to localized development of large rhizospheric resistances, a more intensive view of the nature of root resistance is warranted.

It is generally agreed that the course of water movement from the soil cylinder surrounding the root to the root xylem is a reasonably direct radial path and that longitudinal movement of water along the root axis either in the surrounding soil or in the cortex of the root is quite limited. This is primarily due to the fact that resistance to water flow within the root xylem is considerably less than through the surrounding soil or the cortex of the root tissues (Cowan and Milthorpe, 1968). Newman (1974) reviewed the few available studies which address the subject of resistance along the xylem pathway of roots. He concluded that most studies suggest very small xylem resistances. However, for some plants, if roots extended a meter or more from the stem base, an appreciable longitudinal resistance might occur, hence giving rise to a phenomenon of decreasing water uptake at distances further from the stem base. Passioura (1972) also recently reported rather high xylem resistance in wheat roots. Models such as that of Nimah and Hanks (1973), which incorporate a small xylem resistance term and thereby predict greater uptake of water in the upper part of the soil profile, seem to correspond rather well with observed patterns of soil moisture depletion.

Although there is certainly enough evidence to suggest that xylem resistance cannot be ignored, particularly over great distances, water movement through the root xylem is still considered to be by far the path of least resistance when compared with alternatives such as longitudinal movement through the root cortex or in the surrounding soil (Newman, 1974). In the following discussion, the course of water movement as being radial from the soil to the root xylem and then longitudinal along the root xylem will be assumed.

Returning to the consideration of Eq. (1), components of the plant liquid phase resistance to water flow, R_r and R_{sh}, are difficult to experimentally partition since it is generally not possible to measure directly Ψ^r, or Ψ^x. Most often, the total plant resistance term is calculated based on transpiration rates and the drop in water potential between the shoot tissues and either the Ψ^s or a calculated Ψ^r based on approximations of Eq. (7) (Lawlor, 1972; Newman, 1969a, 1974; Hansen, 1974a, b).

The exact radial pathway of water movement from the root epidermis to the root xylem has been reviewed in some detail by Weatherley (1970) and Newman (1974) and will not be discussed in detail here. It is quite apparent, however, that at some point most of the water must traverse some cytoplasm and cross at least two membranes. This is usually thought to take place at the endodermis (Weatherley, 1970; Brouwer, 1965).

Despite the quantitative importance of R_r, the nature and exact magnitude of this resistance under field conditions are still poorly understood. The hydrologically optimized perennial plant root system depicted earlier calls for a uniform permeability throughout the length of the root system, thereby allowing an even and efficient extraction of water throughout the rooted zone, provided the root density is reasonably uniform. Yet, available evidence suggests that this is not the case. Permeability of roots to water seems to be dependent on a number of factors and furthermore exhibits a pronounced non-Darcy behavior.

Permeability can be expected to be highly dependent on temperature, the dependence of which Kramer (1940) showed to be much greater than that predicted by changes in the viscosity of water alone. In fact, at temperatures below 15°C, a Q_{10} value on the order of 3 has been reported for maize roots (Brouwer, 1965). Factors which adversely affect normal metabolism such as respiration inhibitors also decrease permeability (Brouwer, 1965).

A particular characteristic which deviates from Darcy behavior is the apparent increase in permeability with increasing flux rate, such as would occur when transpiration rates increase due to either decreasing stomatal diffusion resistance or increasing vapor pressure deficit in the atmosphere. Figure 3 depicts this characteristic pattern. Recently, however, Boyer (1974) has presented evidence to suggest that although there is some flux-dependent permeability behavior of the detached root system, at very low flux rates the primary variable resistance component in the liquid phase transport between the root-soil interface and the evaporating surfaces of the leaves, resides in the leaves rather than in the root system. Nevertheless, he did acknowledge the root resistance term to be still the dominant component of the total plant liquid-phase resistance at moderate to high transpiration rates.

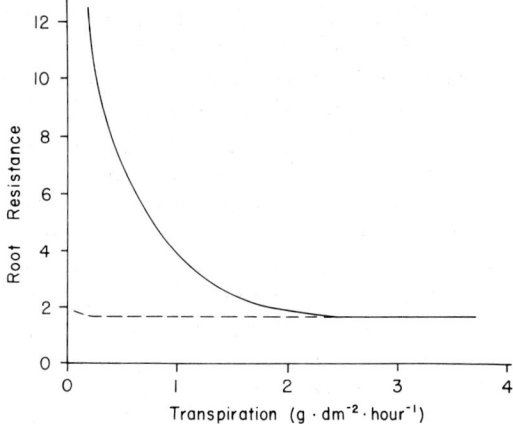

Fig. 3. The relationship between root resistance and flux rate for cotton roots. *Solid line:* resistance of living roots. *Dashed line:* resistance of heat-killed root systems. (After Stoker and Weatherley, 1971)

A further component of non-Darcy behavior is exhibited when the driving potential for water flux through the root system is changed. Kuiper and Kuiper (1974), for example, have recently reported that after the driving potential had been abruptly changed, permeability increased slowly until a steady-state situation was reached after 15 to 60 min. They described this as a temperature-dependent self-induction phenomenon perhaps associated with cytoplasmic streaming.

Permeability has also been reported to exhibit a marked diurnal periodicity with the maximum at midday which may be as much as five times higher than that during the night-time low (Parsons and Kramer, 1974). This cyclic variability in permeability appeared to be in response to signals from the shoot system.

When a periderm is formed, a reasonably continuous suberin sheath encases the root. Entry of water is then confined to lenticels, breaks, wounds, or other disruptions in the periderm. Addoms (1946) clearly showed that these disruptions were the avenues of water entry into suberized roots by observing the pathway of dyes in water solutions when root segments were held under tension. All water entry could be traced to some type of disruption in the periderm. This patchy pattern of water absorption would certainly corroborate the very high variance of permeability among root segments observed by Kramer and Bullock (1966).

The highly variable nature of the root resistance term as suggested by this brief discussion must be kept in mind in reviewing the few extant data for quantitative expressions of root permeability of various species. For the questions being addressed in this chapter, the most significant variables with respect to the R_r term are root age and associated degree of suberization, and the spatial heterogeneity of water uptake along the length of the root system. Certainly, as Newman (1974) indicated, the acceptable experimental data base concerning root permeability, and especially comparisons of suberized and unsuberized roots, are indeed very meager. From the brief foregoing discussion of such evidence, some tentative hypotheses concerning root permeability might be put forth: (1) Although water uptake is certainly not restricted to the young, unsuberized root tips, absorption per unit length is much greater than in the older, suberized portions of the root system. (2) In the older, suberized portion of the roots, decreased permeability will be very contingent on the exact manner of suberin deposition. When a periderm is involved, a reasonably complete suberin cylinder is formed around the root and water uptake only occurs in very localized areas such as lenticels, wounds, cracks, etc. (3) Because in the course of radial movement of water to the xylem, cytoplasm and membranes must be traversed at some point, environmental factors which affect metabolism, such as temperature, are expected to greatly influence permeability. (4) The root resistance term for any segment of root cannot be considered as a constant but will vary with flux rate and may also exhibit self-induction behavior and perhaps diurnal periodicity.

3. Development of Localized Rhizospheric Resistances: A Model

The thesis supported by Newman (1969a, b, 1974), Lawlor (1972), Williams (1974), and Hansen (1974a, b) that R_s is generally unimportant until Ψ^s approaches about -15 atm is predicated on the assumption that most of the root system is involved in water absorption and that permeability along the length of the system is reasonably uniform. As discussed earlier, all these authors used available data on rooting density in the basic formulations of Cowan (1965) and Gardner (1960) to reach these conclusions. They did, however, acknowledge that if not all elements of the root system are functioning in water uptake, appreciable R_s might develop in soil at Ψ^s greater than -15 atm. Although data are scant, enough evidence is available to suggest that the highest permeability per unit length usually occurs in the young, unsuberized root segments and that in the older, suberized portions of the root system entry of water will be highly dependent on the patterns of suberin deposition. Furthermore, where a periderm or any continuous suberin sheath surrounds the root, water uptake will be confined to lenticels or other disruptions in the suberin layer. This depiction of nonuniform water uptake by the root system might suggest a much more complicated model of water flow from the soil to the root and the development of localized areas of appreciable rhizospheric resistance before the bulk soil moisture potential has approached -15 atm.

A model to illustrate quantitatively the resistances to water uptake in a root system where root permeability is not uniform and where localized rhizospheric resistances might develop is based on a simple parallel water flux scheme depicted in Fig. 4. Water flow in the two legs of the circuit representing radial flow from the soil to the root xylem through the suberized and unsuberized segments of the root system are inversely proportionate to the resistances encountered. The total flux is

$$E = {}^sE + {}^uE = \frac{\Psi^x - \Psi^s}{{}^sR_r + {}^sR_s} + \frac{\Psi^x - \Psi^s}{{}^uR_r + {}^uR_s} \tag{10}$$

where ^uE and ^sE are the fluxes through the unsuberized and suberized segments of the root system (cm·d^{-1}), ^sR_r and ^uR_r are root resistances and ^sR_s and ^uR_s are the rhizospheric resistances for the suberized and unsuberized root segments, respectively (d^{-1}). In this equation, Ψ are in units of cm water column. Because of the proximity of the suberized and unsuberized portions of the root system, the mean bulk soil water potential Ψ^s would be the same for soil surrounding the suberized and unsuberized root sections. Since the resistance to water flow in the xylem is generally recognized to be negligible over short distances, water potential in the xylem, Ψ^x, in the suberized and unsuberized root sections would be also the same.

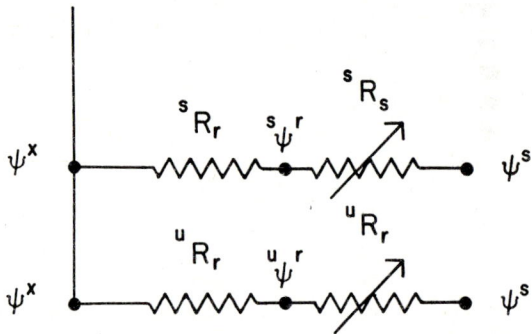

Fig. 4. Parallel flux scheme representing the course of water movement from the bulk soil of water potential, Ψ^s, to the xylem of the root at water potential, Ψ^x, through the unsuberized and suberized portions of the root system. Rhizospheric resistances are denoted as ^sR_s and ^uR_s for the suberized and unsuberized portions of the root system, respectively; and ^sR_r and ^uR_r are the respective resistances to radial flow of water from the root-soil interface to the root xylem. Water potentials at the root-soil interface of the suberized and unsuberized portions of the root system are represented by $^s\Psi^r$ and $^u\Psi^r$

The magnitude of the rhizospheric resistances, ^uR_s and ^sR_s would be contingent on the water extraction rate, Q, according to Eq. (8) and therefore also dependent on the magnitude of resistances to radial flow through the root, ^sR_r and ^uR_r. For a parallel circuit such as this, the reciprocal of the total resistance in the soil-root system would be the sum of the reciprocals of the resistances in the suberized and unsuberized portions of the system; hence

$$\frac{1}{R_{total}} = \frac{1}{^sR_r + ^sR_s} + \frac{1}{^uR_r + ^uR_s}$$

$$= \frac{1}{\dfrac{^s\varrho}{\Gamma_s} + \dfrac{c[\exp(-b\bar{\Theta}_s) - \exp(-b^s\Theta_r)]}{^sE}} + \frac{1}{\dfrac{^u\varrho}{\Gamma_u} + \dfrac{c[\exp(-b\bar{\Theta}_s) - \exp(-b^u\Theta_r)]}{^uE}}$$

(11)

where $^s\varrho$ and $^u\varrho$ are the resistances per unit root length (cm·d) for the suberized and unsuberized root segments, respectively, and Γ_s and Γ_u are the total lengths of the functional suberized and unsuberized roots in the one-dimensional system. The constants c and b are employed in relating soil moisture content, Θ, to matric potential, Ψ, as

$$\Psi = c \exp(b\Theta). \tag{12}$$

Water content at the root-soil interface, Θ_r, is derived from Eq. (8). In this equation, Q is taken as $^sE/z \cdot P_s$ and $^uE/z \cdot P_u$ for the suberized and unsuberized segments of the root system, respectively, where z is the depth of the rooted zone and P_s and P_u are the respective fractional proportions of suberized and unsuberized root length in the system. In order for $\bar{\Theta}_s$ or Ψ^s at the outer radius, r_2, of the cylinder supplying water to the root to be equivalent for the suberized and unsuberized portions of the root system, there will be a tendency for the unsuberized roots which are extracting water at a greater rate per unit length to withdraw this moisture from a cylinder of greater radius than is the case for water extraction by the suberized portions. The radii of the cylinders, surrounding the suberized and unsuberized root sections in Eq. (8), $^sr_2$ and $^ur_2$, are related by

$$^sr_2/^ur_2 = {^sQ}^{1/2}/{^uQ}^{1/2}. \tag{13}$$

If r_2 for the suberized portion of the root systems, $^sr_2$, is taken as 1/2 of the distance between neighboring root elements, as is usually done (Passioura and Cowan, 1968), the value of r_2 for the unsuberized portion of the root system, $^ur_2$, will often tend to be considerably greater. A value of $^ur_2$ which exceeds half the average distance between neighboring root elements would result in competition for moisture between adjacent root elements rather than exclusive use of water by a given root element within the prescribed soil cylinder. This, in effect, would lead to an estimation of even greater rhizospheric resistance. This competition will not be addressed in this model. Therefore, predictions of ^uR_s will be conservative when these values of $^ur_2$ based on $^sr_2 = 1/(\pi L_v)^{1/2}$ are employed in Eq. (8).

The diffusivity term, D, to be employed in Eq. (8) in the model has been taken as

$$D = m \exp(g\Theta) \tag{14}$$

where the constants m and g are employed in relating soil moisture content, Θ, to diffusivity D.

Often the value of Θ in Eq. (8) is taken as that in the bulk soil (e.g. Lawlor, 1972); however, since soil moisture content is depleted most severely closer to the root (see Fig. 1), and hence the primary resistance to water transport also develops within the soil cylinder as water converges toward the root, D has been calculated with Θ in Eq. (14) as $(\Theta_r^{1/2} \cdot \bar{\Theta}_s^{1/2})$ as was done by Passioura and Cowan (1968). Eq. (14) must then be solved simultaneously with Eq. (8).

For operation of the model under the conditions that Ψ^s as well as Ψ^x are the same for the suberized and unsuberized segments of the root systems, Eqs. (10) and (11), with its component equations, must be solved simultaneously.

To illustrate how the model predicts the development of appreciable localized rhizospheric resistances, the following parameters have been applied to represent a realistic though not extremely severe situation. A Yolo light clay has been considered where b is -25.7, c is $-2.67 \cdot 10^5$ cm in Eq. (11), and m is $4.14 \cdot 10^{-2}$ cm$^2 \cdot$d^{-1} and g is 25.7 in Eq. (14) (Cowan, 1965). Coarser-textured soils would exhibit a much more pronounced decrease of k and D with decreasing Θ, as discussed earlier. The root radius, r_1, has been taken as 0.01 cm, and the radius of the soil cylinder for the suberized root elements, $^s r_2$, as one-half the mean distance between root elements. Mean rooting density per volume of soil, L_v, has been set as 1 cm$\cdot$cm^{-3} over the total rooted zone, z, of 100 cm. This represents a reasonable rooting density which is two- to three-fold greater than has been reported for some woody species and yet some grass species may have as much as ten-fold higher rooting densities (Newman, 1969a, 1974; Lawlor, 1972; Kramer, 1969).

As previously mentioned, information concerning the magnitude of $^s R_r$ and $^u R_r$ are, indeed, meager. For the purposes of the present discussion, permeability of the suberized root system has been assumed as 10% of that of the unsuberized roots per unit length (i.e. $^s\varrho/^u\varrho = 0.1$) based on values given by Kramer and Bullock (1966) and Newman (1974). The percentage of unsuberized roots in the diffuse root system, P_u, again based on Kramer and Bullock (1966) has been taken as 4%. The magnitude of the total liquid phase plant resistance R_p has been taken as $5 \cdot 10^3$ days, as was assumed by Cowan (1965). This value has been shown to be of the proper order of magnitude when compared to available data for several species (Newman, 1969a; Lawlor, 1972). Naturally, R_p will be expected to vary as a function of several environmental and plant factors. Nevertheless, it does appear to stand as a reasonable indication of the order of magnitude of this resistance. In this model all of the liquid phase resistance has been assumed to reside in the R_r term and water flux rates which have been employed are in the range where data for several plant species indicate that R_p is generally not greatly influenced by flux rate (Stoker and Weatherley, 1971; Boyer, 1974; Janes, 1970).

The root resistance has then been partitioned between $^s R_r$ and $^u R_r$ according to the previously stated assumptions that the unsuberized portion of the root system possesses a ten-fold greater permeability per unit length but constitutes only 4% of the root system by using Eq. (11) and neglecting the rhizospheric resistances. In this manner, it is not necessary to know the exact magnitude of root resistance per unit length, ϱ, or the actual length of roots, Γ. The $^s\varrho/\Gamma_s$ term would then be 7083 d^{-1} and the $^u\varrho/\Gamma_u$ term would be 17,000 d^{-1}. In this illustration, water uptake along the suberized portion of the root system is assumed to occur through a series of very closely-spaced lenticels. Therefore, the uptake configuration would approach a cylinder.

A moderate rate of transpiration, E, of 0.1 mm h^{-1} (Newman, 1974; Slatyer, 1967) has been employed in this illustration. Rates as much as ten times this value might well be sustained for periods of a few hours in nature and would, of course, result in much greater rhizospheric resistances.

Initially, immediately following recharge of soil moisture in the rooted zone, rhizospheric resistances would be negligibly small and the proportion of water absorbed by the suberized portion of the root system would be 71% since this constitutes most of the root system even though permeability per unit length is only 10% of that of the unsuberized root segments. A similar conclusion was reached by Kramer and Bullock (1966). However, as water is taken up by the roots from this finite system, instead of a uniform withdrawal of water from the soil cylinders surrounding each root element, an uneven depletion of moisture in the soil along the root system would occur. Because of the higher permeability of the unsuberized root segments, the soil cylinders surrounding these segments would be depleted of moisture more rapidly, leading, in turn, to development of a higher localized ^uR_s. When Ψ^s has been depleted to -3 atm, although ^sR_s would still be negligible (90 days, or 1% of $^sR_s + ^sR_r$) ^uR_s would be $1.9 \cdot 10^4$ days and would exceed the magnitude of ^sR_r if the same transpiration rate is sustained and no root elongation occurs. In this situation, the proportion of water uptake by the suberized portion of the root system would increase to 84%. By the time Ψ^s reaches -8 atm, though ^sR_s would still be only 10% of $(^sR_s + ^sR_r)$, ^uR_s would be 78% of $(^uR_s + ^uR_r)$ and 91% of the water uptake would occur in the suberized portion of the root system. It is clear that in the absence of root elongation, if the same rate of transpiration is maintained, appreciable localized rhizospheric resistance would develop around the unsuberized root elements, forcing a greater proportion of the water uptake to occur in the suberized portion of the root system. This leads to a much different conclusion than the calculations of Newman (1969a), Lawlor (1972), Williams (1974) and others. For example, under the same conditions of E, total R_p, L_v and z, rhizospheric resistance would not equal the magnitude of the root resistance term until Ψ^s had decreased to about -19.2 atm if a root system of uniform permeability is assumed, as has been done by these authors.

An attempt to measure Ψ^r on roots of maize seedlings with the use of a thermocouple psychrometer (Fiscus, 1972) has revealed a substantial depression of about 3 atm below the bulk soil Ψ in reasonably moist sand ($\Psi^s > -0.5$ atm). Although there may be some legitimate reservations about how the measuring device altered Ψ of the root-soil interface, the results might well be in part reflecting the development of a localized rhizospheric resistance and hence corroborate the import of this model.

The hypothetical situation portrayed thus far by the model is useful to illustrate that if unsuberized root elements were substantially more permeable than suberized segments, as the few available data indicate, and transpiration were maintained at a constant rate, appreciable localized rhizospheric resistances could develop in the absence of root extension. Unsuberized root segments do not, however, remain at one location in the soil. As described earlier, new root segments do not remain unsuberized for more than a few days to a couple of weeks and on the average suberization keeps pace with the rate of new root extension. Thus, the length of time, t_u (days), that a segment of unsuberized root element is present at a given location should be $t_u = l_u / U$, where l_u is the length of the unsuberized root zone (cm), and U is the rate of root extension $(cm \cdot d^{-1})$, when extension and suberization are considered to proceed at equivalent rates. Therefore, instead of continued development of an increasingly severe ^uR_s as Ψ^s decreases, unsuberized elements in most cases would probably be present in the same location for no more than a few days and would hence alleviate some of the accretion of resistance by extension. The significance of root extension for these unsuberized root elements would thus seem apparent.

If the unsuberized elements of the root system are actively growing and thereby maintaining t_u at only a few days, the proportion of water absorbed by the unsuberized fraction of the root system could in some circumstances exceed that predicted from the proportion of unsuberized root segments and the relative permeability of the suberized and unsuberized segments (e.g. 29% from the example discussed earlier). This would be expected under long drying cycles if ^sR_s increased at a greater rate than ^uR_s.

If permeability per unit length of the suberized root segments is only a small fraction of that of the unsuberized segments, it might seem that the lower effective uptake rate, Q, would always result in a much lower rhizospheric resistance. However, this may not always be the case depending on the exact geometry of water uptake along the suberized segments. As discussed earlier, for many suberized roots, water uptake apparently occurs primarily through lenticels or other breaks or cracks in the suberized layer rather than in an evenly distributed pattern. Furthermore, the frequency of points of entry of water along the root is expected to vary considerably between species, as indicated by the work of Addoms (1946). For example, she found that suberized root segments of *Pinus echinata* did not possess lenticels and water uptake occurred primarily at the sites of small wounds or breaks. The periderm formation around branch roots was also very well sealed against water entry. However, for a species such as *Liriodendron tulipifera* she reported that lenticels were frequent and that water would enter through these pores as well as breaks around the bases of branch roots, and, of course, through any breaks or wounds in the periderm as well. This anatomical evidence combined with the high variance in permeability among root segments measured by Kramer and Bullock (1966) suggests that the sites of water uptake may be very unevenly distributed and root segments of appreciable length might be relatively impermeable. Although Kramer and Bullock did not specifically state the length of the root segments they were using, based on the type of technique employed, segments of a minimum length of 1 cm would be anticipated. Since they report for some species that some root segments showed essentially no permeability, widely-spaced points of water entry in suberized roots might be reasonably anticipated.

If water entry occurs essentially at a series of widely-spaced points, then a spherical zone of water uptake would occur with movement converging toward a point in a manner analogous to the movement of water toward the central axis of a soil cylinder, as described earlier. Figure 5 is a depiction of patterns of water uptake for a root system with actively growing roots and a suberization pattern where water entry is restricted to widely-spaced disruptions of the suberin layer. For the spherical uptake configuration, the analogue of Eq. (7) is

$$\frac{\partial \Theta}{\partial t} = \frac{1}{r^2} \frac{\partial}{\partial \Theta}\left(r^2 D \frac{\partial \Theta}{\delta r}\right) \tag{15}$$

and its steady rate solution analogous to Eq. (8) is

$$\Theta_r = \bar{\Theta}_s - \frac{Q}{3D} \cdot \frac{r_2^6}{r_1(r_2^3 - r_1^3)} \left[1 - \frac{9r_1}{5r_2} + \frac{r_1^3}{r_2^3} - \frac{r_1^6}{5r_2^6} \right] \tag{16}$$

(Lommen, personal communication). In this case, r_1 and r_2 refer to radii of the sphere, analogous to radii in the cylindrical model. However, in this case, r_1 is not taken as the root radius since water is not converging toward a small sphere. Instead, it represents a small pore or break on the root approaching a point sink, but as is the case with the cylindrical model, r_1 must be considered to be of finite size. The results of the employment of the spherical and cylindrical models are depicted in Fig. 6, in which case Ψ^s is taken as -5 atm.

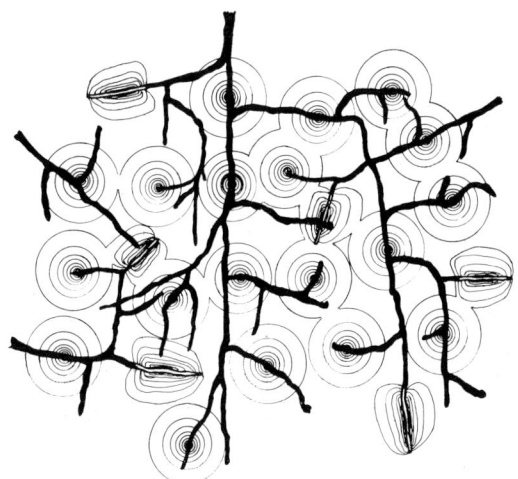

Fig. 5. Depiction of patterns of water uptake around an established perennial plant root system. For the unsuberized root segments, water uptake pattern is in a cylindrical form of increasing radius towards the basal region of the root since the roots are actively growing. The pattern of uptake in the suberized portion of the root system is here considered as a series of widely-spaced points and water flow is in a spherical configuration

These functions have been calculated for two values of Q, 0.0024 and 0.0012. For the cylindrical model, the root diameter, r_1, is 0.01 cm, and for the spherical model the value of r_1 is taken as 0.001 cm. When the outer radius of the cylinder or sphere, r_2, exceeds approximately 0.5 cm, the depression of Ψ increases dramatically for the spherical case and thus illustrates the potential for severe ${}^sR_s$ to develop if a purely spherical model should apply and if the radii of the spheres are appreciable.

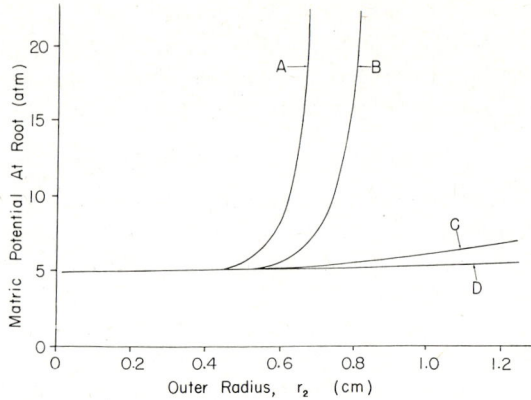

Fig. 6. Soil matric potential at the root-soil interface as a function of the magnitude of the outer radius of a cylinder or sphere of water uptake, r_2. This is represented for two rates of water uptake, 0.012 and 0.0024 cm^3·cm^{-3}·d^{-1} for a spherical, A and B, and a cylindrical model, C and D. For the cylindrical case, r_1 is taken as 0.01 cm, and for the spherical model, the value of r_1 is taken as 0.001 cm

When $^sr_2$ exceeds 0.8 cm, even at low values of Q, substantial localized rhizospheric resistances could develop around the suberized root elements. Such values of r_2 are reasonable if the points of entry of water are well-spaced in a root system of moderate density such as $L_v = 1.0$. For example, at Q of only 0.001 (cm^3·cm^{-3}·d^{-1}), Ψ^s of -5 atm and $^sr_2$ of 0.8 cm, $^s\Psi^r$ would be depressed to -8.5 atm; for the same conditions and $^sr_2$ of 0.9, Ψ^r would be -23.5 atm. These large decreases in Ψ between the bulk soil and the root-soil interface at such low flux rates denote very large rhizospheric resistances. (This value of Q; i.e. 0.001 cm^3·cm^{-3}·d^{-1}, would occur if only 40% of the total water uptake occurred in the suberized portion of the root system, according to provisions of the model (Fig. 4 and Eqs. (10) and (11)) when the suberized root fraction constitutes 96% of the root system and $E = 0.24$ cm·d^{-1}). The development of large rhizospheric resistances around the suberized portion of the root system would tend to cause a greater proportion of water uptake to occur in the unsuberized root elements according to the parallel flux model (Fig. 4), which would, in turn, accentuate the importance of root extension under long drying cycles.

V. Conclusions

The significance of root extension into previously unexplored regions of the soil profile by annual plants or establishing perennial species seems to be of clear advantage facilitating water acquisition, particularly if the soils are only partially wetted and therefore D is small.

The more intriguing question to which the bulk of this chapter has addressed itself relates to the significance of new root extension within the rooted zone of an established perennial plant. A simple depiction of an optimized perennial plant root system would be one of uniform permeability throughout the root length and an evenly-spaced rooting pattern. In this case, extension of new root elements would seem to be of no advantage except to replace root elements lost to death and decay. For the optimized system, sizable rhizospheric resistances would not develop until the soil dried to the point that Ψ^s approached -15 atm unless the root system were extremely sparse or the soil of extremely coarse texture. However, perennial plant root systems do not seem to follow this optimized strategy. New root element formation usually occurs throughout most of the active growing season for the shoot system and there is some indication that this new root formation is of value in water absorption. Furthermore, the root system does not maintain a uniform permeability.

A case is presented (Section IV-3) to suggest that root extension in the rooted zone is of advantage in at least partially evading localized rhizospheric resistances. The discussion and model development suggest a re-evaluation of the consensus (Newman, 1969a, 1974; Lawlor, 1972; Hansen, 1974a, b; Williams, 1974) that rhizospheric resistances are not important until Ψ^s within the rooted zone approaches low levels of -12 to -15 atm. Localized development of appreciable rhizospheric resistances may, indeed, be anticipated at moderate rooting densities when the soil is still relatively moist (e.g. -3 atm). Furthermore, because of the potential for localized rhizospheric resistance development, the importance of root extension within the rooted zone of established perennial plants can also be envisaged. The observation that new root element formation within the rooted zone of a perennial plant species seems to be spatially well distributed at a given depth in the profile so that new growing root tips seldom occur in close proximity (Fernandez and Caldwell, 1975) would seem to fit well with the envisaged scheme of root extension and water uptake. The experimental evidence upon which much of this discussion and model development has been based is, indeed, meager, particularly with respect to root resistances and the nature of water entry along older, suberized portions of the root system. Nevertheless, what evidence is available certainly warrants the depiction of the root system as being of nonuniform permeability. Certainly, a need for experimental work in this area should be recognized.

Although a reasonably plausible case has been made to support the hypothesis that new root element extension within the rooted zone is important to overcome localized rhizospheric resistances in even relatively moist soil, the larger question of the adaptive strategy of perennial plant root systems has not been addressed. From an energetic standpoint, the annual replacement of a significant proportion of the fine root system, the deposition of suberin in root tissues following a mere one to two weeks after formation, and the loss of inorganic as well as organic constituents to the soil as ephermeral root elements die, are all rather difficult to justify when roots are viewed as purely instruments of water absorption in the soil. Although suberization of root surfaces might play a role in slowing or preventing water loss to very dry portions of the soil profile or perhaps aid in the protection against microbes or belowground herbivores, convincing arguments to justify the teleologic concept of energetic efficiency are still to be conceived. Root functions apart from water absorption undoubtedly must also be taken into account in any holistic concept of root system strategies. Nevertheless, from both a physiologic and an ecologic point of view the questions are hardly trivial when it is considered that half to three-quarters of the productivity takes place belowground in such diverse systems as a mixed deciduous forest, a shortgrass prairie, a mesic fescue meadow, and a shrub steppe.

Acknowledgement. Paul Lommen provided the derivation of Eqs. (15) and (16), and Herman Wiebe, Richard Holthausen and Douglas Johnson reviewed the manuscript. The assistance of these colleagues is gratefully acknowledged.

References

Addoms, R. M.: Entrance of water into suberized roots of trees. Plant Physiol. **21**, 109–111 (1946).
Bhar, D. S., Mason, G. F., Hilton, R. J.: *In situ* observations on plum root growth. J. Am. Soc. Hort. Sci. **95**, 237–239 (1970).
Boyer, J. S.: Water transport in plants: mechanism of apparent changes in resistance during absorption. Planta (Berl.) **117**, 187–207 (1974).
Brouwer, R.: Water absorption by the roots of *Vicia faba* at various transpiration strengths. II. Causal relation between suction tension, resistance and uptake. Proc. Kon. Ned. Akad. Wetensch. (C) **56**, 129–136 (1953).
Brouwer, R.: Water movement across the root. In: The state and movement of water in living organisms (ed. G. E. Fogg). Symp. Soc. Exp. Biol. **19**, pp. 131–149. Cambridge: University Press 1965.
Caldwell, M. M., Camp, L. B.: Belowground productivity of two cool desert communities. Oecologia (Berl.) **17**, 123–130 (1974).
Cowan, I. R.: Transport of water in the soil-plant-atmosphere system. J. Appl. Ecol. **2**, 221–239 (1965).
Cowan, I. R., Milthorpe, F. L.: Plant factors influencing the water status of plant tissues. In: Water deficits and plant growth, Vol. 1 (ed. T. T. Kozlowski), pp. 137–193. New York: Academic Press 1968.
De Puit, E. J., Caldwell, M. M.: Seasonal pattern of net photosynthesis of *Artemisia tridentata*. Am. J. Botany **60**, 426–435 (1973).
Elfving, D. C., Kaufmann, M. R., Hall, A. E.: Interpreting leaf water potential measurements with a model of the soil-plant-atmosphere continuum. Physiol. Plantarum **27**, 161–168 (1972).
Fernandez, O. A., Caldwell, M. M.: Phenology and dynamics of root growth of three cool semi-desert shrubs under field conditions. J. Ecol. **63**, 703–714 (1975).
Fiscus, E. L.: *In situ* measurement of root-water potential. Plant Physiol. **50**, 191–193 (1972).
Gardner, W. R.: Dynamic aspects of water availability to plants. Soil Sci. **89**, 63–73 (1960).
Gardner, W. R.: Availability and measurement of soil water. In: Water deficits and plant growth, Vol. 1 (ed. T. T. Kozlowski), pp. 107–135. New York: Academic Press 1968.
Hansen, G. K.: Resistance to water transport in soil and young wheat plants. Acta Agr. Scand. **24**, 37–48 (1974a).
Hansen, G. K.: Resistance to water flow in soil and plants, plant water status, stomatal resistance and transpiration of Italian ryegrass, as influenced by transpiration demand and soil water depletion. Acta Agr. Scand. **24**, 83–92 (1974b).
Harris, G. A., Wilson, A. M.: Competition for moisture among seedlings of annual and perennial grasses as influenced by root elongation at low temperature. Ecology **51**, 530–534 (1970).
Hayard, H. E., Blair, W. M., Skaling, P. E.: Device for measuring entry of water into roots. Botan. Gaz. **104**, 152–160 (1942).

Head, G. C.: Shedding of roots. In: Shedding of plant parts (ed. T. T. Kozlowski), pp. 237–293. New York: Academic Press 1973.

Honert, T. H. van den: Water transport in plants as a catenary process. Disc. Faraday Soc. **3**, 146–153 (1948).

Janes, B. E.: Effect of carbon dioxide, osmotic potential of nutrient solution, and light intensity on transpiration and resistance to flow of water in pepper plants. Plant Physiol. **45**, 95–104 (1970).

Kramer, P. J.: Root resistance as a cause of decreased water absorption by plants at low temperatures. Plant Physiol. **15**, 63–79 (1940).

Kramer, P. J.: Plant and soil water relationships. New York: McGraw-Hill 1969.

Kramer, P. J., Bullock, H. C.: Seasonal variations in the proportions of suberized and unsuberized roots of trees in relation to the absorption of water. Am. J. Botany **53**, 200–204 (1966).

Kuiper, F., Kuiper, P. J. C.: Permeability and self-induction as factors in water transport through bean roots. Physiol. Plantarum **31**, 159–162 (1974).

Lang, A. R. G., Gardner, W. R.: Limitation to water flux from soils to plants. Agron. J. **62**, 693–695 (1970).

Lawlor, D. W.: Growth and water use of *Lolium perenne*. I. Water transport. J. Appl. Ecol. **9**, 79–98 (1972).

Newman, E. I.: Resistance to water flow in soil and plant. I. Soil resistance in relation to amounts of root: theoretical estimates. J. Appl. Ecol. **6**, 1–12 (1969a).

Newman, E. I.: Resistance to water flow in soil and plant. II. A review of experimental evidence on the rhizosphere resistance. J. Appl. Ecol. **6**, 261–272 (1969b).

Newman, E. I.: Root and soil water relations. In: The plant root and its environment (ed. E. W. Carson), pp. 363–440. Charlottesville: Univ. Press Virginia 1974.

Nimah, M. N., Hanks, R. J.: Model for estimating soil water, plant, and atmospheric interrelations. II. Field test of model. Soil Sci. Soc. Am. Proc. **37**, 528–532 (1973).

Parsons, L. R., Kramer, P. J.: Diurnal cycling in root resistance to water movement. Physiol. Plantarum **30**, 19–23 (1974).

Passioura, J. B.: The effect of root geometry on the yield of wheat growing on stored water. Australian J. Agr. Res. **23**, 745–752 (1972).

Passioura, J. B., Cowan, I. R.: On solving the non-linear diffusion equation for the radial flow of water to roots. Agr. Meteor. **5**, 129–134 (1968).

Schulze, E.-D., Lange, O. L., Koch, W.: Ökophysiologische Untersuchungen an Wild- und Kulturpflanzen der Negev-Wüste. II. Die Wirkung der Außenfaktoren auf CO_2-Gaswechsel und Transpiration am Ende der Trockenzeit. Oecologia (Berl.) **8**, 334–355 (1972).

Slatyer, R. O.: Plant-water relationships. New York: Academic Press 1967.

Speidel, B., Weiss, A.: Zur ober- und unterirdischen Stoffproduktion einer Goldhaferwiese bei verschiedener Düngung. Angew. Botan. **46**, 75–93 (1972).

Speidel, B., Weiss, A.: Untersuchungen zur Wurzelaktivität unter einer Goldhaferwiese. Angew. Botan. **48**, 137–154 (1974).

Stoker, R., Weatherley, P. E.: The influence of the root system on the relationship between the rate of transpiration and depression of leaf water potential. New Phytologist **79**, 547–554 (1971).

Taylor, H. M., Klepper, B.: Water uptake by cotton roots during an irrigation cycle. Australian J. Biol. Sci. **24**, 853–859 (1971).

Weatherley, P. E.: Some aspects of water relations. Advan. Botan. Res. **3**, 171–206 (1970).

Williams, J.: Root density and water potential gradients near the plant root. J. Exp. Botany **25**, 669–674 (1974).

B. Resistance to Water Flow in the Roots of Cereals

E. L. GREACEN, P. PONSANA, and K. P. BARLEY

I. Introduction

Given access to soil water at high potential the root system can generally meet evaporative demand; but under adverse conditions with relatively dry soil and high evaporative demand, supply often falls below demand and yield suffers. To extract water from the soil at a rate sufficient to match demand, gradients in water potential have to develop to overcome resistances to flow located in the soil and in the plant itself. This chapter deals with the water transfer properties of plant roots and root systems and how these modify water supply in cereals.

II. Anatomy of Cereal Roots

The root system of cereals, as described by Troughton (1962), consists of two distinct parts: the seminal roots, consisting of three to five axes in wheat, arising from the embryo, and the nodal or adventitious roots arising from the basal nodes of the stem. The seminal roots will be described and significant differences in the anatomy of nodal roots will be noted.

The epidermis is a continuous layer of elongated cells, the outer walls of which are slightly cutinised. Many of these cells develop root hairs, which tend to be relatively long lived in cereals. Walter and Barley (1974) found that 10 to 30% of the total root length of field grown wheat was hair-bearing. Whether all zones bearing hairs retain their effectiveness as organs of water absorption is doubtful.

The cortex, 200 μm thick in the seminal axes of wheat, consists of six to eight layers of thin walled cells with small intercellular spaces. These spaces are not likely to transport water in the field since, having a radius of 2 μm, they will drain at suctions less than 1 bar. In the older zones of the wheat root (>30 days) the cortex tends to shrivel, while the vascular tissue retains its usual characteristics. It is likely that the vapor gap created by the loss of the cortex renders this part of the root ineffective in water uptake. In the younger sections of the root Cole and Alston (1974) observed 60% reduction in the diameter of the root as the water potential of the root decreased from -5 to -10 bar. This may also lead to a break in liquid continuity with the soil. On the inside the cortex is bounded by the endodermis, a single layer of closely fitting cells with thin outer walls and, in older zones, with thickened and strongly suberised inner walls; the radial walls are also thickened and suberised especially where they adjoin the inner walls. Clarkson et al. (1973) recognise three stages of development of the endodermis in barley: primary differentiation and elongation; a secondary stage of laying down the Casparian strip in all endodermal cells at distances greater than 5 mm from the apex; and a tertiary stage commencing with the suberisation of the entire wall. This tertiary stage is delayed in certain cells, the passage cells, adjacent to the protoxylem cells. Thus, while tertiary suberisation may begin within 8 cm of the apex, the entire endodermis does not become thickened until much later (20 to 30 cm from the apex). In the nodal roots unsuberised cells may extend to 30 to 60 cm from the apex. In the first order laterals similar developmental stages are found, but they occur much closer to the tip.

The vascular cylinder is bounded on the outside by the pericycle, a layer of cells that are slightly elongated in the radial direction, continuous except opposite the protoxylem groups, where, in barley, the protoxylem pole cells abut the endodermis. The centre of the stele in seminal roots of wheat is occupied by a single pitted metaxylem vessel, 40 to 110 μm in diameter; occasionally there may be two centrally placed metaxylem vessels. The diameter of the central vessel usually increases with distance from the base of the axis, until it decreases suddenly several cm from the apex (Ponsana, 1975). In wheat 7 or 8 xylem strands each contain a metaxylem vessel approximately 10 μm in diameter.

The stele of the nodal axes differs from that of the seminals in that there are several centrally placed metaxylem vessels. In wheat the number of these decreases from seven at the base of the axis to one near the tip; the diameter of individual vessels is approximately 50 μm.

III. Zone of Water Absorption

Considerable uncertainty exists about the fraction of the total root length effective in water uptake in field crops. Almost all the significant work has been done on young root systems if not young root members, often with a technique where the root is tested under water. Under this condition results are reasonably consistent: the zone of maximum absorption in axes begins some five mm from the tip, where the elements of the stele are well differentiated but tertiary suberisation has not begun. Clarkson et al. (1973) give a value for the rate of uptake for a transpiring barley plant of $0.11 \text{ cm}^2 \text{ day}^{-1}$ (cm^3 per cm root length per day) for the apical 7 cm of the seminal axis, which fell rapidly to $0.024 \text{ cm}^2 \text{ day}^{-1}$ with the onset of tertiary suberisation. This rate was maintained for 50 cm behind the apex. For laterals the rate of water uptake per unit length was uniform at $0.022 \text{ cm}^2 \text{ day}^{-1}$ up to 1 to 2 cm from the apex; near the base (>5 cm) it decreased to $0.01 \text{ cm}^2 \text{ day}^{-1}$. These rates are only roughly comparable as no indication was given of the suction differences across the roots. A similar pattern of uptake by young root axes was obtained for beans by Brouwer (1953) and for wheat by Hansen (1974).

The extent of the absorbing zone varies between species and in the one species with the conditions of growth. In the young (6 cm) first order laterals of barley above we can assume that the "low-resistance" zone extends to at least 1.5 cm behind the tip compared to 6 to 8 cm in seminal axes. Using these values and output from Lungley's (1973) model for long axes with first order laterals spaced at 0.4 cm intervals, the low resistance zones comprise 0.25 of the total root length. Barley (1970) considered that root hairs may be important in water uptake as a means of maintaining liquid continuity with the soil. Walter and Barley (1974) and Ponsana (1975) used the length of hair-bearing root per unit volume of the soil L_H rather than the total length of roots per unit volume of soil L_V (Fig. 1) in water uptake studies. They obtained values of L_H ranging from 0.1 to 0.3 L_V. Nevertheless the walls of root hairs persist for a time after the cells have lost turgor so that not all hair-bearing zones are operative; effective rooting density in cereals is likely to be less than L_H, say from 0.05 to 0.20 L_V.

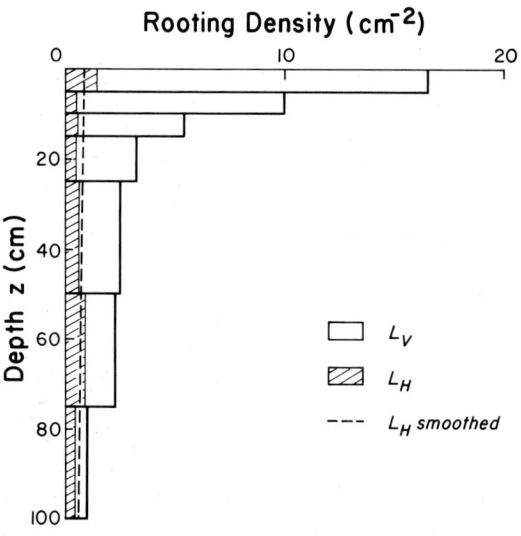

Fig. 1. Rooting density (L_V: total length of roots, L_H: length of hair-bearing roots, per unit volume of the soil) of wheat at successive depths at flowering for an intermediate rate of seeding (S_2)

Claims have been made that even highly suberised regions are effective in water uptake (see also Caldwell, this volume Part 2:A). For example, Kramer (1969) concludes that plants without unsuberised roots can absorb water readily. Also absorption of water by suberised roots has been observed where the roots are submerged in water. When roots are submerged, water may flow through breaks, for example at junctions of laterals with the endodermis. Furthermore, it seems unlikely that radial conduction through a suberised endodermis would offer any advantage. Firstly, as the first order laterals on the seminal axes of cereals are commonly spaced at intervals of 0.4 cm or less, the radial conductance within the stele may be fully taxed by flow from the metaxylem of the laterals to the metaxylem of the axis. Secondly, it is important that some barrier exists inside the points of junction of laterals with the axis to prevent air from entering the metaxylem of the axis following injury to the laterals. Thirdly, a high resistance to the flow of water through the endodermis in proximal zones has the advantage of decreasing the risk of water loss to drier layers of soil.

IV. Forces Causing Flow of Water

Van den Honert (1948), using an Ohm's law analogue of water flow in the plant, distinguished two parts of the flow pathway in the root, the radial pathway from the epidermis to the xylem, and the axial pathway along the xylem. The radial pathway in the soil is often added when, for a steady flow,

$$E = \frac{\Psi_m^s - \Psi_m^r}{R_s^A} = \frac{\Psi^r - \Psi^x}{R_\beta^A} = \frac{\Psi_h^x - \Psi_h^b}{R_\alpha^A} \qquad (1)$$

where E, the evaporation rate per unit ground area (cm day^{-1}) is the flow of water caused by differences in matric potential Ψ_m (bar) between the bulk of the soil s and the root surface r, in the water potential $\Psi = \Psi_m + \Psi_\pi$, where Ψ_π (bar) is the osmotic potential, between the root surface r and in the xylem x, and in the hydraulic potential Ψ_h (bar) between the xylem x in the absorbing zone and at the base of the root b; $\Psi_h = \Psi_p + \Psi_g$, the sum of the pressure and gravitational potentials. R_s^A, R_β^A and R_α^A (cm^{-1} day bar) are resistances to flow per unit ground area in the soil and in the radial β and axial α pathways in the root.

Depending on the nature of the flow path different components of Ψ will operate. Since there is no semi-permeable barrier to the movement of solutes within the soil, any gradient in Ψ_π will be ineffective, and liquid flow of water to the root will be driven solely by differences in Ψ_m. Also there are no semi-permeable barriers in the axial flow path within the xylem, where water and solutes move under a gradient of hydraulic potential. For flow f (cm^3 day^{-1}) in a single vertical root element of length L a change in gravitational potential Ψ_g is involved and vertical flow can be expressed as

axial resistance

$$f = -K_\alpha(\Delta\Psi_p + \Delta\Psi_g)/\Delta L = -K_\alpha \Delta\Psi_h/\Delta L \qquad (2)$$

where Ψ_h is the hydraulic potential and K_α (cm^4 day^{-1} bar^{-1}) is the axial conductance. In rapidly transpiring plants the hydrostatic pressure in the xylem is thought to be negative; the negative pressure arises from a matric potential that develops at the sites of evaporation in the leaves, and is conveniently expressed as Ψ_m.

Flow of water J_w (cm² day⁻¹, cm³ per cm length of root per day) in the radial pathway from the epidermis to the xylem vessels may be expressed as

$$J_w = -K_\beta(\Delta\Psi_m + \Delta\Psi_\pi) \tag{3}$$

where J_w is the influx to the xylem per unit length of root and K_β (cm² day⁻¹ bar⁻¹) is the radial conductance. As discussed above $\Delta\Psi_m$ is induced by a reduction in Ψ_m in the walls of the leaf mesophyll; Ψ_m may fall to low values, usually about -15 bar, but frequently even lower than this. At these values of Ψ_m the xylem is subjected to large mechanical loads.

The second term in Eq. (3) describes the flow of water driven by any difference in osmotic potential between the root surface and the xylem. Between the epidermis and the xylem a semi-permeable barrier occurs that allows the root to behave as an osmometer. Salts, and in herbaceous plants almost exclusively salts (van Overbeek, 1942), are pumped into the xylem on the inner side of this barrier, increasing the solute concentration above that of the external medium. The resulting flow is sometimes manifested as guttation or, in detopped roots, as exudation; it indicates a positive pressure potential or "root pressure" within the xylem.

While root pressures above atmospheric are clearly indicated by water exuding from excised roots, there are conflicting opinions about the importance that osmotic gradients have during transpiration (Brouwer, 1965; Kramer, 1969). Locher and Brouwer (1964) showed that the root systems of 21 day old maize plants, which took up water at the rate of 39 cm³ day⁻¹ during transpiration in light, exuded water at the rate of 29 cm³ day⁻¹ when excised near the base of the stem. However, it is uncertain whether or not the rate of exudation after detopping is a good measure of osmotically induced flow in intact plants. A better method of determining the osmotic flow is described by House and Findlay (1966), who worked with excised roots of maize. They measured J_w (cm² day⁻¹) and J_s, the radial salt flux per unit length of root into the xylem (μ osmol cm⁻¹ day⁻¹), when excised roots from 8 to 14 cm long were immersed in solutions containing various concentrations C_0 (m osmol l⁻¹) of KCl. The conductivity K_β was obtained from the slope of the regression of J_w on $\Delta\Psi_\pi$, where

$$\Delta\Psi_\pi = RT(C_x - C_0). \tag{4}$$

R is the gas constant, T is the absolute temperature, and C_x (m osmol l⁻¹) is the osmolarity of salt in the exudate. From Eq. (3) the osmotically induced component of flow is the product $K_\beta \Delta\Psi_\pi$.

Assuming that $C_0 = 2$ m osmol l⁻¹ for the soil solution, and using the values of K_β and J_s obtained from House and Findlay (1966) for maize roots ($K_\beta = 0.02$ cm² day⁻¹ bar⁻¹, $J_s = 1.0$ μ osmol cm⁻¹ day⁻¹), we find that, when $J_w = 0.02$, 0.05, and 0.10 cm² day⁻¹, the osmotic component can account for 100, 17, and 4% of J_w. We note here that for roots in field soils J_w rarely exceeds $0.01\,\text{cm}^2\,\text{day}^{-1}$ (Walter and Barley, 1974; Greacen and Hignett, 1976). However House and Findlay used high sugar, low salt roots and conducted their experiment at a relatively high temperature ($25°C$).

Using detopped 15-day-old wheat seedlings having seminal axes 8 cm long, Lundegårdh (1944) obtained $K_\beta = 0.01$ to $0.02\,\text{cm}^2\,\text{day}^{-1}\,\text{bar}^{-1}$. In later work with similar seedlings Lundegårdh (1950) found $J_s = 1.2\,\mu$ osmol $\text{cm}^{-1}\,\text{day}^{-1}$ at $C_0 \geq 2\,\text{m}$ osmol l^{-1}. As several centimetres of stem base were retained on the excised root system used by Lundegårdh, the value of K_β may have been reduced below the true value by resistance located in the base of the stem. If we take the higher value of K_β ($0.02\,\text{cm}^2\,\text{day}^{-1}\,\text{bar}^{-1}$) then the osmotic component of flow appears to be similar to that in maize.

Data of Locher and Brouwer (1964) on salt and water uptake suggest that the osmotic component for 21-day-old detopped barley seedlings is relatively small compared to that for maize. Again these results may be influenced by resistance located in the base of the stem.

Further data are clearly needed for roots of cereals sampled at different stages of growth, before we can reach general conclusions about the importance of the osmotically induced component of flow. The importance of $\Delta\Psi_\pi$ during transpiration in wheat in the field depends not only on J_s and J_w but also on the magnitude of R_α and R_β. It is with these resistances that we are next concerned.

V. Resistance to Flow

1. Radial Pathway

Radial resistance R_β refers to the resistance to flow from the epidermis to the lumen of the xylem vessels; it is defined as $R_\beta = 1/K_\beta$ where K_β, the radial conductance, is defined by Eq. (3). The nature of the radial pathway has been described by Slatyer (1967), Ginsburg and Ginzberg (1970), and Tanton and Crowdy (1972). The most widely held view is that most of the flow proceeds via the free space of the cell walls, except at the endodermis where it has to cross the protoplast and its boundary membranes. It has generally been assumed that the endodermis represents the main barrier to water movement, but this has recently been questioned by Dainty (1969): House and Findlay (1966) have deduced from response times of flow to changes in C_0 that, in young roots where the endodermis is still in the secondary stage, the solute barrier referred to in the previous section is located at the walls of the xylem vessels containing symplasm. House and Findlay consider that the site of salt absorption and the main barrier to water movement in the young roots are probably identical. This barrier and resistance to flow in the free space make up the radial resistance in the highly conducting poorly suberised zone of young root members, but as shown by Graham et al. (1974) R_β increases sharply as the endodermis enters the tertiary stage.

There are relatively few measurements of K_β, particularly for cereals, and these have mostly been confined to measurements on axes. Potometers were used by Brouwer (1953) to measure J_w for a single root in solution at several rates of transpiration. Brouwer found Ψ^x by determining the value of C_0 and hence Ψ_π for the external solution that just prevented flow. He obtained values of K_β for the absorbing zone of the radicle of broad bean varying from 0.01 to 0.10 cm^2 day^{-1} bar^{-1}. K_β may alter when C_0 is changed suddenly, and this may have affected Brouwer's results. With excised roots Ψ_π^x can be measured on the exudate; here the question arises whether Ψ_π^x for the exudate is the same as Ψ_π^x at the site of water absorption. Hansen (1974) used a potometer to supply water to 1 cm lengths of young, seminal axes of wheat. He measured leaf water potential Ψ^l. Hansen obtained a uniform value of K_β of 0.05 cm^2 day^{-1} bar^{-1} over the length 1 to 5 cm from the apex. His method ignored any change in water potential associated with vertical flow in the xylem. Measurements have also been made of $\bar{K}_\beta$ for sets of roots of various plants. Cox (1966) applied a suction to the base of two detopped root systems of young wheat plants, and obtained $\bar{K}_\beta$ values of 0.2 and 0.4 cm^2 day^{-1} bar^{-1}. His values are considerably higher than those of Brouwer or Hansen above and leakage may have occurred.

Other methods for finding K_β involve measurement of exudation from excised roots in a stirred aerated osmoticum. As mentioned previously it is important in these methods that R_β is the only significant resistance to flow; where the base of the stem has been retained, it may offer an unknown resistance so that the measured value of K_β is lower than the value for roots alone. Newman (1973) measured $\bar{K}_\beta$ for young plants of several species using the solution change method of Arisz et al. (1951). $\bar{K}_\beta$ was calculated from the change in the rate of exudation from an excised root system after C_0 had been changed suddenly. Values of $\bar{K}_\beta$ were as follows: broad bean, 0.0013; maize, 0.0024; tomato, 0.0043 cm^2 day^{-1} bar^{-1}. These values are about one tenth of those found for the sub-apical zone of single roots as reported above. Possibly a large resistance was located in the base of the stem. Alternatively, the results suggest that only one tenth of the total root length was absorbing water rapidly. For maize the coefficient of variation of $\bar{K}_\beta$ was small compared with that for root length. This is consistent with either (1) relatively uniform absorption of water over the entire root surface—the view favoured by Newman (1969)—or (2) a relatively uniform fraction of total root length being concerned with water absorption. It is interesting to note here that Hackett and Rose (1972) predict that the mean length per root member remains constant or nearly so during growth.

2. Axial Pathway

The axial resistance to flow in a root member R_α is defined here as $R_\alpha = 1/K_\alpha$ where K_α is defined by Eq. (2).

In woody plants axial flow is confined to the xylem (Slatyer, 1967; Kramer, 1969). Normally the same applies to herbaceous plants and cereals. The intercellular spaces in the cortex drain at $\Psi_m \leq -1$ bar, and so they would not conduct water except perhaps in very wet soils. When the roots are submerged in water the intercellular spaces may become filled with water. Kozinka and Luxova (1971) concluded that 25% of the total axial flow in seminal axes of maize could move through the cortical tissue when water was forced along short lengths of axes at positive pressures.

Although it is often held that $R_\alpha^A \ll R_\beta^A$, R_α^A is not always small. Wind (1955) calculated the resistance to vertical flow in the roots of perennial grasses from an estimate of the number of vertical roots and of the radius of the xylem vessels in a representative root member. Wind defined $r_e = \sqrt[4]{\sum_1^n r_i^4}$; that is r_e is the radius of a single tube equivalent to a set of n tubes in parallel. He assumed that resistance to flow in the vessels was twice that calculated from the Poiseuille equation, and concluded that below 20 cm depth, when the water table was at 50 cm depth, the resistance to vertical flow in the roots exceeded the resistance to vertical flow in the soil.

Passioura (1972), using data on the geometry of the root systems of cereals taken from the literature, assumed that, under dry conditions, only the radicle and two other seminal axes per plant would penetrate to a depth of 40 cm or more. Following Emerson (1954) he assumed that the resistance in the central metaxylem vessel was twice that calculated from its radius using the Poiseuille equation. For a planting density of 150 m^{-2} he calculated that a pressure drop along the seminal axes of 16 bar would be required to move water vertically from a depth of 40 cm to the ground surface to meet a transpiration rate E of 1 cm day^{-1}.

The assumption made by these workers that resistance to flow in the metaxylem vessels was twice that obtained from the Poiseuille equation for a tube of identical r_e was derived from measurements made on a few grass roots at low suctions by Emerson (1954). Emerson measured xylem radii for both basal and apical cross sections. Ponsana (1975) calculated r_e for both of these sections and combined the r_e values using an harmonic mean as follows:

$$r_H = \sqrt[4]{n / \sum_1^n \frac{1}{r_{ei}^4}} \tag{5}$$

Using Emerson's data Ponsana found that for *Lolium perenne* and *Dactylis glomerata* the measured resistance was from two to three times that calculated from the Poiseuille equation and r_H, whereas for *Phleum pratense* the measured and theoretical values were similar.

Ponsana (1975) found that for seminal and nodal axes of wheat roots grown in the field or in solution culture, and for axes and first order laterals of *Pisum sativum*, resistances were generally greater than those calculated from the Poiseuille equation ($R_\alpha = 1.3$ to 2.3 R_p where R_p is the resistance calculated from the Poiseuille equation). In Fig. 2 (Ponsana, 1975) ln K_α (cm^4 day^{-1} bar^{-1}) is given as a function of ln r_e for the seminal axes of wheat, var. Halberd, data being pooled for the seedling and flowering stages. However R_α for seminal axes of wheat at the dough stage was significantly less than R_p.

Among conditions required for Poiseuille flow are that the tube is rigid, smooth and circular in cross section and that there is no slip on the walls. These conditions may not be met for water flow in xylem vessels. The vessels are not smooth since perforated septa occur at intervals along their length.

however the septa are widely spaced and the central pore is only slightly narrower than the rest of the vessel. The radius of the vessel is independent of suction at low suctions but this may not be so at high suctions; Greacen (unpublished data) found that R_α increased by 30% when a pressure difference of 2 bar was applied across the xylem wall. The xylem vessels are sometimes elliptical in cross section and they taper towards the base of the root. The conductance of such tubes can be obtained analytically: Preston (1938) showed that for an elliptical tube with axes l_1 and l_2 the flow rate can be obtained with little error by substituting a value of $r = (l_1 l_2)^{1/2}$ in the Poiseuille equation, provided $l_1/l_2 < 1.3$. All the factors mentioned so far result in R_α greater than R_p. It is more difficult to account for positive deviations. Perhaps water may slip

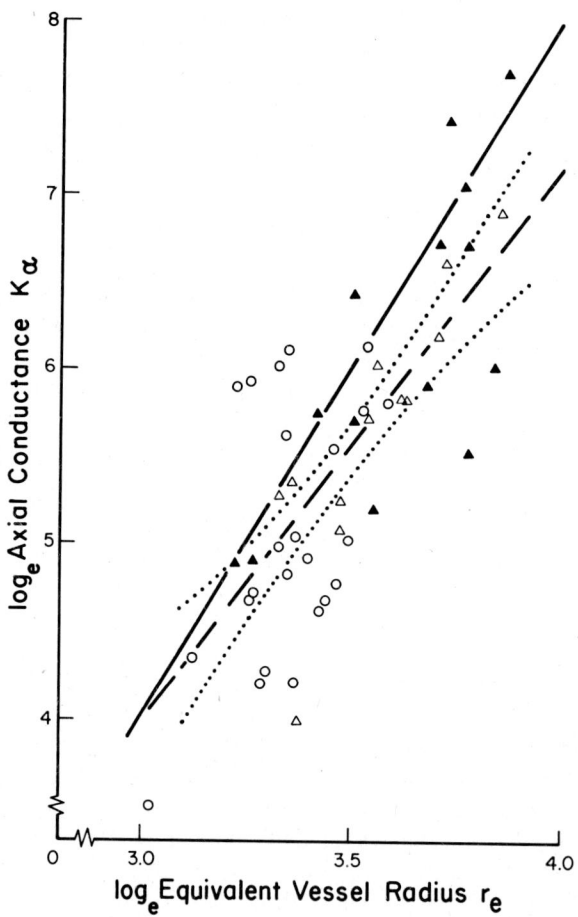

Fig. 2. $\log_e K_\alpha$ (cm^4 day^{-1} bar^{-1}) as a function of $\log_e r_e$ (μm) for segments of seminal axes of wheat cv. Halberd at the seedling stage and at flowering. ○ 18 days after transplanting, △ flowering stage (nutrient solution), ▲ flowering stage (field), —— Poiseuille equation, ----- regression line for data, ····· confidence limits (P, 0.05)

on the walls of the xylem; Barley (unpublished) has found by staining with Sudan III that the walls of the metaxylem in the older zones of seminal axes become impregnated with lipid.

VI. Effect of Root Resistance on Withdrawal of Water from the Soil

Earlier attempts to predict the pattern of water uptake were based on the assumption that root resistance was negligible. For example, in a well known paper, Gardner (1964) postulated that:

$$\lambda = (\Psi_m^s - \Psi_m^r)/\Lambda_s \tag{6}$$

where λ (day^{-1}) is the rate of depletion of volumetric water content due to withdrawal of water by roots, and Λ_s (day bar) is the resistance offered by the soil. This approach is limited in value, as it does not consider the possible contribution of root resistance. Gardner expressed Λ_s as

$$\Lambda_s = 1/B k_s L_V \tag{7}$$

where k_s (cm^2 day^{-1} bar^{-1}) is the unsaturated hydraulic conductivity of the bulk of the soil and B is a constant given to a first approximation as $2\pi/\ln(r_2/r_1)$; r_2 is the radius of the hollow cylinder of soil delivering water to unit length of root of radius r_1 and is calculated as $(\pi L_V)^{-0.5}$.

He then predicted depletion patterns for hypothetical root distributions: one of these, L_V decreasing logarithmically with depth, is shown in Fig. 3.

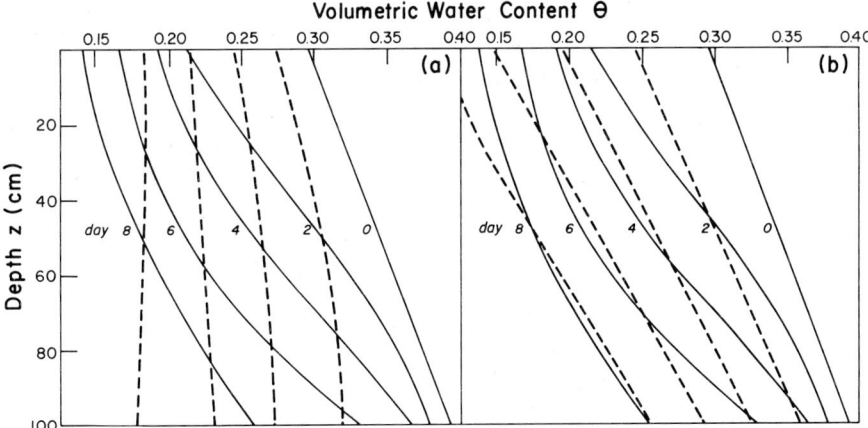

Fig. 3a and b. Calculated soil water depletion curves for Pachappa sandy loam for a logarithmic distribution of L_V ——— (Gardner, 1964) and a linear distribution of L_H ----- as in Fig. 1 (Ponsana, 1975) for (a) $K_\beta = \infty$, and (b) $K_\beta = 0.005$ cm^2 day^{-1} bar^{-1}. Ψ_h was assumed uniform initially at -0.120 bar

The logarithmic distribution approximates the distribution of rooting density L_V of wheat under field conditions but it contrasts strikingly with the distribution of hair-bearing roots L_H (Ponsana, 1975). The latter is probably a more realistic description of the distribution of absorbing roots. In Fig. 3a the water depletion curves obtained using a smoothed distribution of L_H (see Fig. 1), again assuming negligible root resistance, are compared with Gardner's results for the logarithmic distribution. The depletion pattern, dominated by the hydraulic properties of the soil, is reversed, with the greater part of the water being withdrawn from the deeper layers of the soil. If we include R_β in the model, using $R_\beta = 200\,\text{cm}^{-2}$ day bar, the depletion pattern, shown in Fig. 3b, is now dominated by the hydraulic properties of the roots. This is not surprising as the resistance of the soil per unit length of root is only 0.01 R_β at the lowest value of Ψ_m^s in the profile. The depletion pattern now approaches that obtained for Gardner's logarithmic distribution. Inclusion of R_α would bring the depletion curves into still closer agreement, but any agreement is coincidental.

In seeding rate experiments with cereals Kirby (1970) and Walter and Barley (1974) found that, even after the canopy had achieved complete cover, the rate of withdrawal of water from the subsoil was greatest at an intermediate rate of seeding. Kirby hypothesised that this was caused by differences in rooting density in the subsoil. Walter and Barley estimated that at their low rate of seeding (10 kg ha^{-1}) resistance to flow in the relatively few root axes would restrict uptake of physically available water. They also found that the roots penetrated deepest and dried the soil to the greatest depth at the intermediate rate of seeding. Passioura (1972) controlled the number of seminal axes per wheat plant by amputation, and found that water was conserved for use during grain development by reducing the number of axes. Greacen and Hignett (1976) describe a water balance model in which the rate of water uptake from the subsoil is limited by the total length of roots in the subsoil. This allows good simulation of the soil water regime over a wide range of seasonal conditions.

Ponsana (1975) established relations between r_e and K_α for seminal and nodal axes of wheat at different stages of growth. He then measured root geometry of the crop at different seeding rates. This involved tube sampling to obtain measurements of L_V and L_H, and pin-board sampling to obtain counts of the vertical root members. Estimates of L_V and L_H yielded results similar to those described by Walter and Barley (1974) and will not be further described here.

The pin-boards yielded prisms of soil 60 cm × 15 cm × 110 cm deep. After removing the soil by soaking and washing, counts were made of the vertical roots. Segments of a sample of vertical roots were then taken at depths 30, 60 and 90 cm, and cross sectioned. The seminal and nodal axes were distinguished by the anatomy of the xylem, and vertical first order laterals were distinguished from axes by their radius. Vessel radii were measured with an eyepiece graticule at × 100. Results are given in Table 1a and b.

Root counts made at the base of the plants at flowering showed an average of 5 seminal roots per plant, and 55, 30 and 15 nodal roots per plant at $S_1 = 27$, $S_2 = 137$, $S_3 = 595$ plants per m^2. A few nodal roots were found to penetrate deeper than 90 cm at all seeding rates; the largest number of deep nodal roots occurred in S_1, although even here the numbers were small being 13, 5 and

Table 1. Data on root geometry and axial resistance to flow as a function of plant density[a]

(1) Number of axes per square meter at various depths z (cm)

z (cm)	Flowering Stage				Dough Stage			
	S_1[b]	S_2	S_3	L.S.D.	S_1	S_2	S_3	L.S.D.
30	433	433	489	n.s.	614	634	910	n.s.
60	369	396	557	n.s.	521	596	696	n.s.
90	133	303	231	n.s.	157	200	232	n.s.

(2) Mean radius of metaxylem vessels r_A (µm)

30	44.8	41.8	41.7	n.s.	42.3	40.2	36.4	3.0
60	39.7	40.6	39.5	n.s.	34.9	34.8	33.4	n.s.
90	34.4	33.7	35.5	n.s.	33.1	33.2	29.3	n.s.

(3) R_z^A (bar cm^{-1} day) or pressure drop (bar)[c] to supply $E=1$ cm day^{-1} from below depth specified.

30	1 (1)	1 (1)	1 (1)		1 (1)	1 (1)	1 (1)	
60	3 (2)	3 (2)	2 (2)		2 (2)	1 (2)	2 (2)	
90	21 (14)	7 (6)	9 (6)		10 (14)	8 (11)	9 (14)	

[a] Ponsana (1975).
[b] The three seeding rates S_1, S_2 and S_3, 10, 55 and 280 kg ha^{-1}, gave plant densities of 27, 137 and 595 plants m^{-2}.
[c] Bracketed values assume Poiseuille flow; values without brackets corrected for measured deviations from Poiseuille flow.

2% of the total number of axes at 30, 60 and 90 cm depth. Few first order laterals had a vertical component to their path exceeding 10 cm. At the highest plant density only 1 seminal axis per plant had penetrated below 30 cm, while the number of nodal axes at 30 cm depth or greater was negligible. This is surprising, particularly as the supply of water and nutrients appears to have been adequate until flowering. Another unexpected result is the uniformity of axis distribution particularly at the flowering stage; at 30 and 60 cm depth the density of axes was close to 400 m^{-2} or 1 axis per 25 cm^2; at the dough stage this value increases to approximately 600 m^{-2}. Owing to variability, significant differences in axis density between plant densities were not established, but there tended to be fewer axes at S_1 at each depth.

It was not possible to trace the individual roots down the pin board, and Ponsana calculated the radius r_A of a tube representative of the axes found at any particular depth i as $r_A = \sqrt[4]{\sum_1^n r_i^4/n}$, where n is the number of axes sampled at that depth. The effective radius for the depth interval, 0 to 90 cm for example, was then calculated as the harmonic mean:

$$r_H = \sqrt[4]{\frac{3}{\frac{1}{r_{A,30}^4} + \frac{1}{r_{A,60}^4} + \frac{1}{r_{A,90}^4}}}. \tag{8}$$

The axial resistance per unit ground area R_α^A, assuming Poiseuille flow, is given by

$$R_\alpha^A = \frac{\Delta \Psi_h}{E} = \frac{8\eta z}{N \pi r_H^4} \tag{9}$$

where η is the viscosity ($=1.16 \times 10^{-13}$ bar day at $20°C$), z (cm) is the depth from below which water is being withdrawn, and N (cm^{-2}) is the number of axes per unit ground area. Values of R_α^A, representing the drop in hydraulic head (bar) from depth $z=i$ to $z=0$ for a transpiration rate of 1 cm day^{-1}, when water is being withdrawn exclusively from below 30, 60 and 90 cm depth, are given in Table 1c. These values may be compared directly with R_β^A the radial root resistance per unit ground area where $R_\beta^A \equiv 1/K_\beta L_A$, where L_A is now the length of absorbing root per unit area below the depth under consideration. R_α^A and R_β^A may both be compared with R_s^A the radial resistance of the soil per unit ground area defined as

$$R_s^A = \frac{\ln(r_2/r_1)}{2\pi k_s L_A}. \tag{10}$$

To give an example: R_α^A for flow from below a depth of 60 cm at the S_2 seeding rate from the data in Table 1c is 3 cm^{-1} day bar. If we estimate the length of hair-bearing roots per unit ground area from 60 to 120 cm as $0.6 \times 60 = 36$ cm (Fig. 1), and assume $K_\beta = 0.005$ cm^2 day^{-1} bar^{-1}, we obtain a value of $R_\beta^A = 5.6$ cm^{-1} day bar. This shows that R_α^A is comparable with R_β^A for this root geometry at a depth of 60 cm. On the other hand soil resistance R_s^A for Pachappa sandy loam, for instance, does not approach these values until Ψ_m^s approaches -10 bar.

VII. Conclusions

The water supply to crops depends on the geometry and hydraulic properties of the root system. Improvements in sampling and measurement techniques and the development of numerical simulation of the growing root system are providing detailed descriptions or predictions of root geometry. Concurrently more quantitative information is being obtained on the resistance to the flow of water in the root.

Most of the water withdrawn from the soil by cereals is likely to be absorbed by first order laterals, as these members contribute from 60 to 80% of the total root length. Water is absorbed most rapidly from the soil in the hair-bearing zone, where suberisation of the endodermis has not proceeded beyond the secondary stage.

Water flows across the root in response to gradients in matric and osmotic potential. The osmotically induced flow has generally been regarded as small in freely transpiring plants, but this is not always so. For example in maize the osmotic flow remains the major term even when $J_w = 0.03$ cm^2 day^{-1}.

Existing methods of measuring K_β require the root to be submerged in solutions, and the results may often over estimate the in situ conductivity of basal zones in unsaturated soils. Osmotic methods are preferable because, in the absence of a gradient in hydrostatic pressure, solution is less likely to be forced into spaces within the cortex normally occupied by air. However the question arises as to whether K_β differs for flows induced by gradients in osmotic or hydrostatic pressure. The usual range of values of K_β is from 0.002 to 0.02 cm^2 day^{-1} bar^{-1}.

Values of R_α have been obtained by direct measurement or calculated from the radius of the lumen of the metaxylem vessels assuming Poiseuille flow. Measured values of R_α are usually from two to three times the values predicted from the Poiseuille equation, but some exceptions have been found. Depending on xylem geometry K_α in cereals ranges from 10^2 to 10^3 cm^4 day^{-1} bar^{-1}.

Except in soils of extremely low water content water depletion curves are determined primarily by the root distribution and the hydraulic properties of the roots. When calculating R_β^A we need to know the vertical distribution of L_H; this can be obtained by standard methods (Walter and Barley, 1974). Similarly, when calculating R_α^A we need to know the vertical distribution of root axes, this can be obtained with pin-board techniques (Ponsana, 1975). R_β^A and R_α^A may both be important, particularly when water is being withdrawn exclusively from the deeper subsoil.

There is some evidence that the density of vertical roots does not vary greatly over a wide range of seeding rates but the main determinants are not yet known. Models based on the physics of root growth (Greacen, 1976) may prove useful in predicting root geometry as a function of soil and climatic environment.

References

Arisz, W. H., Helder, R. J., Nie, R. van: Analysis of the exudation process in tomato plants. J. Exp. Botany **2**, 257–297 (1951).

Barley, K. P.: The configuration of the root system in relation to nutrient uptake. Advan. Agron. **22**, 159–201 (1970).

Brouwer, R.: Water absorption by the roots of *Vicia faba* at various transpiration strengths. I, II. Proc. Kon. Ned. Akad. Wet. C **56**, 106–115, 129–136 (1953).

Brouwer, R.: Water movement across the root. Symp. Soc. Exp. Biol. **19**, 131–149 (1965).

Clarkson, D. T., Robards, A. W., Jackson, S. M.: The structure of barley roots in relation to the transport of ions into the stele. Agr. Res. Council, Letcombe Lab. Ann. Rep. 1972, 5–7 (1973).

Cole, P. J., Alston, A. M.: Effect of transient dehydration on absorption of chloride by wheat roots. Plant Soil **40**, 243–247 (1974).

Cox, E. F.: Resistance to flow of water through the plant. Ph. D. Thesis, Nottingham, U. K. 1966.

Dainty, J.: The water relations of plants. In: Physiology of plant growth and development (ed. M. B. Wilkins), pp. 421–451. London: McGraw-Hill 1969.

Emerson, W. W.: Water conduction by severed grass roots. J. Agr. Sci. **6**, 147–159 (1954).

Gardner, W. R.: Relation of root distribution to water uptake and availability. Agron. J. **56**, 41–45 (1964).

Ginsburg, H., Ginzberg, B. Z.: Radial water and solute flows in the roots of *Zea mays*. J. Exp. Botany **21**, 580–592 (1970).

Graham, J., Clarkson, D. T., Sanderson, J.: Plant root systems and their relationship to the soil. Agr. Res. Council, Letcombe Lab. Ann. Rep. 1973, 9–12 (1974).
Greacen, E. L.: Soil-plant mechanisms and models—Water. In: Soil factors in crop product in a semi-arid environment (eds. J. S. Russell, E. L. Greacen), Monograph Australian Soil Sci. Soc. Brisbane: Univ. Queensland Press (in press, 1976).
Greacen, E. L., Hignett, C. T.: Water balance model and supply index for wheat in South Australia. CSIRO Div. of Soils Tech. Paper No. 27 (in press, 1976).
Hackett, C., Rose, D. A.: A model of the extension and branching of a seminal root of barley, and its use in studying relations between root dimensions. Australian J. Biol. Sci. **25**, 669–679 (1972).
Hansen, G. K.: Resistance to water transport in soil and young wheat plants. Acta Agr. Scand. **24**, 37–48 (1974).
Honert, T. H. van den: Water transport in plants as a catenary process. Faraday Soc. Discuss. **3**, 146–153 (1948).
House, C. R., Findlay, N.: Water transport in isolated maize roots. J. Exp. Botany **17**, 344–354 (1966).
Kirby, E. J. M.: Evapotranspiration from barley grown at different plant densities. J. Agr. Sci. Camb. **75**, 445–450 (1970).
Kozinka, V., Luxova, M.: Specific conductivity of conducting and nonconducting tissues of *Zea mays* root. Biologia Plantarum (Praha) **13**, 257–266 (1971).
Kramer, P. J.: Plant and soil water relationships: a modern synthesis. New York: McGraw-Hill 1969.
Locher, J. Th., Brouwer, R.: Preliminary data on the transport of water, potassium and nitrate in intact and bleeding maize plants.Jb. Inst. Biol. Scheik. Onderz. LandbGewass, 41–49 (1964).
Lundegårdh, H.: Bleeding and sap movement. Ark. Botan. **31 A**, 2 1–56 (1944).
Lundegårdh, H.: The translocation of salts and water through wheat roots. Physiol. Plantarum **3**, 103–151 (1950).
Lungley, D. R.: The growth of root systems. A numerical computer simulation model. Plant Soil **38**, 145–159 (1973).
Newman, E. I.: Resistance to water flow in soil and plant. I. Soil resistance in relation to amounts of root: theoretical estimates. J. Appl. Ecol. **6**, 1–12 (1969).
Newman, E. I.: Permeability to water of the roots of five herbaceous species. New Phytologist **72**, 547–555 (1973).
Overbeek, J. van: Water uptake by excised root systems of tomato due to non-osmotic forces. Am. J. Botany **29**, 677–683 (1942).
Passioura, J. B.: The effect of root geometry on the yield of wheat growing on stored water. Australian J. Agr. Res. **23**, 745–752 (1972).
Ponsana, P.: Drainage and water uptake terms in the water balance. Ph. D. Thesis. Adelaide, Australia 1975.
Preston, R. D.: The contents of the vessels of *Fraxinus americana* L., with respect to the ascent of sap. Ann. Botany (London) **2**, 1–21 (1938).
Slatyer, R. O.: Plant-water relationships. London: Academic Press 1967.
Tanton, T. W., Crowdy, S. H.: Water pathways in higher plants. II. Water pathways in roots. J. Exp. Botany **23**, 600–618 (1972).
Troughton, A.: The roots of temperate cereals (wheat, barley, oats and rye). Mim. Publ. Commonw. Bur. Past. Field Crops, No. 2, 1–91 (1962).
Walter, C. J., Barley, K. P.: The depletion of soil water by wheat at low, intermediate and high rates of seeding. Proc. 10th. Intern. Congr. Soil Sci. (Moscow) **1**, 150–158 (1974).
Wind, G. P.: Flow of water through plant roots. Neth. J. Agr. Sci. **3**, 259–264 (1955).

C. Soil Water Relations and Water Exchange of Forest Ecosystems

P. BENECKE

I. Introduction

Trying to determine the water exchange of ecosystems in quantitative terms requires the acceptance of basic principles which describe in mathematical statements the relationships that govern the water exchange between the ecosystem and its adjacent systems—the atmosphere, the underground and possibly the neighboring ecosystems—as well as between the internal "compartments" of the ecosystem (Fig. 1). Such a generally accepted basic principle is the equation of continuity, also known as the law of conservation of matter. Another fundamental principle is Darcy's law. Both laws are explained in Sect. III. The combined form of these two laws allows in particular to determine the rates of water transfer within the soil and the corresponding rates of water storage or depletion. The application of the equation of continuity without sink or source terms implies that we have to deal with a constant total amount of recycling water.

To describe the turnover, one can start with the precipitation. Part of it wets the above-ground vegetation and evaporates after the rain ceases. This part is called interception. The rain not intercepted, reaches the ground by three ways: either by throughfall, canopy drip or stem-flow. After reaching the ground the precipitation water tends to infiltrate the soil. Depending on whether or not the "infiltrability" (Hillel, 1971) of the soil is exceeded, the rainfall water will accumulate on the soil surface and, under certain conditions, may give rise to surface runoff.

The infiltrating water increases the amount of water stored in the soil pores and provides in this form or as slowly moving soil water the reservoir from which the plant roots extract water. Excess water seeps vertically or laterally downward. It may recharge groundwater aquifers and will eventually emerge in wells (Fig. 2), provided that it is not withdrawn from aquifers by pumps of water plants.

The fate of the water that enters the soil depends to a high degree on two physical properties of the soil: the water characteristic or water retention curve (pF-curve) and the water conductivity. Both functions generally vary from soil horizon to soil horizon. The water retention curve reflects that the total pore space of a soil consists of pores with a continuous diameter distribution

(Fig. 8). The strength of the forces holding the water and resisting its movement increases strongly with decreasing pore diameter.

If air and water are both present in the pore system, it is the air that occupies the coarser pores. This condition is called "unsaturated". In a "saturated" soil the total pore space is occupied by water. At least 10% of the total soil volume should be occupied by air in the main root zone in order to provide sufficient oxygen supply (Kramer, 1969; Black, 1968; Wesseling and van Wijk, 1957; Finck, 1969).

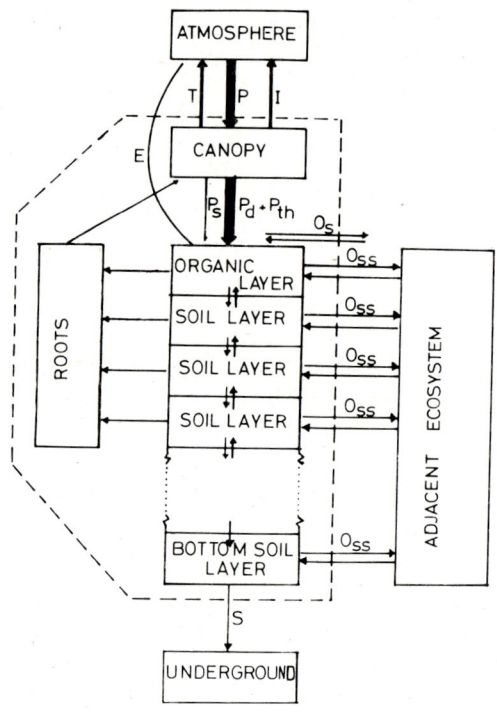

Fig. 1. Compartment model of a soil-plant ecosystem; P precipitation, T transpiration, I interception, P_s stemflow, P_d canopy drip, P_{th} throughfall, O_s surface run-off, O_{ss} subsurface flow, S deep seepage. The arrows indicate possible directions of water fluxes

The already mentioned equation of continuity relates the changes in water storage and water movement. The rates of water movement depend, according to Darcy's law, on the water conductivity and the hydraulic gradient. The water conductivity in turn is strongly dependent on the degree of water saturation (Fig. 9).

The function of the soil is stressed in this chapter in appreciation of the pivotal role the soil plays in the water exchange of plant–soil ecosystems. The soil can be regarded as a kind of hydraulic switch which principally adjusts the water fluxes in favor of the plants. "Principally" means within the given possibilities which are dependent on the water retention curve and the water conductivity. The principles mentioned so far need not be restricted to forest ecosystems; but forest ecosystems constitute in various ways a particular case in the water cycle. In the temperate zone they are frequently located at less favorable sites characterized by high precipitations and relatively low tempera-

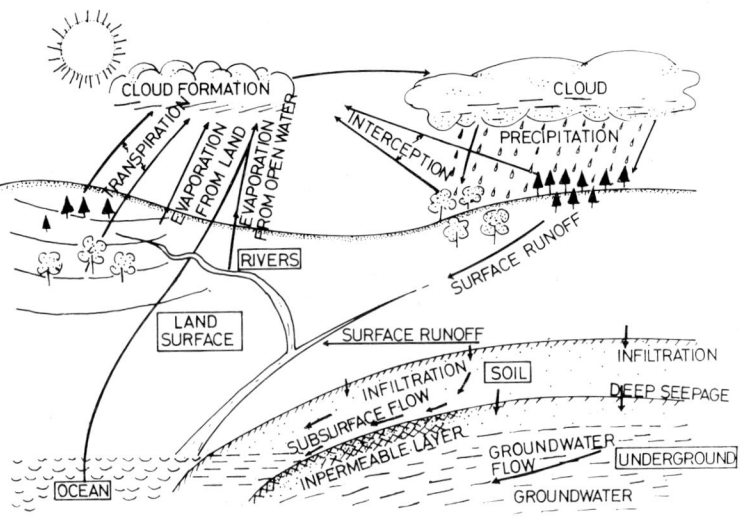

Fig. 2. The water cycle

tures. The soils are often shallow, stony and situated in sloped or mountainous areas. Another important feature of forests in this respect is the development of large amounts of living and dead organic matter, in particular, the formation of a characteristic organic top layer. The structure of the standing crop is in general quite heterogeneous, thus causing an irregular input flux of precipitation water into the soil. This irregularity may be further increased by an irregular extraction by the growing root system and the heterogenities of the soil. All of this makes it difficult to obtain significant measurements for the determination of the components of the water balance equation (Sect. II.1). On the other hand it is the location of forests in the regions of higher precipitation that makes them important in water management. Nowadays, with water shortages being a widely known problem, a new dimension has been added to the economic meaning of forests. The forester may well be confronted with the problem whether to give the timber production or the "water production" higher priority. In order to be able to base his decision on reliable arguments, quantitative results on the water exchange behavior of forest ecosystems are more needed today than ever.

II. Water Balance

1. The Water Balance Equation

For the quantitative description of the water exchange of an ecosystem with the atmosphere, the underground and adjacent ecosystems, the water balance equation is often used [Eq. (1)]. It equates the difference of all input and output fluxes (negative sign) with the change in soil water storage:

$$P - I - O_s - O_{ss} - E - T - S = \Delta R \tag{1}$$

(P = precipitation; I = interception; O_s, O_{ss} = surface and subsurface run-off; E, T = evaporation, transpiration; S = deep seepage; ΔR = change in storage).

The form of Eq. (1) is not entirely satisfactory because it does not show the "dynamic" character of its components. This is to say that all of the components are time-dependent variables. Consequently they have to be integrated over the balance period, which suggests the following form

$$\int_{t_1}^{t_2} (P - I - O_s - O_{ss} - E - T - S) dt = \int_{t_1}^{t_2} \int_{z_b}^{z_0} \frac{\partial \Theta}{\partial t} dz \, dt \tag{2}$$

(Θ = water content by volume; z_b, z_0 = vertical coordinate at the bottom (b) and the top (0) of the soil profile; t = time).

Since the soil is further divided into subcompartments (Fig. 1) the term ΔR has to be specified for each of the subcompartments and thus is also a function of space. The same may be true for the subsurface flow O_{ss}.

In order to solve the water balance equation only one unknown is allowed. The remaining terms have to be measured or determined in another way. Some of the terms in Eqs. (1) and (2), though, may frequently be neglected, for instance, there will be not surface of subsurface run-off in level areas and, due to the organic top layer, surface run-off will seldom occur in forests, not even in sloped areas.

The term ΔR loses importance with the increasing length of the balance period. It may with some justification be neglected, if the period is one or several years and is carefully timed in order to start and end at the same degree of water saturation of the soil profile.

The three evaporation terms, interception, surface-evaporation and transpiration require energy (about 590 cal per cm^3 water) and thus contribute little to the balance during the cold season. In particular, transpiration is absent for about half of the year in moderate climates in the case of deciduous forests. Interception on the other hand does not occur during precipitation-free periods. Evaporation, too, may be neglected during these periods, at least in the presence of an intact organic top layer that after drying at the surface decreases the rate of capillary rise to the surface to a negligeable amount. Since precipitation is also absent, the water balance equation is greatly simplified in such periods, allowing the changes of water storage to be equated with the transpiration and the deep seepage only.

Thus it becomes obvious that the water balance equation may be stated in a number of different ways depending on site properties, season, length of the balance period, and particularly the aim of the investigations. Correspondingly, the equipment may be quite different. In general, requirements regarding time resolution and sensitivity are greater the shorter the balance period is.

It may be mentioned here (and is further discussed in Sect. IV) that in connection with the computer simulation of the water exchange of ecosystems the short-term field observations become more pronounced, since they primarily serve to prove or disprove the validity of the mathematical model. Only if the model is verified by field observations should it be used to infer long-range predictions on the water exchange behavior of an ecosystem.

In the following subsections the physical nature of the components of the water balance equation is discussed in more detail. Closely associated are the ways and means of their determination, which, because of limited space, can be discussed only in a general way. Results are cited only for illustrating the order and relative importance of the components.

2. Input Variables

a) **Precipitation.** According to Eqs. (1) and (2) the precipitations are the only input variable of the water budget of an ecosystem. Artificial irrigation and lateral water flow above (surface flow) or within (subsurface flow) the soil may be mentioned as two other input forms. Upon arriving the canopy the rain is fractioned:

$$P = I + P_t + P_d + P_s \tag{3}$$

(P_t = throughfall; P_d = canopy drip; P_s = stemflow).

The rain is changed qualitatively and quantitatively during the passage through the canopy. It is diminished by the amount of interception and is changed from rather even to very irregular spacial distribution. Only the throughfall has about the same properties as the rain above the canopy. The canopy drip reaches the soil surface with great spacial variability as can be seen from Table 1. Locally, the flux intensity may exceed the rain in the open whereas other spots receive much less rain.

b) **Stemflow.** Still more extreme is the concentration of the rain by means of the stemflow, though great differences between different tree species exist (Fig. 3). A rainstorm may produce well above 500 l of water flowing down the stem and entering the soil at a single small area at the bottom of the stem of a dominant beech tree (Weihe, 1970). Some results of the percentage of stem flow can be found in Table 2. The results of Table 2 agree closely with those of Eidmann (1959).

c) **Throughfall and Canopy Drip.** The generally great spacial variability of the rain beneath the canopy requires numerous normal rain gages or such with extremely big catch openings, in order to determine the representative rate of the input flux. Since it is hardly possible to distinguish between throughfall

Table 1. Rainfall (in mm) in the open, canopy drip (including throughfall but not including stemflow), error of the mean and variability. 30 gages were used under each beech and spruce canopy. The opening, the beech and the spruce site are each about 400 m apart (Solling, Germany)

Solling 1969		Rainfall open	Beech (120 years)			Spruce (95 years)		
Month	Day		Canopy drip	Error of the mean	Variability (% st. dev.)	Canopy drip	Error of the mean	Variability (% st. dev.)
5	2	4.8	3.4	0.1	19.6	2.4	0.2	48.8
5	5	1.0	0.9	0.0	25.2	0.5	0.09	91.7
5	7	19.4	15.3	0.4	14.5	18.7	0.5	14.2
5	10	6.7	4.9	0.1	13.8	3.9	0.3	44.4
5	12	4.5	2.6	0.1	25.2	2.0	0.1	27.1
5	16	8.0	5.2	0.2	19.3	5.8	0.3	25.3
5	17	1.7	0.5	0.0	54.6	0.3	0.05	91.1
5	18	9.7	6.5	0.3	20.9	5.6	0.3	27.7
5	19	5.0	3.2	0.1	24.4	3.1	0.2	40.7
5	21	20.3	12.3	0.3	11.1	15.0	0.3	10.9
5	22	1.0	0.3	0.0	92.1	0.4	0.06	86.3
5	27	25.0	18.2	0.6	18.2	20.1	0.8	22.1
5	30	14.1	8.1	0.3	17.5	8.0	0.4	25.2
5	31	9.8	2.0	0.2	44.6	1.3	0.2	71.0
6	2	18.0	18.3	0.8	24.3	18.3	0.9	26.2
6	5	24.5	15.5	0.6	20.3	16.2	0.6	20.0
6	9	9.8	7.3	0.2	17.5	8.3	0.3	20.3
6	16	25.3	25.1	0.8	17.5	27.3	1.2	24.6
6	20	20.3	14.5	0.8	28.2	17.0	0.8	26.9
6	30	27.8	20.9	1.1	28.2	21.1	1.2	29.8
7	9	15.8	10.7	0.5	27.0	8.6	0.6	40.1
7	12	20.6	12.8	0.5	21.3	15.5	0.5	18.3
7	14	2.5	0.6	0.1	63.1	0.5	0.06	66.6
7	27	3.3	0.9	0.1	56.9	0.6	0.13	108.3
7	30	3.1	1.8	0.1	42.4	0.7	0.13	94.8

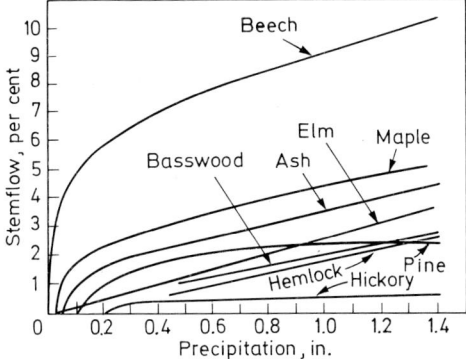

Fig. 3. Percentage of precipitation stemflow for different species in relation to precipitation per storm. (After Kittredge, 1948)

and canopy drip, both are generally put together (Table 1) under the term "canopy drip".

The spacial variability and the percentage of the water input to the soil depend on various factors. Among them are kind, number per area, age and management (particularly thinning) of the trees, season, rain intensity and duration, wind (velocity and direction). Table 2 gives an example how the rainfall is fractioned upon passing through the canopies of a beech and a spruce stand.

Many authors have been dealing with the spacial rainfall distribution beneath the canopy, particularly in order to find out an adequate measuring system (Weihe, 1968; Delfs et al., 1958; Ziemer, 1968; Lang, 1970). From soil moisture measurement (Table 3; Eschner, 1967) it appears that there are gradients along radial lines originating from the stem. They have opposite directions under beech and spruce. The reason obviously lies predominantly in the arrangement of the branches. They are rising in the case of the beeches, thus leading the water on their smooth surfaces to the stem, whereas the falling branches of spruce lead the rainwater to the periphery, where it drips off, leaving practically no water for the stemflow.

Attempts have been made to express the percentage of the canopy drip as a function of meteorological and stand conditions. Lang (1970) found a quantitative relationship but points out that it may be valid only for the specified site conditions. Furthermore, the relationship may be valid only for a short time, because the measurements were taken after the stand was thinned. Ordinarily the trees tend to regain a better closure of the canopy within a few years, which will alter the relationship between rainfall in the open and canopy drip. The general pattern of this relationship is reflected in Fig. 5 and is discussed further in the next subsection.

So far rainfall has been considered as one precipitation form. Fog is another form, though of limited significance in the water budget. Fog may be accumulated on the trees and contribute significantly to the budget under two conditions. Border areas of forests exposed to fog may collect enough fog to produce

Table 2. Rainfall in the open and under the canopy, stemflow and interception for spruce and beech: Solling 1968/69 (Benecke and Mayer, 1970)

1968/1969	Rain in the open (mm)	Beech, 120 years			Spruce, 95 years		
		Rain under canopy (mm) (%)	Stem-flow (mm) (%)	Inter-ception (mm) (%)	Rain under canopy (mm) (%)	Inter-ception (mm) (%)	Stem-flow[a]
May	96.2	71.9 75%	12.1 12%	12.2 13%	59.5 62%	36.7 38%	
June	125.5	88.0 70%	16.9 14%	20.6 16%	80.1 64%	45.4 36%	
July	109.7	70.9 65%	17.3 15%	21.5 20%	59.7 54%	50.0 46%	
Aug.	98.6	64.2 65%	11.0 11%	23.4 24%	57.6 58%	41.0 42%	
Sept.	138.3	98.9 72%	18.8 13%	20.6 15%	102.7 74%	35.6 26%	
Oct.	112.1	79.4 71%	18.0 16%	14.7 13%	92.1 82%	20.0 18%	
Nov.	32.8	32.6 99%	8.6 27%	− 8.4 −26%	35.4 108%	− 2.6 − 8%	
Dec.	29.6	39.4 133%	5.2 18%	−15.0 −51%	41.2 139%	− 11.6 − 39%	
Jan.	75.6	62.2 82%	11.4 15%	2.0 3%	58.0 77%	17.6 23%	
Febr.	79.8	70.0 88%	2.0 2%	7.8 10%	63.8 80%	16.0 20%	
March	76.1	63.2 83%	12.5 16%	0.4 1%	61.2 80%	14.9 20%	
April	134.1	109.7 82%	24.5 18%	− 0.1 0%	115.3 86%	18.8 14%	
Total	1108.4	850.4 76%	158.3 14%	99.7 9%	826.6 75%	281.8 25%	

[a] less than 1%

extra stemflow and canopy drip. Table 4 gives an example. The other condition is freezing temperatures with fog. This leads to the formation of rime on the trees that later may melt and thus produce canopy drip and stemflow, especially during bright weather. It is for this reason that Table 2 contains "negative" interception in Nov. and Dec., 1968.

Table 3. Soil–water–tension response to the distance from the stem. (Data are averaged over time, June to Oct. 1968, depths and plots, in cm water column; Benecke and Mayer, 1971)

	Area of crown projection near to the stem	Canopy drip	Outside crown projection
Beech	−42.3	−58.8	−43.6
Spruce	−94.2	−64.8	−47.6

Table 4. Rainfall (mm) in the open, and in the forest, Odaigahara, Japan, 1922. (After Hirata, 1929, in: Penman, 1963)

		April	May	June	July	Aug.	Sept.	Oct.	Total
No fog	R_0	56	27	307	26	167	49	100	732
	R_f	42	21	209	19	118	36	68	512
Fog	R_0	265	182	56	469	100	230	606	1909
	R_f	293	167	55	517	81	216	450	1780

Far more important is the contribution of snow to the water budget of forests because they are often located at higher altitudes where snowfall is frequent.

The behavior of snow is quite different from that of rain. Far larger amounts may be intercepted during the snowfall than would be the case with rain, but the "intercepted" snow is not or only to a small degree evaporated. Generally most of it is transferred to the soil surface by various ways. Either it melts subsequently or it slips in whole portions to the ground. Thus snow interception is not a genuine interception with subsequent evaporation but a temporary storage which is sooner or later converted in its major portion into an "input variable" for the soil and only in a minor portion into an "output variable" in the sense of interception.

Another important difference as compared with rain is the accumulation of snow above the soil surface prior to entering the soil. By considering the snow layer as a temporary storage compartment, it could be treated in the same way as a soil compartment (Fig. 1; Sect. II.3 and II.4). But unfortunately the principles of water movement and water release are not the same. This makes it extremely difficult to balance the input, the changes of the amount of snow and the output either by release at the bottom or by evaporation from the top. (The interested reader can find plenty of literature on this subject.

Examples are: Mitscherlich, 1971; Brechtel and Zahorka, 1971; Brechtel, 1970, 1971; Heiseke, 1974; de Quervain and in der Gand, 1967; Garstka, 1964).

Rounding up this subsection with respect to the water exchange of forest ecosystems it may be emphasized that the incoming rain is diminished by the interception and converted into an irregularly distributed flux. This flux must be measured with adequate equipment because there is no generally valid relationship which would allow to infer the flux intensity from meteorological and site parameters, nor is it possible to measure the interception directly.

3. Output Variables

Except for the interception, all the remaining output terms depend on the soil, thus emphasizing the focal role of the soil in the water exchange.

a) Interception. The interception is often labeled as "unproductive" evaporation because it does not contribute to plant growth. This makes it desirable to keep interception as low as possible in order to increase the amount of water available for transpiration or groundwater recharge.

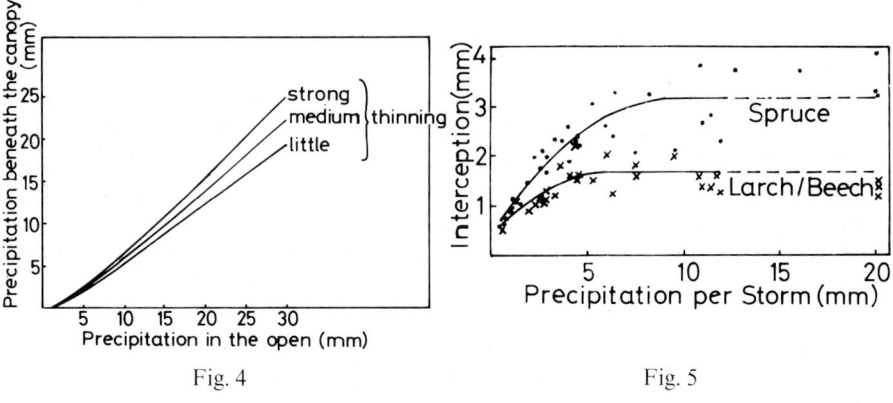

Fig. 4. Canopy–throughfall vs rainfall in the open for stands of Douglas fir with different degrees of thinning. The differences increase relatively and absolutely with increasing rainfall in the open. (After Mitscherlich, 1971)

Fig. 5. Interception of a spruce and a larch/beech stand in relation to the amount of precipitation per storm. The age of both stands is about 45 years; both were rigerously thinned prior to the measurements. (After Lang, 1970)

Substantial amounts of the precipitation can indeed be contributed to interception as can be seen from Figs. 4 and 5, or Table 5. The fraction of the interception in relation to the amount of precipitation per storm depends on the tree species (Tables 2 and 5, and Fig. 5) and the density of the stand (Fig. 4). These relationships provide means to control the interception to some degree.

Particularly wide spacing of plantations and/or well-timed and rigorous thinning may help to keep interception low (Hibbert, 1967).

Figure 5 exhibits at the same time some of the dynamics of interception. It generally starts out with a high rate which gradually slows down, finally reaching zero. At this point the interception capacity is completely filled up and no further interception occurs. After the rain has ceased, the evaporation of the intercepted water will begin. Should a new rain occur before the evaporation is completed, then this time the interception would be correspondingly lower. Figure 5 furthermore shows that there is a difference between the interception capacity and the amount of rain necessary to fill up this capacity completely. For spruce and an annual rainfall of about 600 mm Weihe (1970) found, for example, 3 mm interception capacity, and 6 mm to fill this capacity completely. For 800 mm annual precipitation the figures are 4 and 8 mm. Beech was independent of the annual rainfall with only 0.6 and 1.2 mm being the corresponding figures. Clearly, this difference is due to the fact that some of the rain reaches the ground right from the beginning and that this portion gradually increases until all of the rain, either falls through the openings of the canopy, or dripps, from the canopy, or flows down the stems.

b) Surface Run-Off. All the precipitation water not contributed to interception reaches and infiltrates the soil as long as the input rate does not exceed its "infiltrability" (Hillel, 1971). Once this happens, the water begins to accumulate on the soil surface and may give rise to surface run-off and possibly erosion and flooding in the valleys, provided the rainstorm yields enough water and the area is sloped.

Surface run-off is relatively unlikely under forest cover, mainly because of the organic top layer that itself provides a high storage capacity for water and furthermore shields the underlying mineral soil. Without this protective cover the mineral soil may be puddled by the direct impact of the rain drops that in turn may sharply decrease the infiltrability. Preventing, or at least greatly reducing, the danger of surface run-off is one of the important protective functions of forests.

Table 5. Summer interception by spruce and beech: Sauerland. (After Eidmann, 1959)

Rainfall range (mm)	Spruce I %	Beech I %
0–1	82	72
1–2	63	55
2–3	55	25
3–5	47	18
5–10	33	16
10–15	30	16
15–20	26	17
20–	24	18

c) Subsurface Flow. As long as the infiltrability of the soil exceeds the input rate, all of the rainwater enters the soil. Because all or almost all of the pore space is of capillary dimensions (Fig. 8), the capillary forces play a dominating role. They counteract the gravitational forces tending to keep the water in place. But in general this is only partly achieved, because the capillary forces themselves mostly exhibit differences at different points of the soil, thus giving rise to the establishment of gradients. As is explained further in Sect. III the capillary forces cause a soil water potential that is commonly referred to as the "matric potential" of the soil water. Other notations are frequent, in particular "capillary potential", "soil water tension", and "soil water suction". Matric potential—denoted by Ψ_m—is used throughout this chapter.

Correspondingly, the gravitational forces cause the gravitational potential of the soil water, denoted by the symbol Ψ_g. The sum of the matric potential and the gravitational potential forms the hydraulic potential and it is the gradient of the hydraulic potential that causes the water to move. Thus water retention and water movement are closely interconnected as can also be seen from the general flow Eq. (11).

According to Darcy's law [Eq. (6)] the water flow in the soil is in the direction of the negative hydraulic gradient. Its rate is proportional to this gradient by a coefficient called soil water conductivity or permeability. The conductivity depends on structural properties. Often the upper layers have a higher conductivity than lower layers. To make things more complicated, it must be added that the conductivity is not a constant but depends on the water saturation of the soil, becoming lower the more desaturated the soil is.

This brief introduction makes it possible to understand when subsurface flow will occur. It requires sloped areas, soils with decreasing conductivity in deeper layers, and a high water saturation. In level soils the infiltrating rain will increase the moisture content in a horizontal layer evenly, thus allowing the buildup of hydraulic gradients only in vertical direction. But in a sloped soil a drop of the gravitational potential in the same layer is due to its inclination that causes the soil water to move laterally within the layer. This movement may take any direction between almost vertical and parallel to the surface.

Subsurface flow may become the only non-evaporative output variable in sloped areas with shallow soils over impermeable bedrock or clay. Such an area is particularly suited for catchment investigations, since the total run-off can be measured in weirs at the exit of the valley. Figure 6 is an example that shows the order of the seasonal changes. Bloemen (1974) worked on run-off prediction of watersheds on the base of observations of groundwater levels and run-off values.

The theoretical treatment of subsurface flow is very difficult and no analytical solutions have yet been derived for field conditions. But much work is currently extended (Freeze, 1971) employing numerical methods. Since many forests are located in sloped areas, solutions and results in this field are urgently needed.

d) Deep Seepage. In a physical sense deep seepage is to be distinguished from subsurface flow only by its vertical direction and there are transition stages where both notations apply. Its ecologic as well as its economic significance lies in the fact that this water moves more and more out of the reach of the roots, thus being lost for the vegetation but possibly gained for the supply of springs by recharging groundwater aquifers. It should be pointed out that deep seepage has the last priority of all output variables, this is to say that all the others have to be satisfied first, before deep seepage takes place. This accounts for the generally large seasonal amplitude, of which Fig. 6 gives an example. Under humid conditions the seasonal changes are superimposed by short-range amplitudes caused by heavy or continued rainfall (Figs. 15 and 16).

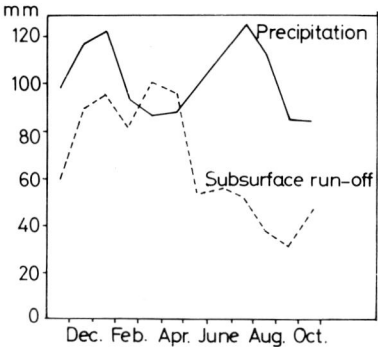

Fig. 6. Precipitation and subsurface run-off on a monthly base, averaged over 15 years (1951–1965). Catchment, 80 ha, with spruce. (After Friedrich et al., 1968)

Since deep seepage is very often determined by use of lysimeters, a brief comment may be appropriate, because a systematical error can easily be introduced and lead to invalid results. Lysimeters are normally large containers filled with the undisturbed or repacked soil. They are often about 1.5 m deep, thus allowing for sufficient rooting depth. The critical installment is a plate at the bottom above which the water runs off laterally to a registrating device. This plate interrupts the hydraulic continuity between the soil and the underground, a fact that in most cases changes the storage–seepage conditions in the direction of higher storage in the soil in the container as compared with the reference soil. This, of course, may make the results rather meaningless.

An attractive alternative is the use of tensiometers, which measure the matric potential at a certain point. By installing them at different levels, especially beneath the rooting depth, and determining independently the conductivity at the same depth the seepage rate may be calculated from Darcy's law [Eq. (6)]. Another way is the mathematical simulation that is discussed in Sect. IV.

e) **Evaporation.** Evaporation consists of two processes: the moving upward of the water through the soil profile to the surface and evaporating from the surface. The upward movement follows the same principles as downward or lateral movement. A hydraulic gradient as driving force is established by depletion of the upper layers either by root extraction or by surface evaporation. Thus it becomes evident that in the top region of the soil profile, which under forest is formed by the organic top layer, evaporation may occur without preceding water movement as is the case with interception. This has lead to the notation of "litter-interception" as an output term that connects interception and surface evaporation. As is the case with interception there is no way for direct measurement and even indirect measurement is difficult to accomplish. Thus, in general, litter interception is regarded as part of the surface evaporation.

As has already been pointed out surface evaporation plays a minor role under forest cover. This is true partly because most of the incoming radiation energy is absorbed by the canopy and perhaps one or more understories and partly because of the water conductivity characteristic of the organic layer, which ordinarily does not allow an appreciable amount of water to move to the surface. Clearly, under these circumstances, the evaporating plane may move down from the surface into the profile. But this, too, is of limited effect because the vapor exchange within the pore systems of the soil is small because of small vapor-pressure gradients. It must be admitted, though, that the rate of vapor exchange or movement within the soil is subject to controversial opinions, which cannot be discussed here. Rounding up this point it may be concluded that most of the surface evaporation under forest with an intact litter layer occurs in the form of "litter-interception".

Once the surface has dried off, the water conductivity in the top region becomes so low that, despite steep hydraulic gradients vertically upward, no appreciable water movement to the surface goes on. Under the spruce forest (referred to in Tables 1, 2 and 3 and Fig. 16) in a small isolated plot where no transpiration was possible and rainfall was shielded off a hydraulic gradient equal to zero developed and remained unaltered the whole summer, thus indicating that despite low tension of the soil water no evaporation took place.

f) Transpiration. The discussion of the transpiration will be limited in this context to the role of the plants as conductive medium between soil and atmosphere, the physiologic regulation of transpiration is discussed in Part 3 of this volume. A basically simple but conceptually important step in this respect is to consider the system soil–plant–atmosphere as a hydraulic continuum. This unified system has been called the "SPAC" by Philip (1966), see Hillel (1971). It is discussed in detail by Oertli in Part 1:C of this volume (see also Caldwell, this volume Part 2:A). Within this system the water transfer is obeying the same law which is principally expressed as Darcy's law [Eq. (6)] and can be stated in more general terms by saying that the rate of water movement is proportional to a gradient as driving force by a transfer-coefficient. The gradient represents ultimately an energy gradient, which in this system is realized as a pressure gradient. The initial point is in the soil in the form of the matric potential or soil water suction (negative pressure) and the terminal point is in the atmosphere as vapor pressure. The overall potential difference is in general very high in the daytime (a hundred or even several hundreds of bars) and low during the night. Most of this enormous pressure difference is used up between the walls of the substomatal cavities and the external air where the water is converted from the liquid into the vapor state (Fig. 7).

It is remarkable that the role of the plant in the water exchange is rather passive as long as the evaporative demand of the atmosphere can be met (Penman, 1963; Hillel, 1971). When sufficient water supply in the root zone is provided, the plants transpire at a rate that is dominated by the energy available at the transpiring surfaces rather than by plant physiological requirements. Only from the moment the soil water availability to plants becomes limited, do the plants begin to regulate the transpiration flux through their stomata. Thus soil water availability becomes a crucial question in the water exchange of ecosystems. It is known from experience that, once the matric potential in the root zone falls below -15 bars, most of the plants will wilt irreversibly (Slatyer, 1967). This has led to the notation of "field capacity", meaning the portion of the soil water that is held by surface forces against the action of gravity. The field capacity is further subdivided into plant-available water and dead water (Fig. 8), the latter having a matric potential < -15 bars. But evidently it is not the binding force that limits the water availability, but the greatly restricted mobility and accessibility of the soil water. It would take <1 cal to release 1 cm^3 of water held with 15 bars as compared to 580 cal necessary for its evaporation. With the water supply becoming limited, water conductivity drops rapidly (Fig. 9). Steep hydraulic gradients around the roots may nevertheless allow for adequate water flux to the roots and the growing roots themselves continue to invade new regions with less exhausted water supply. But even this combined effect eventually becomes insufficient. This point is marked by a matric potential of about -15 bars in the bulk soil, which means that the potential of the soil water may be much lower in the immediate neighborhood of the roots.

Having defined the portion of the soil water that is not available to plants, the question remains if there is a gradually decreasing availability with decreasing potential. Answers to this question are still controversial. Figure 10 shows the three principal versions. One shows that an unrestricted water supply is possible only at matric potentials above about 1 bar, the second one assumes a medium stand between the first and the third, which concludes that water supply is adequate until the -15 bar potential is reached. Though this decade-old controversy can not yet be considered as settled, evidence is growing that favors the medium version, but in a more differentiated way, as demonstrated in Fig. 11. Briefly stated, these results indicate that the transpiration rate is dependent primarily on two factors: the matric potential (in the root zone) and the rate of potential evaporation (as a function of the meteorological conditions) (Richards and Wadleigh, 1952; Denmead and Shaw, 1962; Shaw and Laing, 1965; Gardner, 1965a, b; Waggoner, 1965; Cowan, 1965; Item, 1974).

4. Infiltration, Storage, and Depletion

While discussing the output variables the action of the soil was compared to a hydraulic switch. In fact it can be considered a selfregulating system the functioning of which, in an ecologic sense, depends on its physical properties and dynamics. The basic features are discussed in the following section.

Recalling (Fig. 8) that about half of the total soil volume consists of pores ranging in diameter from $>50 \,\mu\text{m}$ to $<0.2 \,\mu\text{m}$, it is realized that most of the soil water is subject not only to gravitational forces but also to capillary forces.

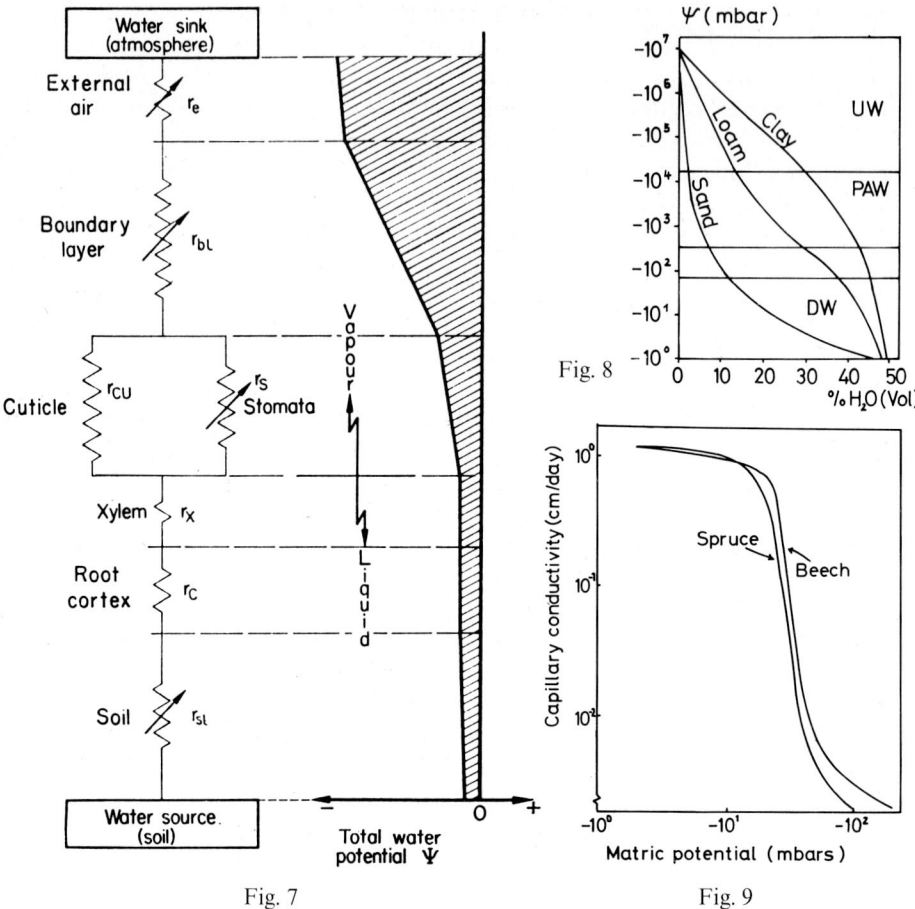

Fig. 7

Fig. 8

Fig. 9

Fig. 7. (*Left*) Illustrating the various soil, plant and atmosphere resistances to the transpiration stream. Whilst no resistance is fixed, those shown with an arrow are particularly variable. (*Right*) Showing the decreasing total potential of water as it moves from soil through a plant to the atmosphere (not to scale). (After Rose, 1966)

Fig. 8. Water characteristic curves for three typical soils. (*UW* unavailable water or "dead water", *PAW* plant-available water held by the soil against gravity with its lower limit dependent on the depth of a zero-tension-plane, *DW* water that drains off within two or three days, free drainage downwards provided, Ψ matric potential (or tension) of the soil water in cm water column)

Fig. 9. Capillary conductivity as a function of the matric potential for the least permeable horizons (sandy loam of very high bulk density, very stony—85 to 140 cm depth) of the soils of Figs. 15 and 16

Equation (5b) says that the capillary forces are inversely proportional to the pore diameter, and Fig. 8 shows the so-called soil water characteristic or the relationship between soil water content and the capillary forces (here expressed

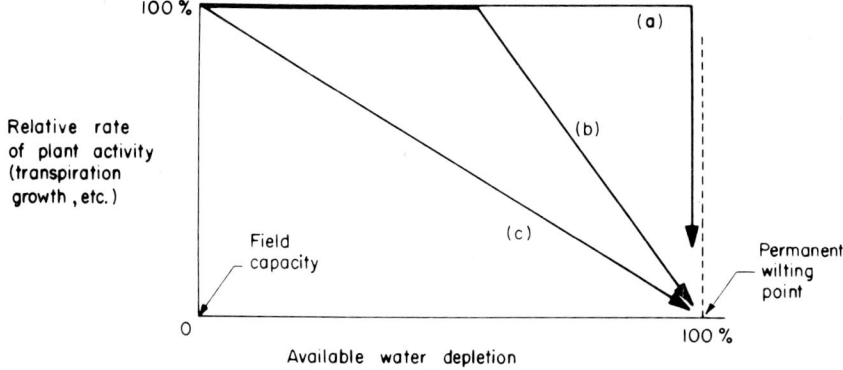

Fig. 10. Three classical hypotheses regarding the availability of soil water to plants: *a* equal availability from field capacity to wilting point; *b* equal availability from field capacity to a "critical moisture" beyond which availability decreases; and *c* availability decreases gradually as soil moisture content decreases. (After Hillel, 1971)

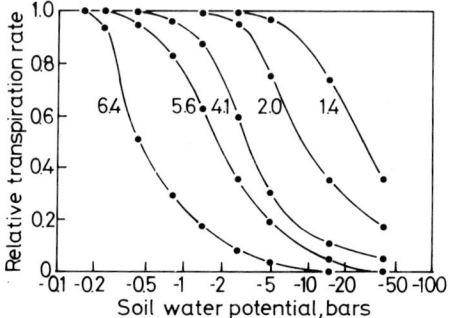

Fig. 11. Transpiration rate as a fraction of that with the soil at field capacity, plotted against matric potential. The curves represent this relation for days on which the transpiration rates at field capacity had the values shown at each curve in units of mm/day. (After Denmead and Shaw, in: Rose, 1966)

as matric potential) for three soils of different, typical texture. The combined action of the capillary and gravitational forces results in the hydraulic potential [Eq. (7)] of the soil water. Figure 12 illustrates this relationship with the condition that the hydraulic potential is the same at any point of a soil with a groundwater table. According to Darcy's law [Eq. (6)] water movement takes place only if a hydraulic (potential) gradient not equal to zero exists. Since the hydraulic gradient $d\Psi/dz$ (Ψ = hydraulic potential, z = vertical coordinate) is equal to zero at any point in Fig. 12, we are dealing with the particular case of a static equilibrium. No water movement takes place and all forces balance each other. In nature this situation is unlikely, because the input and output variables, as already discussed, always cause disturbances that initiate fluxes by means of which the system tries to return to the balanced status.

Figure 12 may serve to give an idea of what is meant by "field capacity", defined as that amount of water held by capillary forces. This amount is represented in Fig. 12 by the straight line that shows the matric potential as a function of depth. Two important conclusions can be drawn. First, the "field capacity" is not confined by a fixed value but depends on the vertical distance of the point (or layer), under consideration, from the groundwater table (where the matric potential is zero). Consequently the field capacity in the upper soil decreases with increasing depth of a zero-tension-plane. Because of the increasingly reduced conductivity (Fig. 9), its becomes almost meaningless to refer to zero-tension-planes deeper than about 3 m under moderate–humid conditions. Hence, the limit of field capacity is at a maximum matric potential or tension of $-1/3$ bar, but frequently assumes higher values. The second conclusion is that field capacity in terms of water content is highly dependent on the texture of the soil. Figure 8 shows the great differences. The total field capacity decreases with increasing coarseness of the texture, whereas the plant-available field capacity assumes the highest values with medium-textured soils.

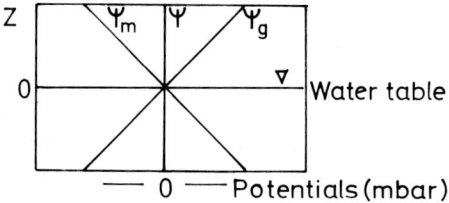

Fig. 12. Soil water suction (matric potential) Ψ_m, gravitational potential Ψ_g, and hydraulic potential $\Psi = \Psi_m + \Psi_g$ in a soil containing a free water table under the condition of static equilibrium

If in Fig. 12 rain was allowed to enter the soil, this would initiate two processes that can best be visualized by going back to the compartment model (Fig. 1). The rain is considered as a flux through the surface of the top compartment. It enters the compartment and hence increases the water content and consequently the hydraulic potential (as can be concluded from Figs. 8 and 12). This in turn establishes a hydraulic potential gradient between the first and the next deepest compartment which initiates a water flux from the top compartment to the next deepest one. This same procedure is then repeated between the second and the third compartment and so on. Hence, considering a certain time interval $t_1 - t_0 = \Delta t_1$, each compartment receives an input flux through its upper boundary plane and releases an output flux through its lower boundary plane. If the incoming flux is greater than the outgoing flux, water will be stored in that compartment. Conversely, if the outgoing flux exceeds the incoming flux, depletion of the originally stored water occurs. Eventually it may happen that both fluxes are equal and then no change of the water content of that compartment takes place though a—in this case constant—flux is passing through the compartment (steady-state-flux).

At the end of the first time interval Δt_1 some or all of the compartments have a different water content as compared to time $t=t_0$. This new water content distribution sets the initial conditions for the next time interval Δt_2 at the end of which there will again be another water content distribution. So time-step after time-step is governed by the outlined rules. Clearly, thinking in discrete differences with respect to both, the thickness of the compartments and the length of the times-steps is an auxiliary device that only approximates the real situation which could only be described adequately by infinitesimally thin compartments and infinitesimally short time-steps.

The question remains, how the rate of flux through the boundary planes of two adjacent compartments is established. According to Darcy's law this rate depends on the hydraulic gradient, and the water conductivity, which in turn is a function of the water content. A more detailed discussion on this subject may be found in Van der Ploeg and Benecke (1974a, b) in which fluxes in two- and three-dimensional flow problems are also treated. The equations describing the outlined dynamics are stated in the next section. In particular, Eq. (11), which combines the equation of continuity [Eq. (10)] with Darcy's law [Eq. (6)], is generally used for work on flow problems in soils.

With the knowledge of the basic relationships that govern water storage and water movement in the soil one can imagine that infinite cases of infiltration are possible, depending on the input function, the hydrologic soil properties, the vegetation and the meteorologic conditions.

A particular case occurs if rainfall continues over a longer period—say two weeks—at about the same daily rate. Provided evapotranspiration does not interfere, a quasi-steady-state infiltration develops, this is to say, input and output (deep seepage only in this case) are the same. Figure 13 shows the tension profiles for different flow rates as calculated from the integrated form of Darcy's law for a layered field soil (dashed lines). Also shown are the results of tensiometer measurements (solid lines; Benecke, 1972). The tension profiles reflect the differences in water conductivity of the different soil layers.

Such a steady-state condition, however, is an exception. Far more frequent is a sequence of rain periods and redistribution periods. Figure 14 shows a series of tension profiles representing the moisture dynamics before and after a heavy rainstorm for the same soil as in Fig. 15. The general pattern is the development of an infiltration "wave" that moves with decreasing velocity and decreasing peak downward through the profile.

If transpiration and evaporation are taken into account, the dynamics of water storage, redistribution and depletion become more complicated, but do not require new principles to be understood or dealt with. Since both are output variables, they decrease the water content and, hence, the hydraulic potential at the soil surface or root surface, respectively. This establishes gradients that cause the water to move into the areas where the extraction takes place by exactly the same principles as described previously in this section. Some of the ways in which these principles are expressed in mathematical statements (for computer simulation) are discussed in Sect. IV.

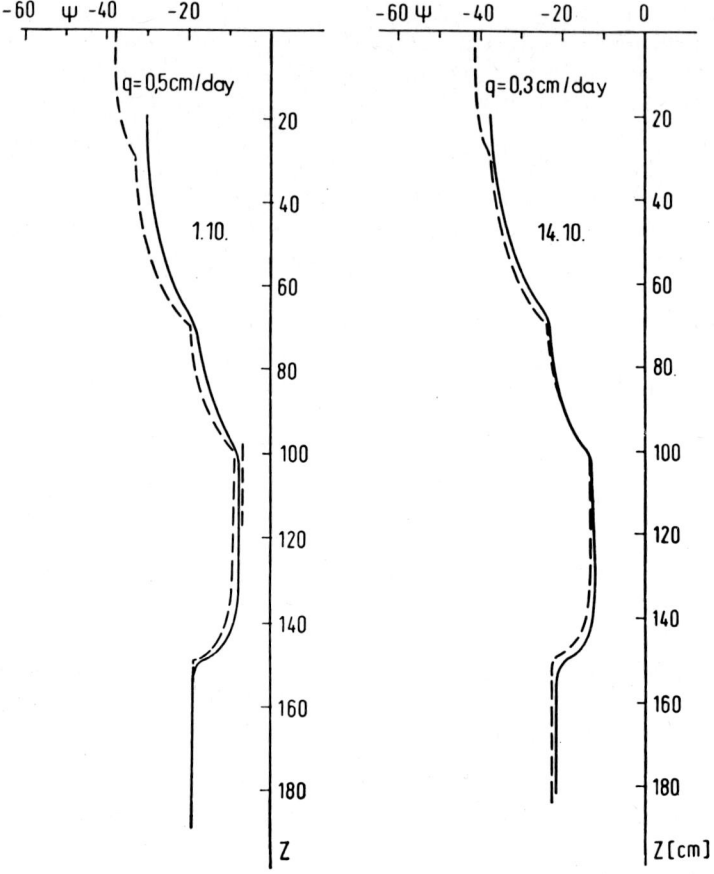

Fig. 13. Comparison of the practically constant matric potential profiles (solid lines) in a soil (referred to also in Fig. 15) consisting of five different horizons, during a period with rather uniform rainfall distribution and neglectable evapotranspiration as measured by tensiometers and the calculated (dashed lines) matric potential profiles for constant values of q as defined in Darcy's law [Eq. (9)]. (After Benecke, 1972)

III. Fundamental Equations and Principles

Only a few basic relationships need to be kept in mind for a principal understanding of the hydrologic processes that have been discussed so far, as well as for the mathematical simulation of the water exchange of ecosystems which is the subject of the next section.

The law of capillarity states that water rises in a cylindrical capillary tube until an equilibrium is established between the surface forces at the liquid: solid: air-contact and the weight of the elevated water column in the tube. Hence

$$2r\pi\sigma\cos\Theta = r^2\pi\varrho hg \qquad (4a)$$
$$h = 2\sigma\cos\Theta/r\varrho g \qquad (4b)$$

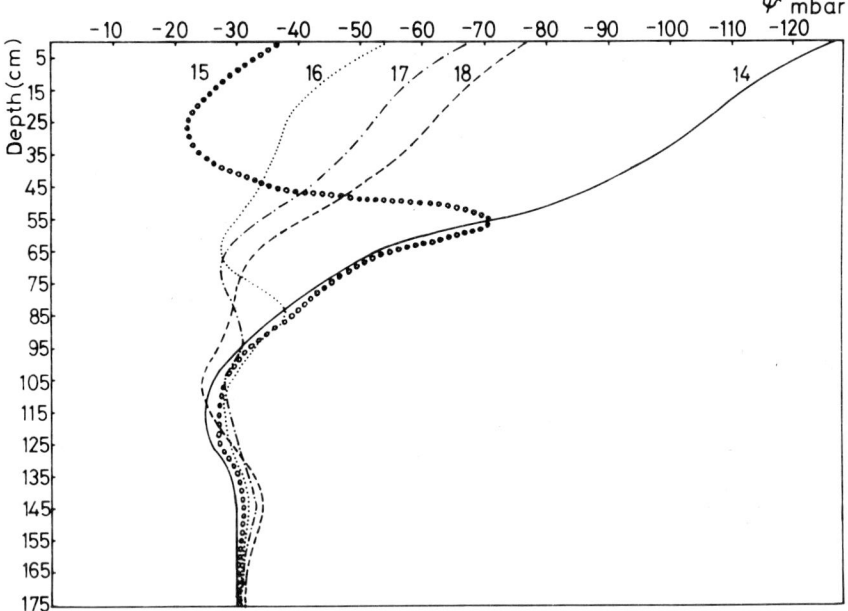

Fig. 14. Matric potential (Ψ) profiles before *(14)* and after *(15–18)* a 2.95 cm rainstorm of 3 h. The numbers indicate days of the month of June 1969. Evapotranspiration was about 0.3 cm/day. Soil and stand (beech) are the same as in Fig. 15

where h is the length of the water column, r the capillary radius, σ the surface tension (73 dyn/cm), Θ the wetting angle[1] (generally taken as zero), ϱ the water density and g the gravitational acceleration. Since the water "hangs" in the capillary, the negativ pressure or tension directly beneath the meniscus is equal to the upward directed surface forces. Hence—if Ψ_m is the tension per unit square area

$$\Psi_m r^2 \pi = 2 r \pi \sigma \quad \text{for } \Theta = 0 \tag{5a}$$

and

$$\Psi_m = 2\sigma/r \quad (\text{dyn/cm}^2). \tag{5b}$$

Frequently the tension Ψ_m is not expressed in the pressure unit dyn/cm² but in cm water column. Substituting Eq. (5a) into Eq. (4a) shows that this is in accordance with the underlying equilibrium of forces. Equation (5b) establishes the inverse proportionality between the tension or matric potential of the soil water and the pore diameter. Thus, the force necessary to withdraw a unit volume of water from the soil depends on the size of the pores that

---

[1] Θ is also used in this article to denote the volumetric water content.

hold the water. This relationship is used to determine the soil water characteristic experimentally, as shown in Fig. 8.

Clearly, since soil pores are not cylindrical tubes, the term "diameter" indicates an idealization. The fact that real soil pores are of rather irregular shape is reflected in a considerable hysteresis that affects the relationship between the matric potential Ψ_m and the water content Θ. Hence Ψ_m is not a unique but a multi-valued function of Θ. This greatly complicates the theoretical treatment of unsteady, unsaturated flow problems.

The hydraulic potential of the soil water in the unsaturated zone is predominantly composed of two components: the matric potential and the gravitational potential. The matric potential is caused by surface forces of the soil particles and results in the tension of the soil water. It is defined in terms of specific energy (energy per unit mass, unit volume, or unit weight). The gravitational potential is the work necessary per unit weight (or unit volume or unit mass) to elevate a (small) amount of water from a reference level to the specified height. The interested reader will find a full discussion and complete definition of the soil water potentials in almost any soil physics textbook, for example in Rose (1966) or Hillel (1971).

According to Darcy's law

$$q = -k(\Theta) \frac{d\Psi}{ds} \quad \left(\frac{cm^3}{cm^2 \, min}\right) \tag{6}$$

(q is the amount of water flowing per unit time perpendicular to a square unit area in the direction of the decreasing hydraulic potential Ψ; s is an arbitrary space coordinate; $k(\Theta)$ is the conductivity coefficient which is dependent on the water content Θ).

Darcy's law may be specified for horizontal and vertical water movement. Observing that

$$\Psi = \Psi_m + \Psi_g \tag{7}$$

(Ψ is the hydraulic potential, often also called "total water potential"; Ψ_m the matric potential, and Ψ_g the gravitational potential)

$$q = -k(\Theta) \frac{d\Psi_m}{dx} \tag{8}$$

for horizontal water movement, and

$$q = -k(\Theta) \left[\frac{d\Psi_m}{dz} + 1\right] \tag{9}$$

for vertical water movement (x is a horizontal and z a vertical coordinate).

Darcy's law describes a constant flux q of water through the soil. The more general and frequent case in the unsaturated zone, however, is a nonstationary flux that combines water movement and water storage or depletion. Referring to the discussion in Sect. II, the equation of continuity applies

$$\frac{dq}{dx} = -\frac{d\Theta}{dt} \qquad (10)$$

(q as defined in Darcy's law and Θ = volumetric water content). It says that the change of flux along some distance x is equal to the change of water content during that time interval, and can best be understood if one considers the flux through a (small) unit volume of soil during a unit time interval. A full derivation is given, for instance, in Kirkham and Powers (1972). This flux may not be in the x-direction but in any direction. In this case the three dimensional components of the flux are considered. Substituting the corresponding right-hand expressions of Darcy's law for Eq. (9) gives the general soil water flow equation for three-dimensional water movement

$$\frac{\partial}{\partial x}\left(k_x \frac{\partial \Psi_m}{\partial x}\right) + \frac{\partial}{\partial y}\left(k_y \frac{\partial \Psi_m}{\partial z}\right) + \frac{\partial}{\partial z}\left(k_z \frac{\partial \Psi_m}{\partial z}\right) + \frac{\partial k_z}{\partial z} = \frac{\partial \Theta}{\partial t}. \qquad (11)$$

If the flux is vertical, the first two terms on the left-hand side disappear

$$\frac{\partial}{\partial z}\left(k_z \frac{\partial \Psi_m}{\partial z}\right) + \frac{\partial k_z}{\partial z} = \frac{\partial \Theta}{\partial t}. \qquad (12)$$

In this form the flow equation is frequently used since many field investigations are carried out in level sites. The simulation of the water exchange of an ecosystem described in the next section is also based on this equation.

IV. Simulation of Evapotranspiration and Percolation

Referring again to the basic discussion in Sect. III and a study of the equations of the previous section lead to the conclusion that a full mathematical description of the water dynamics in the soil should be possible—at least under certain restrictions.

Since Eqs. (11) and (12) are non-linear partial differential equations, analytical solutions are known only for idealized conditions. But with modern computer facilities being widely available today, the employment of numerical methods becomes more and more attractive. Particularly helpful are the newly developed "superpositioned" computer languages like CSMP[2] (de Wit and van Keulen, 1972). They possess many build-in functions and routines that allow the user

---

[2] Continuous System Modeling Program (IBM, 1972).

to work with Eq. (12) even with limited mathematical background. More-dimensional flow problems, may also be attacked with CSMP (van der Ploeg and Benecke, 1974b) but will not be considered here.

In setting up a dynamic model one may start by considering three consecutive compartments and use Eq. (12) after changing it into difference form. The procedure has already been discussed on p. 118. Initially, only the water content distribution must be known since the matric potential and consequently the potential gradient, as well as the conductivity are dependent on the water content. Under unsteady conditions the water content of the compartments changes during the first time-step, resulting in a new water content distribution at the end of that time-step. In this way new "initial" conditions are established for the next time-step and so on. Boundary conditions that must be known are the input function (time rate of water input into the soil) and the drainage conditions at the bottom of the soil profile. Furthermore the soil water characteristic (water content: matric potential relationship) and the conductivity function (water content: water conductivity relationship) must be known for each of the soil layers separately. Figure 15 shows the results of a simulation period of about a month. Evapotranspiration is neglected due to the season in the case

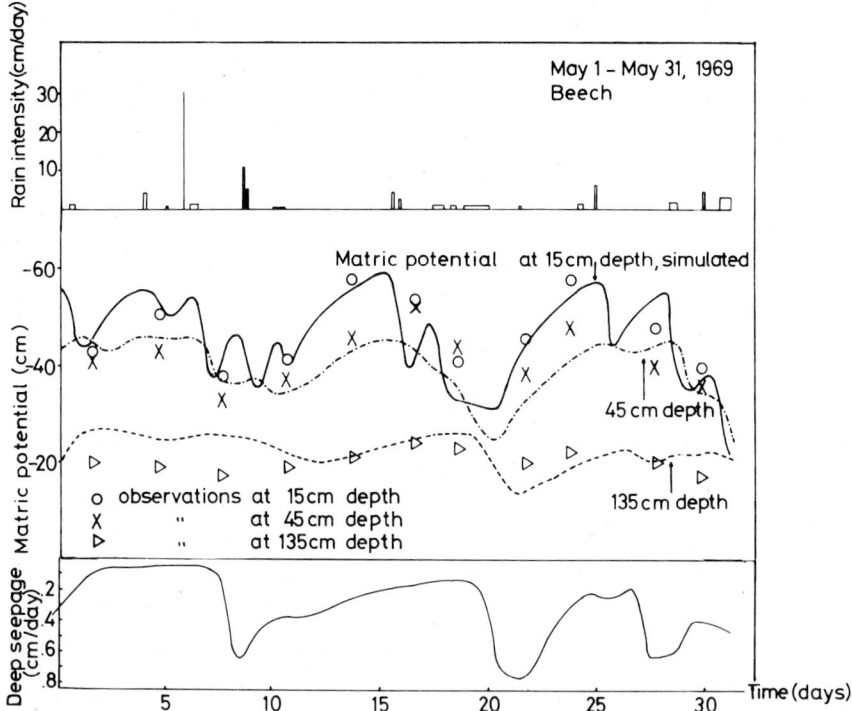

Fig. 15. Simulation of the matric potential in different depths as compared to measured values (mean values of about 40 tensiometers per depth). Rainfall intensity and distribution beneath the canopy (including stemflow) are shown on top of the figure. The total rainfall amounts to 12.27 cm in May, 1969. Deep seepage as determined by the model is depicted at the bottom. It totals 9.27 cm. Evaporation neglected

of beech. The response of the matric potential to the water input (rainfall beneath the canopy—including stemflow) is damped out and delayed with increasing depth. Also the dynamics of redistribution are well reflected in the figure. The deep seepage—or deep percolation—is shown at the bottom of the figure.

For comparison the same simulation was carried out for a spruce stand at a distance of 400 m from the beech stand. Figure 16 shows the results. This time evapotranspiration had to be included. An iterative procedure was used by first assuming daily evapotranspiration rates that were built into the model by adding "sink" terms in different soil depths according to the root distribution (see the subsequent discussion). The daily evapotranspiration rates were corrected until the simulated and the measured values of matric potential agreed satisfactorily.

Table 6 shows the water balance terms of the beech and the spruce stand of Figs. 15 and 16 not only for the displayed period of May 1969, but for most of the year 1969. Simulation has been used to facilitate a high time resolution (30 min) and the field measurements served to prove the correctness of the

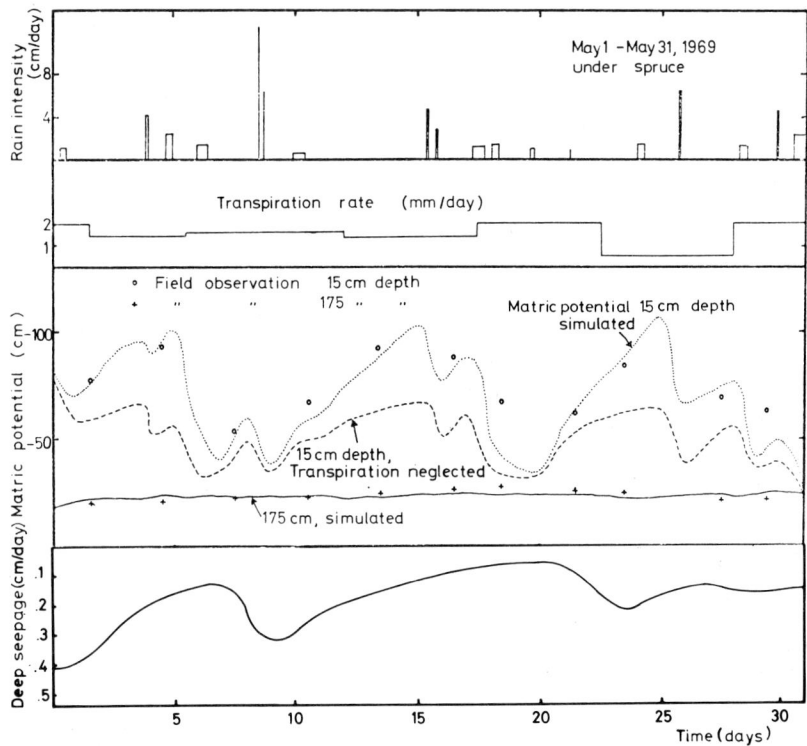

Fig. 16. Simulation of matric potential as in Fig. 15. Under spruce poor agreement between simulated and measured values was observed when evaporation was neglected (dashed line). But with the shown transpiration rate, agreement was good (dotted line). Here the total rainfall is 10.47 cm (due to higher interception as compared to the beech stand in Fig. 15), the total evaporation is 6.25 cm and the total deep seepage 5.41 cm. Both stands are comparable with respect to soil and meteorological conditions

Table 6. Water balances of a beech and a spruce stand of the "Solling" for different time periods in 1969

Water balance components	Beech Jan. 1–Oct. 31	Spruce	Beech Jan. 1–March 31	Spruce	Beech April 1–April 30	Spruce	Beech May 1–May 31	Spruce
	89.63 cm		23.15 cm		13.41 cm		15.31 cm	
Precipitation in the open	10.61	23.94	0.55	4.29	0.00	1.88	2.25	5.70
Interception	79.02	65.69			13.42	11.53	13.06	9.61
Canopy drip + stemflow			6.50	4.34	−8.54	−5.34		
Snow[a]			14.10	14.52				
Water input into the Soil[b]	−2.82	−1.58	0.22	1.92	0.90	0.45	1.12	0.91
Change in Stored Soil Water	49.92	41.23	14.32	12.60	20.06	16.42	10.67	4.55
Deep seepage	31.92	26.04	2.00[c]		1.00[c]		1.27	4.15
Evapotranspiration	42.53	49.98	2.55	4.29	1.00	1.88	3.52	9.85

Water balance components	Beech May 31–June 30	Spruce	Beech July 5–Aug. 5	Spruce	Beech July 15–Aug. 1	Spruce		
	12.07 cm		5.46 cm		0.83 cm			
Precipitation in the open	1.16	3.95	1.49	2.71	0.42	0.76		
Interception	10.91	8.12	3.97	2.75	0.41	0.07		
Canopy drip + stemflow	−1.16	−1.06	−6.85	−4.70	−7.04	−4.55		
Change in stored soil water	4.00	6.70	0.56	0.76	0.26	0.34		
Deep seepage	8.07	2.48	10.26	6.69	7.19	4.28		
Evapotranspiration	9.32	6.48	11.75	9.40	7.61	5.04		

[a] Net increase of the water equivalent of the snow cover.
[b] Calculated as "Precipitation in the open" minus "Interception" minus "Snow" minus the evaporation of the snow.
[c] Assumed value.

model. The results provide detailed information on the seasonal differences with respect to the water turnover of these two important tree species in Germany. Extrapolating over the whole year 1969, which can be regarded as quite a "normal" year (with the restriction, that an unusually high portion of the precipitation arrived as snow), leads to the conclusion that the total evaporation (interception + transpiration + evaporation from the soil surface) is about 10 cm less in case of spruce as compared to beech. On the other hand the seasonal amplitude of the deep seepage is larger under beech. It must be emphasized that these results are valid only for the site conditions of the stands investigated and that in particular, different physical soil conditions may lead to entirely different results.

Adding the previously mentioned "sink" term to Eq. (12) gives

$$\frac{\partial}{\partial z}\left(k_z \frac{\partial \Psi}{\partial z}\right) + A(z,t) = \frac{\partial \Theta}{\partial t} \qquad (13)$$

which is Eq. (2) of Nimah and Hanks (1973a).

In Fig. 16, $A(z,t)$ was determined by trial and error. Nimah and Hanks attempt a more analytical approach, defining $A(z,t)$

$$A(z,t) = \frac{\left[\Psi^{\text{root}} + (\text{RRES} \cdot z) - (z,t)\right] \cdot \text{RDF}(z) \cdot k_z}{\Delta x \cdot \Delta z} \qquad (14)$$

[Ψ^{root} is the hydraulic potential in the root at the soil surface and RRES is a root resistance term equal to $1 + \text{Rc}$; Rc is a flow coefficient in the root system, assumed to be 0.05, $\Psi(z,t)$ is the hydraulic potential in the bulk soil. RDF is a root distribution term, k_z the conductivity in z-direction. Δx is the distance between the plant root and the point in the bulk soil where $\Psi(z,t)$ is measured].

In a further publication (Nimah and Hanks, 1973b) the model was tested in the field to predict water content profiles, evapotranspiration, water flow from or to the water table, root extraction and root water potential at the surface under transient conditions. The test was carried out with alfalfa. Reasonable comparison was found, though the authors point out, that it was poor immediately after irrigation and at its best 48 h later. Using Eqs. (13) and (14), Fazilat (1975), also found good comparisons between predicted and measured data as depicted in Fig. 17. Undisturbed soil columns (40 cm edge length, 120 cm vertical length) were prepared from the same soil as in Figs. 15 and 16. They were planted with grass *(Lolium multiflorum)*. Grass was used rather than trees for conducting the preliminary experiments because of its favorable properties (quick growth, equal root distribution, repeated harvesting). Since the results, as shown in Fig. 17, are encouraging, attempts will be made to utilize Eq. (14) in order to determine the evapotranspiration of forests stands. Other authors who worked on root models are, for excample, Gardner (1960), Feddes and Rijtema (1972).

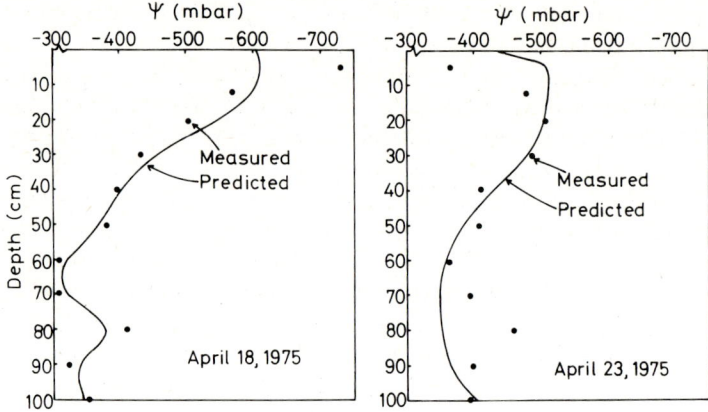

Fig. 17. Comparison of matric potential profiles as measured and predicted for grass (*Lolium multiflorum*) at the beginning and after 5 days of subsurface irrigation (2.6 mm/day). Irrigation is slightly higher than total evaporation thus causing some increase in stored soil water

V. Conclusions

The water balance equation [Eq. (2)] is the basis for a quantitative determination of the water exchange between forest ecosystems and its adjacent systems, the atmosphere, the underground and—in sloped areas—the neighboring ecosystems. Actually the water balance equation is a special form of the equation of continuity, which means that the difference between input and output of water is equal to the amount of change of water stored in the ecosystem.

Considering the whole ecosystem as just one "black" box is not satisfactory because there are six different output terms, three of which are in vapor form and three in liquid form: interception, surface evaporation, transpiration, surface run-off, subsurface flow and deep seepage. It need not be stressed that a strong wish exists to understand the functioning of the water dynamics of ecosystems in order to be able to manage these output terms in an economic and beneficial way.

Consequently, the aim of this chapter is to describe the functioning of forest ecosystems with respect to their water exchange. This is done by first subdividing the ecosystem in a number of compartments and—where necessary—subcompartments, according to their specific functions pertaining to water storage and water transfer. Next the principles and laws that govern the water storage in a compartment and the transfer to a neighboring one are discussed. It is the meaningful utilization of these principles that provides the very tool for controlling and managing the transfer, storage and output of ecosystems. But such activities make sense only if physical laws are known, which quantitatively describe the internal and external exchange processes. Since the soil plays a major role, the self-controlled water dynamics of the soil have received much attention. It is felt that the utilization of the progress that has been achieved in theoretical soil physics in the past years allows a new and prospective approach to solve water exchange problems of ecosystems.

No attempt has been made to provide a more or less complete list of values for the components of the water balance equation for different climatological and soil conditions. For this the reader is referred to Rutter (1968), who has collected all the available data and compiled them in a large table. Rutter and many other authors keep stressing the point that there are limits with respect to the validity and comparability of the data due to different methods, concepts, systematical errors (which are always more or less unavoidable) and the like. But the enormous amount of published data does allow, of course, the inference of characteristic data for certain site conditions. The challenge now is to improve methods—also to introduction of new methods—in order to provide reliable and detailed results for exactly specified site conditions, this is to say, for a unique ecosystem.

Still another point should be mentioned. Utilizing the outlined principles and laws and combining them in a mathematical model allows the prediction of the future behavior of an ecosystem with respect to its water exchange, provided the necessary boundary conditions, as well as some soil and root functions, are known. It need not be emphasized that such predictions are invaluable for the shaping, keeping and management of a landscape.

References

Benecke, P.: Die Ermittlung der Tiefensickerung aus Pseudogleyen. In: Pseudogley and gley (eds. E. Schlichting, U. Schwertmann), Trans. Com. V and VI, Intern. Soc. Soil Sci., pp. 443–452. Weinheim/Bergstr.: Verlag Chemie 1972.

Benecke, P., Mayer, R.: Wasserhaushaltsuntersuchungen im Solling. Mitt. Arb. Kreis "Wald und Wasser" **5**, 71–78 (1970).

Benecke, P., Mayer, R.: Aspects of soil water behavior as related to beech and spruce stands—some results of the water balance investigations. In: Integrated experimental ecology. Ecological Studies, vol. 2 (ed. H. Ellenberg), 153–163. Berlin-Heidelberg-New York: Springer 1971.

Black, C. D.: Soil-plant relationships. 2nd Ed. London-Sidney: Wiley and Sons 1968.

Bloemen, G. W.: On the evaluation of parameter values in water balance models. Tech. Bull. **92**, Inst. Land Water Management Res., Wageningen (1974).

Brechtel, H. M.: Schneeansammlung und Schneeschmelze im Wald und ihre wasserwirtschaftliche Bedeutung. "gwf"-wasser/abwasser **111**, 377–379 (1970).

Brechtel, H. M.: Einfluß des Waldes auf Hochwasserabflüsse bei Schneeschmelzen. Wasser und Boden **23**, 60–63 (1971).

Brechtel, H. M., Zahorka, H.: Beeinträchtigt die Umwandlung von Buchen- in Fichtenbestände die wasserwirtschaftliche Funktion des Waldes? Allgem. Forstz. **26**, 147–150 (1971).

Cowan, I. R.: Transport of water in the soil-plant-atmosphere system. J. Appl. Ecol. **2**, 221–239 (1965).

Delfs, J., Friedrich, W., Kiesekamp, H., Wagenhoff, A.: Der Einfluß des Waldes und des Kahlschlages auf den Abflußvorgang, den Wasserhaushalt und den Bodenabtrag.— Ergebnisse der ersten 5 Jahre der forstlich-hydrologischen Untersuchungen im Oberharz (1948–1953). "Aus dem Walde", Mitt. Nds. Landesforstverw. (Hannover) **3** (1958).

Denmead, O. P., Shaw, H. R.: Availability of soil water to plants as affected by soil moisture and meteorological conditions. Agron. J. **54**, 385–390 (1962).

Eidmann, F. E.: Die Interception in Buchen- und Fichtenbeständen. C. R. Ass. Int. Hydrol. Sci., Hannover Symp. **1**, 5–25 (1959).

Eschner, A. R.: Interception and soil moisture distribution. In: Intern. Symp. on Forest Hydrology (eds. W. E. Sopper, H. W. Lull), pp. 191–199. New York, London: Pergamon Press Ltd. 1967.
Fazilat, M.: Der Wasserentzug der Pflanzenwurzeln und seine Modellierung. Diss. Göttingen 1975.
Feddes, R. A., Rijtema, P. E.: Water withdrawal by plant roots. Tech. Bull. **83**, Inst. Land Water Management Res., Wageningen (1972).
Finck, A.: Pflanzenernährung in Stichworten. Kiel: Ferd. Hirt-Verlag 1969.
Freeze, R. A.: Three-dimensional, transient, saturated-unsaturated flow in a groundwater basin. Water Resourc. Res. **7**, 347–364 (1971).
Friedrich, W., Liebscher, H., Rudolph, R., Wagenhoff, A.: Forstlich-hydrologische Untersuchungen in bewaldeten Versuchsgebieten im Oberharz.—Ergebnisse aus den Abflußjahren 1951–1965. "Aus dem Walde", Mitt. Nds. Landesforstverw. (Hannover) **7** (1968).
Gardner, W. R.: Dynamic aspects of water availability to plants. Soil Sci. **89**, 63–73 (1960).
Gardner, W. R.: Dynamic aspects of soil water availability to plants. Ann. Rev. Plant Physiol. **16**, 323–342 (1965a).
Gardner, W. R.: Soil water movement and root absorption. In: Plant environment and efficient water use (ed. Pierre, Kirkham, Pesek, Shaw), pp. 127–149. Madison, Wisc.: Am. Soc. Agron. and Soil Sci. Soc. Am. Publisher 1965b.
Garstka, W. U.: Snow and snow survey. In: Handbook of applied hydrology (ed. Ven Te Chow), pp. 10:1–10:57. New York, St. Louis, San Francisco: McGraw-Hill 1964.
Heiseke, D.: Schneedecke und Wasserhaushalt der Steilen Bramke im Oberharz 1962/63. "Aus dem Walde", Mitt. Nds. Landesforstverw. (Hannover) **22**, 140–171 (1974).
Hibbert, A. R.: Forest treatment effects on water yield. In: Intern. Symp. Forest Hydrology (ed. Scopper, Lull), 527–543. New York and London: Pergamon Press 1967.
Hillel, D.: Soil and water. New York and London: Academic Press 1971.
IBM: System/360 Continuous System Modeling Program, User's Manual. Program Nr. 360A-CS-16X, 5th ed. New York: IBM Corp. Tech. Publ. Dep. (1972).
Item, H.: A model for the water regime of a deciduous forest. J. Hydrology **21**, 201–210 (1974).
Kirkham, D., Powers, W. L.: Advanced Soil Physics. New York: Wiley and Sons 1972.
Kittredge, J.: Forest influences. New York: McGraw-Hill 1948.
Kramer, P. J.: Plant and soil water relationships. New York: McGraw-Hill 1969.
Lang, W.: Ökologische und hydrologische Untersuchungen in verschieden stark durchforsteten Fichten- und Lärchenbeständen des Schwarzwaldes. Dissertation Freiburg (W-Germany) 1970.
Mitscherlich, G.: Wald, Wachstum und Umwelt. Vol. 2: Waldklima und Wasserhaushalt. Frankfurt: J. P. Sauerländer's Verlag 1971.
Nimah, M. N., Hanks, R. J.: Model for estimating soil water, plant, and atmospheric interrelations: I. Description and sensivity. Soil Sci. Soc. Am. Proc. **37**, 522–527 (1973a).
Nimah, M. N., Hanks, R. J.: Model for estimating soil water, plant, and atmospheric interrelations: II. Field test of model. Soil Sci. Soc. Am. Proc. **37**, 528–532 (1973b).
Penman, H. L.: Vegetation and hydrology. Techn. Column. **53**. Harpenden, England: Commonwealth Agr. Bur. 1963.
Philip, J. R.: Plant water relations: some physical aspects. Ann. Rev. Plant Physiol. **17**, 245–268 (1966).
Ploeg, R. R. van der, Benecke, P.: Simulation of one dimensional moisture transfer in unsaturated, layered field soils. Göttinger Bodenk. Ber. **30**, 150–169 (1974a).
Ploeg, R. R. van der, Benecke, P.: Unsteady, unsaturated, n-dimensional moisture flow in soil: A computer simulation program. Soil Sci. Soc. Am. Proc. **38**, 881–885 (1974b).
Quervain, M. R. de, Gand, H. R. in der: Distribution of snow deposit in a test area for alpine reforestation. In: Forest hydrology (eds. W. E. Sopper, H. W. Lull), pp. 233–239. New York and London: Pergamon Press 1967.
Richards, L. A., Wadleigh, C. H.: Soil water and plant growth. In: Soil physical conditions and plant growth (ed. B. T. Shaw). New York and London: Academic Press 1952.
Rose, C. W.: Agricultural physics. London: Pergamon Press 1966.

Rutter, A. J.: Water consumption by forests. In: Water deficits and plant growth. II. Plant water consumption and response (ed. T. T. Kozlowski), pp. 23–84. New York and London: Academic Press 1968.

Shaw, R. H., Laing, D. R.: Moisture stress and plant response. In: Plant environment and efficient water use (ed. Pierre, Kirkham, Pesek, Shaw), pp. 73–94. Madison, Wisc.: Am. Soc. Agron. and Soil Sci. Soc. Am. Publisher 1965.

Slatyer, R. O.: Plant-water relationships. London-New York: Academic Press 1967.

Waggoner, P. E.: Decreasing transpiration and the effect upon growth. In: Plant environment and efficient water use (eds. Pierre, Kirkham, Pesek, Shaw), pp. 49–72. Madison, Wisc.: Am. Soc. Agron. and Soil Sci. Soc. Am. Publisher 1965.

Weihe, J.: Zurückhaltung von Regenniederschlägen durch Buchen und Fichten. Allgem. Forstz. **23**, 86–90 (1968).

Weihe, J.: Warum noch immer Interzeptionsuntersuchungen im Wald? Mitt. Arb. Kreis "Wald und Wasser" **5**, 10–22 (1970).

Wesseling, J., Wijk, W. R. van: Soil physical conditions in relation to drain depth. In: Drainage of agricultural lands (ed. J. N. Luthin), Agronomy, vol. 7, pp. 461–472. Madison, Wisc.: Soc. Agronomy Publ. 1957.

Wit, C. T. de, Keulen, H. van: Simulation of transport processes in soils. Wageningen: Centre for Agr. Publ. Document. 1972.

Ziemer, R. R.: Soil moisture depletion patterns around scattered trees. U.S. Forest Ser. Res. Note PSW-**166.**, U.S. Dept. Agr. (1968).

Part 3
Transpiration and Its Regulation

Preface

Transpiration is an unavoidable evil for most plants. Certainly, water loss affects the energy budget of the leaves; it decreases leaf temperature and may thus prevent leaf tissues from overheating (see Gates, Part 3:A); the transpiration stream is also necessary for mineral transport. However, in most cases a plant does not profit directly from high transpiration rates demanded by dry air. Water loss can usually even be described as an evil because it may produce water deficits within the living tissues, affecting metabolic activity, and because it involves the danger of injury and even death by desiccation. Water exchange with the atmosphere is unavoidable since the plant needs a gaseous exchange pathway with its environment for production. Most of the diffusive transport of CO_2 and O_2, as well as H_2O, takes place through the stomata, the only pathway of low diffusion resistance in the leaves and stems. Consequently, the plant must transpire in order to be able to pursue photosynthesis; thus, water loss is the necessary by-product of assimilatory CO_2 uptake.

Because normally during the light period the gradients in CO_2 partial pressure between the intercellular air spaces of the mesophyll and the ambient air are much smaller than the gradients in water vapor partial pressure, inward fluxes of CO_2 are lower than outward fluxes of H_2O. The water-use efficiency, that is the ratio between the mass of transpired water per mass of dry matter produced, in herbaceous plants (see Larcher, 1975) ranges between 74 (C_4 plant *Atriplex hymenelytra*; Pearcy et al., 1974) and 840 (C_3 plant *Medicago sativa*) and may even increase above this value. Since water supply is usually restricted, especially during the periods which favor photosynthetic activity, plants must try to utilize water as effectively as possible. This is a key problem for plant establishment and existence particularly in arid and semi-arid habitats. Higher plants have developed various refined strategies in order to meet with these requirements. It is impossible to give a comprehensive review of all these mechanisms within the scope of this book (see Stocker, 1956; Walter, 1960; Kramer, 1969). In Part 3 of this volume an attempt is made to give the reader insight into some important aspects of water loss and its regulation in plants, with special emphasis on ecological problems. The different options of photosynthetic pathways as metabolic adaptations to improve water-use efficiency are treated in Part 5 of this volume. Another Part (6) is devoted to productivity and yield of plants and plant communities in relation to water conditions and water use. The phytogeographical consequences of the task of "tacking adroitly between thirst and starvation" (as Otto Stocker has formulated it; see Larcher, 1975), especially for plants of dry habitats, are made clear in Part 7.

Preface

Transpiration is a process involving physical mechanisms as well as biological properties of leaves. The fundamental biophysical aspects of water loss are treated in Part 3:A by Gates. He formulates the energy balance equation of a leaf and shows how the energy budget is determined by radiation, convection, and evaporation. Special attention is paid to the influence of water loss on leaf temperatures.

The rate of transpiration of a leaf under given meteorological conditions depends upon its total diffusion resistance to water vapor, which consists of two separate terms, namely the cuticular and the stomatal resistance. Although the cuticular transpiration in general is much lower than maximum stomatal water loss, it gains increasing importance in understanding plant water loss control. The cuticle which covers the above-ground organs of all higher plants and the thalli of certain mosses is not fully impermeable to water. Nevertheless, it is an effective protection against excessive water loss. The properties of cuticular membranes, especially their water permeability and the mechanisms of cuticular water transport are the subject of Part 3:B by Schönherr. Early hypotheses of cuticular transpiration are reported, and the concept of the polar pathway is explained by experimental evidence. One consequence of the presence of polar pores in the cuticular membranes is the fact that the cuticular diffusion resistance increases with decreasing water content of the membranes. The importance of this fact on plant water loss needs further investigation.

The main controlling instruments for transpirational water loss are the stomata. Most of the transpired water escapes through them, and they are the valves which can be regulated most sensitively according to the requirements of water balance and photosynthesis. The stomata are versatile sensory devices that react to internal and environmental factors. Since the stomata are so decisively important for water loss, as well as for photosynthetic productivity of plants, much interest has been focused in recent years on their physiology, on the mechanism of guard cell motion, and on its control (see Lange, 1972, 1975). An enormous amount of information has been accumulated. Interpretation of stomatal responses and their biochemical background becomes more and more complicated, and we are still far from a complete understanding of their mechanism. Our present knowledge on stomatal action is summarized in recent comprehensive reviews (Raschke, 1975; Meidner and Willmer, 1975; see also Meidner and Mansfield, 1968; Stålfelt, 1956). In the present volume, only two chapters can be devoted to the functions of stomata. In Part 3:C Levitt presents his consistent concept of the physiological basis of stomatal responses. Then Hall et al. (Part 3:D) report current perspectives of stomatal responses to environmental factors such as humidity, temperature, and light intensity. They stress ecological consequences and explain how stomatal responses determine the diurnal courses of water loss and net photosynthesis of different plants under natural conditions. These aspects are continued in the chapter by Waring and Running (Part 3:E), in respect to the over-all water relations of a whole tree, thus integrating the first three parts of this volume. Part 3:E explains quantitatively by means of a mathematical model how water balance and water status of a plant are the result of water uptake, water storage, and transpiration, and how the water fluxes through the different compartments of the plant are mediated by environmental conditions.

References

Kramer, P. J.: Plant and soil water relationships: A modern synthesis. New York: McGraw-Hill 1969.
Lange, O. L.: Wasserumsatz und Stoffbewegungen. Fortschr. d. Botanik **34**, 91–112 (1972).
Lange, O. L.: Plant water relations. Progress in Botany **37**, 78–97 (1975).
Larcher, W.: Physiological plant ecology. Berlin-Heidelberg-New York: Springer 1975.
Meidner, H., Mansfield, T. H.: Physiology of stomata. London: McGraw-Hill 1968.
Meidner, H., Willmer, C.: Mechanisms and metabolism of guard cells. Current advances of plant science, Commentaries **17**, 1–15 (1975).
Pearcy, R. W., Harrison, A. T., Mooney, H. A., Björkman, O.: Seasonal changes in net photosynthesis of *Atriplex hymenelytra* shrubs growing in Death Valley, California. Oecologia (Berl.) **17**, 111–121 (1974).
Raschke, K.: Stomatal action. Ann. Rev. Plant Physiol. **26**, 309–340 (1975).
Stålfelt, M. G.: Die stomatäre Transpiration und die Physiologie der Spaltöffnungen. In: Pflanze und Wasser. Handbuch der Pflanzenphysiologie, Vol. III (ed. O. Stocker), pp. 351–426. Berlin-Göttingen-Heidelberg: Springer 1956.
Stocker, O. (ed.): Pflanze und Wasser. Handbuch der Pflanzenphysiologie, Vol. III. Berlin-Göttingen-Heidelberg: Springer 1956.
Walter, H.: Grundlagen der Pflanzenverbreitung I. Standortslehre. Stuttgart: Eugen Ulmer 1960.

A. Energy Exchange and Transpiration

D. M. Gates

I. Introduction

Water loss from a plant leaf is a process involving physical mechanisms and biological properties of the leaf. In principle, transpiration is a simple process, but in actuality it is an extremely complicated phenomenon because of the complexities of leaf properties and behavior. It is clear that certain conditions must exist for water to be transpired from a leaf to the environment. Water must be available, energy must be available in order to convert liquid water to vapor, and there must exist a water vapor gradient from inside to outside the leaf which drives the gas diffusion process. A review of earlier research concerning transpiration and leaf temperatures was given by Gates (1968).

II. Gas Diffusion

The rate of transpiration E in g cm^{-2} s^{-1} is equal to the rate at which water molecules diffuse out of the leaf and is described analytically as

$$E = \frac{d_l^s(T_l) - \text{rh}\, d_a^s(T_a)}{r_l + r_a} \tag{1}$$

where $d_l^s(T_l)$ and $d_a^s(T_a)$ are the saturation water vapor densities in the leaf and in the air respectively. They are each a function of the temperature and are expressed in g cm^{-3}. The leaf and air temperatures are T_l and T_a respectively (rh = relative humidity). The diffusion pathway from the wet cell walls within the leaf through the substomatal and stomatal cavities to the leaf surface offers a resistance r_l to the flow of water vapor molecules, and a boundary layer of viscous air adhering to the leaf surface offers an additional resistance r_a to flow. These resistances are expressed in s cm^{-1}. The leaf resistance r_l varies enormously from some minimum value when the stomates are fully open to a maximum value which may be large or nearly infinite when the stomates are closed. Minimum resistances may be as low as 0.5 s cm^{-1} to as high as 10 or 20 or more, depending on the plant species and its growth history. The leaf resistance must be considered as a dynamic quantity which for a given leaf may change constantly with environmental conditions. The leaf resistance is a function of the relative humidity, temperature of the leaf, light intensity, water potential in the leaf, osmotic potential of the stomates, and other factors (see Hall et al., this volume Part 3:D).

The external boundary layer resistance r_a is also a dynamic quantity which varies constantly with environmental conditions. There is always a boundary layer of air which adheres to a leaf surface as the result of viscosity. Across the boundary layer there are gradients of temperature, moisture, carbon dioxide, oxygen, and wind speed. The boundary layer is both laminar and turbulent, depending upon the wind conditions. The larger the leaf the greater is the thickness of the boundary layer. The greater the wind speed the thinner is the boundary layer. Its resistance to water vapor diffusion is described by

$$r_a = K_2 \frac{W^{0.20} D^{0.35}}{V^{0.55}} \qquad (2)$$

where D and W are the leaf dimensions in the direction of the wind and at right angles to the wind respectively. K_2 is a proportionality constant which is between 26 and $35 \cdot 10^{-3}$ depending on the leaf size, and is such that r_a is in min cm^{-1} if the wind speed V is in cm s^{-1}. Details are given in Gates and Papian (1971).

III. Energy Balance

The energy balance for a leaf involves processes of heat exchange by radiation, convection, and evaporation. If the energy consumed by photosynthesis is small and is neglected to a first approximation

$$Q_a - \varepsilon \sigma T_l^4 - K_1 (V/D)^{0.5} (T_l - T_a) - L(T_l) E = 0 \qquad (3)$$

where Q_a is the total radiation absorbed by the leaf, $\varepsilon \sigma T_l^4$ is the radiation emitted by the leaf, and $L(T_l)$ is the latent heat of water which is a function of the leaf temperature. It is approximately 580 cal g^{-1} at 30°C. The constant K_1 varies from about 10 to $16 \cdot 10^{-3}$ and is such that with V in cm s^{-1}, D in cm, $T_l - T_a$ in °C, the convection term is in cal cm^{-2} min^{-1}. Details are given in Gates and Papian (1971). The Stefan-Boltzmann constant $\sigma = 8.13 \cdot 10^{-11}$ cal cm^{-2} min^{-1} °C^{-4} and the emissivity $\varepsilon = 0.95$.

Using these relatively simple expressions one can derive all of the necessary information concerning leaf temperatures and rates of water loss. The net radiation absorbed by a leaf is the difference between the radiation absorbed and the radiation emitted by the leaf. The net radiation gained or lost by a leaf is only exchanged by the processes of convection and evaporation, assuming that the photosynthetic process uses a negligible amount.

IV. Transpiration

Convection will cool or warm a leaf depending upon whether the leaf temperature is greater than or less than the air temperature. The transpiration rate depends upon the amount of energy available after the net radiation is shared with convection. Hence, one can write

$$L(T_l)E = NR - C \qquad (4)$$

where NR is net radiation and C is convection. If $T_l > T_a$ in the convection term, then an amount of energy less than the net radiation is available for the evaporation of liquid water to vapor. If $T_l < T_a$, then the convection term is negative and heat is added to the leaf and more energy is available for evaporation. From Eqs. (1), (2), and (4) one can write

$$E = \frac{d_l^s(T_l) - \mathrm{rh}\, d_a^s(T_a)}{r_l + K_2 \dfrac{W^{0.20} D^{0.35}}{V^{0.55}}} = \frac{NR - C}{L(T_l)}. \qquad (5)$$

All environmental factors affecting the rate of water loss from a leaf are contained in Eq. (5). These include the radiation absorbed by a leaf, the air temperature, wind speed, and relative humidity. Equation (5) also contains all of the plant factors affecting the water loss from a leaf. These include the leaf resistance, the leaf size, and implicitly in the radiation term the absorptivity to incident radiation. A leaf arrives at a specific leaf temperature such that it is in energy balance according to Eq. (3) for any given set of environmental conditions and plant properties. At each specific leaf temperature, a leaf will have a particular transpiration rate dependent upon its resistance and leaf size.

V. Wind Speed Influence

For a given set of environmental conditions, one can ask whether an increase of wind speed will increase or decrease the rate of water loss from a leaf? The answer to this question is not always obvious. Sometimes transpiration will increase, at other times decrease, and sometimes there is no change with increased wind speed. Shown in Fig. 1 is an illustration of the dependence of transpiration rates with wind speed for various amounts of radiation absorbed by a leaf and for the conditions shown. It is clear that when the amount of radiation absorbed is low, an increase of wind speed produces an increase of the transpiration rate, and when it is high, the transpiration rate decreases with increased wind. There is always an intermediate condition when wind speed has no affect on the transpiration rate.

An increase of wind speed always affects several quantities in the energy budget relationship for a leaf. It affects the convection term and the boundary layer resistance in the diffusion term for water vapor. If the leaf temperature is greater than the air temperature, then an increase of wind speed produces increased convective cooling according to the square root of the wind speed. Increased convective cooling drops the leaf temperature which in turn drops the water vapor density in the leaf's substomatal cavity. This reduces the vapor density difference between the inside and the outside of the leaf and, therefore,

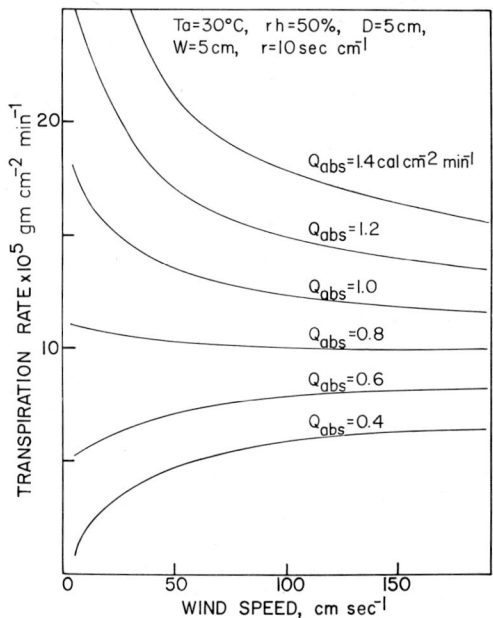

Fig. 1. Transpiration rates of a leaf of dimensions (D, W) 5·5 cm and diffusion resistance (r) 10 s cm^{-1} as a function of the wind speed for an air temperature (T_a) of 30°C and relative humidity (rh) 50% at various amounts of absorbed radiation (Q_{abs})

would drop the transpiration rate. However, there is a contrary force acting to increase the rate of water loss because the boundary layer resistance decreases with increased wind speed. If the leaf resistance is relatively large, then the change in boundary layer resistance is a relatively small change out of the total resistance, and the vapor density effect will dominate. If, however, the radiation absorbed by a leaf is small, then the leaf temperature may be less than the air temperature, convective exchange will put heat into the leaf, and an increase of wind speed will increase convective heating. This will increase the total amount of energy available to a leaf, which in turn will increase leaf temperature which will increase the substomatal water vapor density and increase the water vapor gradient. The increased wind speed will decrease the boundary layer resistance and this, combined with the increased vapor gradient, will cause an increase in the transpiration rate. In between these two situations, any combination of these events may occur and the transpiration rate will increase, decrease, or not change at all with wind speed.

VI. Leaf Temperature Affected by Transpiration

There is absolutely no doubt that the evaporation of water from a leaf must affect the leaf temperature. It is only a question as to how much? The solution to the energy budget relationship given as Eq. (3) gives the exact answer to this question. If the environment of a leaf is properly evaluated and the properties of a leaf are known, then the transpiration rate and leaf temperature are determined. In Fig. 2 is shown the dependence of the transpiration rate and leaf temperature as a function of the leaf resistance and the air temperature for a moderate amount of absorbed radiation, still air, and a relative humidity of 50%. The leaf size is 5·5 cm. It is seen that for conditions below the dotted line the leaf temperature is over the air temperature, and above the dotted line the leaf temperature is under the air temperature. If the amount of radiation absorbed by the leaf is reduced from $1.0\,\mathrm{cal\,cm^{-2}\,min^{-1}}$ to $0.6\,\mathrm{cal\,cm^{-2}\,min^{-1}}$ and all other conditions are the same, one gets the diagram shown in Fig. 3. The line separating the "over-air" temperature from the "under-air" temperature response has shifted to much lower temperatures. It is clear that the air temperature at which this transition occurs depends strongly on the amount of radiation absorbed by the plant leaf. Also, since the dotted line is not parallel to lines of constant air temperature, it is a function of the diffusion resistance to water vapor.

A mathematical prediction of the "over-air", "under-air" leaf temperature phenomena can be seen by linearizing the energy budget equation and solving for the temperature difference between leaf and air. If $\Delta T = T_l - T_a$, one gets

$$Q_a = \varepsilon \sigma T_a^4 + 4\varepsilon \sigma T_a^3 \Delta T + K_1 \left(\frac{V}{D}\right)^{0.5} \Delta T + LE. \qquad (6)$$

Solving for ΔT, one gets

$$\Delta T = \frac{Q_a - \varepsilon \sigma T_a^4 - LE}{4\varepsilon \sigma T_a^3 + K_1 \left(\dfrac{V}{D}\right)^{0.5}}. \qquad (7)$$

Hence, $\Delta T = 0$ when $Q_a = \varepsilon \sigma T_a^4 + LE$. If the leaf resistance $r_l = \infty$, then $\Delta T = 0$ when $Q_a = \varepsilon \sigma T_a^4$. This is the intercept of the dotted line with the abscissa. For all other values of the leaf resistance (lower values) the dotted line will slope towards lower air and leaf temperatures. This shows that whether a leaf is an "under-air" or "over-air" temperature plant is primarily a function of the environmental conditions for which it is observed and also is a function of its transpiration rate. For a given plant species with leaves of specific characteristics, the leaf temperature is "over-air" or "under-air" temperature depending on the amount of radiation absorbed, the air temperature, and the transpiration rate.

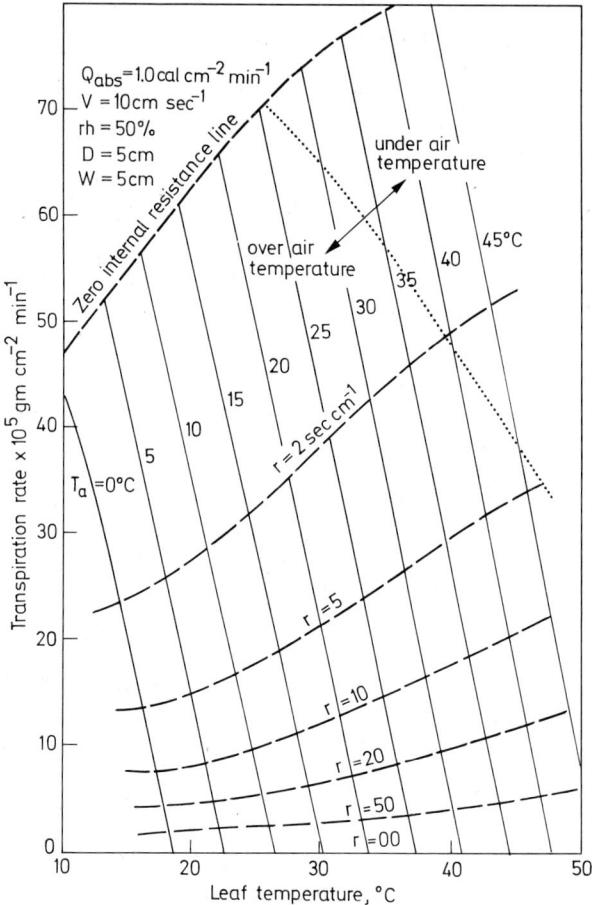

Fig. 2. Transpiration rates and leaf temperatures as a function of the air temperature (T_a) and diffusion resistance (r) for a leaf of dimensions (D, W) 5·5 cm in still air (wind speed V) at a relative humidity (rh) of 50% and an amount of absorbed radiation (Q_{abs}) of 1.0 cal cm^{-2} min^{-1}

Lange (1959) reported field observations of "under-air" and "over-air" leaf temperature responses of plants growing in Mauretania. Shown in Fig. 4 is the leaf temperature of *Zygophyllum* cf. *fontanesii* in comparison with the air temperature. The intensity of radiation was high and, although the amount of radiation absorbed was not measured, it was probably at least 1.0 cal cm^{-2} min^{-1}. According to Fig. 2, one would expect that at these air temperatures (25 to 30°C) this plant would respond in an "over-air" leaf temperature condition. If the amount of radiation diminished greatly, this plant would have responded in an "under-air" leaf temperature condition.

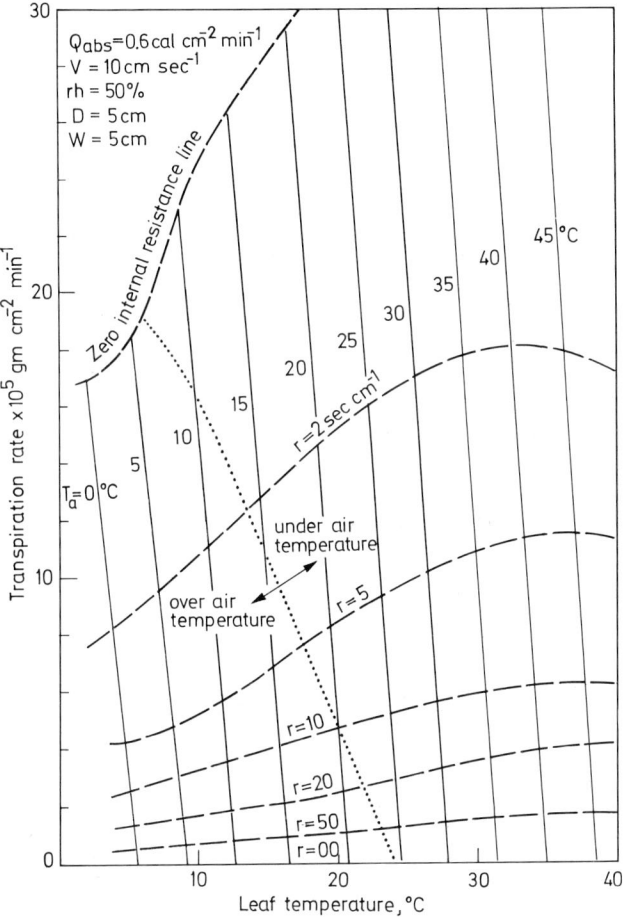

Fig. 3. Transpiration rates and leaf temperatures as a function of the air temperature (T_a) and diffusion resistance (r) for a leaf of dimensions (D, W) 5·5 cm in still air (wind speed V) at a relative humidity (rh) of 50% and an amount of absorbed radiation (Q_{abs}) of 0.6 cal cm^{-2} min^{-1}

In order to demonstrate the influence of transpiration on leaf temperature, Lange (1959) measured the leaf temperature of *Citrullus colocynthis* before and after severing the leaf from the plant during very hot conditions. These measurements are shown in Fig. 5. The temperatures of two plant leaves were monitored when the air temperature was from 40 to 52°C. Under these environmental

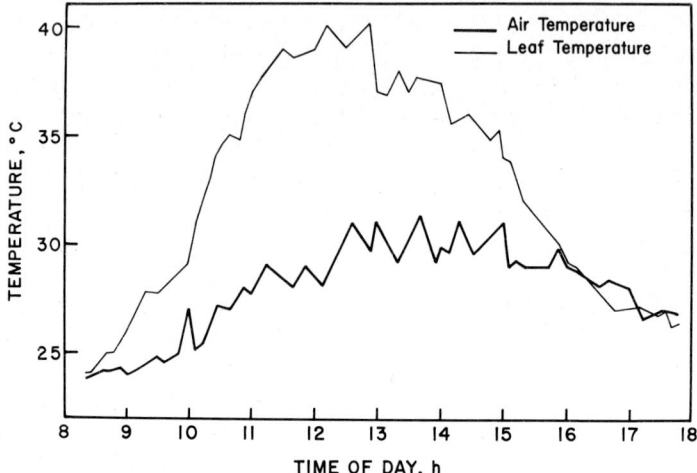

Fig. 4. Leaf temperature for *Zygophyllum* cf. *fontanesii* and the air temperature as a function of the time of day for an intense radiation environment in Mauretania. (Redrawn from Lange, 1959)

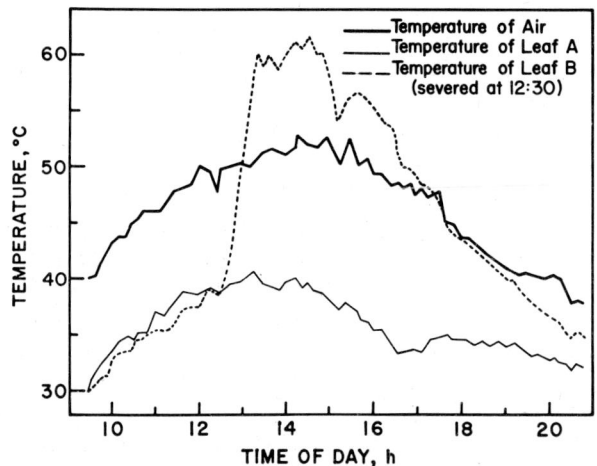

Fig. 5. Leaf temperature for *Citrullus colocynthis* and the air temperature as a function of the time of day for an intense radiation environment in Mauretania. The leaf temperature of an intact leaf is shown in comparison with a severed leaf. (Redrawn from Lange, 1959)

conditions, the plant leaves were in an "under-air" temperature condition, as one would predict from Fig. 2. At 12:30 pm one of the leaves was severed, thereby cutting off the supply of water and causing $LE=0$ or the equivalent of $r_1 = \infty$. The severed plant leaf quickly went from an "under-air" leaf temperature condition to an "over-air" leaf temperature condition. In fact, the leaf temperature increased by 22°C with the removal of the water supply. This response is consistent with Fig. 2. However, the relative humidity of the desert air in Mauretania was probably much drier than the 50% relative humidity used to calculate Fig. 2. In fact, if the relative humidity is very low, the lines of constant air temperature are less steep and slant more from left to right since the influence of transpiration on leaf temperature is stronger. If the appropriate properties of the *Zygophyllum* cf. *fontanesii* and *Citrullus colocynthis* were known and all environmental factors were measured, the leaf temperature behavior of these plants in Mauretania would be in complete agreement with theoretical prediction.

During field work in the Sierra Nevada Mountains of California during August, 1963, Gates et al. (1964) made a similar discovery to that of Lange for plants of the genus *Mimulus*. These are plants which generally grow in wet soil along stream banks, drainage ditches, and alpine snow melt fields. Two species were involved in the field observations—*Mimulus lewisii* growing at elevations above 2000 m and *Mimulus cardinalis* growing at lower elevations. A variety of these species and their hybrids were growing in some of the Carnegie Institution's transplant gardens at Mather and Timberline. Measurements were made of the leaf temperatures during clear days between 10:00 and 15:00 h. Shown in Fig. 6 are the leaf temperatures at the following sites: Priest's Grade, 400 m; Mather, 1400 m; Carlin, 1520 m; Smoky Jack, 2300 m; Warren Creek, 2700 m; Timberline station and slope garden, 3000 m; and Timberline talus slope, 3200 m. At the high elevations above 2300 m air temperatures were from 15 to 22°C and leaf temperatures were generally 20 to 31°C, with exceptions for shade leaves (indicated by subscript s) and clouded conditions (indicated by subscript c). At elevations below 2300 m air temperatures were from 22 to 37°C. The striking thing is that when the air temperature was 37°C most *Mimulus* leaf temperatures were between 29 and 35°C. When air temperatures changed by 17 to 22°C, the leaf temperatures generally changed by about 5°C, but there were exceptions. However, when taking into account different leaf orientation and exposure, it is not surprising that there should be a considerable spread of leaf temperatures in a population of leaves. At the time these observations were made, it was a great surprise to find such homeostasis among plant leaves in the field.

Field conditions were simulated in the laboratory at the Department of Plant Biology, Carnegie Institution, Stanford. Many very different plants were observed in the growth chamber and leaf temperatures and transpiration rates measured for a wide range of chamber temperatures. All plants exhibit the phenomenon of having "over-air" leaf temperature behavior at low chamber temperatures and "under-air" leaf temperatures at high chamber temperatures. At levels of moderate illumination (1700 fc and 0.14 cal cm^{-2} min^{-1}) the crossover from "above-air" to "under-air" leaf temperatures occurred at 28°C, while at higher illumination (2100 fc and 0.56 cal cm^{-2} min^{-1}) it occurred at around 45°C. The behavior of a particular plant species depended on its diffusion resistance, leaf size, and absorptivity. There is no doubt if each leaf property was known all field and laboratory observations could be precisely accounted for by the theoretical analysis.

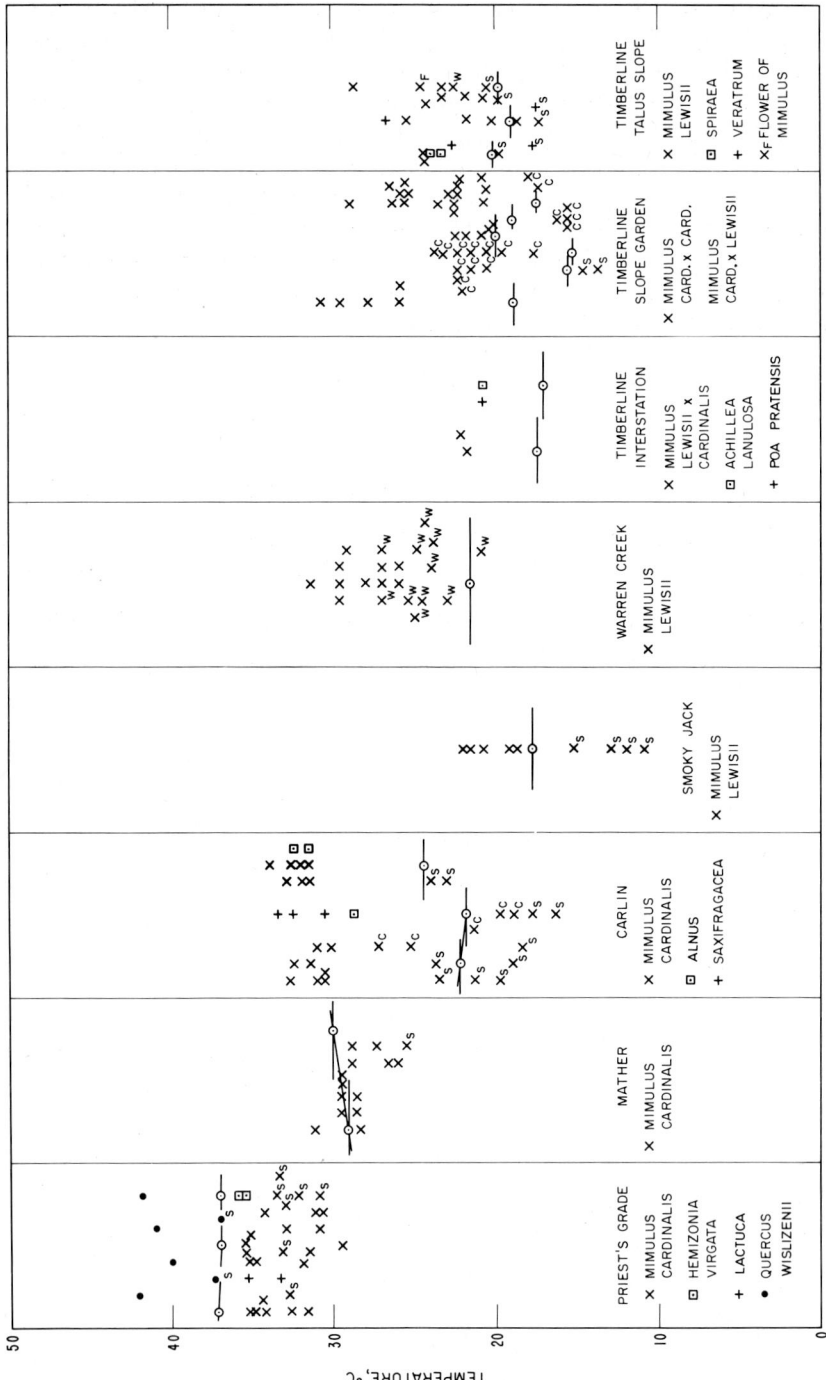

Fig. 6. Leaf temperatures in the field of *Mimulus cardinalis*, *Mimulus lewisii*, and of hybrids, as well as of other selected plant species along a transect from Priest's Grade at 400 m elevation to timberline at 3 200 m during midday in the Sierra Nevada Mountains on a cloudless day. Air temperatures at each site are given as —⊙—

VII. Conclusions

The physical environment of plants is complex and highly variable with time and space. Various characteristics of plants, such as leaf geometry and diffusion resistance, change with time and with environmental conditions. Nevertheless, water loss from plant leaves can be completely understood both qualitatively and quantitatively by energy budget analysis. However, to do so in specific instances for plants growing in the field requires careful measurements of all pertinent environmental factors and plant parameters. Such factors include the amount of radiation, both shortwave and longwave, incident on a leaf, air temperature, relative humidity, and wind speed in the vicinity of the leaf. Plant parameters required include absorptivity to radiation, leaf size, and diffusion resistance to water loss as a function of time.

References

Gates, D. M.: Transpiration and leaf temperature. Ann. Rev. Plant Physiol. **19,** 211–238 (1968).

Gates, D. M., Hiesey, W. M., Milner, H. W., Nobs, M. A.: Temperature of *Mimulus* leaves in natural environments and in a controlled chamber. Carnegie Institution of Washington Year Book **63,** 418–430 (1964).

Gates, D. M., Papian, L. E.: Atlas of energy budgets of plant leaves. London-New York: Academic Press 1971.

Lange, O. L.: Untersuchungen über Wärmehaushalt und Hitzeresistenz mauretanischer Wüsten- und Savannenpflanzen. Flora **147,** 595–651 (1959).

B. Water Permeability of Cuticular Membranes

J. Schönherr

I. Introduction

The plant cuticle is a lipid membrane that covers all above-ground organs of the Spermatophyta and Pteridophyta (Esau, 1969). The thalli of certain members of the Hepaticeae are likewise covered by a cuticle (Schönherr and Ziegler, 1975). Since the cell walls of the intercellular air spaces are also lined with a cuticle (Ursprung, 1925; Scott, 1950), it would appear that in the above plant species a cuticle is generally formed at all plant-air interfaces.

The main function of the cuticle is to serve as a barrier to minimize water loss from plant cells to the surrounding atmosphere. However, the cuticle is not impermeable to water, and under severe water stress it cannot prevent wilting of plants. For this reason the water permeability of cuticles has been the subject of extensive research, especially during the first half of this century (see Stålfelt, 1956, for a review). When radioisotopes became available the emphasis shifted to the study of cuticular permeability to pesticides, insecticides, herbicides, growth regulators and mineral nutrients (for a comprehensive review see Hull, 1970). It was found that the permeability of cuticles to lipophilic molecules was proportional to the oil–water partition coefficient of the permeating species (Darlington and Cirulis, 1963; van Overbeek, 1956; Hull, 1970). The results obtained with polar permeants were confusing, often contradictory and a general description for the transport of polar molecules across the cuticle was not arrived at.

Ectodesmata were once suggested to serve as pathways for polar molecules across the cuticle and the outer epidermal wall (Franke, 1967). These structures were shown to be fixation artefacts (Schönherr and Bukovac, 1970). In the following it will be shown that the pore membrane model developed by physiologists (Pappenheimer et al., 1951; Paganelli and Solomon, 1957; Nevis, 1958; Solomon, 1968) provides a consistent description of the permeability of cuticles to water and other small polar molecules.

II. Cuticular Transpiration
Early Observations and Hypotheses

In an attempt to judge the efficiency of the protection which the cuticle provides against water loss Seybold (1929) and Kamp (1930) compared the total transpiration of leaves (when stomata are open) with cuticular transpiration (transpiration of leaves with stomata closed). This ratio was found to be much larger in xerophytic plant species and to depend on leaf morphology. The ratio varied between 2 and 8 for leaves having a thin lamina and increased with increasing leaf thickness (expressed as the ratio leaf surface/leaf fresh weight) to a figure of 45 for *Laurus nobilis*. A similar trend is observed when the absolute values of cuticular transpiration are compared (Pisek and Berger, 1938).

Cuticular transpiration of the hygrophytic *Impatiens noli-tangere* was greater by a factor of 85 than that of the xerophytic *Pinus silvestris*. Kamp (1930) tried to relate these differences to the thickness of the cuticles but found no good correlation. This may be due to the following inadequacies and difficulties:

(a) Transpiration rates were usually expressed per unit fresh weight. The differences between species are reduced considerably if cuticular transpiration is referred to unit leaf surface (Kamp, 1930; Pisek and Berger, 1938).

(b) Unambiguous cuticular transpiration rates can only be measured using leaf surfaces free of stomata. When whole leaves are used the transpiration rates observed under conditions when stomata are believed to be closed are taken to represent cuticular transpiration. As pointed out by Stålfelt (1956), this assumption is most likely erroneous.

(c) It would be very surprising if cuticle thickness was the sole factor determining the water permeability of cuticles. The fact that cuticular transpiration rates of young leaves tend to be lower than those of older ones (Kamp, 1930; Pisek and Berger, 1938) demonstrates that factors other than cuticle thickness are involved in the regulation of water permeability of cuticles.

Since the cuticle is permeable to water it is generally assumed that it swells in the presence of water and that water permeability depends on its water content. There exist, however, opposing opinions as to whether or not cuticular transpiration rates are positively or negatively correlated with the water content of the cuticle. The membrane model usually adopted is a pore membrane with the pore size depending mainly upon the swelling in the presence of water. Early studies were concerned with the effect of relative humidity of the atmosphere on swelling of the cuticle and cuticular transpiration (Livingston and Brown, 1912; Seybold, 1929). A moisture saturation deficit was thought to cause deswelling and therefore an increased frictional resistance in the cuticle, resulting in reduced transpiration rates (incipient drying).

Later, Härtel (1947, 1951) observed that a pretreatment of the leaves with buffer solutions affected the rate of cuticular transpiration. These results were taken to indicate an effect of pH and ions on swelling and pore size of the cuticle. In order to be able to account for the observed pH-dependence of cuticular transpiration with a maximum around pH 7, Härtel hypothesized that cuticular transpiration was controlled by an ampholyte with an isoelectric point around

7. In contrast to the earlier workers it was assumed that the swelling caused an increased diffusional resistance in the cuticle because water "bound" to polar groups and ions associated with fixed charges of the pore walls would reduce the pore area available for the diffusion of "free" water.

It should be realized that these attempts to explain the rates of cuticular transpiration observed on the basis of swelling of cuticles, are highly speculative and mutually exclusive. This conflict cannot be resolved at present because (a) there are no reliable data available on the swelling of cuticles and (b) when experimenting with whole leaves there is the uncertainty of whether or not it is the water permeability of cuticles which is actually measured or some other variable instead.

A new approach was therefore adopted. Using isolated cuticles of the upper leaf surface of *Citrus aurantium* L., which are free of stomata and trichomes, and a transport apparatus that permitted accurate control of all experimental variables, unambiguous flux determinations were made and tested for their consistency with predictions derived from a hypothetical pore membrane model.

III. The Concept of the Polar Pathway Through Lipid Membranes

1. General Description

The description of the permeability properties of biological membranes in terms of parallel pathways for lipid soluble and lipid insoluble molecules has its root in a study by Collander and Barlund (1933). The concept of the equivalent pore as a description of the path taken by lipid insoluble molecules was developed by Pappenheimer et al. (1951), Koefoed-Johnson and Ussing (1953), Renkin (1954), Durbin (1960) and Solomon (1968). A treatment on the basis of irreversible thermodynamics describing the permeability properties of biological membranes by phenomenological coefficients and their translation into frictional quantities, which provide informations about membrane structure and the mechanism of transport, is credited to Kedem and Katchalsky (1958, 1961). These theories have been tested with various artificial membranes (Robbins and Mauro, 1960; Ginzburg and Katchalsky, 1963; Hanai and Haydon, 1966; Cass and Finkelstein, 1967; Lakshminarayanaiah, 1967) and were successfully applied to the analysis of the membrane structure of erythrocytes (Paganelli and Solomon, 1957; Solomon, 1968).

Basically two observations have led to the development of the polar pore concept. (a) The permeability of lipid membranes to polar molecules, which are insoluble in the membrane matrix, was found to depend on the size of the permeating species. The permeability decreased rapidly with increasing size of the molecules and the membranes were impermeable to polar molecules that exceeded a certain critical size. This observation can be explained if the membrane contains water-filled pores through which the solute molecules permeate. The membrane is impermeable to polar solute molecules of a size larger than that of the cross-section of the pores. The sizes of the pores can be estimated from the measured permeabilities (Renkin, 1954; Durbin, 1960; Beck and Schultz, 1970). (b) The water permeability of membranes under a gradient of osmotic or hydrostatic pressure was often much greater than could be accounted for by diffusion. This observation can be explained if the water transport takes place through pores in the membranes.

Two different mechanisms of transport can be visualized: viscous flow (the relative motion of adjacent portions of a liquid) and diffusion (the relative motion of the liquid molecules). When passage through a membrane can take place only by dissolution of the water molecules in the membrane, water transport takes place by diffusion alone (Cass and Finkelstein, 1967; Hanai and Haydon, 1966). When a lipid membrane contains channels filled with water the relative contributions of viscous flow and diffusion to the total transport depends on the size of the channels relative to the size of the water molecules (Robbins and Mauro, 1960). If a pressure gradient is imposed across a membrane containing aqueous channels whose radius is greater by orders of magnitude than the radius of the water molecules, viscous flow alone is of primary importance. With decreasing pore radius, the contribution of diffusion to the total flux will increase. When the pore radius is approximately twice that of the water molecules the two mechanisms of transport are of equal magnitude (Ticknor, 1958). Of course, in the absence of a gradient of hydrostatic or osmotic pressure, transport is by diffusion, no matter what the radius of the pore is. The dimensions of the pores can be estimated from the permeability coefficients for diffusion and hydrodynamic transport (Paganelli and Solomon, 1957; Nevis, 1958).

2. Principles of Experimental Procedures

The upper cuticular membrane of *Citrus aurantium* L. leaves which has no stomata was isolated enzymatically or chemically and inserted between the two compartments of a transport apparatus. When diffusion was to be measured the two compartments were filled with identical buffer solutions. Isotopically labeled water (THO), urea (^{14}C) or glucose (^3H) were added to one compartment and the rate of appearance of the radio-isotopes was measured under steady state conditions in the opposite compartment. The hydrodynamic permeability of the membranes was determined using identical buffers in both compartments plus a solute in one compartment. The water flux caused by the osmotic pressure difference was measured directly using a calibrated capillary connected to the compartment containing the solute. Care was taken to maintain constant temperature and pH. All fluxes measured were controlled by the permeability of the membranes and were not affected by unstirred layers. Details of the experimental procedures are given elsewhere (Schönherr, 1976). It should be pointed out that the waxes of the cuticles used had been extracted, as these studies were primarily concerned with the structure of the cutin-matrix. To denote this difference the term cuticular membrane will be used when referring to these cuticles devoid of waxes.

3. Diffusion of Water as Criterium for the Existence of Polar Pores

The flux by diffusion J_d (mol s^{-1}) of water across a porous membrane is given by Fick's law

$$J_d = -D A_w \frac{dC}{dx} \tag{1}$$

where C is the concentration (mol cm^{-3}) of isotopically labeled water at any point x in the membrane, D is the diffusion coefficient (cm^2 s^{-1}) of water in bulk liquid filling the pores of the membrane and A_w is the area of the membrane permeable to water, that is, the total pore area (cm^2). Integrating Eq. (1) one gets

$$J_d = -\frac{D A_w}{\Delta x} \Delta C = -P_d \Delta C \tag{2}$$

where the differential has been replaced by the difference taken across the membrane, ΔC being the concentration difference of isotopically labeled water and Δx represents the total path length through the membrane, which is probably greater than the membrane thickness because of tortuosity (Kedem and Katchalsky, 1961). Since A_w and Δx are usually not known they are combined with D to give the permeability coefficient for diffusion, P_d (cm^3 s^{-1}).

Using tritiated water (THO) it is found that the water permeability of cuticular membranes is pH-dependent. Between pH 3 and pH 11 an almost 5-fold increase of the permeability coefficient is observed (Table 1). Since D depends very little on the pH the increase in permeability can be attributed to a proportional increase in the diffusion area per unit diffusion path ($A_w/\Delta x$). By neglecting tortuosity the fractional pore area (total area of the water-filled pores (A_w)/total area of the membrane (A_m)) can be calculated. Kedem and Katchalsky (1961) have shown that the fractional pore area is numerically equal to the fractional volume of water in the membrane (volume of water in the membrane/total

Table 1. The effect of pH of the external solutions on water diffusion across isolated *Citrus aurantium* leaf cuticular membranes

pH	P_d	$A_w/\Delta x$[a]	A_w/A_m[b]
	cm^3 s^{-1}	cm	
3.0	1.14·10^{-5}	0.47	1.25·10^{-4}
5.0	1.80·10^{-5}	0.74	1.96·10^{-4}
7.0	2.74·10^{-5}	1.12	2.98·10^{-4}
9.0	3.48·10^{-5}	1.43	3.80·10^{-4}
11.0	5.45·10^{-5}	2.23	5.94·10^{-4}

[a] $A_w/\Delta x = P_d/D$; $D = 2.44 \cdot 10^{-5}$ cm^2 s^{-1} taken from Lakshminarayanaiah (1967).
[b] $A_w = (A_w/\Delta x)\Delta x$; $\Delta x = 2.66 \cdot 10^{-4}$ cm, calculated from the dry weight of the cuticular membrane, tortuosity neglected. Data from Schönherr (1976).

volume of the membrane). Hence, the pH-dependent increase in water permeability is the consequence of an increased water content (swelling) of these membranes. As water content of the membranes and water permeability are directly proportional, Härtel's hypothesis (1947) is incompatible with these data.

The data in Table 1 are also in contrast to those reported by Härtel (1947), as no permeability maximum was found around pH 7. This, however, was not to be expected since tomato fruit cuticular membranes were recently shown to be cation exchangers with the density of fixed charges increasing with increasing pH up to pH 12 (Schönherr and Bukovac, 1973). Between pH 3 and pH 9 —COOH groups and above pH 9 phenolic hydroxyl groups fixed to the membrane matrix dissociate. With increasing degree of dissociation of these groups, the polymer takes up water and swells (Howe, 1952; Katchalsky, 1954; Gregor et al., 1955, 1956). Due to ion pair formation —COOH groups are ionized and hydrated only to a minor extent. When neutralized with monovalent alkali metal ions association between —COO$^-$ and M$^+$ is less close than with H$^+$ and the polymer swells because of (a) the electrostatic repulsion of fixed charges of equal sign, (b) the volume occupied by the counter ions and their hydration shells, and (c) the tendency of the highly concentrated pore fluid to dilute itself (osmotic pressure difference between the pore solution and external solution). Thus, the observed pH-dependence of the water permeability demonstrates the existence of negatively fixed charges of the weak acid type in these cuticular membranes.

4. The Relation of Water Diffusion to Osmotic Flow as an Index of Equivalent Pore Dimensions

The thermodynamic equation for volume flow J_v (cm^3 s^{-1}) as a function of applied pressure head ΔP (atm) and osmotic pressure difference $\Delta \pi$ (atm) across a membrane of total pore area A_w is (Kedem and Katchalsky, 1958)

$$J_v = L_p A_w \Delta P - \sigma L_p A_w \Delta \pi. \tag{3}$$

With cuticular membranes which are too fragile to withstand large hydrostatic pressures it is convenient to determine the hydrodynamic (or filtration) coefficient L_p (cm^3 dyn^{-1} s^{-1}) at $\Delta P = 0$ using a solute to which the membrane is impermeable. Then the reflection coefficient σ is equal to one and Eq. (3) becomes

$$J_v = -L_p A_w \Delta \pi = -L_p A_w RT \Delta C_s \phi \tag{4}$$

where ΔC_s is the concentration difference of solute across the membrane and ϕ is the mean osmotic coefficient which corrects for non-ideal behavior. Introducing $\Phi_w = J_v / \bar{v}$ in Eq. (4) the flux of water Φ_w (mol s^{-1}) is given by

$$\Phi_w = -\frac{L_p A_w RT}{\bar{v}} \Delta C_s \phi = -P_f \Delta C_s \phi \tag{5}$$

where $\bar{v}$ is the partial molar volume of water (18 cm³ mol⁻¹) and P_f the so-called filtration (or osmotic) permeability coefficient (cm³ s⁻¹). From the slope volume vs time, Φ_w is readily obtained and P_f is calculated from Eq. (5).

The osmotic water permeability of cuticular membranes increased with increasing pH in the same degree as diffusion, such that the ratio P_f/P_d remained essentially constant at approximately 2.60 (Table 2). The fact that P_f is greater than P_d is generally accepted as criterium for the existence of polar pores in membranes (Pappenheimer et al., 1951; Koefoed-Johnson and Ussing, 1953; Robbins and Mauro, 1960; Cass and Finkelstein, 1967; Solomon, 1968). It is therefore concluded that these cuticular membranes are in fact porous.

Table 2. Hydrodynamic and diffusion permeability of isolated *Citrus aurantium* leaf cuticular membranes as function of pH

pH	P_f	P_d[a]	P_f/P_d
	cm³ s⁻¹	cm³ s⁻¹	
3.0	$0.68 \cdot 10^{-4}$	$2.56 \cdot 10^{-5}$	2.66
6.0	$1.08 \cdot 10^{-4}$	$4.16 \cdot 10^{-5}$	2.60
9.0	$1.80 \cdot 10^{-4}$	$7.30 \cdot 10^{-5}$	2.47

Data from Schönherr (1976).
[a] $\Delta C_s = 0.25$ molal raffinose.

Nevis (1958) has used P_f and P_d values to estimate the average pore radius. The diffusion flux is given by Fick's law [Eq. (2)] where the total pore area A_w is composed of n pores of mean area πr^2. According to Poiseuille's law

$$\Phi_w = \frac{n\pi r^4}{8\eta\bar{v}} \cdot \frac{\Delta P}{\Delta x} = P_f \Delta C_s \quad (\Delta P = RT \Delta C_s) \tag{6}$$

where η is the viscosity of the solution and P_f includes the diffusional flow. From Eqs. (2) and (6) it follows that

$$r^2 = \frac{8\eta D\bar{v}}{RT} \cdot \frac{P_f}{P_d}. \tag{7}$$

Substituting the literature values for the constants and accounting of the diffusional component in P_f Nevis (1958) wrote for the average pore radius (r_N):

$$r_N = 3.6 \left[\frac{P_f - P_d}{P_d} \right]^{1/2}. \tag{8}$$

Application of Eq. (8) to the present data results in an average pore radius of 4.52 Å between pH 3 and pH 9 with no significant effect of the pH value (Table 3).

In the derivation of Eq. (8) it is assumed that circular pores of uniform cross-section traverse the membrane in perpendicular direction and that Poi-

Table 3. Estimation of the average pore radius in isolated *Citrus aurantium* leaf cuticular membranes

pH	$A_w/\Delta x$	J_v	L_p	r_N [a]	r_P [b]
	cm	cm^3 s^{-1}	cm^3 dyn^{-1} s^{-1}	Å	Å
3.0	1.13	$3.07 \cdot 10^{-7}$	$4.96 \cdot 10^{-14}$	4.64	4.58
6.0	1.70	$4.80 \cdot 10^{-7}$	$7.84 \cdot 10^{-14}$	4.55	4.71
9.0	2.99	$8.25 \cdot 10^{-7}$	$1.33 \cdot 10^{-13}$	4.36	4.61

Data from Schönherr (1976).
[a] Calculated from Eq. (8), using the P_f/P_d values of Table 2.
[b] Calculated from Eq. (9).

seuille's law holds for these very narrow capillaries. None of these assumptions, save the last one, is made in a derivation due to Paganelli and Solomon (1957) who also take into account steric hindrance at the entrance of the pore and frictional resistance with the pores felt by the water molecules. Their equation

$$r_p = -a \left[2a^2 + \frac{8\eta L_p}{A_w/\Delta x} \right]^{1/2} \tag{9}$$

which includes the radius of the water molecules ($a = 1.97$ Å) has been very useful in estimating the equivalent pore radii of a number of artificial (Lakshminarayanaiah, 1967) and natural (Paganelli and Solomon, 1957) membranes with pore radii down to 4 Å. Equation (9) results in an equivalent pore radius of 4.63 Å for cuticular membranes (Table 3); the pore radius being again pH-independent.

Paganelli and Solomon (1957) pointed out the difficulties inherent in the assumptions of bulk values for the diffusion coefficient and the viscosity of water when it is contained within the membrane. They introduced the term: equivalent pore radius, to stress the operational nature of this description; "a radius equivalent to the pore radius of an ideal membrane containing uniform, circular pores in which diffusion and bulk flow may be described by the equations of Fick and Poiseuille" (Solomon, 1968).

5. Diffusion of Polar Non-electrolytes and the Reflection Coefficient as Index of Equivalent Pore Radius

The empirical nature of Eqs. (8) and (9) and the assumptions made in their derivation necessitate an independent test of the pore-size estimates so obtained. A test procedure often employed is the use of polar probe molecules of known size which permeate through the water-filled pores of the membrane. By reference to Table 4 it is seen that the cuticular membranes discriminate between polar molecules on the basis of their size. Even though the permeability to glucose was low, the fact that glucose did penetrate indicates that there must be some pores larger than 4.44 Å in radius, the size of the glucose molecule. The rapid decrease in apparent diffusion area with increasing molecular weight can be attributed to molecular exclusion at the entrance of the pore (the molecule must pass through the opening without striking the edge) and to friction between the solute molecules and the pore walls (Renkin, 1954; Beck and Schultz, 1970). Since glucose penetrated the membranes while sucrose did not, it is concluded that the pore radius must be somewhere between 4.4 and 5.5 Å, which is in good agreement with the pore size estimated using Eqs. (8) and (9).

A second independent test criterium is the reflection coefficient. The volume flux per unit concentration gradient across cuticular membranes increased in the order urea < glucose < sucrose = raffinose. From these flux measurements reflection coefficients can be calculated. They increase with increasing solute radius and reach the limiting value of 1.0 for sucrose and raffinose (Table 5) indicating that the membranes were impermeable to these solutes. The reflection coefficients were pH-independent, hence pore size was not affected by pH. Again these data are in excellent agreement with the pore size estimated using Eqs. (8) and (9).

Table 4. Permeability of isolated *Citrus aurantium* leaf cuticular membranes to THO, ^{14}C-urea and ^{3}H-glucose at pH 9

Molecule	Molecular[a] radius	P_d	D^b	$A/\Delta x$
	Å	cm^3 s^{-1}	cm^2 s^{-1}	cm
THO	1.97	$4.35 \cdot 10^{-5}$	$2.44 \cdot 10^{-5}$	1.783
Urea	2.64	$3.76 \cdot 10^{-7}$	$1.38 \cdot 10^{-5}$	0.027
Glucose	4.44	$1.98 \cdot 10^{-8}$	$0.67 \cdot 10^{-5}$	0.003

Data from Schönherr (1976).
[a] Molecular dimensions taken from Renkin (1954) and Longsworth (1953).
[b] Diffusion coefficients in water taken from Lakshminarayanaiah (1967) for THO and Beck and Schultz (1970) for urea and glucose.

Table 5. Reflection coefficients for the transport of various solutes across isolated *Citrus aurantium* leaf cuticular membranes at different pH values

Solute	Solute[a] radius	Reflection coefficients[b]		
		pH 3.0	pH 6.0	pH 9.0
	Å			
Raffinose	6.54	1.00	1.00	1.00
Sucrose	5.55	1.00	1.00	0.98
Glucose	4.44	0.94	0.94	0.95
Urea	2.64	0.78	0.75	0.81

Data from Schönherr (1976).
[a] Molecular dimensions taken from Beck and Schultz (1970).
[b] Reflection coefficients calculated according to Eq. (4). L_p was determined using 0.25 m raffinose, J_v was referred to unit membrane area, not A_w. σ is usually defined under the condition of zero volume flow. Since this was not the case here, the values given above are not true thermodynamic values, but the deviations are small as the membranes are very impermeable.

Since both P_f and P_d increase with increasing pH while the pore radius is pH-independent one is driven to the conclusion that the increase in permeability is due to an increase in the number of pores. Neglecting tortuosity the number of pores increases from $5.1 \cdot 10^{10}$ at pH 3 to $15.8 \cdot 10^{10}$ at pH 9 (Table 6).

Table 6. The effect of pH on total pore area and number of pores in isolated *Citrus aurantium* leaf cuticular membranes

pH	$A_w/\Delta x$	A_w^a	r_P	Number of pores/cm$^{2\,b}$
	cm	cm^2	Å	
3.0	1.13	$3.39 \cdot 10^{-4}$	4.60	$5.1 \cdot 10^{10}$
6.0	1.65	$4.95 \cdot 10^{-4}$	4.60	$7.5 \cdot 10^{10}$
9.0	3.50	$1.05 \cdot 10^{-3}$	4.60	$15.8 \cdot 10^{10}$

Data from Schönherr (1976).
[a] $A_w = (A_w/\Delta x)\Delta x$; $\Delta x = 3 \cdot 10^{-4}$ cm, tortuosity neglected.
[b] Number of pores $= A_w/r_P^2 \pi$.

IV. Conclusions

On the basis of the theory and the data presented, the following conclusions are drawn with regard to structure of the cuticular membranes and the mechanism of water transport:

The cuticular membranes swell in the presence of water. Swelling increases with increasing pH because of the presence of —COOH groups fixed to the matrix. Thus, the assumptions of earlier workers (Livingston and Brown, 1912; Seybold, 1929) are confirmed experimentally and it is shown that water permeability is directly proportional to the water content of the cuticular membranes, in contrast to Härtel's (1947) assumption.

The ratio osmotic permeability coefficient (P_f)/diffusion permeability coefficient (P_d) is greater than one. This is taken as evidence for the existence of polar pores in these membranes. Under a gradient of osmotic pressure water transport will be by both diffusion and viscous flow.

The pore radius estimated from the ratio P_f/P_d is 4.5 to 4.6 Å. Diffusion of polar probe molecules and the reflection coefficients indicate that the pores are larger than the radius of glucose (4.44 Å) and smaller than that of sucrose (5.55 Å). Thus, three independent methods yield consistent results.

Even though the permeability of cuticular membranes to water (both diffusion and viscous flow) increases with increasing pH, the equivalent pore radius does not. Hence, the increase in permeability is due to an increase in the number of pores per unit membrane area. This indicates that pores lined with —COOH groups of either high or low acid strength exist in the membranes. At low pH values (pH 3 to pH 6) only pores lined with —COOH groups of relatively low pK_a exist. As the pH is increased new pores lined with weakly acidic —COOH groups come into being.

The polar pores are dynamic structures. They do not exist in dry cuticular membranes but develop only on hydration. Hence, they cannot be demonstrated by current methods of electron microscopy.

These results cannot be extrapolated to natural (non-extracted) cuticles as it is not known how the waxes which had been extracted affect swelling and membrane structure. The procedures used in this study which provide information about membrane structure from simple flux determinations should nevertheless be helpful in answering this question.

References

Beck, R. E., Schultz, J. S.: Hindered diffusion in microporous membranes with known pore geometry. Science **170**, 1302–1305 (1970).
Cass, A., Finkelstein, A.: Water permeability of thin lipid membranes. J. Gen. Physiol. **50**, 1765–1784 (1967).
Collander, R., Barlund, H.: Permeabilitätsstudien an *Chara ceratophylla*. II. Die Permeabilität für Nichtelektrolyte. Acta Botan. Fennica **11**, 1–14 (1933).
Darlington, W. A., Cirulis, N.: Permeability of apricot leaf cuticle. Plant Physiol. **38**, 462–467 (1963).
Durbin, R. P.: Osmotic flow of water across permeable cellulose membranes. J. Gen. Physiol. **44**, 315–326 (1960).
Esau, K.: Pflanzenanatomie. Stuttgart: Gustav Fischer 1969.
Franke, W.: Mechanism of foliar penetration of solutions. Ann. Rev. Plant Physiol. **18**, 281–301 (1967).
Ginzburg, B. Z., Katchalsky, A.: The frictional coefficients of the flows of non-electrolytes through artificial membranes. J. Gen. Physiol. **47**, 403–418 (1963).
Gregor, H. P., Hamilton, M. J., Becher, J., Bernstein, F.: Studies on ion exchange resins. XIV. Titration, capacity and swelling of methacrylic acid resins. J. Phys. Chem. **59**, 874–881 (1955).
Gregor, H. P., Hamilton, M. J., Oza, R. J., Bernstein, F.: Studies on ion exchange resins. XV. Selectivity coefficients of methacrylic acid resins toward alkali metal cations. J. Phys. Chem. **60**, 263–267 (1956).
Hanai, T., Haydon, D. A.: The permeability to water of bimolecular lipid membranes. J. Theoret. Biol. **11**, 370–382 (1966).
Härtel, O.: Über die pflanzliche Kutikulartranspiration und ihre Beziehung zur Membranquellbarkeit. Sitz.-Ber. Akad. Wiss. Wien, math. nat. Kl. Abt. I **156**, 57–86 (1947).
Härtel, O.: Ionenwirkung auf die Kutikulartranspiration von Blättern. Protoplasma **40**, 107–136 (1951).
Howe, P. G.: Studies on ion exchange. Doct. Dissert., Univ. London (England) (1952).
Hull, H. M.: Leaf structure as related to absorption of pesticides and other compounds. Residue Rev. **31**, 1–155 (1970).
Kamp, H.: Untersuchungen über Kutikularbau und kutikuläre Transpiration von Blättern. Jb. Wiss. Botan. **72**, 403–465 (1930).
Katchalsky, A.: Problems in the physical chemistry of polyelectrolytes. J. Polymer Sci. **22**, 159–184 (1954).
Kedem, O., Katchalsky, A.: Thermodynamic analysis of the permeability of biological membranes to non-electrolytes. Biochim. Biophys. Acta **27**, 229–246 (1958).
Kedem, O., Katchalsky, A.: A physical interpretation of the phenomenological coefficients of membrane permeability. J. Gen. Physiol. **45**, 143–179 (1961).
Koefoed-Johnson, V., Ussing, H. H.: The contribution of diffusion and flow to passage of D_2O through living membranes. Acta Physiol. Scand. **28**, 60–76 (1953).
Lakshminarayanaiah, N.: Studies with thin membranes. III. Measurement of water permeabilities of parlodion membranes formed by the dip technique. J. Appl. Polymer Sci. **11**, 1737–1754 (1967).

Livingston, B., Brown, W. H.: Relation of the daily march of transpiration to variation in water content of foliage leaves. Botan. Gaz. **53**, 309–317 (1912).
Longsworth, L. G.: Diffusion measurements, at 25°, of aqueous solutions of amino acids, peptides and sugars. J. Am. Chem. Soc. **75**, 5705–5709 (1953).
Nevis, A. H.: Water transport in invertebrate peripheral nerve fibres. J. Gen. Physiol. **41**, 927–958 (1958).
Overbeek, J. van: Absorption and translocation of plant growth regulators. Ann. Rev. Plant Physiol. **7**, 355–372 (1956).
Paganelli, C. V., Solomon, A. K.: The rate of exchange of tritiated water across the human red cell membrane. J. Gen. Physiol. **41**, 259–277 (1957).
Pappenheimer, J. R., Renkin, E. M., Borrero, L. M.: Filtration diffusion and molecular sieving through peripheral capillary membranes. Am. J. Physiol. **167**, 13–46 (1951).
Pisek, A., Berger, E.: Kutikuläre Transpiration und Trockenresistenz isolierter Blätter und Sprosse. Planta (Berl.) **28**, 124–155 (1938).
Renkin, E. M.: Filtration, diffusion and molecular sieving through porous cellulose membranes. J. Gen. Physiol. **38**, 225–243 (1954).
Robbins, E., Mauro, A.: Experimental study of the independence of diffusion and hydrodynamic permeability coefficients in collodion membranes. J. Gen. Physiol. **43**, 523–532 (1960).
Schönherr, J.: Water permeability of isolated cuticular membranes. I. The effect of pH and cations on diffusion, hydrodynamic permeability and size of polar pores in the cutin matrix. Planta (Berl.) **128**, 113–126 (1976).
Schönherr, J., Bukovac, M. J.: Preferential polar pathways in the cuticle and their relationship to ectodesmata. Planta (Berl.) **92**, 189–201 (1970).
Schönherr, J., Bukovac, M. J.: Ion exchange properties of isolated cuticular membrane: Exchange capacity, nature of fixed charges and cation selectivity. Planta (Berl.) **109**, 73–93 (1973).
Schönherr, J., Ziegler, H.: Hydrophobic cuticular ledges prevent water entering the air pores of liverwort thalli. Planta (Berl.) **124**, 51–60 (1975).
Scott, F. M.: Internal suberization of plant tissues. Botan. Gaz. **111**, 378–394 (1950).
Seybold, A.: Die pflanzliche Transpiration. Berlin: Springer 1929.
Solomon, A. K.: Characterization of biological membranes by equivalent pores. J. Gen. Physiol. **51**, 335s–364s (1968).
Stålfelt, M. G.: Die cuticuläre Transpiration. In: Handbuch der Pflanzenphysiologie Vol. III, (ed. W. Ruhland), pp. 342–350. Berlin-Göttingen-Heidelberg: Springer 1956.
Ticknor, L. B.: On the permeation of cellulose membranes by diffusion. J. Physiol. Chem. **62**, 1483–1485 (1958).
Ursprung, A.: Über das Eindringen von Wasser und anderen Flüssigkeiten in Interzellularen. Beih. Botan. Zentrbl. **41**, 15–40 (1925).

C. Physiological Basis of Stomatal Response

J. LEVITT

I. Introduction

1. Functions and Ecological Adaptations of Stomata

Stomata fulfill three major functions in the physiology of the plant. (1) They permit the entrance of CO_2 into the green leaf at a rate sufficient to support an adequate rate of photosynthesis for the normal growth and development of the plant. They have become adapted to this function by opening in response to low CO_2 concentrations inside the leaf. (2) They permit the entrance of O_2 into the leaf at a rapid enough rate to support an aerobic respiration sufficient to provide the metabolic needs of the leaf. They have become adapted to this function by opening in response to low O_2 concentrations inside the leaf. (3) If these two were the only functions of the stomata they would fulfill them by remaining constantly open. Accompanying the uptake of CO_2 and O_2, however, there must also be an increased loss of water vapor, as long as the water potential of the external atmosphere is below that of the leaf. When such a loss of water approaches the point of water stress injury, the third function of the stomata is to close, reducing the rate of water loss to that due to cuticular transpiration—a small fraction of stomatal transpiration. They have become adapted to this function by means of a guard-cell structure which leads to closure when loss of water from these cells lowers their turgor pressure to a sufficient degree. Recent evidence has revealed another adaptation which permits them to close even before the evaporative loss of water is sufficient to lower the guard cell turgor (Lange et al., 1971). In this case, closure must be due to an osmotic loss of water, again followed by loss of guard cell turgor. This adaptation is not yet understood. Today, the need for a fourth function has arisen—to prevent or decrease the entrance of pollutants into the leaf. Whether such an adaptation will be developed, remains to be seen.

The evidence for the first adapation leaves no doubt as to its existence, although the mechanism does not always function, due to reversal by the third adaptation. The evidence for the second adaptation is less conclusive. Indirect evidence is the night-opening of stomata in spite of a high CO_2 concentration inside the leaf. This has long been known to occur in many plants (Loftfield, 1921). More direct evidence is the induction of opening in the dark by submersion under oil, or water, or reducing the O_2 tension (Scarth et al., 1933). More recent evidence is the failure of stomata to close in an O_2-free atmosphere (Akita and Moss, 1973), and the opening of stomata in the dark in a N_2 atmosphere (Arntzen et al., 1973). The existence of the third adaptation has been amply and repeatedly established.

2. Biochemical Idiosyncrasies

The physiologist is concerned with the biochemical mechanisms which lead to these adaptations, and therefore to the opening and closing of stomata. These appear, at first sight, to differ basically from those occurring in the other leaf cells. In opposition to the mesophyll cells, for instance, the guard cells accumulate their maximum starch content and achieve their minimum turgor when they close at night instead of during daylight. In partial explanation of this difference, it is proposed (see below) that the guard cells differ from the mesophyll cells in another fundamental way. They assimilate C via the Calvin cycle while closing in the dark (or weak light) rather than when open in bright sunlight. They apparently achieve all three of these differences by possessing much higher activities of certain enzymes than are found in the other green cells of leaves. Specifically, they compress the two different carboxylating systems which occur in two different kinds of cells in C_4 plants, into two different locations within each guard cell. They also resemble the other green cells of CAM[1] plants in possessing an active PEP carboxylase system, but it is active primarily in daylight in the guard cells, at night in the CAM plants.

II. Biochemical Processes Leading to Movement

The basic concepts of stomatal mechanisms and the evidence for them have already been discussed (Levitt, 1974). The proposed sequence of events can be better understood from the following diagrammatic representations (Figs. 1–4).

1. Photoactive Opening

This occurs when leaves with stomata closed in the dark are illuminated. The series of reactions leading to stomatal opening are set in motion by the initial steps of photosynthesis—the photoelectron transport leading to $NADP^+$ reduction, the accompanying proton transport leading to ATP formation (Fig. 1). This occurs in all the green cells of the leaf, mesophyll as well as guard cells, and initially leads to photosynthetic C-assimilation via the Calvin cycle. At a certain point the $[CO_2]$ inside the leaf drops to a value too low to support net photosynthesis, and C-assimilation via the Calvin cycle essentially ceases.

Since the NADPH and ATP are essentially no longer being used up photosynthetically, they will accumulate to a maximum, and so will the ΔpH between

---
[1] For explanation of abbreviations see Notation, p. 167

the thylakoids and the stroma of the chloroplast [Fig. 2 (1–3)]. The guard cells differ from the mesophyll cells by their possession of two carboxylating systems, PEPC as well as RuDPC. Just where the PEPC is located has not been clearly demonstrated. According to Edwards et al. (1974) it is in the cytoplasm of the C_4 mesophyll cells, according to Rathnam and Das (1974) in the chloroplast envelope. In Fig. 2, the second of these findings is arbitrarily adopted for the guard cells, although the first would fit in equally well. Due to the ΔpH, the rise in pH of the stroma must raise the $[HCO_3^-]$ there by about two orders of magnitude [Fig. 2 (4)]. This increased $[HCO_3^-]$, together with the stroma pH of 8.0 to 8.5 which is optimum for PEPC activity (von Willert, 1974), will activate the PEPC-controlled carboxylation of PEP to $R(COO^-)_2$ [Fig. 2 (5)]. As it is used up in this reaction, the PEP is replaced by breakdown of the starch or other (CH_2O) previously accumulated in the dark [Fig. 2 (6)]. The H^+ ions from the $R(COOH)_2$ are now available for exchange with K^+ ions of the subsidiary cell, at the expense of some of the ATP formed photosynthetically, and in the presence of an H^+, K^+, ATPase [Fig. 2 (7)].

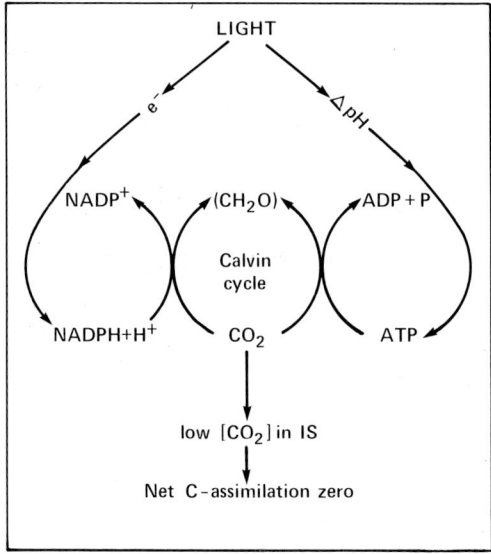

Fig. 1. Reactions leading to photoactive stomatal opening. Stomata closed in the dark, leaf then exposed to the light. Stage 1. Biochemical reactions in all the green cells

This completes the photoactive process per se, and the hydropassive process now takes over. The accumulation of $R(COO^-)_2$ and K^+ ions lowers the guard cell Ψ_s, leading to endosmosis of water, increase in turgor pressure, and stomatal opening. Since photoactive opening depends on the hydropassive process, this explains apparent discrepancies, such as an apparent lack of a relation between opening and $[CO_2]$, which is sometimes reported.

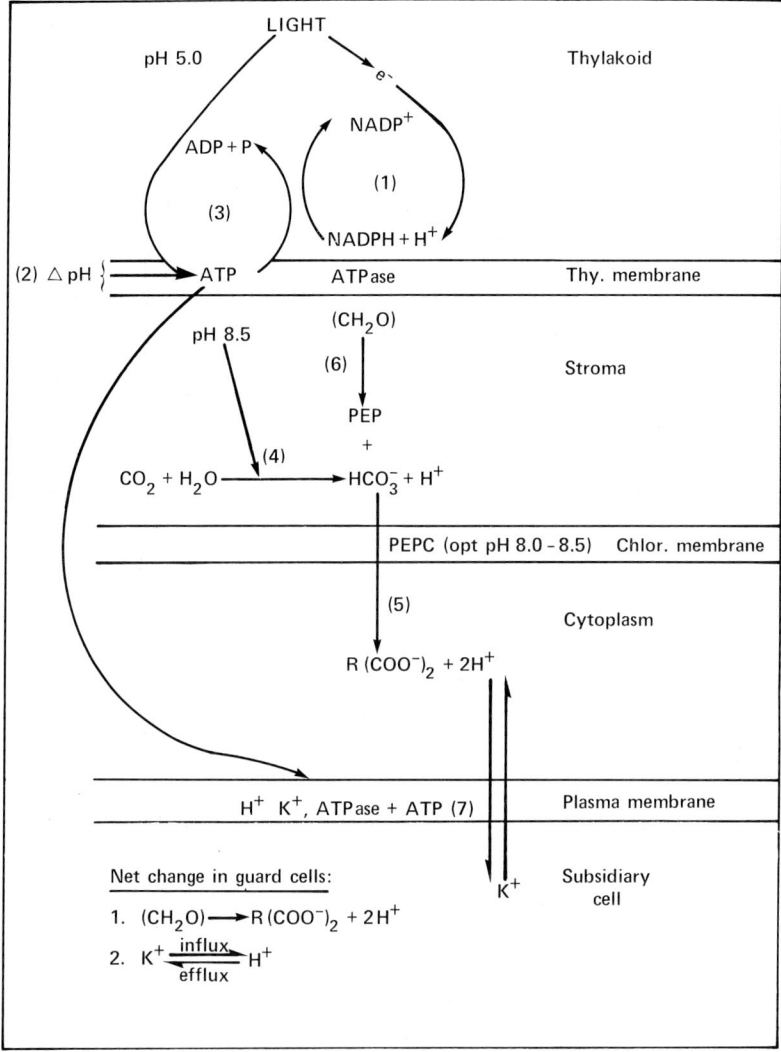

Fig. 2. Reactions leading to photoactive stomatal opening. Stage 2. Biochemical reactions in the guard cells. See text for further details

2. Scotoactive Closing

This occurs when leaves with open stomata in the light are darkened. The cessation of photosynthesis by the green leaf cells, leads to a rapid rise in $[CO_2]$ of the intercellular spaces due to a diffusion from the external atmosphere through the still open stomata and cessation of CO_2 assimilation. The attainment of maximum $[CO_2]$ may require a few minutes since Calvin cycle C-assimilation continues for a short time in the dark at the expense of the accumulated NADPH and ATP. In the guard cells, photoelectron transport will also cease instantly on darkening, and with it will disappear the ΔpH, the high [HCO_3^-] relative to [CO_2], the favorable pH for PEPC activity, and, therefore, the conditions for R(COOH)$_2$ synthesis and for K$^+$ accumulation. But stomatal closing requires more than the cessation of the reactions leading to stomatal opening. The solutes which accumulated in the light must disappear. This disappearance may be brought about as follows.

Due to the above rise in [CO_2] of the intercellular spaces, and the shift in equilibrium within the guard cells from HCO_3^- to CO_2, the [CO_2] in the guard cell chloroplasts rises to a level sufficient for Calvin cycle C-assimilation, at the expense of the NADPH and ATP accumulated in the previous light period [Fig. 3 (1)]. Unlike the mesophyll cells, which have at their disposal only the NADPH and ATP accumulated in the previous light period, and which must, therefore, cease C-assimilation as soon as these small reserves are used up, the guard cells can regenerate these substances as follows.

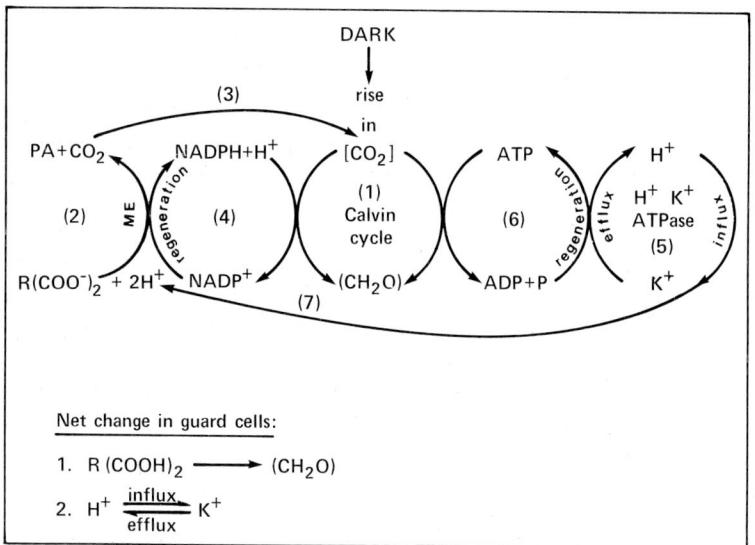

Fig. 3. Reactions leading to scotoactive stomatal closing. Stomata open in the light, leaf then exposed to the dark. See text for further details

(a) They contain a malic enzyme which decarboxylates the R(COOH)$_2$ in the presence of the NADP$^+$ liberated by the Calvin cycle, and regenerates the NADPH which can then again help drive the Calvin cycle, in combination with the CO_2 released by the malic enzyme [Fig. 3 (2–4)]. (b) They contain an ATPase which dephosphorylates ATP in the light, the energy released in this process driving the K$^+$ ion uptake in exchange for H$^+$ ions [Fig. 3 (5)]. In the dark, the reverse process will occur due to the high concentration of these light-accumulated K^+ ions and to the above loss of the $R(COO^-)_2$ counterions by decarboxylation and assimilation to (CH_2O). There is, therefore, both a diffusion potential (of K^+ ions) and an electric potential (due to the consequent excess of cations over anions) available to supply the energy required for ATP synthesis [Fig. 3 (6)]. This occurs as the K^+ ions diffuse out in exchange for incoming H^+ ions [Fig. 3 (7)], which immediately lose their charge by combining with $R(COO^-)_2$ ions. This process continues until all the $R(COOH)_2$ is decarboxylated and converted to (CH_2O), and therefore no further NADPH can be made available for Calvin cycle C-assimilation.

This completes the scotoactive process, per se, and the hydropassive process takes over. The Ψ_s of the guard cells rises as the $R(COOH)_2$ is converted to starch or other (CH_2O) and K^+ ions leave the cell in exchange for H^+ ions which become part of the $R(COOH)_2$ molecule. The rise in Ψ_s leads to exosmosis of water from the guard cells, loss of turgor, and closing of the stomata.

3. Scotoactive Opening

This occurs in many, if not all, leaves after the stomata have remained closed in the dark for some time (2–3 h or more, depending on the temperature). Little is known about the process, other than the above-mentioned results of Scarth et al. (1933). The concept to be discussed is, therefore, purely speculative, as a basis for suggesting an experimental attack.

Scotoactive closing results in the accumulation of starch or some other (CH_2O) in the chloroplasts of the guard cells. Scotoactive opening is accompanied by loss of this (CH_2O) in a series of metabolic changes which conceivably occur as follows. In the guard cells, the (CH_2O) must be slowly converted to a substrate for the respiratory breakdown which occurs in all living cells. It, therefore, must be in equilibrium with a small amount of (CH_2O)—P [Fig. 4 (1)]. During the normal glycolytic breakdown (the first stage of respiration which normally occurs in the cytoplasm) PEP is formed [Fig. 4 (2)] as an intermediate in the formation of pyruvate, which is the pivot substance for the aerobic breakdown in the mitochondria [Fig. 4 (3)]. In the accompanying conversion of respiratory energy to ATP, there is a proton-transport from the mitochondria into the cytoplasm [Fig. 4 (4)]. After the stomata are closed for some time, an O_2 deficiency will arise in the leaf cells, slowing down the aerobic process, and initiating alcoholic fermentation [Fig. 4 (5)]. The diversion of pyruvate to reaction (5) will result in a rise in cytoplasmic pH due to a decreased proton transport from the mitochondria and to a decrease in CO_2 evolution. In spite of the latter decrease to presumably 1/3, the $[HCO_3^-]$ will rise due to the rise in pH [Fig. 4 (6)] and the PEPC will be activated both by this rise in $[HCO_3^-]$ and because its pH optimum is 8.0 to 8.5. Consequently, some PEP will be diverted to form $R(COO^-)_2$ [Fig. 4 (7)]. This will be followed by K^+ uptake from the subsidiary cell [Fig. 4 (8)]. Finally, the hydropassive process again takes over leading to stomatal opening. The products of anaerobic respiration will then be respired aerobically to CO_2 and H_2O.

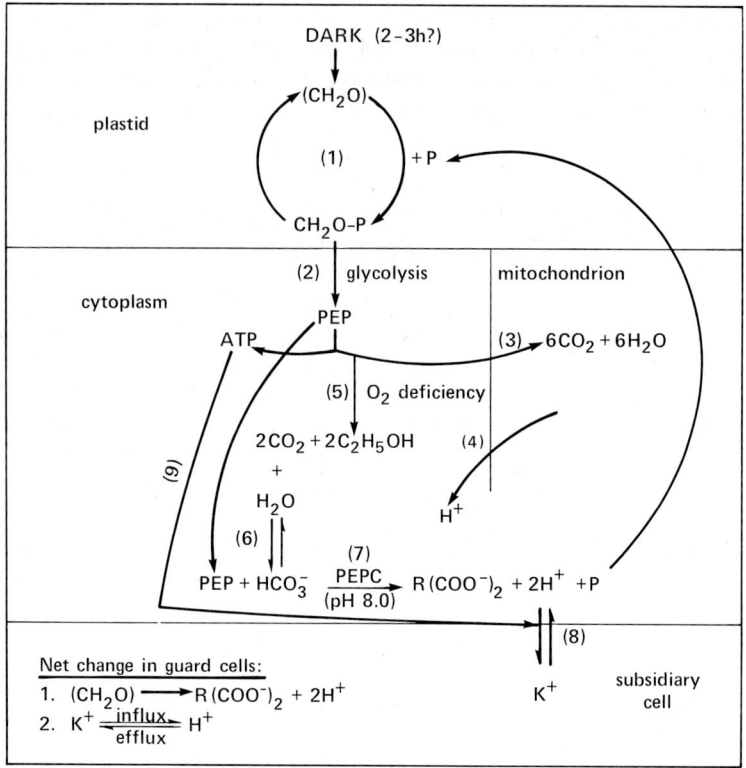

Fig. 4. Reactions leading to scotoactive opening. Stomata closed in the dark, leaf then maintained in the dark for at least 2–3 h, at a high enough temperature. See text for further details

It must be pointed out that scotoactive opening is a slow process, usually requiring hours, as opposed to photoactive opening or scotoactive closing which occur in minutes. This is in agreement with its dependence on respiration, which is a much slower process than photosynthesis—particularly the photoelectron transport of the latter.

On the basis of the above concept, scotoactive opening can only be temporary; for as soon as the stomata open, O_2 diffuses into the leaf and reverses the processes which led to scotoactive opening. Again, the closing will be slow, because it depends on the slower respiratory process. In CAM plants, on the other hand, scotoactive opening may persist for longer periods during the dark period and closing may persist for the major part of the daylight period. When the stomata operate in this way, the CAM plants must differ from other plants by maintaining parallel metabolic changes in the mesophyll cells and guard cells during both day and night. This similarity between the two kinds of cells in CAM plants is undoubtedly dependent on the presence of a highly active PEPC in the mesophyll as well as in the guard cells. However, the reversal in CAM plants of the normal pattern of stomatal movement is most likely controlled by hydropassive water movement into and out of the guard cells, rather than by photoactive and scotoactive processes per se, since the whole significance of the reversal is as an adaptation to dry climates.

III. Conclusions: Ability of the Mechanism to Explain the Known Facts

Growth hormones have repeatedly been shown to control stomatal movement, at least under certain conditions. This control can readily be explained on the basis of the above schemes. Both fusicoccin and auxin, for instance, accelerate K^+ absorption by coupling it with proton extrusion (Marré et al., 1974). They, therefore, mimic the proton transport of photosynthesis which sets in motion the biochemical changes leading to photoactive opening. Similarly, the effects of inhibitors are readily explained. The inhibition of opening by DCMU and FCCP (Pallaghy and Fischer, 1974) is to be expected, since both of these substances prevent the development of ΔpH in the light (Brinckmann and Lüttge, 1972).

Notation

Ψ_s	= osmotic potential
[ ]	= concentration
PEP	= phosphoenol pyruvate
PEPC	= phosphoenol pyruvate carboxylase
RuDP	= ribulose diphosphate
RuDPC	= ribulose diphosphate carboxylase
CAM	= Crassulacean acid metabolism
$R(COOH)_2$	= dicarboxylic acid
(CH_2O)	= carbohydrate
$(CH_2O)-P$	= carbohydrate phosphate
NADPH	= reduced nicotine adenine diphosphate
$NADP^+$	= oxidized nicotine adenine diphosphate
ATP	= adenosine triphosphate
ATPase	= adenosine triphosphate phosphohydrolase
H^+, K^+, ATPase	= H^+ ion, K^+ ion-activated ATP phosphohydrolase
DCMU	= 3-(3',4-dichlorphenyl)-1,1-dimethylurea
FCCP	= carbonyl-cyanide-p-trifluoromethoxy phenyl hydrazone
PA	= pyruvate
ME	= malic enzyme
IS	= intercellular space

References

Akita, S., Moss, D. N.: The effect of an oxygen-free atmosphere on net photosynthesis and transpiration of barley (*Hordeum vulgare* L.) and wheat (*Triticum aestivum* L.). Plant Physiol. **52**, 601–603 (1973).

Arntzen, C. J., Haugh, M. F., Bobick, S.: Induction of stomatal closure by *Helminthosporium maydis* pathotoxin. Plant Physiol. **52**, 569–574 (1973).

Brinckmann, E., Lüttge, U.: Vorübergehende pH-Änderungen im umgebenden Medium intakter grüner Zellen bei Beleuchtungswechsel. Z. Naturforsch. **27**, 277–284 (1972).

Edwards, G. E., Guttierez, M., Ku, S. B., Kanai, R.: Intracellular localization of enzymes of the C_4 pathway in mesophyll cells of C_4 plants. Plant Physiol., ann. suppl. **1974**, 28 (1974).

Lange, O. L., Lösch, R., Schulze, E.-D., Kappen, L.: Responses of stomata to changes in humidity. Planta (Berl.) **100**, 76–86 (1971).

Levitt, J.: The mechanism of stomatal movement—once more. Protoplasma **82**, 1–17 (1974).

Loftfield, J. V. G.: The behavior of stomata. Carnegie Inst. Wash. Pub. No. 314. Wash. D.C. 1921.

Marré, E., Lado, P., Rasi-Caldogno, F., Colombo, R., De Michelis, M. I.: Evidence for the coupling of proton extrusion to K^+ uptake in pea internode segments treated with fusicoccin or auxin. Plant Sci. Letters **3**, 365–379 (1974).

Pallaghy, C. K., Fischer, R. A.: Metabolic aspects of stomatal opening and ion accumulation by guard cells in *Vicia faba*. Z. Pflanzenphysiol. **71**, 332–344 (1974).

Rathnam, C. K. M., Das, V. S. R.: Role of carbonic anhydrase in C-4 photosynthesis. Plant Physiol., ann. suppl. **1974**, 28 (1974).

Scarth, G. W., Whyte, J., Brown, A.: On the cause of night opening of stomata. Trans. Roy. Soc. Can. (Ser. 3) **27**, 115–117 (1933).

Willert, D. J. von: Der Säurestoffwechsel in Abhängigkeit von osmotischem Wert und NaCl-Belastung. Abstracts. Tagung Deut. Botan. Ges., Würzburg, 103 (1974).

D. Current Perspectives of Steady-state Stomatal Responses to Environment

A. E. HALL, E.-D. SCHULZE, and O. L. LANGE

I. Introduction

Plant responses to environment are frequently influenced by stomatal functioning, through effects on plant water use, the development of plant water deficits, net photosynthesis, and temperature relations. In ecophysiology, predicting stomatal responses to environment and relating these responses to the functioning and physical state of plants are important objectives. Qualitative and quantitative differences in stomatal behavior between plant species and ecotypes, or at different phenological stages play important roles in plant performance and adaptation.

In their natural habitats plants are exposed to continually changing environmental conditions. Stomatal functioning is examined in steady-state environments to facilitate the prediction of stomatal responses to these dynamic natural environments. It is assumed in this approach that the diurnal and seasonal responses of stomata can be treated as a sequence of quasi-steady-states. This assumption is convenient in that appropriate dynamic models of stomatal response will be more difficult to develop and are not yet available. Quasi-steady-state models will not be appropriate, however, when sustained dynamic events occur, such as stomatal oscillation. The time-dependent effects of environmental stresses on stomatal responses to environment must also be considered. For example, short-term (less than 1 h) exposure to low or to high temperatures can modify the subsequent responses of stomata, for instance in citrus and sesame (see also Drake and Salisbury, 1972; Crookston et al., 1974). Consequently, steady-state stomatal responses to temperature may vary depending upon the range and sequence of temperatures used in the experiment. Exposure to low carbon dioxide concentrations ($< 50 \mu l\ l^{-1}$) may also modify subsequent stomatal responses (Whiteman and Koller, 1967a; McPherson and Slatyer, 1973) and after-effects of water stress on stomatal responses have been frequently observed (Fischer et al., 1970). Long-term (more than 1 day) effects of stresses on stomatal responses to environment are important in acclimation and can be modeled by making the parameters of steady-state models depend upon environmental prehistory and time. The influences of endogenous rhythms on stomatal aperture (Stålfelt, 1965) can be modeled in a similar manner when the interactions between these rhythms and environment are known.

Steady-state stomatal responses to environmental factors that are important to ecophysiology are examined in this chapter. Additional information may be found in the reviews of Meidner and Mansfield (1968) and Raschke (1975). Precedence is not given to the elucidation of stomatal mechanisms which are discussed by Levitt in Part 3:C of this volume. A reinterpretation of stomatal response to environment is presented that is based upon recent information on the effects of "humidity" on stomata.

II. Measurement of Stomatal Responses to Environment

Any analysis of stomatal response to environment should give consideration to experimental techniques. In ecophysiological studies quantitative descriptions of stomatal influence on gas exchange by plants are required. Diffusive leaf conductances and resistances have both been used as quantitative parameters of stomatal function. Leaf conductance is particularly useful and should be used more frequently, because transpiration, leaf water status and net photosynthesis will often be directly related to conductance, whereas they are inversely related to resistance. Possible misinterpretations of data that were caused by the use of leaf resistance rather than leaf conductance will be pointed out later. Resistances are, however, more convenient than conductances when the effects of a catenary series components are partitioned. The units currently used for conductance and resistance (cm s^{-1} and s cm^{-1}) are theoretically incorrect because the vapor pressure gradient is a more appropriate driving force than the absolute humidity gradient. In this analysis conductance is expressed as flux density per unit difference in relative partial pressure of water vapor between leaf and air (e.g., millimole m^{-2} s^{-1}). Conductances in cm s^{-1} may be converted to these units by multiplying by 424 and 383 at 15°C and 45°C respectively. Flux densities were calculated based upon one leaf surface but using the gas exchange from both sides of the leaves. When average values are determined the arithmetic mean of conductances is appropriate but with resistances the harmonic mean should be used.

The measurement of stomatal function is a good example of the classical phenomenon whereby the act of measurement disturbs the process being measured. In the past, mass flow porometers were used which disturbed the environment of stomata for extended periods and changed stomatal apertures. These changes in aperture were probably due to depletions in CO_2 or to changes in humidity. Some of the data described in this analysis were obtained with diffusion porometers of the type that utilize humidity sensors. Humidity is frequently reduced below ambient levels during measurements with this type of porometer, and it is assumed that over short measurement periods the low humidity does not influence stomatal conductance. This assumption may not always be valid. Porometer systems based upon the humidity compensation principle (Beardsell et al., 1972) may be more appropriate providing the flow of dry air is adjusted so that ambient humidity is maintained at the leaf surface during measurement. Excision of leaves may also produce artifacts, both with and without a supply of water to the cut surface, because it is now apparent that endogenous hormones have relatively short term effects on stomata and may be present in the transpiration stream (Loveys and Kriedemann, 1973).

The experiments described in the following figures were conducted with intact plants in steady-state, gas exchange systems (reviewed by Šesták et al., 1971). In some instances artifacts may occur with these methods due to the enclosure of only part of the plant shoot in the cuvette, except when the cuvette and ambient climates are similar, or to the removal of some leaves while placing the plant in the cuvette. The artifacts associated with the use of single attached leaves have not been adequately defined and they may be insignificant in many cases. However, Hall and Loomis (1972) observed that a leaf in a humid cuvette was turgid, with open stomata and high photosynthetic rates while the remainder of the plant, which was in dry air became wilted due to cold, anaerobic soil. The gas exchange systems used in these studies provided the climatic control needed to determine the separate effects on leaf conductance of individual climatic factors. Leaf conductances were calculated from measurements of transpirational flux densities, climatic and boundary layer conditions. The leaf conductances to water vapor obtained should be appropriate for ecophysiological purposes but they do describe the integrated effects of stomata, and cuticle on both sides of the leaf, and of cell-wall resistances to water vapor. These cell-wall resistances may be small (Fischer, 1968) but it has also been proposed that they are of significant size at high transpirational flux densities (Jarvis and Slatyer, 1970).

III. Steady-state Stomatal Responses to Environment

The environmental factors that are known to have a major influence on stomatal conductance are irradiance, leaf water status, ambient humidity, leaf temperature and carbon dioxide concentration. These factors have direct and interactive effects, and it has been hypothesized that two feedback systems are present which are schematically depicted in Fig. 1 (see also Raschke, 1975). The H_2O feedback system involves stomatal response to leaf water status and indirect effects of stomata on leaf water status through effects on transpiration. Ambient humidity is considered to have indirect effects on stomatal conductance through effects of transpiration on leaf water status. Whether "humidity" has direct effects on stomatal conductance will also be examined. The CO_2 feedback system involves stomatal response to internal CO_2 concentration and effects of stomata on this concentration through effects on the flux of CO_2 into or out of the leaf. Additional interactions are possible between these feedback systems through indirect effects of leaf water status on internal CO_2 concentration via effects on photosynthesis and respiration. Irradiance and temperature could influence the operation of these feedback system in many ways: through indirect effects on internal CO_2 concentration and leaf water status; and through effects on the movement of ions within the leaf. Experimental evidence for the separate and interactive effects of these environmental factors on leaf conductance are examined in the following.

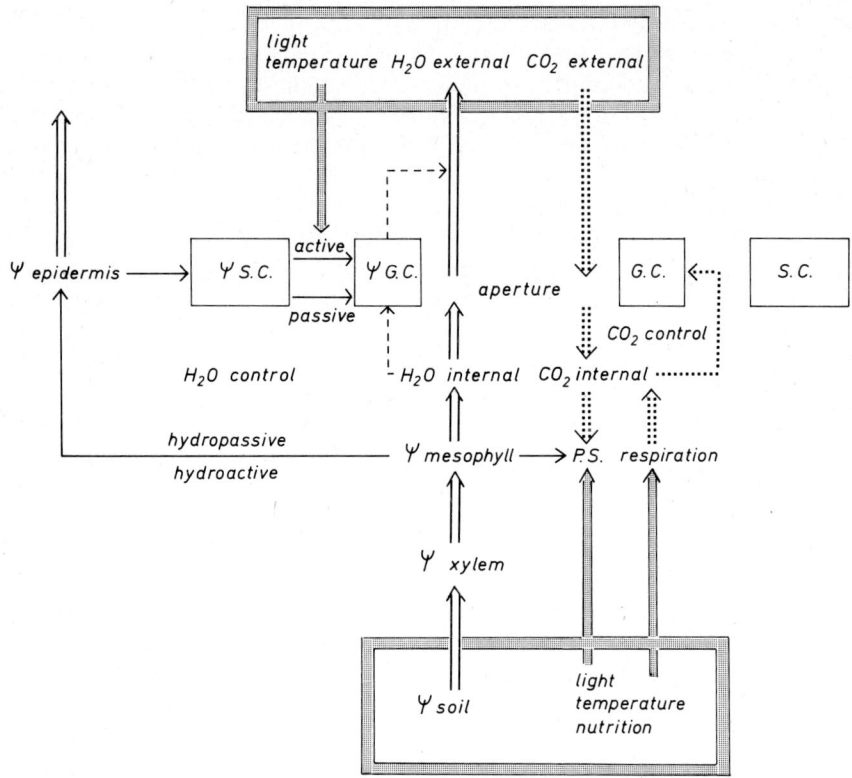

Fig. 1. Schematic description of stomatal response to environment. *White arrows:* flux of water; *dashed arrows:* flux of CO_2; *shaded arrows:* environmental influences; *thin lines with arrows:* the controlling dependencies; G.C. guard cell; S.C. subsidiary cell; P.S. photosynthesis

1. Stomatal Response to Humidity and Plant Water Status

The cause-effect relationships between stomatal response, humidity and plant water status have not been clearly established and are important to ecophysiology. If stomata respond to humidity via a negative feedback system that depends upon changes in the turgor relations of the bulk leaf (Raschke, 1975), then the physiologically active mesophyll cells must experience at least a minimal level of stress for the system to work. In contrast, if stomata respond to humidity either directly or via changes in the water relations of the epidermal cells, then this response system can prevent or reduce the level of water stress in the mesophyll tissue (Schulze et al., 1972) providing the system is compatible with the functioning of the water supply system to the leaves (Camacho-B et al., 1974).

There is abundant evidence that in some conditions stomata remain open until a threshold level of leaf water deficit is reached after which stomata close dramatically (reviewed by Hsiao, 1973). This threshold level of leaf water deficit may be associated with a bulk leaf pressure potential of zero (Kanemasu and Tanner, 1969; Kassam, 1973; Turner, 1974). This indicates that leaves may be subjected to moderate water deficits before stomata respond to changes in bulk leaf water status. In field conditions leaf water potentials may drop to very low levels without reaching the threshold value for stomatal closure (Jordan and Ritchie, 1971; Schulze et al., 1975). The threshold concept may, however, be partially an artifact due to the use of the parameter "resistance" because the relation between conductance and leaf pressure potential appear to be linear over a substantial range of values (Gardner, 1973; McCree, 1974) while resistance responses are curvilinear increasing dramatically at the "threshold" value. Threshold type responses are not always observed, in some cases leaf resistance has been more closely coupled to leaf water potential (Biscoe, 1972). Kassam (1973) postulated that interactions with irradiance may be responsible for the different observations: that high light levels may result in a threshold response, whereas low light levels result in leaf resistance being more closely coupled to bulk leaf water status. The data of Beadle et al. (1973), however, do not support this postulate. Acclimation to drying soil (McCree, 1974) and differences in rate of drying may modify the relationship between leaf resistance and leaf water status.

In numerous studies decreases in ambient humidity have resulted in increases in leaf resistance (e.g. Schulze et al., 1972; Camacho-B et al., 1974) or partial stomatal closure (Macklon and Weatherley, 1965; Lange et al., 1971). Plants growing in lower humidities have been shown to have higher leaf resistances (Slavík, 1973; Beardsell et al., 1973). Also, experiments have been described where stomata did not significantly respond to humidity (Raschke and Kühl, 1969; Hall and Kaufmann, 1975b). In the past it was frequently assumed that partial stomatal closure in dry air was due to increased water deficits in the bulk leaf, caused by increased transpirational water loss. However, several experiments have shown that dry air can cause measurable increases in leaf resistance without measurable changes in bulk leaf water potential (Macklon and Weatherley, 1965; Camacho-B et al., 1974; Watts et al., 1974). These results could be a consequence of extremely close coupling, in a negative feedback system, between leaf resistance and leaf water potential. However, this feedback system includes a dependence of leaf water potential upon the transpiration rate (Cowan, 1972) and this was not observed in these studies.

The results of Schulze et al. (1972) provide more definitive evidence for effects of humidity on leaf resistance that are not due to the classical negative feedback system. With increasing vapor pressure difference between leaf and air, the authors observed decreases in leaf conductance, increases in leaf relative water content and decreases in transpiration (Fig. 2). An over-shoot of the classical negative feedback system could result in non-steady-state reductions in transpiration as stomata closed due to decreases in ambient humidity but as occurred in the experiments of Aston (1973) and Farquhar (1973) these events would have been accompanied by increasing water deficits. The proposal that the observations of Schulze et al. (1972) were steady-state manifestations of direct humidity effects is supported by the reversibility of the responses that they observed. Decreases in water deficit with increases in evaporative demand

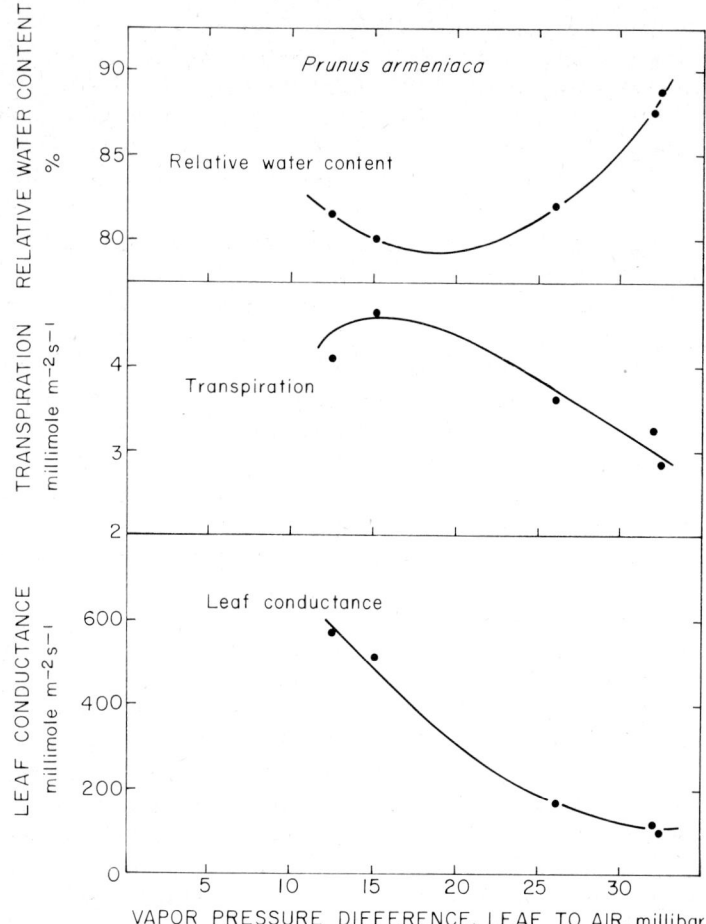

Fig. 2. Steady-state responses by irrigated *Prunus armeniaca* to changes in vapor pressure difference with a constant leaf temperature of 30°C and natural light conditions. (Data from Fig. 1 of Schulze et al., 1972)

were also observed by Tinklin and Bowling (1969). Tendencies for decreases in transpiration with increases in the vapor pressure difference between leaf and air are also apparent in the data of Jarvis and Slatyer (1970) and Hall and Kaufmann (1975a).

The increase in leaf resistance with decreasing humidity may be partially due to increases in internal cell-wall resistance as suggested by Jarvis and Slatyer (1970), but, according to their model, this could not be the case where changes in leaf resistance are not accompanied by increases in transpiration. Concurrent measurements of CO_2 exchange by Hall and Kaufmann (1975b) indicated that mesophyll conductance to CO_2 was fairly constant when leaf conductance to H_2O decreased with decreasing humidity (Fig. 6b). The simplest explanation for these data is that the changes in both leaf conductance and in net photosynthesis were due to changes in stomatal aperture. Ludlow and Wilson (1971) also observed constant mesophyll resistances to CO_2 while leaf resistances to H_2O were increasing due to changes in humidity, but Gale et al. (1966) obtained contrasting results. The observations of Hall and Kaufmann (1975b) indicate that the humidity effect on stomata is not dependent upon changes in carbon dioxide concentration inside the leaf (Fig. 6b).

The mechanism for "direct" stomatal response to humidity is not known and the use of the vapor pressure gradient between leaf and air as the driving force for the response rather than some other variable such as relative humidity should be examined. Stomata may respond to "humidity" if changes in cuticular transpiration result in epidermal water deficits. This could occur if the epidermis is partially isolated, hydraulically, from the mesophyll tissue (Macklon and Weatherley, 1965; Sheriff and Meidner, 1974). Uniform water potentials in the epidermis could cause stomatal closure if the hydraulic capacitance of the guard cells is different from that of the subsidiary cells (unpublished suggestion by Cowan and Farquhar). Alternatively, gradients in pressure potential between guard cells and subsidiary cells would influence stomatal aperture. These gradients could develop due to differential rates of cuticular transpiration between these cells (Seybold, 1961/62; Lange et al., 1971) or differential rates of water supply. In this case a degree of hydraulic isolation between guard cells and subsidiary cells would be needed. With both of these mechanisms cuticular transpiration would be responsible for stomatal response to humidity and the vapor pressure gradient (or vapor pressure difference between leaf and air) would be the appropriate driving force for the response. Another less substantiated possibility is that the epidermis responds directly to ambient relative humidity through effects of the degree of hydration on the mechanical properties of the cell walls.

Interactions between humidity and bulk leaf water status in determining leaf resistance have been considered. Schulze et al. (1974) observed that leaf resistance response to humidity with non-irrigated apricots increased during the dry summer season in the Negev desert. Other observations indicated that with irrigated apricots the response of leaf resistance to humidity did not change during this same period. This suggested that long-term increases in plant water deficits may be responsible for the increased sensitivity of leaf resistance to humidity with the unirrigated plants. Even day-to-day changes in the sensitivity of leaf resistance to humidity have been observed with citrus (Hall et al., 1975). Acclimation of stomatal response to humidity associated with drying soils would have substantial adaptive significance because it would improve water-use-efficiency (Hall and Kaufmann, 1975a) and reduce plant water deficits (Schulze et al., 1972, 1975).

Species differences in the sensitivity of leaf resistance response to humidity were observed by Camacho-B et al. (1974) and Hall and Kaufmann (1975a). The use of resistance can, however, give a misleading impression due to the large changes in conductance with small changes in resistance at low resistance values. Species differences in leaf conductance responses to humidity calculated from the same data differed mainly with respect to absolute conductance values (Fig. 3). This illustrates that the stomata of many species may be responsive

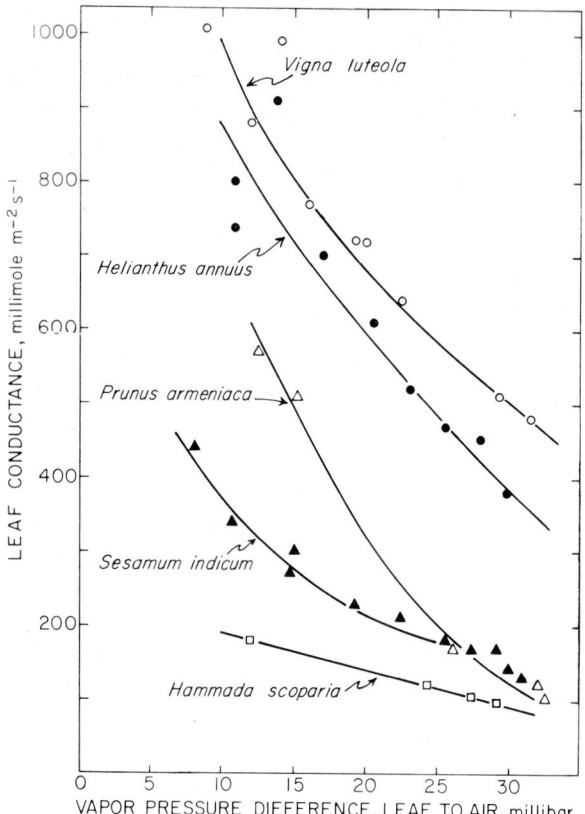

Fig. 3. Calculated leaf conductances from steady-state, gas exchange responses at a leaf temperature of 30°C and moderate to high irradiances with well-watered plants. (Data for *Vigna luteola* from Fig. 9b of Ludlow and Wilson, 1971; *Helianthus annuus* and *Sesamum indicum* from Fig. 6 of Hall and Kaufmann, 1975a; *Prunus armeniaca* from Fig. 1 of Schulze et al., 1972; *Hammada scoparia* data obtained on June 2, 1971 in the Negev desert)

to humidity but that the responses may be difficult to measure where minimal leaf resistances are very small even though the changes in leaf resistance may have substantial effects on transpiration. Differences in stomatal response to humidity may be related to species differences in ecological strategy (Hall and Kaufmann, 1975a) and differences in leaf conductance have been observed among sesame strains in hot, dry climates that may be related to agricultural productivity (Hall and Yermanos, 1975).

At lower irradiances stomatal response to humidity may be more pronounced (Davies and Kozlowski, 1974; Kaufmann, 1976). This could provide an adaptive advantage in some habitats since it may be expected that stomatal closure will have more beneficial effects on water-use-efficiency when photosynthesis is less responsive to the internal CO_2 concentration.

Stomatal response to the vapor pressure gradient may decrease with increasing leaf temperatures with some species (Fig. 4). This interaction between temperature

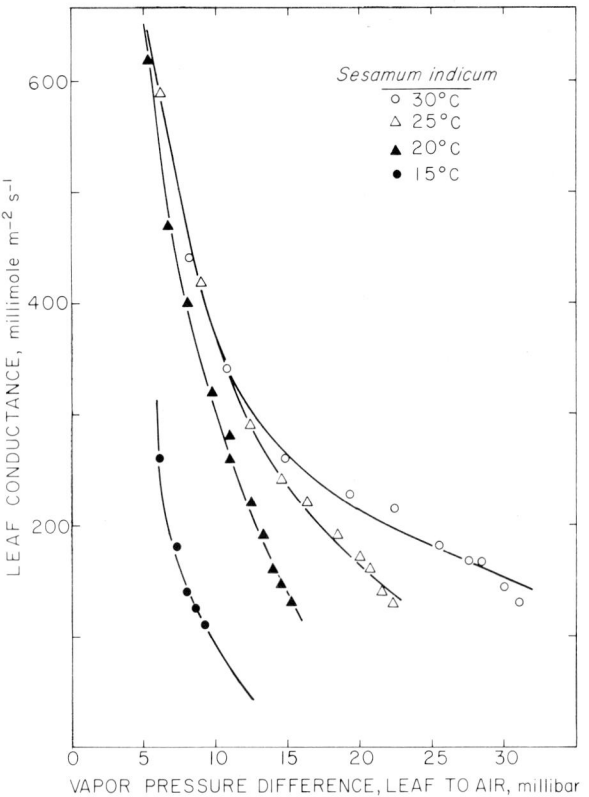

Fig. 4. Calculated leaf conductances from steady-state, gas exchange responses by *Sesamum indicum* to the vapor pressure differences at different temperatures with PHAR of $7.7 \cdot 10^4$ erg cm^{-2} s^{-1}. (Data from Fig. 2 of Hall and Kaufmann, 1975a)

and humidity could provide an adaptive advantage for broad leaved species through the enhancement of evaporative cooling at high air temperatures.

2. Stomatal Response to Temperature

Studies of stomatal response to temperature have yielded contradictory results. Stomatal opening with increasing temperature has been observed (Stålfelt, 1962; Hofstra and Hesketh, 1969; Drake et al., 1970; Drake and Salisbury, 1972; Crookston et al., 1974). Stomatal conductance has also exhibited an optimum response curve to temperature with maximal values at intermediate temperatures (Hofstra and Hesketh, 1969; Sharpe, 1973) and decreases in conductance with increasing temperature have been reported (Heath and Orchard, 1957; Heath and Meidner, 1957; Rees, 1961; Lange et al., 1969; Downes, 1970).

Interactions between temperature and humidity effects on stomata may partially account for these conflicting results. If the vapor pressure external to the leaf is kept constant, the vapor pressure gradient will increase as leaf tempera-

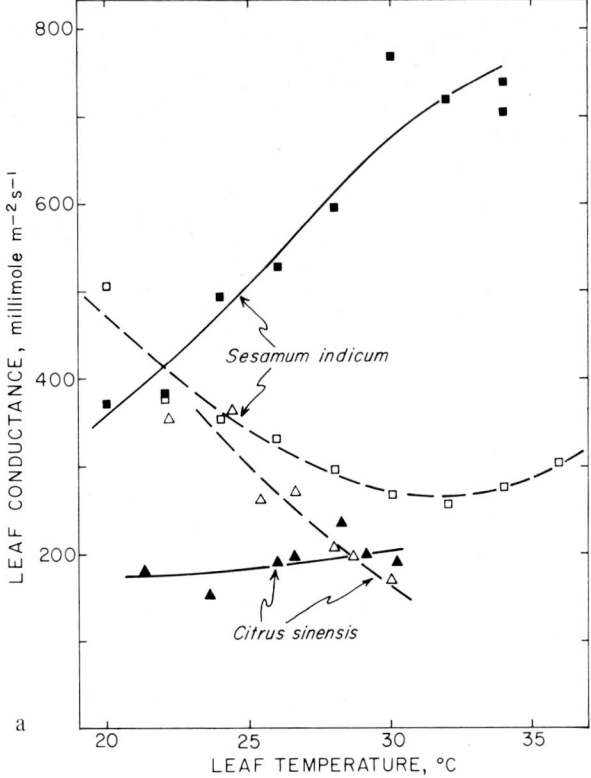

Fig. 5. (a) Effects of leaf temperature on leaf conductance with constant vapor pressure differences of 7 millibars (■) and 15 millibars (▲) and with constant external vapor pressures of 17 millibars (□) and 25 millibars (△). (Data sources as in Fig. 5b). (b) Effects of leaf temperature on leaf conductance with constant vapor pressure differences of 7 millibars with *Sesamum indicum* (data from Fig. 5 of Hall and Kaufmann, 1975b); 5 millibars (△) and 15 millibars (▲) with *Citrus sinensis* (data from Fig. 4 of Hall et al., 1975); and 15 millibars (○) and 40 millibars (●) with *Prunus armeniaca* (data from Fig. 2 of Schulze et al., 1974)

ture is raised and leaf conductance may decrease, whereas with constant vapor pressure gradients conductance may increase with increasing temperature (Fig. 5a). Stomatal response to temperature, with constant vapor pressure gradients, will also depend upon the magnitude of the vapor pressure gradient (Fig. 5b). Species contrasts in stomatal response to temperature are apparent in this figure and have been described by Hofstra and Hesketh (1969) and Schulze et al. (1973). At high temperatures increases in leaf conductance may confer an adaptive advantage due to enhanced evaporative cooling (Schulze et al., 1973). However, at extremely high temperatures leaf conductance may decrease due directly to changes in bulk leaf water status resulting from high transpiration rates, or to changes in endogenous abscisic acid levels resulting from plant water stress (Hiron and Wright, 1973). Leaf conductance may also decrease with increasing temperature when initial plant water deficits are large, due to dry soil (Schulze

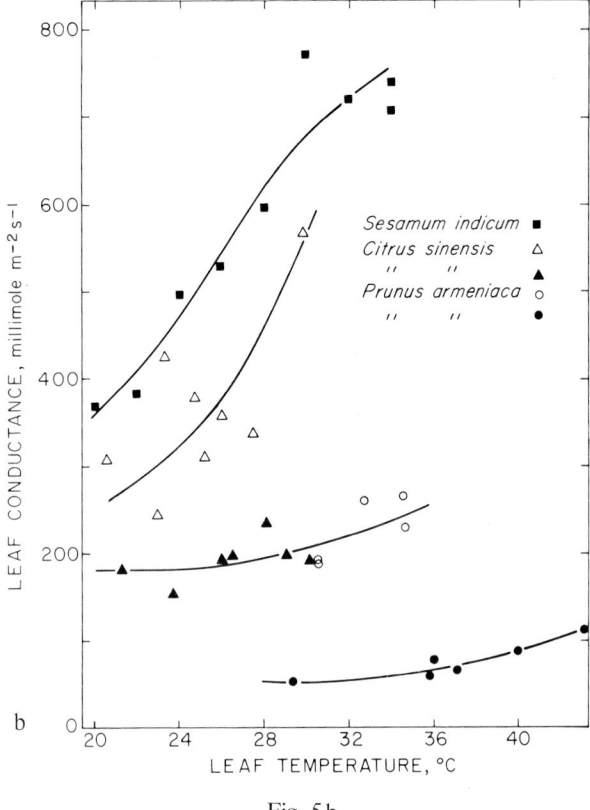

Fig. 5b

et al., 1973). Under water stress, which also induces an increase in the heat resistance of the leaves (Hammouda and Lange, 1962), the restriction of water loss is of greater significance for plant survival than transpirational cooling.

3. Stomatal Response to Carbon Dioxide Concentration

The response of stomata to the carbon dioxide concentration inside the leaf, C_i, has frequently been considered to be important in the control of stomata (Ketellapper, 1963; Meidner and Mansfield, 1968; Raschke, 1975; see also Levitt, this volume Part 3:C). While there is abundant evidence for stomatal closure with increases in C_i it is not clear whether natural variations in C_i have a substantial influence on stomatal behavior (Zelitch, 1969). Many studies have been conducted with levels of CO_2 far above ambient values. A number of reports indicate only small effects of ambient carbon dioxide concentration, C_a, over a natural range of values on leaf resistance with some species in some conditions (Heath and Milthorpe, 1950; Gaastra, 1959; Whiteman and Koller, 1967b; Jones and Mansfield, 1970; Boyer, 1971; Akita and Moss, 1972; Hall and Kaufmann, 1975b). Increases in leaf resistance with decreases in C_a below ambient levels have also been reported (Parkinson, 1968; Kriedemann,

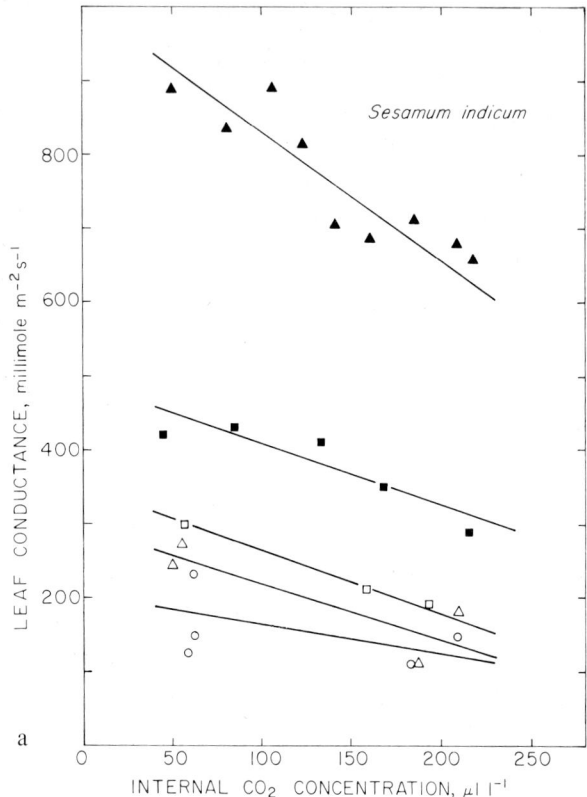

Fig. 6. (a) Leaf conductance response to the calculated internal carbon dioxide concentration with *Sesamum indicum* at constant vapor pressure differences of 5.5 ± 0.5 millibars (▲, ■) and 23.5 ± 2 millibars (□, △, ○), at fluxes of photosynthetically active quanta of 105 ± 25 nano einstein cm^{-2} s^{-1} and leaf temperatures of $26\pm1°C$ (data from Figs. 2 and 3 of Hall and Kaufmann, 1975b and unpublished data of Hall). (b) Effects of the vapor pressure difference on leaf conductance to water vapor and mesophyll conductance to carbon dioxide with *Sesamum indicum* at constant internal carbon dioxide concentrations of $206\pm20\,\mu l\,l^{-1}$ (●) and $56\pm5\,\mu l\,l^{-1}$ (▲) (data from Fig. 3 of Hall and Kaufmann, 1975b)

1971; Ludlow and Jarvis, 1971). Some of these conflicting results may be due to species differences in stomatal response to CO_2 (Pallas, 1965), and stomata of some C_4 species may be more responsive to CO_2 than those of some C_3 species (Ludlow and Wilson, 1971; Akita and Moss, 1972). Stomatal closure with increasing CO_2 over the natural range may also be more pronounced at low light intensities (Gaastra, 1959; Downes, 1971; Akita and Moss, 1972; McPherson and Slatyer, 1973). Hall and Kaufmann (1975b) reported that leaf resistance responded more to Ci in dry air than in humid air and pointed out that this would explain the observations of Heath and Milthorpe (1950). However, in relating these data to stomatal function, conductance responses should be examined. The effect of Ci on leaf conductance was not less in humid

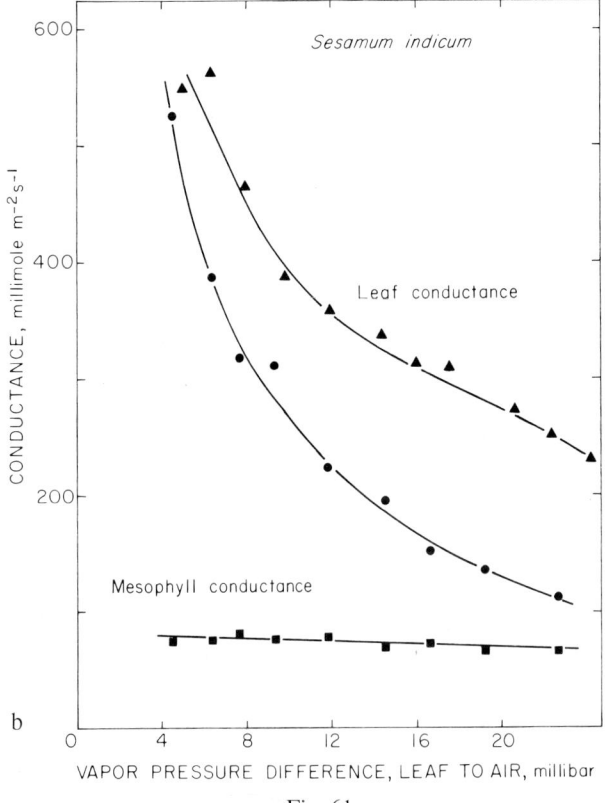

Fig. 6 b

air than in dry air (Fig. 6a). The effects on leaf conductance of changes in Ci over a natural range were, however, much smaller than the conductance response to humidity obtained with the same plants and constant Ci (Fig. 6b).

It has been proposed that midday stomatal closure may be due to high temperature-induced increases in Ci (Meidner and Mansfield, 1968). It has been shown, however, that the increasing vapor pressure gradients that frequently accompany increasing leaf temperatures may cause decreases in leaf conductance (Fig. 5a). Hall and Kaufmann (1975b) also demonstrated in the same experiment that the humidity-induced partial stomatal closure was associated with decreases in Ci. Schulze et al. (1975) observed that under natural variable conditions leaf conductance was not consistently related to Ci (Fig. 7) on a day when stomata were partially closed during the middle of the day (Fig. 8). During the morning stomatal closure was associated with decreasing Ci, whereas, in the late afternoon leaf conductance increased with an almost constant Ci. Small decreases in leaf conductance during the middle of the day were, however, associated with increasing Ci. It is apparent that the proposed role of high Ci in midday closure of stomata should be further investigated. The suggestion that stress-induced, endogenous ABA may sensitize stomata to CO_2 (Raschke, 1975) could help to resolve the controversy concerning the role of internal CO_2 concentration in stomatal responses to environment.

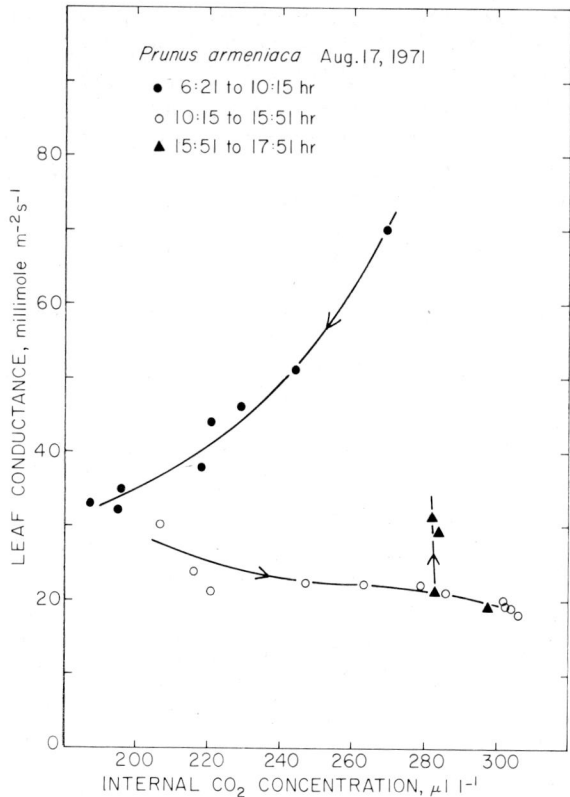

Fig. 7. Relationships between leaf conductance and calculated internal carbon dioxide concentrations with *Prunus armeniaca* during a day (August 17, 1971) whose variation in climate is described in Fig. 8. (Data from Fig. 9 of Schulze et al., 1975)

4. Stomatal Response to Radiation

Stomatal responses to radiation have provided more consistent data than responses to other environmental factors in that stomata usually partially close when the irradiance is decreased and partially open when it is increased. However, hysteresis in radiation-induced stomatal movements and non-steady-state responses to step-changes in radiation do occur at low irradiances. Stomatal responses to radiation are frequently conducted with decreasing irradiances in laboratory studies, consequently these data may not be completely applicable to radiation-induced stomatal opening in nature.

It is frequently assumed that stomata do not respond to decreases in radiation until a low threshold value has been attained. This threshold response is partially due to the use of leaf resistance rather than leaf conductance, and due to the insensitivity of some measurement techniques at high conductance values. Small changes in leaf resistance at low leaf resistance values are hard to resolve yet they can have a substantial influence on plant water loss. Partial stomatal closure with decreases in irradiance may be more pronounced with some species than with others (Downes, 1970), with older leaves than with younger leaves (Kriedemann, 1971; Gee and Federer, 1972) and with upper leaf surfaces than with lower surfaces (Turner, 1970; Kassam, 1973). Stomatal responses to irradiance are also probably influenced by humidity, temperature, leaf water status, and carbon dioxide concentration but these interactions have not been adequately defined. The ecological significance of differences in stomatal response to radiation is discussed by Ludlow, this volume Part 5:D.

IV. Stomatal Responses to Diurnal Changes in Environment

Information concerning steady-state stomatal responses to environment will now be applied to the prediction of stomatal responses to diurnal changes in environment. Data obtained by Schulze et al. (1974, 1975) with non-irrigated apricots during a hot, dry day and a cooler, more humid day in the Negev desert during the summer (Fig. 8) are analyzed. Effects on leaf resistance of leaf temperature and vapor pressure difference had been determined with apricots in the field using different steady-state conditions. A mathematical model of leaf resistance response to temperature and the vapor pressure difference was developed using response curves similar to those for apricot in Figs. 3 and 5b. It is apparent that the predicted values are in good agreement with the observed values except for the afternoon on the hot dry day when the predicted resistances are lower than the observed values (Fig. 8). The tendency for leaf resistances to remain higher than predicted during the afternoon of hot dry days may be due to non-steady-state events such as after-effects of water stress (Fischer et al., 1970) or high temperatures (Bauer, 1972). Changes in irradiance did not appear to be influencing leaf resistance during the day, supporting the concept of a low threshold value for stomatal response to irradiance. Effects of irradiance on leaf resistance were only observed at sunrise and sunset. Changes in leaf resistance were not correlated with changes in xylem pressure potential (Fig. 8) or with changes in Ci (Figs. 7 and 8).

In this instance a simple, steady-state model of stomatal response to environment using only leaf temperature and ambient humidity gave reasonable predictions. In denser canopies and with different species and environments it will often be necessary to include irradiance in models of stomatal response to environment. Also field studies have demonstrated correlations between leaf conductance and leaf water status but it is not clear at this time whether they are due to negative feedback effects or to the modulation of stomatal response to humidity and temperature by long-term changes in plant water status. For example, Running et al. (1975) have reported that daily mean conductance is related to predawn plant moisture status. More complete models of stomatal responses to environment are needed. Many models have been developed but they are based upon functions for stomatal response to CO_2 and plant water status that should be more completely tested.

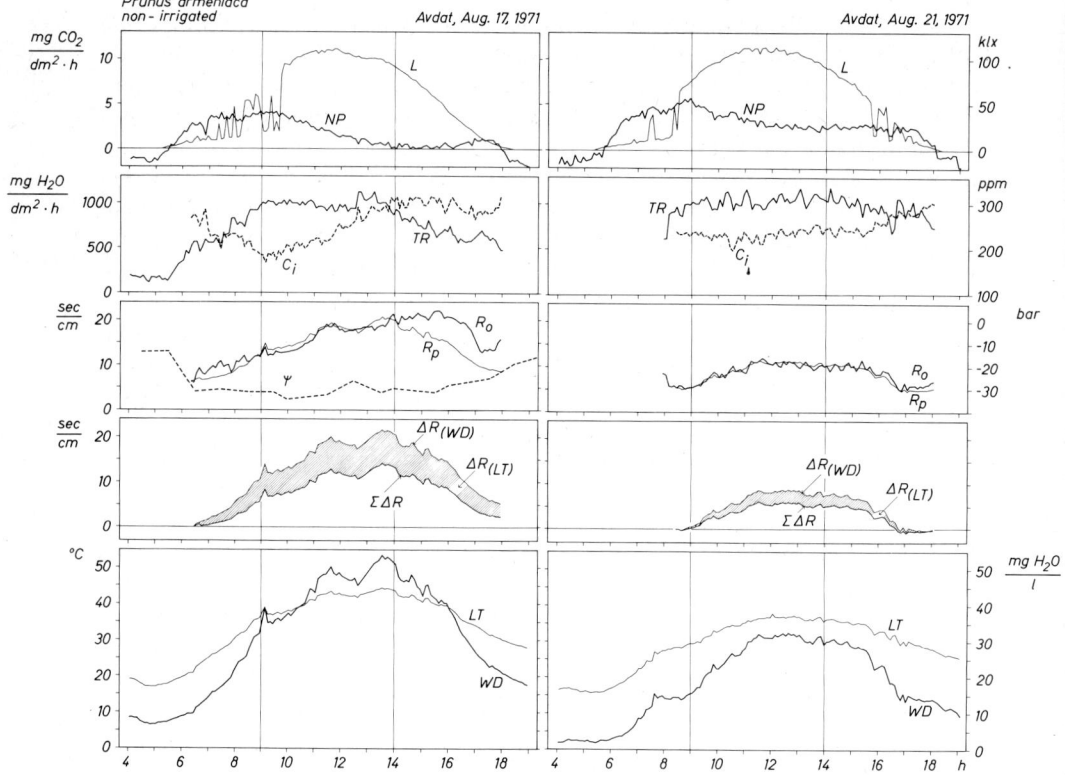

Fig. 8. The daily course of net photosynthesis (NP), light intensity (L), transpiration (TR), water potential (pressure chamber values, Ψ), internal carbon dioxide concentration (C_i), observed (R_o) and predicted (R_p) values of total diffusion resistance, the change in diffusion resistance caused by changes in the water vapor concentration difference between leaf and air ($\Delta R_{(WD)}$), the change in diffusion resistance caused by changes in leaf temperature ($\Delta R_{(LT)}$), and total change in diffusion resistance ($\sum \Delta R$) caused by changes in leaf temperature (LT) and water vapor concentration difference between the evaporating sites in the leaf and the surrounding air (WD) for *Prunus armeniaca*. (Data from Schulze et al., 1974, 1975)

V. Conclusions and Future Research Directions

Optimal stomatal functioning will reflect a compromise between the conflicting requirements for controlling water use and the development of plant water deficits, while maintaining adequate levels of photosynthesis and evaporative cooling. Optimal stomatal functioning will depend upon environmental conditions, physiological, morphological, anatomical and phenological properties, and the ecological strategies of plants. During the evolution of plant ecotypes, complex systems governing stomatal response to both the external and the internal leaf environment have probably developed, that can achieve optimal functioning with different plant and environmental conditions. Integration of the many facets of stomatal response to environment in relation to hypothetical optimal functioning would appear to be a useful direction for future research. Comparative studies with contrasting ecotypes and cultivars that focus on qualitative and quantitative differences in stomatal response to environment in relation to other attributes of the plant and its environment could improve our understanding of plant adaptation and environmental effects on primary productivity.

Ecophysiological aspects of stomatal response to environment can be elucidated by a combination of field and controlled environment studies, and mathematical modeling. Greater care must be taken with respect to plant preconditioning, environmental control, plant disturbance during measurement, and the accurate measurement of parameters such as leaf conductance, especially at large levels of conductance. The interactive effects of different environmental factors on stomatal behavior should be examined since they are important and since they are probably responsible for many of the apparently contradictory results in the literature.

The influence of hormonal relations on stomatal behavior will probably continue to be an active research area. It appears likely that hormones are involved in acclimation and in phenological aspects of stomatal response to environment. The possibility that changes in endogenous hormone levels may modulate stomatal response to environment appears particularly attractive as a control mechanism that has the potential for achieving the complex responses of optimal stomatal functioning.

References

Akita, S., Moss, D.: Differential stomatal response between C_3 and C_4 species to atmospheric CO_2 concentration and light. Crop Sci. **12**, 789–793 (1972).

Aston, M. J.: Changes in internal water status and the gas exchange of leaves in response to ambient evaporative demand. In: Plant response to climatic factors (ed. R. O. Slatyer), pp. 243–247. Paris: Unesco 1973.

Bauer, H.: CO_2-Gaswechsel nach Hitzestress bei *Abies alba* Mill. und *Acer pseudoplatanus* L. Photosynthetica **6**, 424–434 (1972).

Beadle, C. L., Stevenson, K. R., Neumann, H. H., Thurtell, G. W., King, K. M.: Diffusive resistance, transpiration, and photosynthesis in single leaves of corn and sorghum in relation to leaf water potential. Can. J. Plant Sci. **53**, 537–544 (1973).

Beardsell, M. F., Jarvis, P. G., Davidson, B.: A null-balance diffusion porometer suitable for use with leaves of many shapes. J. Appl. Ecol. **9**, 677–690 (1972).

Beardsell, M. F., Mitchell, K. J., Thomas, R. G.: Transpiration and photosynthesis in soybean. Effects of temperature and vapor pressure deficit. J. Exp. Botany **24**, 587–595 (1973).

Biscoe, P. V.: The diffusion resistance and water status of leaves of *Beta vulgaris*. J. Exp. Botany **23**, 930–940 (1972).

Boyer, J. S.: Nonstomatal inhibition of photosynthesis in sunflower at low leaf water potentials and high light intensities. Plant Physiol. **48**, 532–536 (1971).

Camacho-B, S. E., Hall, A. E., Kaufmann, M. R.: Efficiency and regulation of water transport in some woody and herbaceous species. Plant Physiol. **54**, 169–172 (1974).

Cowan, I.: Oscillations in stomatal conductance and plant functioning associated with stomatal conductance: observations and a model. Planta (Berl.) **106**, 185–219 (1972).

Crookston, R. K., O'Toole, J., Lee, R., Ozbun, J. L., Wallace, D. H.: Photosynthesis depression in beans after exposure to cold for one night. Crop Sci. **14**, 457–464 (1974).

Davies, W. J., Kozlowski, T. T.: Stomatal responses of five woody angiosperms to light intensity and humidity. Can. J. Botany **52**, 1525–1534 (1974).

Downes, R. W.: Effect of light intensity and leaf temperature on photosynthesis and transpiration in wheat and sorghum. Australian J. Biol. Sci. **23**, 775–782 (1970).

Downes, R. W.: Adaptation of sorghum plants to light intensity: its effects on gas exchange in response to changes in light, temperature and CO_2. In: Photosynthesis and photorespiration (eds. M. O. Hatch, C. B. Osmond, R. O. Slatyer), pp. 57–62. New York, London, Sydney, Toronto: Wiley Interscience 1971.

Drake, B. G., Raschke, K., Salisbury, F. B.: Temperatures and transpiration resistances of *Xanthium* leaves as affected by air temperature, humidity and wind speed. Plant Physiol. **46**, 324–330 (1970).

Drake, B. G., Salisbury, F. B.: Aftereffects of low and high temperature pretreatments on leaf resistance, transpiration and leaf temperature in *Xanthium*. Plant Physiol. **50**, 572–575 (1972).

Farquhar, G. D.: A study of the responses of stomata to pertubations of environment. Ph. D. Thesis. Canberra City, Australia 1973.

Fischer, R. A.: Resistance to water loss in the mesophyll of leek *(Allium porrum)*. J. Exp. Botany **19**, 135–145 (1968).

Fischer, R. A., Hsiao, T. C., Hagan, R. M.: After-effects of water stress on stomatal opening potential. I. Techniques and magnitudes. J. Exp. Botany **21**, 371–385 (1970).

Gaastra, P.: Photosynthesis of crop plants as influenced by light, carbon dioxide, temperature and stomatal diffusion resistance. Med. Landbouwhogesch. Wageningen **59**, 1–68 (1959).

Gale, J., Kohl, H. C., Hagan, R. M.: Mesophyll and stomatal resistances affecting photosynthesis under varying conditions of soil, water and evaporative demand. Israel J. Botany **15**, 64–71 (1966).

Gardner, W. R.: Internal water status and plant response in relation to the external water regime. In: Plant response to climatic factors (ed. R. O. Slatyer), pp. 221–225. Paris: Unesco 1973.

Gee, G. W., Federer, C. A.: Stomatal resistance during senescence of hard wood leaves. Water Resources Research **8**, 1456–1460 (1972).

Hall, A. E., Camacho-B, S. E., Kaufmann, M. R.: Regulation of water loss by citrus leaves. Physiol. Plantarum **33**, 62–65 (1975).

Hall, A. E., Kaufmann, M. R.: Regulation of water transport in the soil-plant-atmosphere continuum. In: Perspectives of biophysical ecology. Ecological Studies, vol. 12 (eds. D. M. Gates, R. B. Schmerl), pp. 187–202. Berlin-Heidelberg-New York: Springer 1975a.

Hall, A. E., Kaufmann, M. R.: Stomatal response to environment with *Sesamum indicum* L. Plant Physiol. **55**, 455–459 (1975b).

Hall, A. E., Loomis, R. S.: An explanation for the difference in photosynthetic capabilities of healthy and beet yellows virus-infected sugar beets (*Beta vulgaris* L.). Plant Physiol. **50**, 576–580 (1972).

Hall, A. E., Yermanos, D. M.: Leaf conductance and leaf water status of sesame strains in hot, dry climates. Crop Sci. **15**, 789–793 (1975).

Hammouda, M., Lange, O. L.: Zur Hitzeresistenz der Blätter höherer Pflanzen in Abhängigkeit von ihrem Wassergehalt. Naturwissenschaften **21**, 500–501 (1962).

Heath, O. V. S., Meidner, H.: Effects of carbon dioxide and temperature on stomata of *Allium cepa* L. Nature **180**, 181–182 (1957).

Heath, O. V. S., Milthorpe, F. L.: The role of carbon dioxide in the light response of stomata. J. Exp. Botany **1**, 227–254 (1950).

Heath, O. V. S., Orchard, B.: Temperature effects on the minimum intercellular space carbon dioxide concentration "Γ". Nature **180**, 180–181 (1957).

Hiron, R. W. P., Wright, S. T. C.: The role of endogenous abscisic acid in the response of plants to stress. J. Exp. Botany **24**, 769–781 (1973).

Hofstra, G., Hesketh, J. D.: The effect of temperature on stomatal aperture in different species. Can. J. Botany **47**, 1307–1310 (1969).

Hsiao, T. C.: Plant response to water stress. Ann. Rev. Plant Physiol. **24**, 519–570 (1973).

Jarvis, P. G., Slatyer, R. O.: The role of the mesophyll cell wall in leaf transpiration. Planta (Berl.) **90**, 303–322 (1970).

Jones, R. J., Mansfield, T. A.: Increases in the diffusion resistances of leaves in carbon dioxide-enriched atmosphere. J. Exp. Botany **21**, 951–958 (1970).

Jordan, W. R., Ritchie, J. T.: Influence of soil water stress on evaporation, root absorption, and internal water status of cotton. Plant Physiol. **48**, 783–788 (1971).

Kanemasu, E. T., Tanner, C. B.: Stomatal diffusion resistance of snap beans. I. Influence of leaf water potential. Plant Physiol. **44**, 1547–1552 (1969).

Kassam, A. H.: The influence of light and water deficit upon diffusive resistance of leaves of *Vicia faba* L. New Phytologist **72**, 557–570 (1973).

Kaufmann, M. R.: Stomatal response of Engelmann spruce to humidity, light, and water stress. Plant Physiol. **57**, 898–901 (1976).

Ketellapper, H. J.: Stomatal physiology. Ann. Rev. Plant. Physiol. **14**, 249–270 (1963).

Kriedemann, P. E.: Photosynthesis and transpiration as a function of gaseous diffusive resistances in orange leaves. Physiol. Plantarum **24**, 218–225 (1971).

Lange, O. L., Koch, W., Schulze, E.-D.: CO_2-Gaswechsel und Wasserhaushalt von Pflanzen in der Negev-Wüste am Ende der Trockenzeit. Ber. Deut. Botan. Ges. **82**, 39–61 (1969).

Lange, O. L., Lösch, R., Schulze, E.-D., Kappen, L.: Responses of stomata to changes in humidity. Planta (Berl.) **100**, 76–86 (1971).

Loveys, B. R., Kriedemann, P. E.: Rapid changes in abscisic acid-like inhibitors following alterations in vine leaf water potential. Physiol. Plantarum **28**, 476–479 (1973).

Ludlow, M. M., Jarvis, P. G.: Photosynthesis in Sitka spruce [*Picea sitchensis* (Bong.) Carr.] I. General characteristics. J. Appl. Ecol. **8**, 925–953 (1971).

Ludlow, M. M., Wilson, G. L.: Photosynthesis of tropical pasture plants. I. Illuminance, carbon dioxide concentration, leaf temperature and leaf-air vapour pressure difference. Australian J. Biol. Sci. **24**, 449–470 (1971).

Macklon, A. E. S., Weatherley, P. E.: Controlled environment studies of the nature and origins of water deficits in plants. New Phytologist **64**, 414–427 (1965).

McCree, K. J.: Changes in the stomatal response characteristics of grain sorghum produced by water stress during growth. Crop Sci. **14**, 273–278 (1974).

McPherson, H. G., Slatyer, R. O.: Mechanisms regulating photosynthesis in *Pennisetum typhoides*. Australian J. Biol. Sci. **26**, 329–339 (1973).

Meidner, H., Mansfield, T. A.: Physiology of stomata. London: McGraw-Hill 1968.

Pallas, J. E.: Transpiration and stomatal opening with changes in carbon dioxide content of the air. Science **147**, 171–173 (1965).

Parkinson, K. J.: Apparatus for the simultaneous measurement of water vapor and carbon dioxide exchanges of single leaves. J. Exp. Botany **19**, 840–856 (1968).

Raschke, K.: Stomatal action. Ann. Rev. Plant Physiol. **26**, 309–340 (1975).

Raschke, K., Kühl, U.: Stomatal responses to changes in atmospheric humidity and water supply; experiments with leaf sections of *Zea mays* in CO_2-free air. Planta (Berl.) **87**, 36–48 (1969).

Rees, A. R.: Midday closure of stomata in the oil palm *Elaeis guineensis* Jacq. J. Exp. Botany **12**, 129–146 (1961).

Running, S. W., Waring, R. H., Rydell, R. A.: Physiological control of water flux in conifers. A computer simulation model. Oecologia (Berl.) **18**, 1–16 (1975).

Schulze, E.-D., Lange, O. L., Buschbom, U., Kappen, L., Evenari, M.: Stomatal responses to changes in humidity in plants growing in the desert. Planta (Berl.) **108**, 259–270 (1972).

Schulze, E.-D., Lange, O. L., Evenari, M., Kappen, L., Buschbom, U.: The role of air humidity and leaf temperature in controlling stomatal resistance of *Prunus armeniaca* L. under desert conditions. I. A simulation of the daily course of stomatal resistance. Oecologia (Berl.) **17**, 159–170 (1974).

Schulze, E.-D., Lange, O. L., Kappen, L., Buschbom, U., Evenari, M.: Stomatal responses to changes in temperature at increasing water stress. Planta (Berl.) **110**, 29–42 (1973).

Schulze, E.-D., Lange, O. L., Kappen, L., Evenari, M., Buschbom, U.: The role of air humidity and leaf temperature in regulating stomatal resistance of *Prunus armeniaca* L. under desert conditions. II. The significance of leaf water status and internal carbon dioxide concentration. Oecologia (Berl.) **18**, 219–233 (1975).

Šesták, Z., Čatský, J., Jarvis, P. J. (eds.): Plant photosynthetic production—Manual of methods. The Hague: W. Junk 1971.

Seybold, A.: Ergebnisse und Probleme pflanzlicher Transpirationsanalyse. Jh. Heidelberger Akad. Wiss. **1961/62**, 5–8 (1961/62).

Sharpe, P. J. H.: Adaxial and abaxial stomatal resistance of cotton in the field. Agron. J. **65**, 570–574 (1973).
Sheriff, D. W., Meidner, H.: Water pathways in leaves of *Hedera helix* L. and *Tradescantia virginiana* L. J. Exp. Botany **25**, 1147–1156 (1974).
Slavík, B.: Transpiration resistance in leaves of maize grown in humid and dry air. In: Plant response to climatic factors (ed. R. O. Slatyer), pp. 267–269. Paris: Unesco 1973.
Stålfelt, M. G.: The effect of temperature on opening of the stomatal cells. Physiol. Plantarum **15**, 772–779 (1962).
Stålfelt, M. G.: The relation between the endogenous and induced elements of the stomatal movements. Physiol. Plantarum **18**, 177–184 (1965).
Tinklin, R., Bowling, D. J. F.: The water relations of bracken: a preliminary study. J. Ecol. **57**, 669–671 (1969).
Turner, N. C.: Response of adaxial and abaxial stomata to light. New Phytologist **69**, 647–653 (1970).
Turner, N. C.: Stomatal behavior and water status of maize, sorghum, and tobacco under field conditions. II. At low soil water potential. Plant Physiol. **53**, 360–365 (1974).
Watts, W. R., Jarvis, P. G., Neilson, R. E., Beadle, C. H.: Responses of stomata of Sitka spruce to environmental variables. Abstracts, Meeting Intern. Assoc. Plant Physiol., Würzburg, 85 (1974).
Whiteman, P. C., Koller, D.: Species characteristics in whole plant resistances to water vapor and CO_2 diffusion. J. Appl. Ecol. **4**, 363–377 (1967a).
Whiteman, P. C., Koller, D.: Interactions of carbon dioxide concentration, light intensity and temperature on plant resistances to water vapor and carbon dioxide diffusion. New Phytologist **66**, 463–473 (1967b).
Zelitch, I.: Stomatal control. Ann. Rev. Plant Physiol. **20**, 329–350 (1969).

E. Water Uptake, Storage and Transpiration by Conifers: A Physiological Model

R. H. WARING and S. W. RUNNING

I. Introduction

As part of an ecosystem study, addressed are the questions of water utilization by coniferous forests and the influence water has on growth and mineral cycling. Individual trees are an integrated system by themselves. Their water flux can be envisioned as representing an integration of four components: (1) uptake, (2) internal storage, (3) the controls upon the rate of movement of water from one part of the system to another and (4) the atmospheric demand.

The objective of this chapter is to present a conceptual framework for the movement of water through individual trees from the soil to the atmosphere. Conifers are of special interest because they illustrate the importance of internal storage and exhibit year-around physiological controls upon water movement.

The demand for water, as Gates has presented in Part 3:A of this volume is a function of the atmospheric environment and leaf geometry. The movement of water in the soil to the roots and its uptake from different zones has been treated in Part 2 of this volume. The same general principles are assumed to operate on conifers but the relationships are simplified to apply them to the variable environment associated with tall trees with widely distributed roots.

In a previous paper a computer simulation model was constructed to account for daily uptake, storage, and transpiration by conifers (Running et al., 1975). The same basic design applies here but to account for significant changes in diurnal behavior the model's structure has been refined and the resolution changed to hourly. The exact equations for this hourly model await derivation and depend upon experiments requiring simultaneous measurements of important variables.

II. Description of the Model

In Fig. 1, a flow diagram of a water transport system for a tree is presented which includes water flux: (1) from the soil, (2) from internal storage within the plant, and (3) by transpiration to the atmosphere. Table 1 summarizes the data requirements, internal functions, and predicted output of the model.

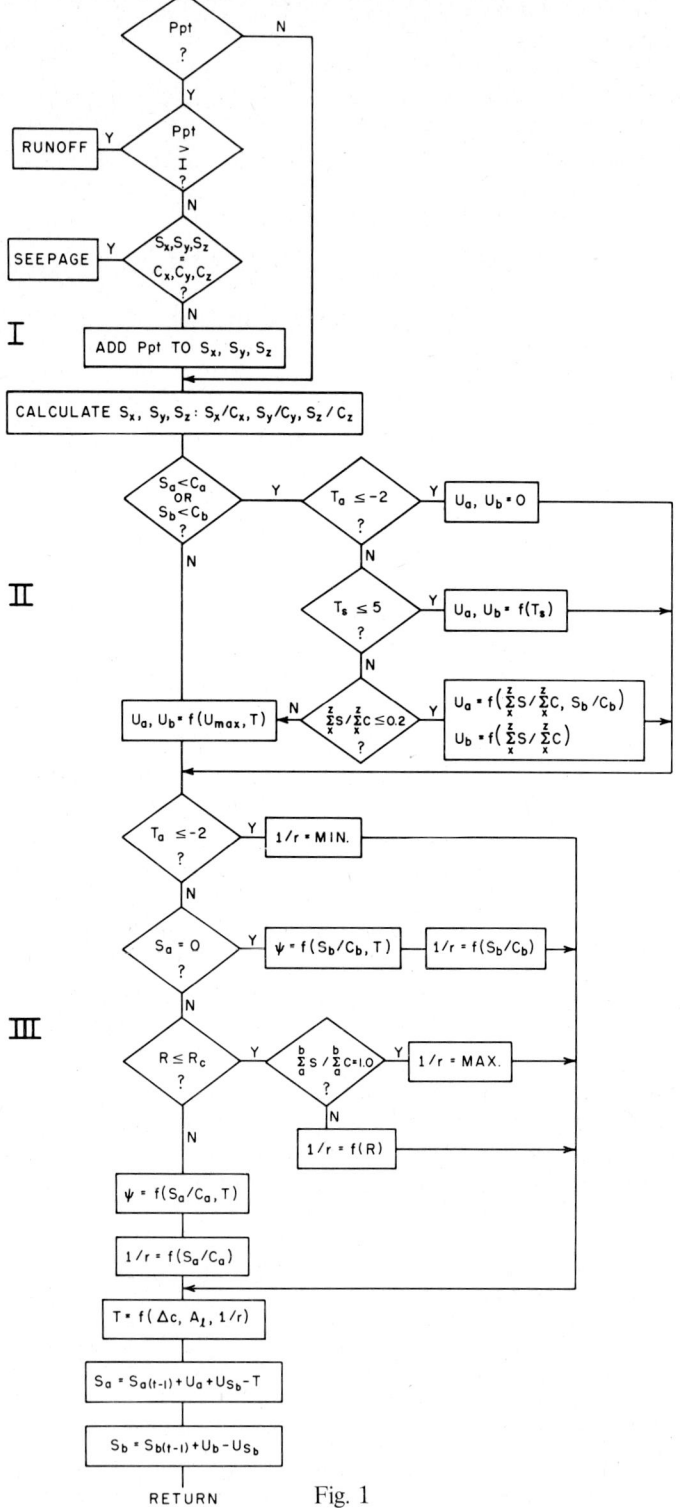

Fig. 1

In the soil root zone, Section I (see Fig. 1), the amount of precipitation (Ppt) entering or leaving is accounted for. Uptake (U_a, U_b) from the various soil zones (S_x, S_y, S_z) is calculated in the second Section (II) where it is routed to a small, but rapidly depletable storage (S_a) or to a larger, more slowly extractable one representing mature sapwood (S_b). Temperature (T_a or T_s) and the supply of water in the root zone (S_x, S_y, S_z) control the rate of uptake that is possible.

The last Section (III) is concerned with how internal water deficits and environment (T_a, Δ_c, R) affect leaf conductance ($1/r$) which, together with foliage area (A_1) and atmospheric demand (Δ_c), determine the rate of transpiration (T). Plant water potential (Ψ) is predicted from knowledge of the transpiration alone when all tissue reserves are full. Otherwise it is a function of the relative water content of the two storage reservoirs (S_a/C_a or S_b/C_b). Following an updating procedure in which water is withdrawn from internal storage, the sequence is repeated.

1. Root Zone Water

Effective precipitation, the amount of water entering the root zone, is a basic input acquired from a general hydrologic model after Sollins et al. (1974). This model considers energy and mass exchange from the canopy, snowpack, and litter.

If the entire root zone is at capacity, removal of water will progress from the upper to lower zones (Woods and O'Neal, 1965). Krygier (1971) has shown from neutron probe moisture measurements that under a coniferous forest of *Pseudotsuga menziesii* (Mirb.) Franco, water is extracted mainly from the upper zones until soil water tensions drop below -2 bars, which corresponds to a supply/capacity ratio of about 0.2. Then the next lower zone begins to supply the majority of water to the trees with the upper zone contributing an exponentially decreasing amount (Running et al., 1975; see Benecke, this volume Part 2:C). For simplicity it is assumed that when water enters the root zone from above, it recharges any depleted zones back to capacity before affecting the next lower.

Water uptake then, is calculated from each soil zone based upon its vertical position, water supply, and the water deficit within the tree. Water uptake must be evaluated because the volume of water available is a function of a tree's rooting volume. With some conifers such as *Pseudotsuga menziesii*, a horizontal projection of the outer limits of the crown may adequately define the lateral

---

◀ Fig. 1. Flow diagram for single tree water flux model. The flow is downward through a decision matrix with yes (Y) or no (N) alternatives. Section I accounts for the flow through or accumulation of water in the rooting zone. Section II treats internal storage deficits in the tree and provides for uptake from the soil or transfer from the sapwood to extensible storage. Air and soil temperature as well as soil water content affect root uptake. In Section III leaf conductance is evaluated as it is affected by low temperatures, low radiation, and internal water deficits. Finally, transpiration by the tree is calculated for the period (1 h) and the soil and internal storages updated before repeating the sequence for the next time step. For symbology refer to Table 1

Table 1. Data requirements and output of a water flux model for a tree

Environmental variables		(Units)
Dew point (for calc. humidity gradient Δc)		°C
Air temperature	T_a	°C
Soil temperature	T_s	°C
Precipitation (effective)	P_{pt}	cm^3 h^{-1}
Time	t	h
Radiation, short wave	R	ly

Required parameters

Soil root zone
Infiltration rate	I	cm^3 h^{-1}
Soil storage capacity	C_x, C_y, C_z	cm^3

Internal storage
Extensible storage capacity	C_a	cm^3
Maximum uptake rate from soil	$U_{a\,max}$	cm^3 h^{-1}
Wood storage capacity	C_b	cm^3
Maximum uptake rate from soil	$U_{b\,max}$	cm^3 h^{-1}
Maximum uptake rate from wood to extensible storage	$U_{Sb\,max}$	cm^3 h^{-1}

Transpiration
Leaf area (all surfaces)	A_1	cm^2
Radiation, critical level	R_c	ly

Required functions

Uptake as a function of soil temperature	$U_a, U_b = f(T_s)$
Uptake as a function of root zone water	$U_a = f(\sum_x^z S / \sum_x^z C, S_b)$
	$U_b = f(\sum_x^z S / \sum_x^z C)$

Extensible storage empty: $S_a = 0$
Leaf conductance as a function of sapwood water	$1/r = f(S_b/C_b)$
Water potential as function of sapwood water and transpiration	$\Psi = f(S_b/C_b, T)$

Extensible storage full: $S_a/C_a = 1.00$
Water potential as function of transpiration	$\Psi = f(T)$

Extensible storage intermediate: $0 > S_a/C_a < 1$
Water potential as function of extensible storage	$\Psi = f(S_a/C_a)$
Leaf conductance as function of extensible storage	$1/r = f(S_a/C_a)$
Leaf conductance as function of light when $R = R_c$	$1/r = f(R)$

Predicted by model

Soil root zone
Runoff		cm^3 h^{-1}
Seepage		cm^3 h^{-1}
H$_2$O Supply	S_x, S_y, S_z	cm^3

Internal storage
H$_2$O Supply	S_a, S_b	cm^3
Uptake	U_a, U_b, U_{Sb}	cm^3 h^{-1}

Transpiration
Leaf conductance	$1/r$	cm sec^{-1}
Transpiration	T	cm^3 H$_2$O h^{-1}
Water potential	Ψ	bars

extension of the roots (Wagenknecht, 1960). Other species such as the pines often exhibit greater lateral root extension which may be approximated by accounting for stand density. Although the depth of rooting depends partly upon soil characteristics, the rooting habits of conifers differ, i. e. that similar-sized trees may tap different volumes of soil (Fowells, 1965; Rutter, 1968). The actual volume of water available in the rooting volume represents only that which can be withdrawn by the roots.

2. Internal Storage

There are two notably different sources of water from within the tree itself. One, represented by fine roots, leaves, phloem, cambium and young xylem cells, shows volumetric change as water enters or leaves; the other is mature wood which shows little dimensional change because it is rigid and the extracted water is replaced at least temporarily by gas (Clark and Gibbs, 1957).

a) Extensible Tissue Reserve. From dimensional measurements of tree stems it is known that water is withdrawn initially from the tissue closest to the sites of evaporation. Then, as the first reserves are depleted, the main sources of supply are from lower and lower down the stem of the plant (Schnock, 1972; Jarvis, 1975; Dobbs and Scott, 1971). Recharge follows the reverse sequence (Doley, 1967). When conditions are favorable, the extensible tissue are refilled overnight (Stewart et al., 1973; Wilson et al., 1953).

Recently Huck et al. (1970) demonstrated that roots may shrink to 60% of their turgid diameter. This represents a change in relative water content of 40%, more than twice that characteristic of diurnal changes in foliage (Clausen and Kozlowski, 1965; Jarvis, 1975). In small trees, the root storage can be of considerable importance, for the small reserves in the foliage and stem are soon exhausted, as demonstrated by shrinkage in dimensions (Lassoie, 1973; Waggoner and Turner, 1971; Jarvis, 1975). In large trees, however, the time required to transfer water from the roots to the foliage is too great to permit roots to contribute much to daily water deficits as shown by isotope studies (Kline et al., 1976; Smith, 1972; Owston et al., 1972).

The water in the above ground extensible tissue reserve may be estimated by severing the stem from the roots and observing the loss in weight until stomata close. On small 1–2 m tall Douglas-fir the reserve defined in this manner represents about 12% of the total water in the tree (Waring, unpublished); very little of this comes from the mature sapwood. Rather, it comes from a change in relative water content observed in the foliage and the 100-cell thick band of living tissue in the branches and stem (Stewart et al., 1973; Jarvis, 1975).

b) Mature Wood Reserve. Seasonal changes in the moisture content of inextensible sapwood are reported in both conifers and hardwoods (Stewart, 1967; Clark and Gibbs, 1957; Markstrom and Hann, 1972; Gibbs, 1958; Chalk and Bigg, 1956).

In a mature Douglas-fir forest it is estimated that more than 6 cm of water is available from the sapwood to supplement storage in the soil. This is equivalent to at least a 10-day requirement for transpiration even in midsummer. On the basis of individual trees, it is estimated that a single 80-m tall, 1.5-m thick Douglas-fir has 4300 l available in just the stem wood (Running et al., 1975), assuming a change in relative water content of 40% (Chalk and Bigg, 1956). The sapwood in the branches would have about half the storage of a comparable volume of stem wood because of its greater density. The roots, which are between 10–40% of the above ground biomass (Santantonio et al., 1976) would contribute proportionately.

The volume of sapwood to heartwood varies with the stem diameter, crown size and species. In general, the genus *Pinus* is noted for its high proportion of sapwood to heartwood. At the other extreme are the cedars and redwood (Lassen and Okkonen, 1969). The amount of sapwood in a tree increases linearly with leaf area or mass (Grier and Waring, 1974; Dixon, 1971) and asymptotically with diameter (Lassen and Okkonen, 1969). Thus an estimate of both the extensible and inextensible tissue reserves is possible without requiring destructive sampling.

Only a fraction of the water in sapwood can be extracted in a day, with the most active exchange occurring in the outer 1–2 cm band. Over a period of a few weeks, however, most of the exchangeable water can be extracted or reabsorbed (Chalk and Bigg, 1956; Gibbs, 1958; Clark and Gibbs, 1957; Fries, 1943). The gas which is introduced into the conducting elements of conifers is sealed off until redissolved by the action of valves (bordered pits) on the sides of tracheids (Gregory and Petty, 1973). The formation and removal of gas from the conducting tissue also occurs under freezing and thawing conditions (Hammel, 1967).

c) Modeling Internal Water Flux. In modeling water flux from the internal storage, it is important to determine whether the extensible reserve is full. If the supply is less than the capacity then the amount of uptake possible from the soil and mature wood must be calculated (Fig. 1). Of course, recharge cannot exceed the internal deficit. A deficit in the extensible reservoir has priority over one in the mature wood. A deficit will develop during the day whenever the rate of transpiration exceeds the rate of recharge from the soil root zone and sapwood. If transpiration ceases at night, and water available in the rooting zone exceeds 20% of soil capacity, then complete recharge would normally occur sometime during the night (Running et al., 1975; Stewart et al., 1973). Below 20% of capacity, uptake from the soil will be reduced exponentially to zero, when all available root zone water is exhausted.

Recharge of the extensible reserve is also affected by low air or soil temperatures. As Zimmermann (1964) has shown, stem temperatures below $-2°C$ stop the ascent of sap in conifers. This corresponds to the temperature when all water in the soil freezes (Nerseova and Tsytovich, 1966). In the root zone, temperatures above freezing may also inhibit uptake (Havranek, 1972; Babalola et al., 1968; Kramer, 1942; Kuiper, 1964). In modeling the relationship, Running et al. (1975) effectively reduced uptake for Douglas-fir from 100% of potential at a threshold soil temperature of $5°C$ to zero uptake at $-2°C$. Again, because of the lag between uptake from the root system and transpiration, large trees must depend more upon their sapwood reserves, and are thus less affected by cold soils, than smaller trees.

The mature sapwood has the same temperature controls on water uptake from the root zone as the extensible reserves; because of its mass and position within the tree trunk its temperature lags behind that of the air and may differ considerably. Due to the relatively low rate of water transfer between sapwood and extensible tissue, however, a small error in estimating sapwood temperature will probably have little effect over a day.

The maximum transfer from sapwood to extensible tissue occurs when the sapwood is filled with water and the extensible reserves have been exhausted. This situation has been observed under clear weather conditions in the spring. As the sapwood dries, the water potential gradient may increase, but the paths of transfer also become more tortuous. From the studies of Chalk and Bigg (1956), and those of Clark and Gibbs (1957), it is estimated that about 5% of the sapwood reserve can be extracted daily. As water is extracted, the remaining reserve is less easily extracted, but an increased water potential gradient partly compensates and may explain the nearly linear decrease in moisture content reported by the above cited authors. Recharge may actually be somewhat faster, particularly following a thaw (Fries, 1943).

The maximum uptake by roots should occur when the sapwood reservoir has been emptied and the soil root zone has recently been recharged. At such a time, uptake reflects considerably more than transpiration. Because the functional conducting tissue changes as water is extracted, it is difficult to obtain accurate water flux estimates from heat-pulse or isotope measurements. These methods only show good agreement with actual water uptake when the conducting area is known and the flow rates are small (Heine and Farr, 1973; Ladefoged, 1960; Doley and Grieve, 1966). The maximum vertical movement reported in conifers is about $2 \, m \, h^{-1}$ (Owston et al., 1972).

3. Transpiration and Its Control

The last part of the water transport system involves the controls on the rate of water movement through the leaves to the atmosphere. The calculation is simplified by assuming that the rate of water vapor transferred from the saturated atmosphere within a leaf to the outside atmosphere is proportional to the gradient in water vapor concentration (Dainty, 1969; Rawlins, 1963; see Gates, this volume Part 3:A).

To estimate transpiration, the water vapor concentration gradient, Δc, between the foliage and the air is estimated first. Errors in estimating Δc are minimized on conifers because the needles are usually within a few degrees of ambient air temperature (Gates, 1968). Next the area of foliage is estimated through the previously mentioned relationship with conducting area (Grier and Waring, 1974; Dixon, 1971). Finally, from experimental observations equations to predict leaf conductance are developed.

In modeling leaf conductance, three major controls are envisioned (cf. also Hall et al., this volume Part 3:D): (1) temperature, (2) water deficits, and (3) light. Changes in guard cell concentration of CO_2, K^+ and ABA are envisioned as responding to the above basic controls so are not treated in the model.

a) **Temperature.** Reed (1968) and Drew et al. (1972) demonstrated that the stomata of both Douglas-fir and Ponderosa pine close if the air temperature is below $-2°C$. Thus, under these conditions, the model assumes minimum leaf conductances. Low soil temperatures may indirectly affect leaf conductance by inhibiting uptake, which in turn may result in a deficit in the extensible tissue.

b) **Water Deficits and Water Potential.** As Jarvis (1975) has emphasized, a large water potential gradient between the roots and leaves does not indicate necessarily that a water deficit exists within the plant. He points out that in a completely rigid flowpath, a potential difference is necessary to drive steady-state flow against the frictional resistance when input and output flow rates are equal. This situation exists in trees when soil water is available and the evaporative demand is not extreme. Lassoie (1973), for example, has reported more than a 10-bar change in the water potential gradient during the day with almost no diameter variation in the stem of a Douglas-fir. If no water deficit develops in the extensible tissue the stomata may remain open and one may find a linear relationship between increasing transpiration and decreasing (more negative) leaf water potential (Landsberg et al., 1975; Jarvis, 1975).

A change in relative water content of the extensible tissue, on the other hand, can directly affect the stomata. Again, there is not a simple relationship with water potential because the cell turgor is greatly affected by changing solute concentration. Even conifers, which are rarely found growing on saline soils, show rather large changes in solute potential depending upon tree height, season, and degree of desiccation (Richter et al., 1972; Larcher, 1972; Hellkvist et al., 1974).

Fortunately, it is possible through the pressure extraction technique developed by Scholander et al. (1965) to measure both water potential and solute potential. Such a procedure was followed by Hellkvist et al. (1974), to provide the solute and water potential curves for Sitka spruce [*Picea sitchensis* (Bong.) Carr.] in relation to relative water content (Fig. 2). In general, these curves tend to be less steep for conifers, such as Douglas-fir, pine and juniper, which are more drought adapted (Scholander et al., 1965; Jarvis and Jarvis, 1963). When the solute potential is equal to the water potential one can be assured that the stomata are completely closed and all transpiration will be cuticular.

c) **Water Potential and Leaf Conductance.** When equilibrium conditions are approached just before dawn, there is a predictable relationship between plant water potential and maximum possible leaf conductance when the stomata first open (Fig. 3). If no further change in the relative water content of extensible tissue occurred, this leaf conductance would exist throughout the day, despite a diurnal decrease in water potential. Under high humidity gradients leaf conductance will decrease slightly, apparently due to changes in guard cell turgor alone (Running, 1976; see Hall et al., Part 3:D of this volume). However, in the normal course of a clear warm day, even when predawn water potentials are less than -5 bars, water can be withdrawn from the extensible tissue, reducing the relative water content and the water potential. When this happens a point is reached, usually around 0.8 relative water content, where stomata begin to close abruptly. On small Douglas-fir and Ponderosa pine the closure threshold

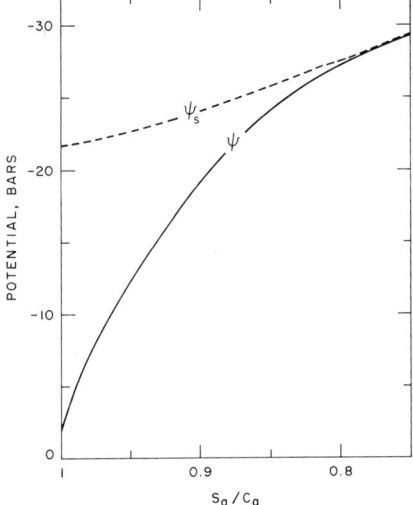

Fig. 2. Change in solute (Ψ_s) and water potential (Ψ) of Sitka spruce with decreasing relative water content of twigs. (After Hellkvist et al., 1974)

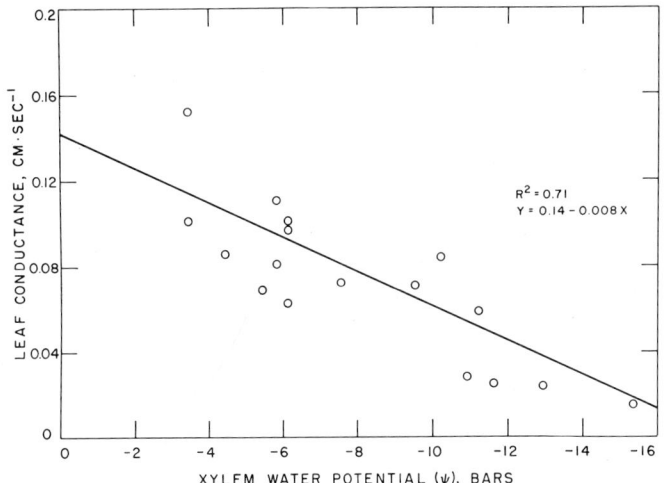

Fig. 3. Maximum morning leaf conductance as a function of predawn plant water potential. Each point represents a canopy average from at least 10 leaf conductance measurements over four age classes of needles on 2–3 m Douglas-fir

in terms of water potential is -20 and -17 bars respectively (Running, 1976; Lopushinsky, 1969).

In the present model (Fig. 1), leaf conductance as well as water potential are assumed to be proportional to relative water content. Once the relative water content has been reduced to a minimum (about 0.5), all of the extensible storage has been depleted and conductance will be minimum, usually around 0.003–0.005 cm sec^{-1}. The water potential under the latter circumstances is a function of the relative water content of the sapwood (S_b/C_b) and will range between -20 to -60 bars.

In the case where the rate of water uptake from the soil is sufficient to maintain full turgor of the extensible tissue, the leaf conductance can approach its maximum of around 0.25 cm sec^{-1} for new foliage of a wide variety of conifers. Water potential then is a simple linear function of transpiration (as stated previously).

d) Light. Low light levels can also affect leaf conductance. The particular level at which stomata begin to close varies. Pioneer plants generally have relatively high light thresholds while advanced successionary species have low (Woods and Turner, 1971). Red pine (*Pinus resinosa* Ait.) has been reported to have a threshold of 0.2 langleys (Waggoner and Turner, 1971) and for Douglas-fir closure begin below 0.1 langley. Thus, the stomata are normally closed when it is dark. However, if the plant rapidly recovers turgor, as illustrated by night-time water potential of around -3 bars, then the stomata of Douglas-fir and a variety of other conifers remain open. Seedlings of *Pinus contorta* var. *murrayana* (Balf.) Engelm., however, remain closed at night even with high water content (Lopushinsky, personal communication).

In the model (Fig. 1, Section III), if radiation (R) is below a critical level, the leaf conductance will be decreased below that predicted from knowledge of relative water content. When there is no internal water deficit, leaf conductance will not be reduced by low light.

The significance of the light effect is that, in a forest, many of the branches are often shaded and near or below the critical level, especially during the winter at latitudes greater than about 40°. This means that not only transpiration but photosynthesis may also be reduced. At night, if stomata remain open, some water will be transpired if the temperature is not at or below dew point. This can account for the occasional low values of water potential observed at night on well-watered soils.

e) Transpiration. Calculation of transpiration is the last major step in the water flux model. It is calculated as a function of the water vapor concentration gradient, Δc, the total leaf area, A_1, and the leaf conductance, $1/r$. Following this, the amount of water in each of the storage reserves is recalculated, and the sequence repeated for the next time step (Fig. 1).

III. Applications

From experience with the daily resolution model (Running et al., 1975) one can expect major differences in water balance depending upon the size of a tree and the extent of its rooting system. For example, from a similar environmental data base, it was calculated that an 80 m Douglas-fir would exhibit major stomatal control as a result of soil drought for only 17 days out of a total of 170 during the growing season (Table 2). A 2 m Douglas-fir, on the other hand would control

Table 2. Simulation results for an 80 m Douglas-fir
May 10–October 27, 1972, Cascade Mountains of Oregon (Running et al., 1975)

Given	
Rooting zone capacity	47,000 l
Extensible storage capacity	50 l
Sapwood storage capacity	4,300 l
Leaf area	$3.74 \cdot 10^7$ cm^2
Climatic data, averaged daily	

Predicted	
Leaf conductance range	0.025–0.008 cm sec^{-1}
Water potential range (at dawn)	-5 to -15 bars
Maximum transpiration rate	1 140 l day^{-1}
Days of reduced transpiration by drought	17 days
Total transpiration (170 days)	66,000 l
Average transpiration (170 days)	390 l day^{-1}

Table 3. Simulation results for a 2 m Douglas-fir
May 10 – October 27, 1972, Cascade Mountains of Oregon (Running et al., 1975)

Given	
Rooting zone capacity	450 l
Extensible storage capacity	0.5 l
Sapwood storage capacity	3.2 l
Leaf area	$3.5 \cdot 10^5$ cm^2
Climatic data, averaged daily	

Predicted	
Leaf conductance range	0.14–0.003 cm sec^{-1}
Water potential range (at dawn)	-2 to -30 bars
Maximum transpiration rate	16.6 l day^{-1}
Days of reduced transpiration by drought	54 days
Total transpiration (170 days)	723 l
Average transpiration (170 days)	4.3 l day^{-1}

transpiration to a fraction of the potential possible for 54 days (Table 3). In these comparisons a greater internal resistance to water is assumed to exist in large conifers (Hudson and Shelton, 1969). This results in generally lower average leaf conductances throughout the day.

From an understanding of the water transport system one can now more fully explain how it is possible for two conifers of the same height to respond similarly under well-hydrated conditions and modest evaporative demand. Yet, when evaporative demand increases and the rooting depth of one is less than the other, further depletion of soil reserves and differing degrees of stomatal control will result in different water balances.

Borchert (1973) suggests a further implication of such a water transfer model in his simulation of rhythmic growth in trees. He proposes that the periodicity of flushing and the foliage size is in a large part a function of internal water deficits that may occur, even under constant environment, when the amount of foliage area exceeds the root absorptive capacity necessary to meet the transpirational demand.

Further, adaptive significance becomes apparent for species with low leaf areas, and large volumes of sapwood growing where drought is frequent. It is clear that differences in leaf conductance exhibited in relation to light, temperature, and moisture stress may help explain the positions that different species fill in forest succession.

IV. Conclusions

The present concept of the water transport system has been modified as research points out deficiencies in previous assumptions. A lag between transpiration and uptake commonly observed in trees requires recognition of internal storage. The fact that foliage and other living tissue lose or regain their turgor faster than exchange is possible from sapwood signifies that at least two kinds of internal storage reservoirs should be considered.

Observations that light, reduction in relative water content, and low temperatures may all affect stomata have been incorporated in a system context. From this standpoint the response of plant water potential has also been reevaluated.

Much still remains unclear or unquantified. The expressed requirements for simultaneous measurements of important variables will require more integrated research to rigorously test and modify the concepts summarized here. It is to be hoped that such research will progress to further the understanding of plant water relations.

Acknowledgements. We are thankful to Drs. W. Emmingham, W. Lopushinsky, W. Ferrell, and J. Rogers for reviewing the original manuscript.

Publication of this work was supported by National Science Foundation Grant GB-20963 to the Coniferous Forest Biome, Contribution No. 162, Ecosystem Analysis Studies.

References

Babalola, O., Boersma, L., Youngberg, C. T.: Photosynthesis and transpiration of Monterey pine seedlings as a function of soil water suction and soil temperature. Plant Physiol. **43**, 515–521 (1968).

Borchert, R.: Simulation of rhythmic tree growth under constant conditions. Physiol. Plantarum **29**, 173–180 (1973).

Chalk, L., Bigg, J. M.: The distribution of moisture in the living stem in Sitka spruce and Douglas fir. Forestry **29**, 5–21 (1956).

Clark, J., Gibbs, R. D.: Studies in tree physiology. IV. Further investigations of seasonal changes in moisture contents of certain Canadian forest trees. Can. J. Botany **35**, 219–253 (1957).

Clausen, J. J., Kozlowski, T. T.: Use of the relative turgidity technique for measurement of water stresses in gymnosperm leaves. Can. J. Botany **43**, 305–316 (1965).

Dainty, J.: The water relations of plants. In: Physiology of plant growth and development (ed. M. Wilkins), pp. 421–452. New York: McGraw-Hill 1969.

Dixon, A. F. G.: The role of aphids in wood formation. J. Appl. Ecol. **8**, 165–179 (1971).

Dobbs, R. C., Scott, D. R. M.: Distribution of diurnal fluctuations in stem circumference of Douglas fir. Can. J. For. Res. **1**, 80–83 (1971).

Doley, D.: Water relations of *Eucalyptus marginata* Sm. under natural conditions. J. Ecol. **55**, 597–614 (1967).

Doley, D., Grieve, B. J.: Measurement of sap flow in a Eucalypt by thermo-electric methods. Australian For. Res. **2**, 3–27 (1966).

Drew, A. P., Drew, L. G., Fritts, H. C.: Environmental control of stomatal activity in mature semiarid site ponderosa pine. Ariz. Acad. Sci. **7**, 85–93 (1972).

Fowells, H. A.: Silvics of forest trees of the United States. Ag. Handbook No. 271. Washington, D.C.: U.S. Dept. Agr. 1965.

Fries, N.: Zur Kenntnis des winterlichen Wasserhaushalts der Laubbäume. Svensk. Botan. Tidskr. **37**, 241–265 (1943).

Gates, D. M.: Transpiration and leaf temperature. Ann. Rev. Plant Physiol. **19**, 211–238 (1968).

Gibbs, R. D.: Patterns in the seasonal water content of trees. In: The physiology of forest trees (ed. K. V. Thimann), pp. 43–69. New York: Ronald Press 1958.

Gregory, S. C., Petty, J. A.: Valve actions of bordered pits in conifers. J. Exp. Botany **24**, 763–767 (1973).

Grier, C. C., Waring, R. H.: Conifer foliage mass related to sapwood area. Forest Sci. **20**, 205–206 (1974).

Hammel, H. T.: Freezing xylem sap without cavitation. Plant Physiol. **42**, 55–66 (1967).

Havranek, W.: Über die Bedeutung der Bodentemperatur für die Photosynthese und Transpiration junger Forstpflanzen und die Stoffproduktion an der Waldgrenze. Angew. Botan. **46**, 101–116 (1972).

Heine, R. W., Farr, D. J.: Comparison of heat-pulse and radioisotope tracer methods for determining sap-flow velocity in stem segments of popular. J. Exp. Botany **24**, 649–654 (1973).

Hellkvist, J., Richards, G. P., Jarvis, P. G.: Vertical gradients of water potential and tissue water relations in Sitka spruce trees measured with the pressure chamber. J. Appl. Ecol. **11**, 637–668 (1974).

Huck, M. G., Klepper, B., Taylor, H. M.: Diurnal variations in root diameter. Plant Physiol. **45**, 529–530 (1970).

Hudson, M. S., Shelton, S. V.: Longitudinal flow of liquids in southern pine poles. Forest Prod. J. **19**, 25–32 (1969).

Jarvis, P. G.: Water transfer in plants. In: Heat and mass transfer in the environment of vegetation. 1974 Seminar of the Intern. Centre for Heat and Mass Transfer, Dubrovnik, 369–394. Washington, D.C.: Scripta Book Co. 1975.

Jarvis, P. G., Jarvis, M. S.: The water relations of tree seedlings. Physiol. Plantarum **16**, 501–516 (1963).

Kline, J. R., Reed, K. L., Waring, R. H., Stewart, M. L.: Direct measurement of transpiration and biomass in coniferous trees. J. Appl. Ecol. **13**, 273–283 (1976).

Kramer, P. J.: Species differences with respect to water absorption at low soil temperatures. Am. J. Botany **29**, 828–832 (1942).

Krygier, J. T.: Comparative water loss of Douglas-fir and Oregon white oak. Ph. D. Thesis, Colorado State Univ. 1971.

Kuiper, P. J. C.: Water uptake of higher plants as affected by root temperature. Med. Landbouwhogesch. Wageningen **64**, 1–11 (1964).

Ladefoged, K.: A method for measuring the water consumption of larger intact trees. Physiol. Plantarum **13**, 648–658 (1960).

Landsberg, J. J., Beadle, C. L., Biscoe, P. V., Butler, D. R., Davidson, B., Incoll, L. D., James, G. B., Jarvis, P. G., Martin, P. J., Neilson, R. E., Powell, D. B. B., Slack, E. M., Thrope, M. R., Turner, N. C., Warrit, B., Watts, W. R.: Diurnal energy water and CO_2 exchanges in an apple *(Malus pumila)* orchard. J. Appl. Ecol. **12**, 659–684 (1975).

Larcher, W.: Der Wasserhaushalt immergrüner Pflanzen im Winter. Ber. Deut. Botan. Ges. **85**, 315–327 (1972).

Lassen, L. E., Okkonen, E. A.: Sapwood thickness in Douglas-fir and five other western softwoods. U.S.F.S., F.P.L. **124**, 1–16 (1969).

Lassoie, J. P.: Diurnal dimensional fluctuations in a Douglas-fir stem in response to tree water status. Forest Sci. **19**, 251–255 (1973).

Lopushinsky, W.: Stomatal closure in conifer seedlings in response to leaf moisture stress. Botan. Gaz. **130**, 258–263 (1969).

Markstrom, D. C., Hann, R. A.: Seasonal variation in wood permeability and stem moisture content of three Rocky Mountain softwoods. U.S.F.S., Res. Note RM-**212**, 1–7 (1972).

Nerseova, F. A., Tsytovich, J. A.: Unfrozen water in frozen soils. Proc. 1st Intern. Permafrost Conf., Nat. Res. Council, Nat. Acad. Sci., 230–234 (1966).

Owston, P. W., Smith, J. L., Halverson, H. G.: Seasonal water movement in tree stems. Forest Sci. **18**, 266–272 (1972).

Rawlins, S. L.: Resistance to water flow in the transpiration stream. In: Stomata and water relations of plants (ed. I. Zelitch), pp. 69–85. Conn. Agr. Exp. Sta. Bull. **664** (1963).

Reed, K. L.: The effects of sub-zero temperatures on the stomata of Douglas-fir. M. S. Thesis, Seattle, Washington 1968.

Richter, H., Halbwachs, G., Holzner, W.: Saugspannungsmessungen in der Krone eines Mammutbaumes *(Sequoiadendron giganteum)*. Flora **161**, 401–420 (1972).

Running, S. W.: Environmental control of leaf water conductance in conifers. Can. J. Forest Res. **6**, 104–112 (1976).

Running, S. W., Waring, R. H., Rydell, R. A.: Physiological control of water flux in conifers: a computer simulation model. Oecologia (Berl.) **18**, 1–18 (1975).

Rutter, A. J.: Water consumption by forests. In: Water deficits and plant growth. II (ed. T. T. Kozlowski), pp. 23–84. New York: Academic Press 1968.

Santantonio, D., Hermann, R. K., Overton, W. S.: Root biomass studies in forest ecosystems. Pedobiologia (in press, 1976).

Schnock, G.: Contenu en eau d'une phytocénose et bilan hydrique de l'ecosystème: Chênaie de virelles. Oecol. Plant. **7**, 205–226 (1972).

Scholander, P. F., Hammel, H. T., Hemmingsen, E. A.: Sap pressure in vascular plants. Science **148**, 339–346 (1965).

Smith, J. L.: Forest soils and the associated soil-plant water regime. In: Symp. on the use of isotopes and radiation in research of soil-plant relationships including forestry, pp. 399–412. Vienna: Intern. At. Energy Agency 1972.

Sollins, P., Waring, R. H., Cole, D. W.: A systematic framework for modeling and studying the physiology of a coniferous forest ecosystem. In: Integrated research in the Coniferous Forest Biome (Proc. AIBS Symp. Conif. For. Ecosyst.) (eds. R. H. Waring, R. L. Edmonds), pp. 7–20. Conif. For. Biome Bull. **5** (1974).

Stewart, C. M.: Moisture content of living trees. Nature (London) **214**, 138–140 (1967).

Stewart, C. M., Tham, S. H., Rolfe, D. L.: Diurnal variations of water in developing secondary stem tissue of Eucalypt trees. Nature (London) **242**, 479–480 (1973).

Wagenknecht, E.: Beiträge zur Kenntnis der Wurzelausbildung verschiedener Bestockungen. Mitt. aus der Staatsforstverwaltung Bayerns **31**, 252–274 (1960).

Waggoner, P. E., Turner, N. G.: Transpiration and its control by stomata in a pine forest. Bull. Conn. Agr. Exp. Sta. **726**, 1–87 (1971).

Wilson, C. C., Boggess, W. R., Kramer, P. J.: Diurnal fluctuations in the moisture content of some herbaceous plants. Am. J. Botany **40**, 97–100 (1953).

Woods, D. B., Turner, N. C.: Stomatal response to changing light by four tree species of varying shade tolerance. New Phytologist **70**, 77–84 (1971).

Woods, F. W., O'Neal, D.: Tritiated water as a tool for ecological field studies. Science **147**, 148–149 (1965).

Zimmermann, M.: Effect of low temperature on the ascent of sap in trees. Plant Physiol. **39**, 568–572 (1964).

Part 4
Direct and Indirect Water Stress

Preface

After the discussion of the general principles of water balance and water uptake by plants, and the regulation of their water balance, the following Parts 4 and 5 are primarily dedicated to critical conditions of water supply and the adaptation of plants to water deficits. In recent years the responses of plants to water stress have been intensively investigated in ecology and physiology, as well as in biochemistry and from the viewpoint of research on cellular ultrastructure.

Part 4 is not intended as a review (see Hsiao, 1973; Kozlowski, 1968a, b, 1972) or as an attempt at a general survey of our present knowledge about water stress in plants (cf. also Levitt, 1972). It is rather aimed at pointing out various modern aspects concerning the efficiency of plants and their ability to survive under drought stress in their habitats. Of course, there exist many species of lower, and some higher, plants which can survive nearly complete desiccation. Many of these plants can even utilize widely differing degrees of hydration for their metabolically active life in terrestrial environments (poikilohydric plants). Many aquatic algae adapt themselves to a wide range of concentrations of their medium by means of osmoregulation (cf. Kluge, Part 4:C). However, the majority of the higher plants, are more or less sensitive to water deficits. Some of the most essential questions, for instance, are: what are the primary symptoms of water stress in the plant? Are they mechanical in nature, due to loss of turgor and spatial changes in the position of the compartments to one another, or are they primarily chemical in nature, due to higher ion concentrations or shifts in the hormonal relations?

These questions, which are dealt with in the following chapters, present two important view points, on first glance seemingly incompatible. The answer cannot be all-encompassing, also for the reason that, depending on the type of water stress and on the morpho-genotype of the plant investigated, the results vary due both to species-specific qualities and to the ontogenetic state of the plant. A further difficulty in the way of a consistent explanation is the comparison of the effects of mild and severe water stress. The question then is whether or not the reaction systems are identical in both cases.

Vieira da Silva (Part 4:A) has been able to observe the influence of water stress in the cellular ultrastructure. He discusses the disruption of certain compartments and the possibilities of destructive enzymes being activated, which are especially injurious during the rehydration of the tissues. After mild water stress the ultrastructure becomes less affected or not at all. Drought-tolerant plants obviously have more resistant compartments and less activation of destructive enzymes than tender plants under the same water stress. While this author tends more to a structural stress concept and also does not exclude turgor change as a possible trigger of the destructive processes, Itai and Benzioni (Part 4:B) express doubt as to the primary role of a mechanical trigger—at least in moderate and transient water stress. Their approach also aims to make use of a general stress concept which consequently excludes special effects of different stresses, as in case of water stress the turgor changes. These authors focus on changes of hormonal balance as the primary effect of water stress, and discuss the possibility that hormones trigger changes of membrane permeability which subsequently cause stomatal response and metabolic changes, or any chain of processes leading to after-effects or injury.

Searching for reaction sites that are sensitive to water stress, it is important to analyze and localize adaptational processes by which the plants can protect themselves against water-stress injury. This problem is discussed particularly in the chapters of Heber and Santarius (Part 4:D), of Kluge (Part 4:C) and of Lee-Stadelmann and Stadelmann (Part 4:E). Most proteins and enzyme systems, as well as nucleic acids, are very resistant and become only reversibly inactive by desiccation, independent of the fact whether they belong to resistant or to sensitive plants. This shows that biochemical systems per se are not sensitive to desiccation. The sensitive sites must be sought in certain structures of the organelles. With leaf tissues it has been demonstrated that sugars, proteins, organic acids and other substances can protect membrane-bound structures and processes found to be sensitive to water stress, from injury by freezing and desiccation. Such protective substances accumulating in plants, for instance during frost hardening, obviously might have different effects (Heber and Santarius, Part 4:D): they can directly interfere with membrane-damaging substances, they can have a structure-preserving effect by specific interaction with membranes, or they can even modify structures. The permeability changes of the cells of pea seedlings might be understood in a similar way (cf. Lee-Stadelmann and Stadelmann, Part 4:E). In this case, increasing drought tolerance is connected with a decrease in water permeability, as was also found with increasing freezing tolerance in cortical tissues of *Malus* (cf. Mittelstädt, 1971), and with increasing permeability for lipophilic nonelectrolytes. However, the problem remains as to the significance of increasing lipophily for drought tolerance and how it comes about. Thus we have arrived again at the initial question of what primary biochemical or mechanical processes lead to a change in the membrane structure.

Although in most of the chapters attention is primarily focused on the responses of cells and their compartments, one must not forget that the plant as a whole comprises a complex of different tissues in different developmental states, that is, different sites and kinds of reactions. For now it can be demonstrated, at least with models, by what combination of series of reactions local or overall water stress influences such factors as the productivity of plants. The chapter of Hsiao et al. (Part 4:F) illustrates to what extent models of such reactions series are feasible today and the difficulties, for instance, of an analysis of the effect of water stress in the root. On the other hand, it can be seen more clearly through complex approach than through any localized approach that very mild water stress which does not yet induce stomatal closure can influence long-term regulative processes and growth, primarily by changes of turgor. On the other hand, it is possible to maintain growth even with a decreasing total water potential of tissues by osmotic adjustment. A long series of effects must be examined for an explanation of water-stress situations, e.g., the interesting phenomenon that *Sorghum*, in general, reacts more flexibly to water stress than maize and consequently yields more grain under the same conditions (cf. Hsiao et al., Part 4:F). There is no doubt that future research will increasingly endeavor to elucidate the whole complex of plant reactions. A prerequisite for this, however, is a good knowledge of all the single phenomena. Only then will it be possible to explain exactly the seemingly trivial statement that plants suffer from water stress, and only then will processes of ecological adaptation to moisture conditions and with this, the preference of certain plants for certain habitats become more easily understandable.

References

Hsiao, T. C.: Plant responses to water stress. Ann. Rev. Plant Physiol. **24**, 519–570 (1973).
Kozlowski, T. T. (ed.): Water deficits and plant growth. I. Development, control and measurements. New York, London: Academic Press 1968a.
Kozlowski, T. T. (ed.): Water deficits and plant growth. II. Plant water consumption and response. New York, London: Academic Press 1968b.
Kozlowski, T. T. (ed.): Water deficits and plant growth. III. Plant responses and control of water balance. New York, London: Academic Press 1972.
Levitt, J.: Responses of plants to environmental stresses. New York, London: Academic Press 1972.
Mittelstädt, H.: Untersuchungen über die zellphysiologischen Grundlagen der Frostresistenz beim Apfel. I. Der Einfluß von Kältebehandlungen auf die Permeabilität. Flora **160**, 195–216 (1971).

A. Water Stress, Ultrastructure and Enzymatic Activity

J. Vieira da Silva

I. Introduction

Water stress can have profound metabolic effects in plants, resulting not only in impaired gas exchange but also in profound alteration of physiological processes, as is presented in the comprehensive review by Hsiao (1973). It indicates that even slight stress could affect such different phenomena as cell growth, wall synthesis, nitrogen and chlorophyll metabolism, and the levels of growth substances.

Most of the literature concerning the effect of water stress on enzymatic activity was reviewed by Todd (1972) and shows clearly that hydrolytic enzymes, and some oxidases frequently increase in activity with stress.

That such effects are linked with structure is an old hypothesis, as stated by Iljin (1935, 1953), Oparin (1937, 1953), Oparin and Kaden (1945), Kursanov (1940, 1946), etc. Water stress would result, according to this hypothesis, in the liberation of enzymes from the lipoprotein complex in a similar way to senescence.

An identical hypothesis was demonstrated by De Duve and Wattiaux (1966) with the discovery of lysosomes in animals, these particles being loaded with hydrolytic enzymes that could be discharged in the cells under abnormal conditions, as in disease. In plants lysosome-like structures have also been found (see review by Gahan, 1973), but autotrophy presents a more complicated situation, since the chloroplasts are known to contain fairly high amounts of acid hydrolases.

On the whole the analysis of the effects of water stress on the structure of organelles and on enzymatic activity seems a promising way to explain the disruption, with dehydration, of such a complex system as the plant cell. The following is a highly personal view of this problem, the arguments being taken from results presented in literature, as well as from recent findings from this author's laboratory.

II. Effects of Water Stress on Hydrolytic Enzymatic Activity

1. Induction of Enzymatic Activity by Water Stress

Activity increases in several enzymes in response to dehydration have been observed by several authors (see extensive tabulation in Todd, 1972). Those increases can usually be found in hydrolytic enzymes and in some oxidases. Some Russian authors, (see review by Mothes, 1956), were among the first to find a link between tissue water content and enzymatic activity, and as early as 1929, Maximov suggested that the study of the physico-chemical nature of the protoplasm would lead to a better understanding of the resistance of plants to desiccation.

The role of protoplasmic structure and its alteration with stress was pointed out by Iljin (1935, 1953) and modern results seem to confirm these early suggestions. Other, Russian authors (in Mothes, 1956), found that some enzymes are linked with the lipoproteic complex and that desiccation can lead to their liberation and an increase in activity.

The similarity between these views and the results of De Duve and Wattiaux (1966) concerning the hydrolytic enzyme complex in the lysosomes and the destructive results of their lysis, is striking. Lysosomes were found in plants, (see review in Vieira da Silva, 1970 and Gahan, 1973), in spherosomes, vacuoles, aleurone grains, Golgi apparatus and lutoids. Nevertheless, they do not seem to be the only organelles containing acid hydrolases, as has been demonstrated with acid phosphatase (Ragetli et al., 1966; Ragetli, 1967) and ribonuclease (Hadziyev et al., 1969) in chloroplasts. The basic lysosomal hypothesis, however, of compartmentation of hydrolytic enzymes in a latent condition followed by wholesale release in the cell with stress, seems attractive in the light of recent research. It also seems that the activation or synthesis of those enzymes could be a response of plant cells to drought.

a) Solubilization of Structure-bound Enzymes. A study of several strains of drought-sensitive *Gossypium hirsutum*, compared with drought-resistant *Gossypium anomalum*, shows that under soil drought conditions, water deficits in the leaves produce an important solubilization of acid phosphatase (Fig. 1) in the drought-sensitive species. Solubilization here represents the enzymatic activity in the supernatant of centrifuged leaf homogenates made under isotonic conditions, as a percentage of total activity obtained after disruption of cellular structures by a detergent (Vieira da Silva, 1970). *G. anomalum*, however does not show any increase in solubilization even with a relative water content of less than 60%. Floating leaf discs over an osmoticum (polyethylene glycol 600) or adding the osmoticum to the root nutrient solution results in the same solubilization picture as does soil drought. Solubilization of acid ribonuclease (Fig. 2), or β-fructofuranosidase, and of α- and β-amylases, parallel the solubilization of acid phosphatase. In *G. anomalum*, however, drought increases the solubilization of β-fructofuranosidase and α- and β-amylases, without increasing acid phosphatase solubilization (Vieira da Silva and Poisson, 1969). Hybrid derivatives between *G. anomalum* and *G. hirsutum*, both $2n+1$ aneuploids for chromosome

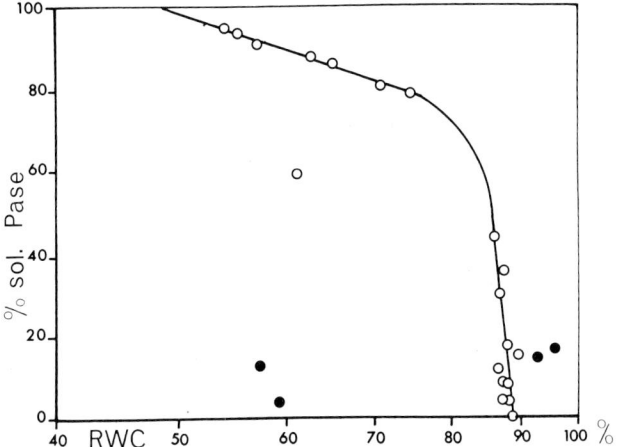

Fig. 1. Relationship between relative water content (*RWC*) of leaf tissue and percentage of solubilization of acid phosphatase (sol. Pase). ○ *G. hirsutum*; ● *G. anomalum*. (From Vieira da Silva, 1970. Courtesy of Gauthier Villars)

3 of *G. anomalum* and euploids derived from this segregating population, exhibit under osmotic stress conditions the same pattern of solubilization of the hydrolytic enzymes β-fructoforanosidase β-amylase and acid phosphatase, as the wild parent *G. anomalum*.

The wild species of Gossypium can show characteristics of either drought avoidance (usually by early leaf abscission) or of drought resistance. The former manifest a high degree of acid phosphatase solubilization with stress [*G. raimondii* (91 %) and *G. thurberi* (78 %)] compared with the later drought-resistant [*G. anomalum* (18.5 %) and *G. australe* (21.6 %)] (Vieira da Silva, 1968, 1969).

Soluble activity of acid phosphatase in *G. hirsutum* can increase more than three-fold within 30 min after the commencement of floating the leaf discs over polyethylen glycol solution of -20 Joules mole^{-1} of osmotic potential (Vieira da Silva, 1969).

It seems therefore that resistance to solubilization of structure-bound hydrolytic enzymes, namely acid phosphatase, accompanies fairly well drought resistance in the genus *Gossypium*.

b) Increase in Global Enzymatic Activity: de novo Synthesis or Activation of Zymogens. Total hydrolytic activity of acid phosphatase, in the leaf tissues of drought-sensitive *G. hirsutum*, i.e. total specific activity after disruption of cellular structures by detergent, can increase in stressed plants to between two and four times that of the controls.

This increase in activity could be the result either of activation of zymogen, or of de novo synthesis of the enzymes. In both cases the increase of activity could be the result of the same molecular form or of isozymes.

Treating the leaf discs before stress with D-actinomycine, cycloheximide or chloramphenicol shows that D-actinomycine has no effect on the level of total phosphatase activity obtained after stress, but that chloramphenicol decreases

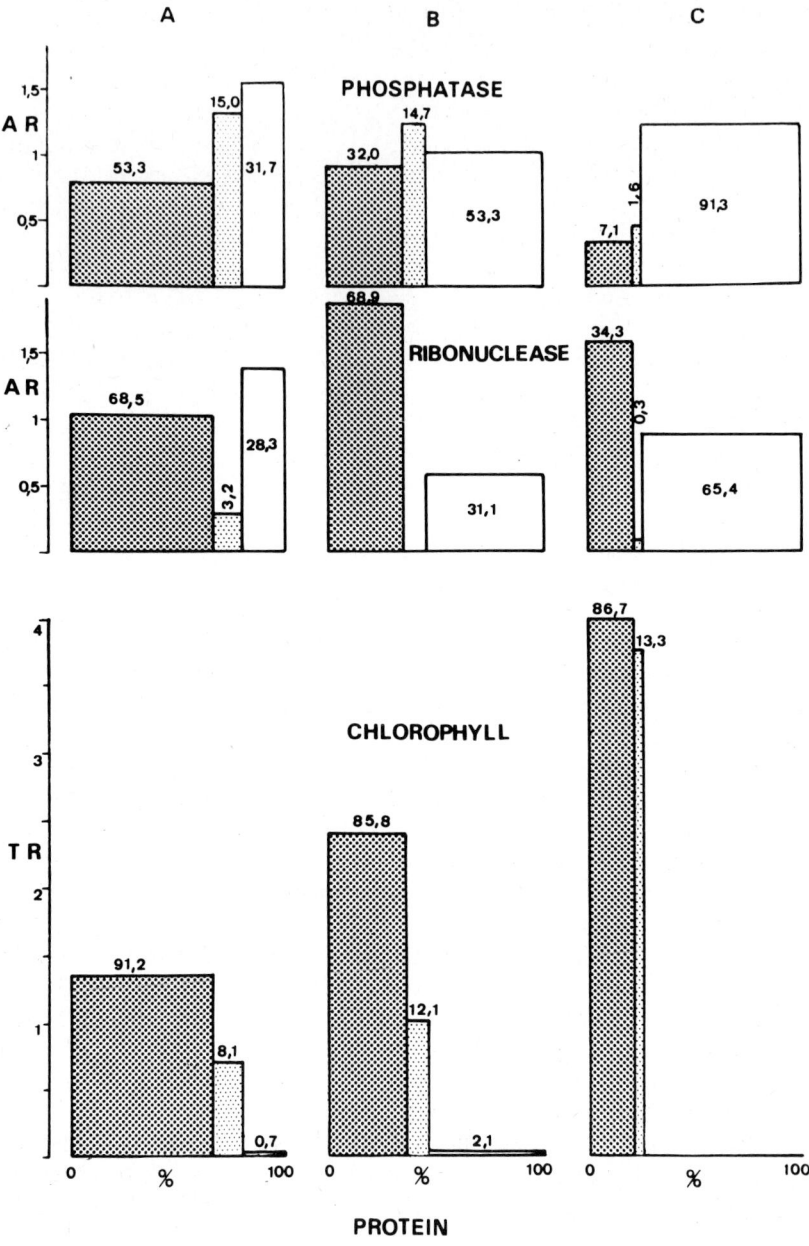

Fig. 2A–C. Distribution of enzyme activity and chlorophyll in three cellular fractions. *Heavy shading:* heavy fraction (3000 g, 1 h); *light shading:* middle fraction (40,000 g, 1 h); *white:* supernatant. The width of each rectangle represents the percentage of protein. Percentage of total activity or of chlorophyll content is indicated in each rectangle. *AR* Relative activity = % activity/% protein; *TR* Relative content = % content/% protein. (From Vieira da Silva, 1970. Courtesy of Gauthier-Villars)

total activity of acid phosphatase between 26% and 32% and of ribonuclease by 54%. Cycloheximide has no effect either in acid phosphatase or ribonuclease, but decreases β-fructofuranosidase activity by 75% (Vieira da Silva, 1970).

Owing to the difficulty in interpreting inhibition experiments that show an effect on protein synthesis, but that cannot confirm whether or not the inhibited synthesis is that of the enzyme studied, ^{14}C-leucine was infiltrated in the discs, and ribonuclease and acid phosphatase were purified by chromatography on DEAE cellulose.

In fact a new peak of activity appears after stress, both in ribonuclease and acid phosphatase chromatographies, but these peaks are not associated with radioactivity. It appears therefore, contrary to the results of Young and Varner (1959), Bagi and Farkas (1967), and McHale and Dove (1968), that water stress in leaf tissue does not produce a de novo synthesis of ribonuclease and acid phosphatase, the increase in activity being probably the result of zymogen activation (Vieira da Silva, 1970). This is confirmed by the fact that increase in activity of one enzyme is strictly proportional to the increase in activity of the other, the activation process being probably of common origin. Presley and Fowden (1965) find that the increase in acid phosphatase in germinating seeds is also the result of an activation of zymogens.

2. Effects of Water Stress in the Cellular Compartmentation of Enzymes

The cellular compartmentation of hydrolytic enzyme has been studied (Vieira da Silva, 1970) by differential centrifugation comparing *G. hirsutum* cultivated in a greenhouse (Fig. 2A) with the same species and variety cultivated at 50% relative humidity (Fig. 2B) and with the plants cultivated at 50% relative humidity and -15 Joules mole^{-1} of osmotic stress in the nutrient solution for 48 h (Fig. 2C).

Much of the enzymatic activity in the leaf tissue of control plants takes place in the heavy fraction, and stress treatment increases the solubilization to the supernatant.

In the control, the protein content in the heavy fraction amounts to 67.9%, in the middle fraction to 11.6%, and to 20.5% in the supernatant. Under atmospheric drought, protein content is 35.0% in the heavy fraction, 11.9% in the middle fraction, and 53.1% in the supernatant. With both atmospheric and soil drought, the protein content of the heavy fraction is 21.5%, of the middle fraction 3.5% and of supernatant 75%.

Even if this situation represents also the effect of homogenate grinding, it shows that the cellular compartments are weaker and can empty themselves, leaking protein to the supernatant and that this protein contains hydrolytic enzymes.

That incubation of chloroplasts in hypertonic solutions could lead to shrinkage of chloroplasts and leakage of ribosomes has already been observed by Filippovich et al. (1967) and this can lead to artifacts when studying the compartmentation of enzymes by sucrose gradient centrifugation.

Therefore the distribution of acid phosphatase and ribonuclease was studied using isoosmotic Dextran-sucrose density gradients, comparing stressed and normal plants of *G. thurberi* (drought-sensitive) and *G. anomalum* (drought-resistant).

Most of the acid phosphatase activity corresponds to the chloroplastic peak in both species, and this peak is void of activity in stressed *G. thurberi*, whereas activity continues in stressed *G. anomalum*. Ribonuclease, however, appears in the nuclear, chloroplastic and microsomal (ribosomes?) peaks in the control treatments, but in the supernatant is stressed *G. thurberi*. Chloroplasts are denser in *G. anomalum* than in *G. thurberi*, and conserve their protein content in stressed condition in the former species.

This compartmentation confirms the results of several authors (Yin, 1945; Nir and Poljakoff-Mayber, 1966; Ragetli et al., 1966; Ragetli, 1967) concerning the localization of most of phosphatase activity in the chloroplasts of leaf tissue and also the works of Hadziyev et al. (1969) and Hsiao (1970) concerning the localization of ribonuclease.

That some hydrolytic activity could be found in the supernatant in some instances could be either an artifact of grinding, or the confirmation of a lysosomal character of the vacuole (Matile, 1969).

However, if the vacuolar sap is extracted twice with an hydraulic press (after freezing and thawing), the sap, amounting to 30 % of cellular water content, contains only 1.25 % of the total acid phosphatase activity (Vieira da Silva, 1970).

Most of studies concerning the lysosomes in plants have been made either in nonphotosynthetic or in etiolated tissues, and many instances of lysosomal activity in plants have been recorded (see review by Gahan, 1973).

One of the most interesting works, that of Pitt and Galpin (1973) with etiolated potato shoots, shows a heavy lysosomal fraction (density 1.10) containing acid phosphatase, as well as a low density (1.07) fraction. In cotton however, *G. hirsutum* chloroplasts sediment at a density of 1.13 in an isoosmotic gradient and it is possible that the heavy fraction observed by Pitt and Galpin (1973) contains proplastides with a density near that of chloroplasts. Nevertheless, in some osmotic stress treatment with cotton (Vieira da Silva, 1970) resulting in an empty chloroplastic compartment, a peak of light particles (density 1.05) can be seen in the density gradient. Since this peak was never observed in the controls, and is accompanied by a chloroplastic compartment void of acid phosphatase, there is a possibility that the light particles originate in this compartment. Also, electron micrographs by Ragetli et al. (1966) and Vieira da Silva et al. (1974) show phosphatase activity to be linked to the lamellae of chloroplasts.

Therefore, the conclusion can be made that the heavy organelles in the green leaf tissue of cotton harbor most of the hydrolytic activity usually associated with lysosomal bodies, and that water stress destroys this compartmentation. In other tissues, the compartments containing hydrolases can probably be different (Pitt, 1975).

III. Effects of Water Stress on the Ultrastructure of the Cell

1. General Influences on the Cell Organelles

Only few studies have been devoted to the influence of water stress on the cell organelles. Nir et al. (1970) observe that after water loss, *Zea mays* root cells contain mitochondria which are almost devoid of cristae, and that the cytoplasm harbors numerous lipid droplets which disappear on rehydration. However, in the experiments of Guillot-Salomon (1972), high tonicity of incubating medium (with mannitol) produces contracted mitochondria with dense stroma. Maize leaf tissue was studied by Giles et al. (1974) who find great differences in the responses of bundle sheath and mesophyll cells to water stress. The former appear to be more resistant to stress than the latter, and with progressing stress, chloroplast ballooning as well as formation for cytoplasmic vesicles and finally (stress - 19 bars), disruption of tonoplast, could be observed. After destruction of the tonoplast, mesophyll cells could not recover. Sorghum cells show a much greater resistance to water-stress damage. In these experiments, however, it is difficult to decide on the greater resistance of bundle sheath cells, because it is possible that they have a higher water potential than mesophyll cells, since the heterogeneity of water content in the leaf cannot be ruled out (Sant-Ibanez, 1974). In the fungi *Pseudoperonospora cubensis* and *Phytophthora infestans* (Cohen et al., 1974), partial deterioration of the mitochondria cristae can also be observed with stress, this effect being less advanced in the drought-resistant species *Ps. cubensis*.

Salinity appears to give results similar to those of water stress, producing in *Atriplex* (Blumental-Goldschmidt and Poljakoff-Mayber, 1968) numerous large lipid droplets in the cell, swelling of chloroplasts and mitochondria and extensive vacuolization with distortion of tonoplast. In cotton, salinity (Gausman et al., 1972) induces shrinkage, lamellae swellings, and condensed mitochondria, with oxalate crystals occurring in the vacuole.

Experiments (Vieira da Silva, 1974; Vieira da Silva et al., 1974; Pham Thi and Vieira da Silva, 1975a) with water-stressed cotton leaf tissue, using polyethylen glycol 600 as the osmotic agent, give more or less the same results. The ultrastructural destructive effects of water stress seem to be different in the drought-sensitive *G. hirsutum* from those in the African wild drought-resistant *G. anomalum*.

Duration of stress treatment seems to be extremely important and the results of the treatment on cellular ultrastructure are summarized in Table 1 and in the electron micrographs presented in Figs. 3–10.

During the first few hours of stress *G. hirsutum* shows plasmolysis (Fig. 5) and contraction of protoplasm, bringing the organelles together (Fig. 4). Lipid droplets are found in chloroplasts (Fig. 4) and vacuole (Fig. 5). Afterwards plasmolysis disappears, but cytoplasmic ribosomes are less frequent and many cytoplasmic vesicles appear (Figs. 6 and 8).

Chloroplasts seem to be the organelles more sensitive to stress. In *H. hirsutum* swelling of thylakoids and disappearance of chloroplastic ribosomes (Fig. 4) are among the earliest effects observable. Some conspicuous spheric bodies, seemingly different from lipid droplets, exist in the cotton chloroplasts. These intrachloroplastic bodies do not seem to be much affected by water stress in *G. hirsutum*. In *G. anomalum*, however, while the effect of stress in the membranes is much less important (Figs. 10–15), and never achieves the destruction by long stress as observed in *G. hirsutum* (Figs. 4–8), these intrachloroplastic bodies seem to increase in number and size considerably after a long stress (Fig. 10). These bodies appear to be linked with the dynamic of chloroplast recovery (results not published).

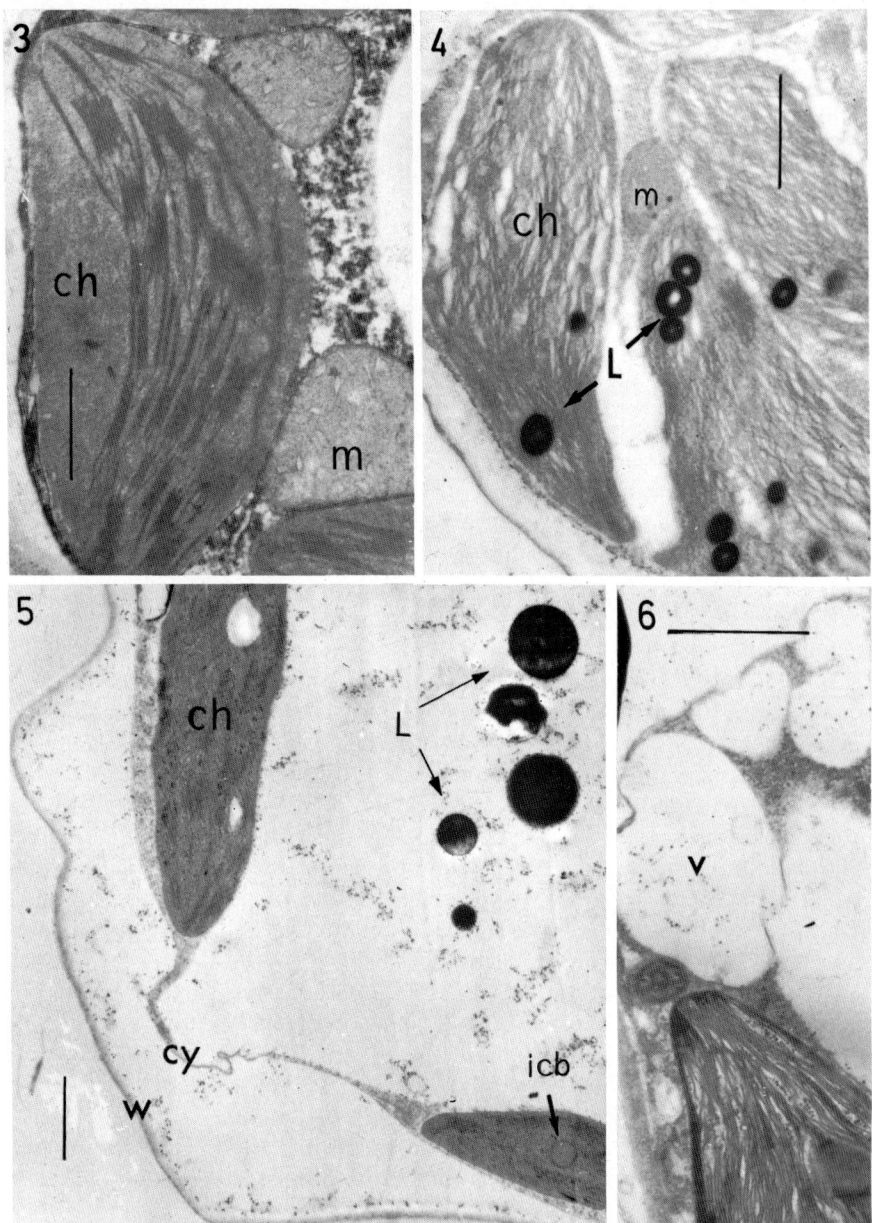

Fig. 3. *G. hirsutum* control, lead post staining
Fig. 4. *G. hirsutum* osmotic stress of -30 Joules mole^{-1} for 2 h
Fig. 5. *G. hirsutum* osmotic stress of -30 Joules mole^{-1} 2 h lead post staining. Plamolysis, cytoplasm detached from cell wall
Fig. 6. *G. hirsutum* osmotic stress of -30 Joules mole^{-1} for 4 h, permanganate post staining

m mitochondria; *ch* chloroplast; *L* lipid droplets; *v* Cytoplasmic vesicles; *icb* intrachloroplastic bodies; *w* cell wall; *cy* cytoplasm. In each figure the bar represents 1 µm length

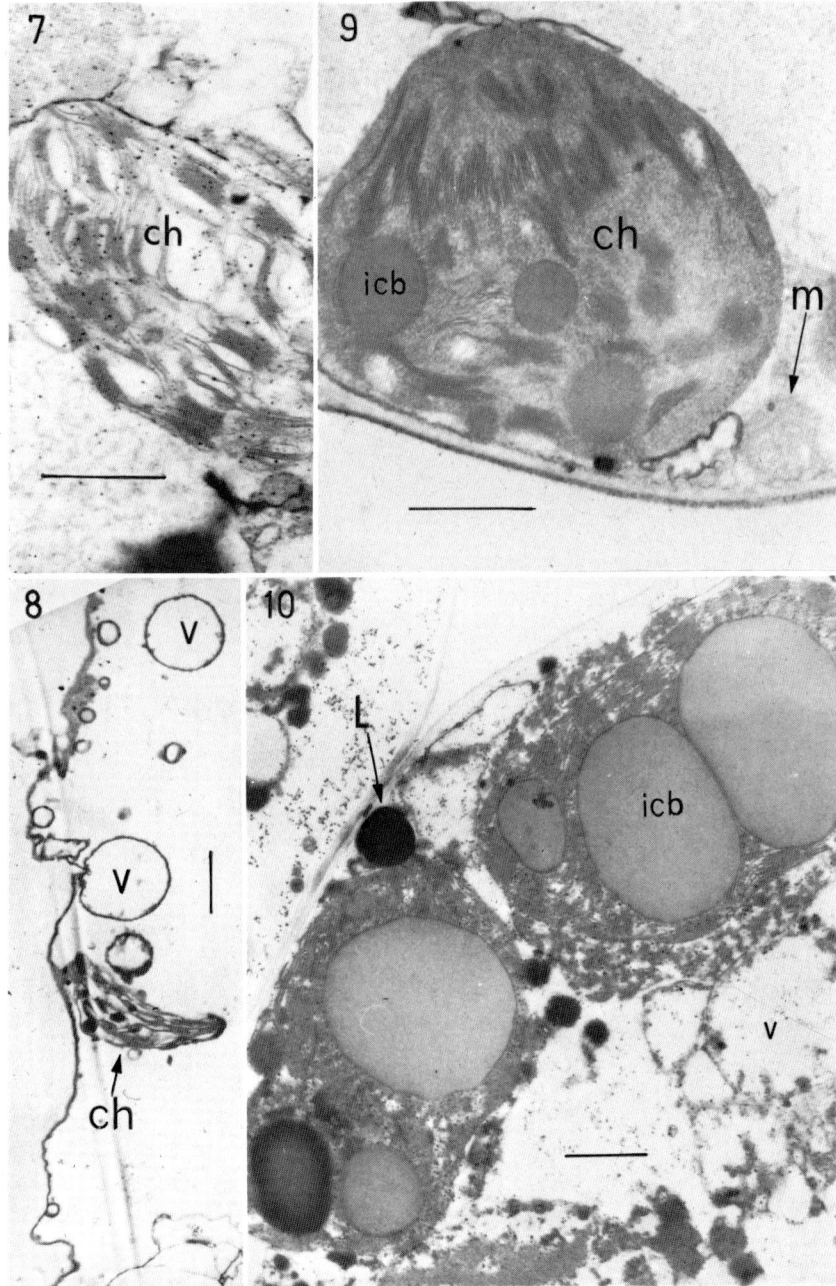

Fig. 7. *G. hirsutum* osmotic stress of -30 Joules mole^{-1} for 24 h lead post staining
Fig. 8. *G. hirsutum* osmotic stress of -30 Joules mole^{-1} for 26 h
Fig. 9. *G. anomalum* control, permanganate post staining
Fig. 10. *G. anomalum*, osmotic stress of -30 Joules mole^{-1} for 24 h, lead post staining
m mitochondria; *ch* chloroplast; *L* lipid droplets; *v* cytoplasmic vesicles; *icb* intrachloroplastic bodies. In each figure the bar represents 1 μm length

Table 1. Effect of length of osmotic stress treatment (polyethylene glycol 600, −30 Joules mole^{-1}) on ultrastructure of drought sensitive (*G. hirsutum*) and drought resistant (*G. anomalum*) cotton species

		2 h	4 h	17–26 h
G. hirsutum	Cytoplasma	Plasmolysis, lipid droplets in the vacuole	Retraction of cytoplasma. Vacuolization started from invaginations of cytoplasmic membranes. disorganization of ergastoplasma.	Cytoplasma highly vacuolized or empty
	Chloroplasts	Grana less organized, sometimes lipid droplets in the stroma	Grana disappears. Swelling of thylakoids. Stroma clear (ribosomes disappear). Osmiophilic droplets.	Disorganized thylakoid fragments in the cytoplasma
	Mitochondria	Swelling of cristae	Cristae rare or lacking entirely. Stroma empty.	Destroyed
	Peroxisomes	No effects	No effects, but less frequent	Disappear
G. anomalum	Cytoplasma	No effects	Plasmolysis, secondary vacuoles	Ergastoplasma disorganized. Cytoplasma clear and highly vacuolized.
	Chloroplasts	No effects	Swelling of thylakoids. Grana remains. Intrachloroplastic bodies digested.	Thylakoids are still observable, but a clear stroma. Lipid droplets. Great number of hypertrophied intrachloroplastic bodies.
	Mitochondria	No effects	Cristae less visible and in some cases clear stroma	Empty
	Peroxisomes	No effects	Rare	Disappear

Mitochondria show the same reactions that were observed by other authors: swelling, decrease of cristae contrast, and finally internal destruction. External membrane seems to resist longer that the internal contents. The peroxisomes do not appear to change with stress treatment, but they become rate.

It seems that drought resistance is tantamount in the cotton to the maintenance of structural integrity in the cell organelles, as shown by *G. anomalum*. The role of intrachloroplastic bodies in resistance and recovery manifestations, needs to be investigated. These bodies seem to be different from lipid droplets and only further research can reveal something about their nature. They are probably either protein (Ames and Pivorun, 1974) or phospholipids.

2. Loss of Ribosomes from the Chloroplasts

That the effect of water stress can change the chloroplastic compartment has already been demonstrated. Leakage and alteration of membranes can result in the loss of enzymatic protein to the soluble phase or, even more strikingly, in the loss of ribosomes from stressed chloroplasts (observed by Philippovitch et al., 1967) incubated under hypertonic conditions.

Results with cotton (Marin and Vieira da Silva, 1972) show that RNA content in the cell is not affected by water loss if it is less than 10% of the maximum cellular water content. The next 10% of water loss, however, decreases the RNA content by 40% and this rate continues linearly till 30% water loss is reached (wilting stage). Decrease of RNA by further water loss is slow.

If ribosomal RNA is taken into account, the 30% water deficit marks the point of sudden increase in ribosomal cytoplasmic content, which is the result of a sudden leakage of ribosomes outside the chloroplast. The chloroplastic ribosome content had already been reduced by 80% with the first 20% water loss, confirming electron microscopy observations. The decrease in the quantity of chloroplastic ribosomes seems therefore to be first the result of nuclease activity (McHale and Dove, 1968; Hadziyev et al., 1969; Hsiao, 1970; Vieira da Silva, 1970; Marin and Vieira da Silva, 1972), and, after wilting, to be the result of leakage from the chloroplasts to the cytoplasm. Loss of turgor in the cell is therefore probably a critical point, as has already been shown by Hsiao (1973).

In revivescent mosses it seems that polyribosomes are conserved during desiccation (Bewley, 1973), and that the site of protein synthesis after recovery is the cytoplasm (Tucker and Bewley, 1974). It remains to be seen if this is the result of a particularly resistant chloroplastic compartment or a peculiarity of the revivescent mosses.

IV. Relationships of Ultrastructural Alteration and Hydrolytic Enzyme Decompartmentation and Activation, with Alteration of Chloroplasts and Mitochondria Metabolism

That the alterations of ultrastructure can bring about enzyme decompartmentation seems to be a generally acceptable hypothesis. Nevertheless it is difficult to imagine that such small decreases of water potential and water content can have such great repercussion on the organelles structure, and given the lipoprotein composition of cellular and organelles membranes, it seems more probable that a biochemical action could bring about such a result. In fact, the results of experiments with cotton (Vieira da Silva et al., 1974) show that water stress activates both acid and alkaline lipases, the former in the mitochondria cristae and between the chloroplast lamellae (Fig. 11), and the latter between the chloroplast lamellae (Figs. 13 and 14). That drought-resistant G. anomalum does not show such an activation (Vieira da Silva, 1974), and appears to conserve a fair degree of ultrastructure integrity with stress, seems to confirm the hypothesis further. Nir et al. (1970) have already suggested that lipids are displaced from their position in the mitochondrial and other cellular membranes.

The effect of water stress on dark respiration seems to differ with species (Brix, 1962) or according to the degree and length of stress (Greenway and Hiller, 1967). Some authors (Schneider and Childers, 1941; Upchurch et al., 1955; Iljin, 1957) find that dark respiration increases with stress. Others (Boyer, 1965; Pinto and Flowers, 1970) observe that respiration decreases with drought. Finally Santarius (1967) finds no effect at all. Mayer and Plantefol (1925) observe with revivescent mosses, that not only respiration decreases with dessication but that the respiratory coefficient doubles, showing the establishment of a new respiratory process. Nir et al. (1970) however, observe that the activity of cytochrome oxidase increases four-fold while oxygen uptake decreases. With cotton, Nguyen Duc and Vieira da Silva (1972) confirm the results of Mayer and Plantefol (1925) but find (Pham Thi and Vieira da Silva, 1975b), by polarographic measures of O_2 uptake, that respiration of stressed G. hirsutum increases slightly during the first hour of treatment and decreases steadily afterwards. With the drought-resistant G. anomalum, however, there is only a slight decrease during a 7-h treatment (-30 Joules mole^{-1}).

These results should be linked with ultrastructural changes observed in mitochondria. Miller et al. (1971) confirm that water stress has a marked effect on the membranes of mitochondria, altering their swelling characteristics. Cohen et al. (1974) observe inhibition of oxygen uptake by stressed fungi sporangia, Zholkevich and Rogacheva (1968) show that drought results in uncoupling of oxidative phosphorylation, and Nir et al. (1970) compare drought action in mitochondria with detergent treatment (Simon, 1958) known to solubilize the lipids.

Fig. 11. G. hirsutum osmotic stress of -18 Joules mole^{-1} for 20 h, incubated for acid ▶ lipase, uranyl acetate post staining
Fig. 12. G. hirsutum control of Fig. 11 incubated for acid lipase, uranyl acetate post staining
Fig. 13. G. hirsutum osmotic stress of -18 Joules mole^{-1} for 20 h, incubated for alkaline lipase, uranyl acetate post staining
Fig. 14. G. hirsutum same as Fig. 13. Shows lead deposits representing alkaline lipase activity, between thylakoids
Fig. 15. G. hirsutum control of Fig. 13, incubated for alkaline lipase, uranyl acetate post staining
(Figs. 11, 12, 13, 15 after Vieira da Silva et al., 1974).

m mitochondria; ch chloroplast; s starch; p peroxisomes; icb intrachloroplastic bodies; acl acid lipase; akl alkaline lipase. In each figure the bar represents 1 μm length, except in Fig. 14

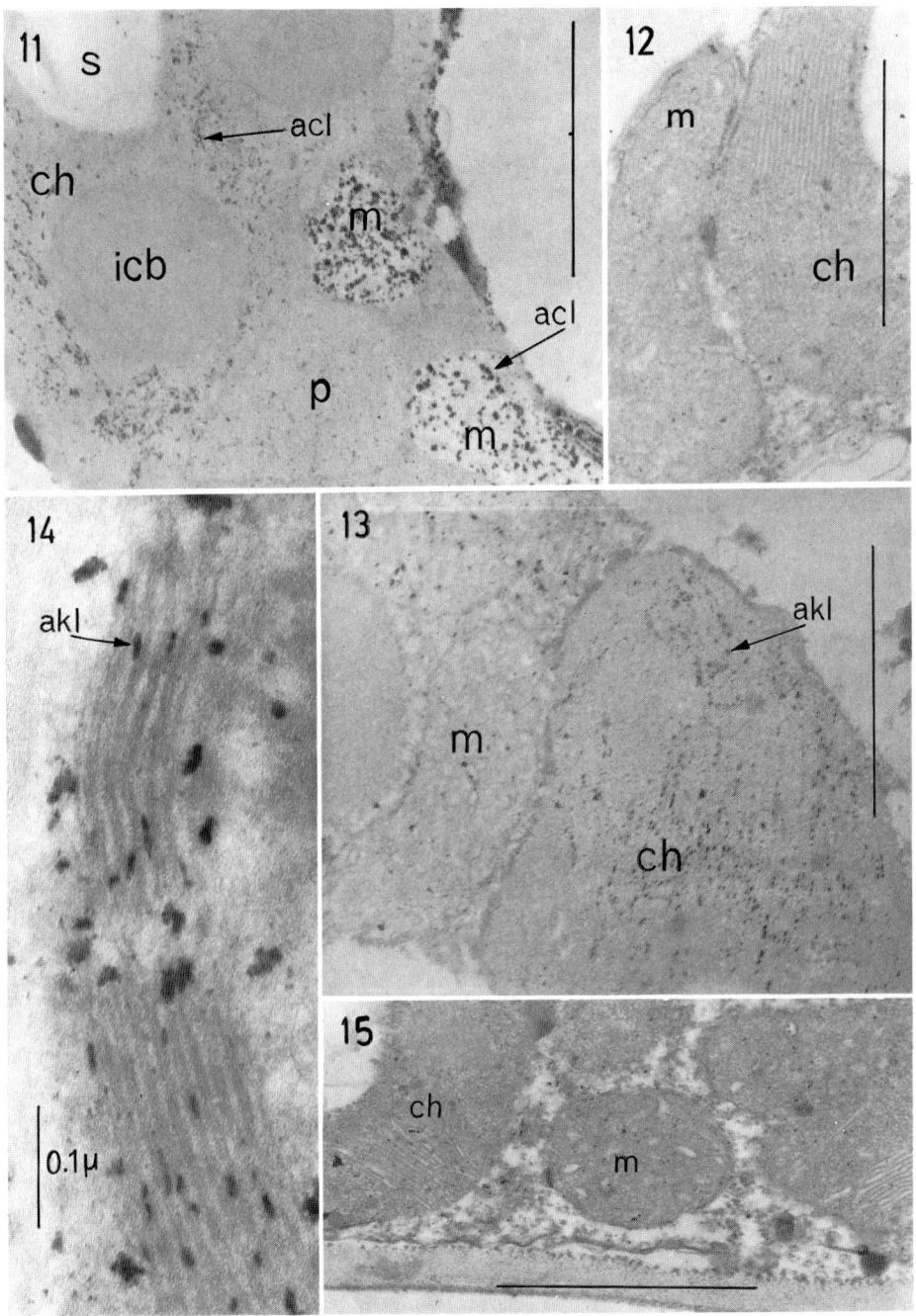

Figs. 11–15

It is easy to link the observed phenomena to alteration of membrane and ultrastructure, and acid lipase activation by water stress in mitochondria.

Photorespiration is decreased by water stress (Nguyen Duc and Vieira da Silva, 1971, 1972) even if glycolate oxidase is not affected. It seems that it is the decarboxylation of glycine in mitochondria that decreases with dessication. Infiltration of stressed discs with glycine could not bring about CO_2 evolution (Pham Thi, 1972).

Enhanced lipase activity (Fig. 11) could lead to reduced glycine decarboxylation. This is supported by the finding that inhibition of the decarboxylation of glycine occurs after lipase treatment of mitochondria (Kisaki et al., 1971). The decrease in photorespiration with drought is much smaller in G. anomalum (Pham Thi and Vieira da Silva, 1975b), which shows no lipase activity in mitochondria (Vieira da Silva, 1974), than in G. hirsutum, which shows lipase activity with drought.

The effects of water stress on photosynthesis (MacCarty and Jagendorf, 1965; Bamberger and Park, 1966; Nir and Poljakov-Mayber, 1967; Fry, 1970; Vieira da Silva and Veltkamp, 1970) appear to be duplicated by the action of lipase (Okayama, 1964; Mantai, 1970; Butler and Okayama, 1971) and the addition of certain unsaturated fatty acids (Bamberger and Park, 1966; Constantopoulos and Kenyon, 1968; Krogmann and Jagendorf, 1959) that could result from lipolytic action. Enhanced lipase activity in water-stressed leaf can therefore account for a significant part of the long-lasting effect of water stress on the chloroplast. Depression of photochemical reactions is known to occur in the presence of free fatty acids (Krogmann and Jagendorf, 1959; Nir and Poljakoff-Mayber, 1967; Boyer and Bowen, 1970; Fry, 1970; Butler and Okayama, 1971). Specifically fatty acids interfere with the Hill reaction.

Inhibition of CO_2 absorption by inorganic phosphate in isolated chloroplasts was observed by Champigny and Miginiac-Maslow (1971).

Inorganic phosphatate liberated by dehydration, has been shown to increase reversibly the CO_2 compensation point. Here also, the drought-resistant G. anomalum does not show such a reversible increase in CO_2 compensation point. The long-lasting effect of water stress that can be attributed to lipase, is smaller in this species than in G. hirsutum.

The same phenomena can be observed (unpublished results) when drought-resistant and drought-sensitive lines of the oil palm Elaeis guineensis are compared. The latter show a greater solubilization of acid phosphatase, and a reversible increase of CO_2 compensation point with stress. The long-lasting effect of stress on CO_2 compensation point is also more pronounced in the drought-sensitive line.

V. Conclusions

Water stress results, in sensitive plants, in a disruption of compartmentation in the leaf cells. Liberated, and previously latent, acid hydrolases can have a profound destructive action, and most of the metabolic effects of water stress could be explained by such destructions.

This situation is similar to that of lysosomes that can liberate acid hydrolases in mechanically damaged, diseased, or senescent tissues (Pitt, 1975). In cotton, however, the chloroplast seems the most critical compartment as far as acid phosphatase is concerned.

Cytochemical and ultrastructural studies with stressed tissue point to the activation of acid and alkaline lipases as the phenomena responsible for the alteration of membranes and for decompartmentation. That drought-resistant plants do not show such an activation, and have fairly resistant compartments is a further argument in this line of thought.

Nevertheless the trigger process in the lipase activation is unknown. It could be a process linked with changes in turgor, and this would explain the effect of slight water stress and dry atmosphere (Bourque and Naylor, 1971), or it could involve a complex metabolic process linked with cyclic AMP, as is proposed in the explanation of lipolysis in mammals (Burns et al., 1972). The effect of plant growth substances as triggers for such process should also not be neglected.

For the present, however, it appears that the observed processes could explain, partially at least, the decrease of CO_2 absorption at the chloroplast level, the reduction of photochemical activity, the reduction of photorespiration and the reduction of dark respiration.

Recovery from water stress is currently under study and it seems that drought-resistant cotton species have a better recovering mechanism than sensitive species (unpublished results). Also it is attractive to think about the increase in intrachloroplastic bodies with prolonged stress in *G. anomalum* (Fig. 10) with respect to such a capacity for recovery.

Iljin's (1935, 1953) hypothesis of destruction of the structure of protoplasm by dehydration and rehydration that results in the death of tissues with water stress, seems at least partially confirmed by present research. In fact, if *G. hirsutum* leaf discs are floated for 4h over a polyethylen glycol solution of -30 Joules mole^{-1}, a slight increase in the CO_2 compensation point occurs, but if the same discs are floated afterwards over water, the increase in turgor disrupts the already weak membrane system, and CO_2 compensation point jumps to 300 ppm and increases quickly thereafter (Pham Thi and Vieira da Silva, 1975b).

It seems that there is a point of no return in the destruction of leaf cell compartments, that recovery is not possible after such serious injuries and that rehydration aggravates the stress damage.

References

Ames, I. H., Pivorun, J. P.: A cytochemical investigation of a chloroplast inclusion. Am. J. Botany **61**, 794–797 (1974).

Bagi, G., Farkas, G. L.: On the nature of increase in ribonuclease activity in mechanically damaged tobacco leaf tissues. Phytochemistry **6**, 161–169 (1967).

Bamberger, E. S., Park, R. B.: Effect of hydrolytic enzymes on the photosynthetic efficiencies and morphology of chloroplasts. Plant Physiol. **41**, 1591–1600 (1966).

Bewley, J. D.: Desiccation and protein synthesis in the moss *Tortula ruralis*. Can. J. Botany **51**, 203–206 (1973).

Blumental-Goldschmidt, S., Poljakoff-Mayber, A.: Effect of substrate salinity on growth and on submicroscopic structure of leaf cells of *Atriplex halimus* L. Australian J. Botany **16**, 469–478 (1968).

Bourque, D. P., Naylor, A. W.: Large effects of small water deficits on chlorophyll accumulation and ribonucleic acid synthesis in etiolated leaves of Jack Bean [*Canavalia ensiformis* (L.) DC.]. Plant Physiol. **47**, 591–594 (1971).

Boyer, J. S.: Effects of osmotic water stress on metabolic rates of cotton plants with open stomata. Plant Physiol. **45**, 229–234 (1965).

Boyer, J. S., Bowen, B. L.: Inhibition of oxygen evolution in chloroplasts isolated from leaves with low water potentials. Plant Physiol. **45**, 612–615 (1970).

Brix, H.: The effects of water stress on the rates of photosynthesis and respiration in tomato plants and loblolly pine seedlings. Physiol. Plantarum **15**, 10 (1962).

Burns, T. W., Langley, P. E., Robison, G. A.: Studies on the role of cyclic AMP in human lipolysis. Advan. Cyclic Nucleotid Res. **1**, 63–85 (1972).

Butler, W. L., Okayama, S.: The photoreduction of C-550 in chloroplasts and its inhibition by lipase. Biochim. Biophys. Acta **245**, 237–239 (1971).

Camefort, H.: Evolution de la structure des plastes pendant la maturation de l'arille de l'If (*Taxus baccata* L.). C. R. Acad. Sci. (Paris) **258**, 1 071–1 020 (1964).

Champigny, M. L., Miginiac-Maslow, M.: Relations entre l'assimilation photosynthétique de CO_2 et la photophosphorylation de chloroplastes isolés. Biochim. Biophys. Acta **234**, 335–343 (1971).

Cohen, Y., Perl, M., Rotem, J., Eyal, H., Cohen, J.: Ultrastructural and physiological changes in sporangia of *Pseudoperonospora cubensis* and *Phytophthora infestans*, exposed to water stress. Can. J. Botany **52**, 447–450 (1974).

Constantopoulos, G., Kenyon, C. N.: Release of free fatty acids and loss of Hill activity by aging spinach chloroplasts. Plant Physiol. **43**, 531–536 (1968).

De Duve, C., Wattiaux, R.: Functions of lysosomes. Ann. Rev. Physiol. **28**, 435–492 (1966).

Filippovich, I. I., Spandararyan, O. A., Svetailo, E. N., Sissakian, N. M.: State of ribosomes in chloroplasts. Dokl. Akad. Nauk SSSR **172**, 1 214–1 217 (1967).

Fry, K. E.: Some factors affecting the Hill reaction activity in cotton chloroplasts. Plant Physiol. **45**, 465–469 (1970).

Gahan, P. B.: Plant lysosomes. In: Frontiers of biology, lysosomes in biology and pathology, Vol. 3 (ed. J. T. Dingle), pp. 68–85. Amsterdam, London: North Holland Publ. Co. 1973.

Gausman, H. W., Baur, P. S., Porterfield, M. P., Cardenas, C.: Effects of salt treatment of cotton plants (*Gossypium hirsutum* L.) on leaf mesophyll cell microstructure. Agron. J. **64**, 133–136 (1972).

Giles, K. L., Beardsell, M. F., Cohen, D.: Cellular and ultrastructural changes in mesophyll and bundle sheath cells of maize in response to water stress. Plant Physiol. **54**, 208–212 (1974).

Greenway, H., Hiller, R. G.: Effects of low water potentials on respiration and on glucose and acetate uptake, by *Chlorella pyrenoidosa*. Planta (Berl.) **75**, 253–274 (1967).

Guillot-Salomon, T.: Variations de ultrastructure de mitochondries isolées de tissus végétaux, en fonction de la tonicité du milieu de suspension. C. R. Acad. Sci. (Paris) **274**, 869–872 (1972).

Hadziyev, D., Mehta, S. L., Zalik, S.: Nucleic acid ribonuclease of water leaves and chloroplasts. Can. J. Biochem. **47**, 273–282 (1969).

Hsiao, T. C.: Rapid changes in levels of polyribosomes in *Zea mays* in response to water stress. Plant Physiol. **46**, 281–285 (1970).

Hsiao, T. C.: Plant responses to water stress. Ann. Rev. Plant Physiol. **24**, 519–570 (1973).

Iljin, W. S.: Die Lebensfähigkeit der Pflanzenzellen in trockenem Zustand. Planta (Berl.) **24**, 742–745 (1935).

Iljin, W. S.: Causes of death of plants as a consequence of loss of water conservation of life in desiccated tissues. Bull. Torrey Botan. Club **80**, 166–177 (1953).

Iljin, W. S.: Drought resistance in plants and physiological processes. Ann. Rev. Plant Physiol. **8**, 257 (1957).

Kisaki, T., Ayako, I., Tolbert, N. E.: Intracellular localization of enzymes related to photorespiration in green leaves. Plant Cell Physiol. (Tokyo) **12**, 267–273 (1971).

Krogmann, D. W., Jagendorf, A. T.: Inhibition of the Hill reaction by fatty acids and metal chelating agents. Arch. Biochem. Biophys. **80**, 421–430 (1959).

Kursanov, A. L.: The reversible effect of enzymes in the living cell (in Russian). Moscow and Leningrad Acad., 1940.

Kursanov, A. L.: The adsorption of enzymes by tissues of higher plants (in Russian). Biochimija **11**, 333–348 (1946).

MacCarty, R. E., Jagendorf, A. T.: Chloroplast damage due to enzymatic hydrolysis of endogenous lipids. Plant Physiol. **40**, 725–735 (1965).

Mantai, K. E.: Some effects of hydrolytic enzymes on coupled and uncoupled electron flow in chloroplasts. Plant Physiol. **45**, 563–566 (1970).

Marin, B., Vieira da Silva, J.: Influence de la carence hydrique sur la repartition cellulaire de l'acide ribonucleique foliaire chez le cotonnier. Physiol. Plantarum **27**, 150–155 (1972).

Matile, Ph.: Vacuoles as lysosomes of plant cells. Biochem. J. **111**, 26 (1969).

Maximov, N. A.: The physiological nature of drought resistance of plants. Proc. Intern. Congr. Plant Sci. (Ithaca) **2**, 1169–1175 (1929).

Mayer, A., Plantefol, L.: Equilibre des constituants cellulaires et forme des oxydations de la cellule. Imbibition et types respiratoires chez les plantes revivescentes. C. R. Acad. Sci. (Paris) **181**, 131–132 (1925).

McHale, J. S., Dove, L. D.: Ribonuclease activity in tomato leaves as related to development and senescence. New Phytologist **67**, 505–515 (1968).

Miller, R. J., Bell, D. T., Koeppe, D. E.: The effects of water stress on some membrane characteristics of corn mitochondria. Plant Physiol. **48**, 229–231 (1971).

Mothes, K.: Der Einfluß des Wasserzustandes auf Fermentprozeß und Stoffumsatz. In: Handbuch der Pflanzenphysiologie, Vol. 3 (ed. W. Ruhland), pp. 656–664. Berlin: Springer 1956.

Nguyen Duc, A. T., Vieira da Silva, J.: Evolution du point de compensation de CO_2 chez le Cotonnier sous l'action de traitements osmotiques. C. R. Acad. Sci. (Paris) **273**, 1291–1294 (1971).

Nguyen Duc, A. T., Vieira da Silva, J.: Action d'un traitement osmotique sur la respiration à l'obscurité et sur l'émission de CO_2 à la lumière chez le *Gossypium hirsutum* L., C. R. Acad. Sci. (Paris) **274**, 3234–3237 (1972).

Nir, I., Poljakoff-Mayber, A.: The effect of water stress on the activity of phosphatases from Swiss chard chloroplasts. Israel J. Botany **15**, 12–16 (1966).

Nir, I., Poljakoff-Mayber, A.: Effects of water stress on the photochemical activity of chloroplasts. Nature **213**, 418–419 (1967).

Nir, I., Poljakoff-Mayber, A., Klein, S.: The effects of water stress on mitochondria of root cells. Plant Physiol. **45**, 173–177 (1970).

Okayama, S.: Effect of lipase-digestion on photochemical activities of isolated chloroplasts. Plant Cell Physiol. (Tokyo) **5**, 145–156 (1964).

Oparin, A. I.: Richtungseinstellung der Invertasewirkung in der lebenden Pflanzenzelle. Enzymologia **4**, 13–23 (1937).

Oparin, A. I.: Variations de l'activité des enzymes dans la cellule végétale sous l'effet des facteurs extérieurs. Bull. Soc. Chim. Biol. **35**, 67–82 (1953).

Oparin, A. I., Kaden, S. B.: The transformation of beta-amylase in germinating wheat seeds (in Russian). Biochimija **10**, 25–36 (1945).

Pham Thi, A. T.: Contribution à l'étude de l'action de la sécheresse sur la photosynthèse et la respiration du cotonnier (*Gossypium hirsutum* L.). Doctorate thesis, Paris-South, 1972.

Pham Thi, A. T., Vieira da Silva, J.: Action d'un traitement osmotique sur l'ultrastructure de feuilles du cotonnier (*Gossypium hirsutum* L. et *G. anomalum* Waw. et Peyr.); C. R. Acad. Sci. (Paris) **280**, 2857–2860 (1975a).

Pham Thi, A. T., Vieira da Silva, J.: Action de deficits hydriques sur la photosynthèse et sur la respiration des feuilles du cotonnier. In: IBP publication (ed. Alexis Moyse), (in press, 1975b).

Philippovitch, I. I., Spandaryan, O. A., Svetailo, E. N., Sissakyan, N. M.: State of Ribosomes in chloroplasts. Dokl. Akad. Nauk SSSR **172**, 1214–1217 (1967).
Pinto, C. M. D., Flowers, T. J.: The effects of water deficits on slices of beet root and potato tissue. J. Exp. Botany **21**, 754–767 (1970).
Pitt, D.: Lysosomes and cell function. London, New York: Longman 1975.
Pitt, D., Galpin, M.: Isolation and properties of lysosomes from dark grown potato shoots. Planta (Berl.) **109**, 233–258 (1973).
Presley, H. J., Fowden, L.: Acid phosphatase and isocitritase production during seed germination. Phytochemistry **4**, 169–176 (1965).
Ragetli, H. W. J.: Virus host interactions, with emphasis on certain cytopathic phenomena. Can. J. Botany **45**, 1221–1234 (1967).
Ragetli, H. W. J., Weintraub, M., Rink, U. M.: Latent acid phosphatase in chloroplasts. Can. J. Botany **44**, 1723–1725 (1966).
Santarius, K. A.: Das Verhalten von CO_2-Assimilation, NADP und PGS-Reduktion und ATP-Synthese intakter Blattzellen in Abhängigkeit vom Wassergehalt. Planta (Berl.) **73**, 228–242 (1967).
Sant-Ibanez, F.: Etude des variations de la teneur en eau des feuilles au moyen d'une méthode rapide et non destructive. Doctorate thesis, Paris VII, 1974.
Schneider, G. W., Childers, N. F.: Influence of soil moisture on photosynthesis, respiration, and transpiration of apple leaves. Plant Physiol. **16**, 565–583 (1941).
Simon, E. W.: The effect of digitonin on cytochrome oxidase activity of plant mitochondria. Biochem. J. **69**, 67–74 (1958).
Todd, G. W.: Water deficits and enzymatic activity. In: Water deficits and plant growth, Vol. 3 (ed. T. T. Kozlowski), pp. 177–216. New York, London: Academic Press 1972.
Tucker, E. B., Bewley, J. D.: The site of protein synthesis in the moss *Tortula ruralis* on recovery from desiccation. Can. J. Biochem. **52**, 345–348 (1974).
Upchurch, R. P., Peterson, M. L., Hagan, R. M.: Effect of soil moisture content on the rate of photosynthesis and respiration in ladino clover (*Trifolium repens* L.). Plant Physiol. **30**, 297–303 (1955).
Vieira da Silva, J.: Le potentiel osmotique du milieu de culture et l'activité soluble et latente de la phosphatase acide dans le *Gossypium thurberi*. C. R. Acad. Sci. (Paris) **267**, 729–732 (1968).
Vieira da Silva, J.: Comparaison entre cinq espèces de *Gossypium* quant à l'activité de la phosphatase acide après un traitement osmotic. Etude de la vitesse de solubilisation et de formation de l'enzyme. Z. Pflanzenphysiol. **60**, 385–387 (1969).
Vieira da Silva, J.: Contribution à l'étude de la résistance à la sécheresse dans le genre *Gossypium*. II. La variation de quelques activités enzymatiques. Physiol. Végétale **8**, 413–447 (1970).
Vieira da Silva, J.: Ultrastructural and enzymatic effects of water stress in cotton. Abstr., 73. Intern. Ass. Plant Physiol., 1st Meet. Würzburg (1974).
Vieira da Silva, J., Naylor, A. W., Kramer, P. J.: Some ultrastructural and enzymatic effects of water stress in cotton (*Gossypium hirsutum* L.) leaves. Proc. Nat. Acad. Sci. **71**, 3243–3247 (1974).
Vieira da Silva, J., Poisson, Ch.: Solubilisation hydrolytiques chez *Gossypium hirsutum*, *G. anomalum* et des dérivés de l'hybridation entre des deux espèces. Can. J. Genet. Cyt. **11**, 582–586 (1969).
Vieira da Silva, J., Veltkamp, J.: Action du potential somotique de la solution nutritive sur la réaction de Hill et la photophosphorylation de chloroplastes de cotonnier. C. R. Acad. Sci. (Paris) **271**, 1376–1379 (1970).
Yin, H. C.: A histochemical study of the distribution of phosphatase in plant tissues. New Phytologist **44**, 191–195 (1945).
Young, J. L., Varner, J. E.: Enzymes synthesis in the cotyledons of germinating seeds. Arch. Biochem. Biophys. **84**, 71–78 (1959).
Zholkevich, V. N., Rogacheva, A. Y.: Ratio of phosphorylation to oxydation in mitochondria from wilting plant tissues. Fiziol. Rast. **15**, 537–545 (1968). (English summary); Biol. Abstr. **50**, 894033 (1969).

B. Water Stress and Hormonal Response

C. Itai and A. Benzioni

I. Introduction

A transient water deficit is a frequent and recurrent phenomenon experienced by most plants. Hence it can be assumed that efficient adaptive mechanisms were developed in the course of evolution which enable plants to cope with water deficits of varying intensities. Evidence is accumulating that these adaptive mechanisms whether rapid, like stomatal movement, or slow, like developmental and morphological adaptation, may involve regulation through changes in the hormone balance of the plant (Hsiao, 1973; Livne and Vaadia, 1972). Plant response to environmental changes requires a complex control system. Even a very simple control system is necessarily made up of several components such as receptor, integrator, modulator, effector and amplifier, and growth regulators may logically be presumed to be part of such a system. The possibility that growth regulators are part of the control mechanism of plant response to water stress is further supported by the following:

1. Water stress evokes concurrent responses in different plant organs which are not directly exposed to the stress. This was found in wheat when stress effects were transmitted to the whole plant even when only part of the root was exposed to 1 bar osmoticum (Meyer and Gingrich, 1964). Similarly, the root/shoot ratio increases owing to higher restriction of shoot growth, irrespective of whether plant shoots are exposed to atmospheric stress or roots are exposed to low water potential in their media.

2. It is known that many developmental changes occur under water stress, e.g. shortening of life cycle, abscission, dormancy and induction of flowering. It is generally assumed that these processes are regulated by a control mechanism of which plant hormones constitute an important part.

3. Moderate water deficits effecting only minute changes in the water potential of plant tissue cause considerable metabolic changes. The fact that a given metabolic process assayed in vitro is not affected by a similar change in water potential, indicates that the effect is not direct and may involve regulating components serving as modulators or effectors.

4. Plant response to renewal of the water supply after the stress, is characterized by "overshoot" and the so-called "after-effects" (Miller and Kramer, 1965; Fischer, 1970). An example of an "overshoot" can be seen in growth rates: rewatered, stressed plants exhibit a growth rate surpassing that of controls. The "after-affect"

is the delay in stomatal opening after turgor is regained by water-stressed plants. These phenomena indicate the existence of oscillation and a "memory", both features of control systems.

5. Plant responses to different stresses such as heat, low temperature, drought, and salinity, have common features, despite the diversity of the stress agents (see also Heber and Santarius, this volume, Part 4:D); this also may point to some regulatory mechanism.

6. The intensity of plant response to water stress varies with plant age and growth conditions (e. g. thermo and photoperiod). These differences indicate the possibility that endogenous hormone levels which are influenced by plant age and its growth conditions, act as modulators of plant response.

In the following chapter we examine the evidence of hormonal changes induced by water stress and the physiological significance of such changes. Three experimental approaches are used. One is the determination of endogenous hormone level during stress; the other is to evoke different plant responses to water stress by manipulating plant hormone levels (application of hormones, preconditioning of plants, use of antihormones); and the third is to try to cause stress symptoms by hormonal treatments.

II. Endogenous Hormonal Changes Due to Water Stress

1. Changes in Cytokinin Activity

Water stress caused by different means effects a reduction in cytokinin (CK) activity. Table 1 shows that CK activity in xylem exudates is decreased by drought (Itai and Vaadia, 1965). Similarly, Browning (1973) found in *Coffea arabica* plants which had previously been subjected to water stress, that the CK level in the xylem sap rose after irrigation. Reduction of the water potential of whole plants by osmotica or salinity produced similar results. Table 1 summarizes the results of three separate experiments in which sunflower plants were treated for two days by adding different amounts of solutes to the culture solution. The decrease in CK content of the exudate is linearly related to the amount of NaCl added to the culture solution (Fig. 1).

Table 1 and Fig. 1 show that various stresses effect a reduction in CK activity. However, one instance has been reported where this effect was not manifested. Mizrahi et al. (1971) found that 24-h salination did not affect CK levels in tobacco leaves. The result of this experiment could have been due

Table 1. Effect of osmotic agents added to culture solutions or withholding irrigation on CK content of exudate. Four separate experiments

Treatment	g l^{-1}	Atm	Kinetin equiv.	% of respective control
Water-stressed	—	—	0.075	30.0
NaCl	3	2.4	0.038	40.0
Mannitol	30	3.6	0.018	50.5
Carbowax 6000	100	1.35	0.087	64.4

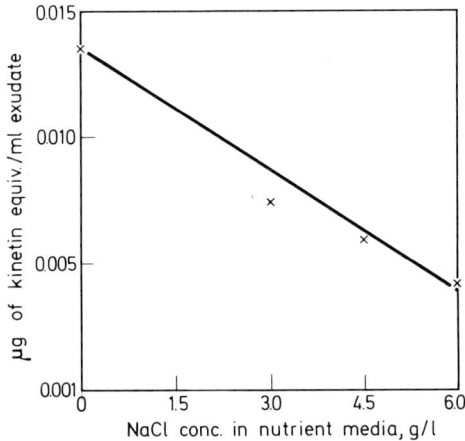

Fig. 1. Kinetin equivalents of fractions obtained from exudate of plants subjected to different concentrations of NaCl. (After Itai et al., 1968)

to assaying leaves of widely differing ages and thus obscuring the changes in CK levels in the stress-responsive leaves.

A decrease in CK content was also found in detached tobacco leaves exposed to atmospheric drought for 15–30 min, during which period they lost 5–10% of their fresh weight (Table 2). Since the shoot depends on the root for its cytokinin supply (Kulaeva, 1962), these data imply that decreased water potential in the leaf increases CK degradation rates. Figure 2 gives the result of an experiment in which ^{14}C kinetin was applied through the petiole before the leaves were air-dried. The results indicate that kinetin metabolism is enhanced during drying. It is significant that, like drought, abscisic acid (ABA) treatment also enhances ^{14}C kinetin metabolism (Fig. 3).

More direct evidence for the effects of ABA on CK levels can be deduced from Table 3. In this experiment, ABA was added to the culture media of 30 bean plants for two days. The plants were topped and the exudate was collected and assayed for CK. The results indicate that ABA indeed affects the level of CK in the exudate.

Table 2. Cytokinin activity in extracts of detached tobacco leaves as a function of drying time (Itai and Vaadia, 1971)

Drying time (min)	Kinetin equiv. $\mu g\ kg^{-1}$ fr. wt.	Activity %
0	43.4	100
15	35.1	81
30	22.9	53

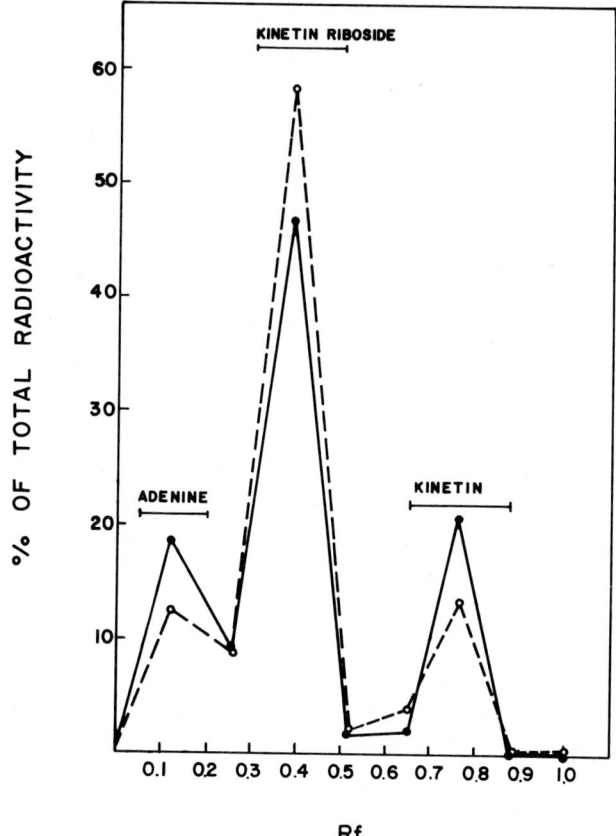

Fig. 2. ^{14}C kinetin metabolism caused by air desiccation of tobacco leaves. The leaf was fed with ^{14}C kinetin, then dried so that it lost about 20% of its initial weight. The leaf was ethanol extracted and the extract separated on TLC using isopropanol : chloroform (4:1 v:v). The result represents one of 4 similar experiments. —●— control ----○---- desiccated (Ze'ev Even-Chen, unpublished)

Table 3. CK concentration in exudate of ABA-treated bean plants

ABA concentration	Exudate per plant	Kinetin equiv.
ppm	ml	$\mu g\ l^{-1}$
0	4.5	283.0
0.01	3.4	192.0
0.10	6.8	84.2

The pattern of recovery of CK levels after stress withdrawal depends upon the nature of the stress applied. In most cases an "overshoot" in the CK level is noted (Fig. 4).

It can be concluded that endogenous CK activity decreases owing to drought, salinity or osmotic stress and that the extent of the decrease correlates with

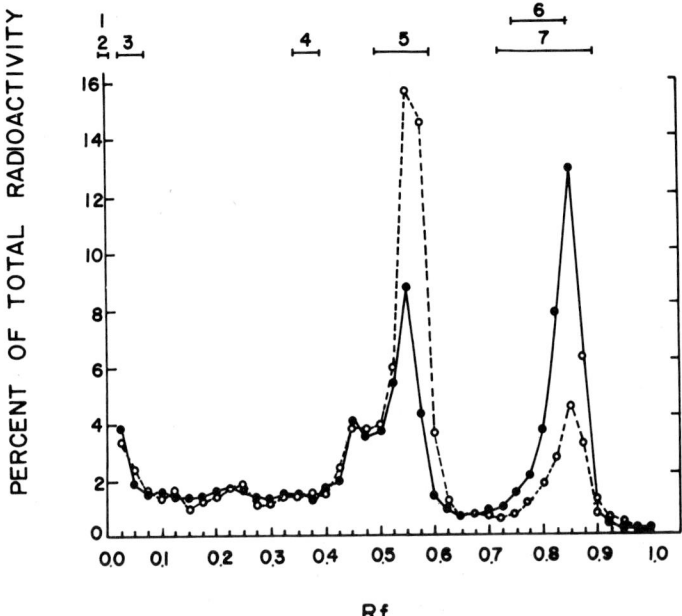

Fig. 3. A chromatogram (developed in butanol:acetic acid:water 4:1:1) showing the effect of ABA on the distribution of radioactivity in the ethanol extracts of *Rumex* leaves after the administration of kinetin $-8^{14}C$. (After Back et al., 1972). ——●—— extract from half-leaf treated with ABA; ----○---- extract from half-leaf treated with water. *1–7* show the standards run in the same system. *1 and 2* ATP, ADP; *3* AMP; *4* Hypoxanthine; *5* Adenine; *6* Kinetin Riboside; *7* Kinetin

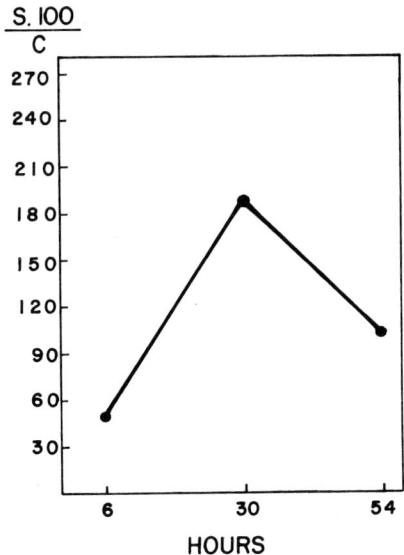

Fig. 4. Changes in levels of CK in exudate during recovery from mannitol stress. (After Itai et al., 1968)

the intensity of the stress. Furthermore, when stress is relieved, CK levels surpass their initial level before reverting to it again. Other stresses such as temperature (Itai and Benzioni, 1974; Skene and Kerridge, 1967; Skene, 1972) and root flooding (Burrows and Carr, 1969) have a similar effect.

2. Abscisic Acid Levels

Numerous reports show that a decrease in water potential results in a marked increase in ABA levels. This increase occurs irrespective of whether the changes in water potential are due to soil-drought (Loveys and Kriedemann, 1973), a desiccating environment (Zabadal, 1974) or osmotic agents such as salinity and mannitol (Fig. 5). Furthermore, there is an indication that ABA levels are correlated with drought resistance. Larque-Saavedra and Wain (1974) show that two *Zea mays* varieties with different degrees of drought tolerance also differ in their endogenous ABA levels. Treatments which elevated ABA levels also increased drought tolerance (Mizrahi et al., 1972).

The increase in ABA content as result of a decrease in water potential is due to an increased rate of ABA synthesis (Milborrow and Noddle, 1970), as well as to a decreased rate of ABA degradation (Table 4). The relative contribu-

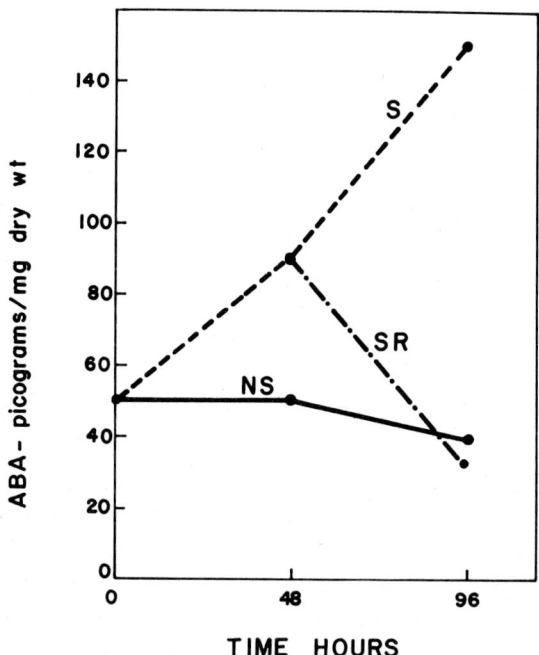

Fig. 5. Effect of return of plants from salinated solution to half-Hoagland medium, on the cis-trans-ABA content of the leaves. *S* salinated plants; *NS* non-salinated plants; *SR* plants salinated for 48h, then returned to half-Hoagland for another 48 h. Material for ABA determination was obtained from three plants from three separate experiments, with similar results. Results of one experiment are presented. (After Mizrahi et al., 1972)

Table 4. Effect of kinetin treatment and wilting on the metabolism of synthetic ^{3}H ABA in tobacco leaves. ^{3}H ABA with or without kinetin was fed through the petiole. Half the leaf was air-dried for 45 min while the other half was kept in a humid chamber. The leaf was methanol-extracted and the extract was separated on TLC. R_f 0.7–0.8 corresponding to synthetic ABA. The relative radio-activity of each R_f fraction was determined. The results are those of one representative experiment out of three similar experiments

Treatment	% total radio-activity				
	Kinetin −	Kinetin +	Control	Dried	Recovered
R_f 0–0.2	51.2	55.8	69.4	62.8	69.3
R_f 0.7–0.8	46.5	41.7	27.1	31.4	28.3

tion of higher synthesis or lower degradation rates to elevated ABA levels due to water stress has not been assessed. Apparently, not only water deficiency but also CK per se can affect ABA degradation (Table 4). Kinetin treatment has the opposite effect to water stress on ABA metabolism. These reciprocal effects of cytokinin and ABA on the level of their respective metabolisms, together with the rise in endogenous ABA levels, should be borne in mind whenever trying to elucidate plant response to water stress.

3. Other Hormones

It is generally accepted that an interaction of all plant hormones is involved in the regulation of plant performance. This would imply that levels of other hormones such as gibberellins, auxins and ethylene are involved in the regulation of plant response to water stress. Despite widespread knowledge of the physiological role of auxins and gibberellins, there is, as far as we know, no evidence of the effect of water stress on the endogenous levels of these hormones.

Indirect evidence of changes in auxin levels can be deduced from the study of the activity of IAA oxidase (Darbyshire, 1971; Mills and Todd, 1973). The results of these two investigations do not agree. The first study reported that stress caused a decline in IAA oxidase activity, while the second found a rise.

More information is available concerning ethylene. The effect of drought upon ethylene has been investigated by El Beltagy and Hall (1974). They showed that ethylene level in the internal space of bean leaves rises in two separate spells. In the first, lasting 24 h, a seven-fold increase in ethylene level was noted: thereafter it returned to almost its initial level. This phase coincided with a rise in WSD from 20% to about 47%. The second phase lasted for nine days, during which the WSD hardly increased at all (only 8%), while the ethylene levels increased slowly but consistently up to four times the initial level.

Similarly, McMichael et al. (1972) reported that ethylene production rates of petioles increased with development of water deficits. In this case too, no apparent quantitative relationship exist between the magnitude of internal water deficits and the rates of ethylene production. Water deficits in orange leaves, imposed by low relative humidity, cause a similar rise in the rate of emanation and endogenous levels of ethylene (Ben-Yehoshua and Aloni, 1974).

These reports on the increase of ethylene levels indicate the possibility that certain metabolic and developmental responses to water stress should be considered in relation to ethylene effects.

III. The Physiological Significance of Hormonal Effects

This section evaluates the physiological significance of the changes in the activity of hormonal factors due to water stress. Since most of the evidence of endogenous changes is related mainly to cytokinins and ABA, only these hormones will be considered.

1. Effects on Membrane Permeability to Water and Solutes

The influence of plant hormones on membrane characteristics has been widely reported. These data permit the assumption that changes in hormone balance in water-stressed plants probably have an immediate effect on membrane parameters such as permeability to water and solutes, selectivity in mineral uptake, and solute translocation. Hormonal changes could also affect membrane composition by changing the pattern of biosynthesis. Feng (1973) and Feng and Unger (1972) showed that kinetin treatment of *Allium cepa* cells increased their permeability to glycerol, thiourea and urea, but not to malonamide and dimethylurea. Studies in fungal cells on the effects of kinetin and energy-linked import of sugars, nucleosides and amino acids show that kinetin inhibited the uptake of these metabolites while enhancing both the release and the uptake of Ca^{++} (Lejohn et al., 1973). ABA and kinetin also affect water permeability: ABA was found to increase permeability, while CK decreased it (Glinka and Reinhold, 1971; Tal and Imber, 1971; Collins and Kerrigan, 1974). Only few data are available on changes in water permeability during water stress (cf. Lee-Stadelmann and Stadelmann, this volume Part 4:E); but it may be assumed that if the root's content of ABA is increased and that of CK decreased, water permeability will probably rise and consequently water balance will improve. ABA and CK also affect mineral uptake and translocation. Collins and Kerrigan (1974) showed that kinetin reduced K^+ and Ca^{++} fluxes to the exudate while ABA increased them (Table 5). The results are contrary to those of Cram and Pitman (1972). Another study by Benzioni et al. (1974) on the uptake and translocation of Na^+ and Cl^-, showed that kinetin caused a reduction of salt transport from the roots to the shoot without affecting salt uptake (Table 6).

The effects of ABA and kinetin on Na/K selectivity in leaves and cotyledons have been closely investigated (e. g. Jacoby and Dagan, 1970; Ilan, 1971; van Stevenick, 1972; Reed and Bonner, 1974). These studies indicate that CK effect a change in the selectivity toward K^+ and Na^+, so that the "affinity" of the cells for potassium is increased while that for sodium is decreased. ABA, on the other hand, inhibits K^+ uptake and causes a preference for Na^+.

These hormonal effects may be relevant to the study of water stress effects on stomatal movements, since K^+ fluxes probably regulate water potential in guard cells.

2. Effects on Stomatal Mechanism and Water Balance

Evidence of ABA as an integral component of the regulation of stomatal movement can be deduced from the study of Tal and his group. They used a wilting mutant of tomato (flacca) in which a one-locus lesion is detrimental to ABA biosynthesis (Tal and Nevo, 1973). As a result, the level of ABA in this mutant remains low even in extreme stress conditions while the stomata do not close, not even in darkness nor when the tissue is plasmolyzed. After ABA treatment, the mutant reverts completely to its normal phenotype (Table 7).

The involvement of ABA in the regulation of stomatal movement is further supported by the work of Kriedemann et al. (1972), who found that stomatal closure begins eight minutes after ABA application and is completed in half an hour. They also found that doubling the endogenous amounts of ABA already initiates the closure. Loveys and Kriedemann (1973) have followed the transpiration rate and ABA levels of vine leaves during the imposition and relief of

Table 5. Ion fluxes into the xylem exudate in the presence of ABA or kinetin (Collins and Kerrigan, 1974)

	K^+ flux	Ca^{++} flux
10^{-6} M ABA	54.9	2.95
10^{-6} M Kinetin	6.0	1.07
Control	39.6	3.30

Table 6. The effect of 0.1 mg l^{-1} kinetin on Na^+ and Cl^- uptake in tobacco plants. Values are given in μ mole·g^{-1} dr. wt. (Benzioni et al., 1974)

	Roots				Shoots			
Treatment	Na	%	Cl	%	Na	%	Cl	%
Control	653	100	1101	100	283	100	726	100
Kinetin	759	116	1165	106	187	66	457	63

Table 7. Concentration of ABA-like substance and water loss in detached leaves of the upper shoot of a normal and a mutant tomato plant (Tal and Imber, 1970; Imber and Tal, 1970)

Plant	ABA concentration mg eq·kg^{-1} fr. wt.	Transpiration mg cm^{-2} for 24 h	
		$-$ABA	$+$ABA
Normal	118.9	190	195
Mutant	11.9	280	188

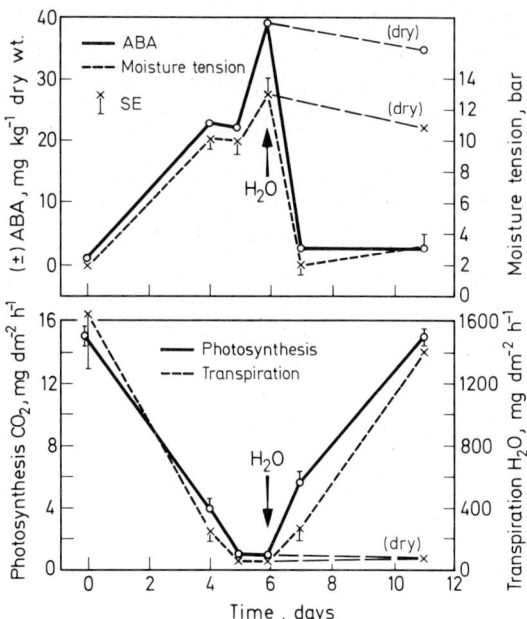

Fig. 6. Changes in leaf water potential, ABA-like inhibitor levels, and gas exchange rates during prolonged moisture stress and subsequent recovery. (After Loveys and Kriedemann, 1973)

water stress. They showed (Fig. 6) that over six days, the water tension increased sixfold, while ABA levels were 44 times higher and that the plants hardly transpired. On rewatering, the ABA level decreased to twice the initial level in one day and maintain this level until the end of the experiment. Transpiration increased at a slower rate than the fall in ABA level and reached the initial level only five days after rewatering. This divergence between transpiration rates and endogenous ABA levels after rewatering, points towards the possible involvement of an additional factor which is probably cytokinin. Details of the "after-effect" and the possible role of the CK/ABA ratio are discussed below.

Since the stomatal mechanism is still not clear (Levitt, 1974 and this volume Part 3:C), the role of ABA in this process cannot yet be defined. There is, however, a clue in the inhibition exerted by ABA on the flux of K^+ into the guard cells (Horton and Moran, 1972). Raschke (1974) suggested that this inhibition is probably due to the enhanced sensitivity of ABA-treated stomata to CO_2. This enhanced sensitivity can inhibit organic acid formation which may affect the movement of K^+. This ABA/CO_2 interaction was demonstrated by Loveys et al. (1973) who found that ABA induces stomatal closure only in the presence of CO_2 while ABA levels are not affected by CO_2. Regrettably, no evidence has so far been brought of any correlation of endogenous CK levels to stomatal opening, although the effect of CK treatment on stomata is well documented (Livne and Vaadia, 1972). It can reasonably be assumed that both ABA and CK affect stomatal mechanism simultaneously. The hormones exert a mutual antagonism when applied to leaves (Horton, 1971; Cooper et al., 1972). This interaction allows the postulation that the "after-effect" is due to a possible role of CK in this system. This concept is plausible since ABA affects stomata linearly to its concentration while kinetin has an optimum curve, and could possibly reach a supra optimal level during the period of "overshoot" in CK levels, after stress release (Fig. 4). Another possibility is that the "after-effect" is the result of an accumulation of phaseic acid, a degradation product of ABA which is known to inhibit photosynthesis (Loveys and Kriedemann, 1974). Finally, the evidence that CK, on the one hand, causes an unfavorable change in physical parameters such as RWC, WSD and turgor, and on the other hand that it decreases stomatal resistance under stress conditions, supports the view that the reduction of CK under stress conditions may play a major role in the regulation of plant response (Mizrahi et al., 1974; Kirkham et al., 1974; Prisco and O'Leary, 1973). The same reasoning holds for the role of ABA, namely, the finding that treatment of stressed plants with ABA and treatments that cause a rise in endogenous ABA levels, always improved plant water balance (Mizrahi et al., 1974). This also points toward the same conclusion regarding the role of ABA under conditions of plant stress.

3. Effects on Metabolism and Developmental Processes

Water stress affects metabolic and developmental processes. A few investigations have dealt with the direct relationship between the endogenous hormonal levels and their effects on these processes. Thus, the most common experimental approach is to study metabolic changes in water-stressed plants as affected by application of hormones.

Senescence is one of the most obvious symptoms of water stress. Leaves from water-stressed plants were placed in a humid chamber and the rate of chlorophyll degradation was determined (Table 8). Kinetin treatment partly arrested chlorophyll degradation and averted the effect of the stress. Cytokinin treatment of a stressed whole plant, enhanced senescence owing to its adverse effect on the water balance of the plant (Mizrahi et al., 1974; Prisco and O'Leary, 1973). In the same study, Mizrahi et al. showed that ABA treatment of water-stressed barley leaves considerably delayed chlorophyll degradation. In plants exposed to a progressive water stress, treatment with ABA kept the chlorophyll level almost unchanged for a month, while in stressed untreated leaves this

Table 8. Effect of salt stress and kinetin treatment on chlorophyll degradation in leaves

	Stress		Control	
	+kinetin	−kinetin	+kinetin	−kinetin
Initial OD	0.253	0.246	0.235	0.226
OD on day 7	0.203	0.162	0.186	0.168
OD day 7/day 1 (%)	80.2	65.8	78.2	74.3

level was maintained for only 19 days, thus indicating that ABA markedly affects survival at plants exposed to drought. In a similar experimental system, Arad et al. (1973) found that RNase activity increases during drought, and that this rise can be arrested by ABA treatment. In this case, CK increases RNase activity in water-stressed plants.

Conversely, Shah and Loomis (1967) found that application of benzyl adenine to water-stressed sugar beets prevented a decrease in RNA content. The cytidylic/uridylic ratio, too, remained unchanged, in contrast with untreated water-stressed plants. Other findings (Benzioni et al., 1967) lend further support to the premise that CK affects protein synthesis. Leaf discs excised from water-stressed plants were kinetin-treated and ^{14}C-leucine incorporation was studied (Fig. 7). The reduced capacity of stressed leaf-discs from NaCl-stressed plants for incorporation of amino acids was partly restored by CK treatment. Similarly, incorporation of ^{14}C-leucine into root segments grown on NaCl media was doubled (Kahane and Poljakoff-Mayber, 1968).

A more morphological and developmental approach was reported by Giles et al. (1974). They found in *Zea mays* a correlation between water potential, endogenous levels of ABA and metabolic processes leading to senescence caused by water stress (Table 9). These data support the assumption that the level of ABA is of primary importance in regulating plant response to water stress. Prisco and O'Leary (1973) assayed the growth and development of red kidney beans in response to salinity stress and found that benzyl adenine exacerbates the injury. These findings support the conclusion that CK application to water-stressed plants has an adverse effect. However, in excised tissue or when transpiration is limited because of favorable environmental conditions (e.g. high relative humidity), CK treatment exerts an opposite effect to that of water stress. This assumption is further supported by a study of germination.

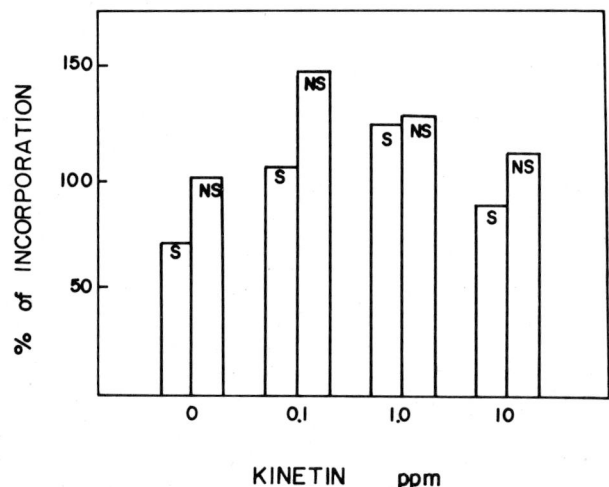

Fig. 7. Influence of treatment with kinetin (*NS*) and stress (*S*) on the incorporation capacity of ^{14}C L-leucine into protein. (After Benzioni et al., 1967)

The germination of lettuce seeds in soil was totally inhibited when water potential decreases to −4 bars. After treatment with kinetin, half the seeds germinated (Kaufmann and Ross, 1970). Another developmental process which can be affected by growth regulators is the induction of flowering. *Coffea arabica* requires both water stress and relief from stress to induce flowering. Both CK and GA levels increased during the last stage of induction, and it was found that application of these hormones could replace renewed watering (Browning, 1973).

In studying the effects of water stress, it is also helpful to consider indirect evidence on hormonal effects on metabolic processes also known to be affected by water stress (Table 10). Water stress may retard growth by reducing biosynthetic processes and enhancing catabolic ones. These processes are also affected by CK and ABA levels. This may indicate that the metabolic and developmental changes due to water stress are regulated by a hormonal shift.

Table 9. Summary of changes in ultrastructure, leaf water potential, relative water content and abscisic acid levels in maize leaves during increasing water stress (Giles et al., 1974)

Day No.	ψ^{leaf} bars	RWC %	ABA ng cm^{-2}	Changes
1 and 2	− 6.0	97	0.5	None. Normal C_4 structure
3	− 10.5	80	7.0	Level of starch in bundle sheath cells reduced. Stomata closed
4	− 13.5	73	7.5	No starch in bundle sheath cells. Cytoplasmic vesicles appeared in bundle sheath and mesophyll cells
7	− 18.5	55	9.0	Bundle sheath chloroplast randomly distributed around cell. Tonoplast breakdown in 25% of mesophyll cells resulting in complete cell disruption

Table 10. Hormonal and water stress effects on metabolic processes

Type of process	Effects of	
	CK	ABA
Enhanced by stress		
Respiration	retardation	?
Nucleic acid degradation	retardation	not conclusive
Proline accumulation	?	enhancement
Amylase activity	not conclusive	?
Sensitivity to pathogens	retardation	?
Chlorophyll degradation	retardation	not conclusive
Retarded by stress		
Photosynthesis	enhancement	retardation
Cell wall synthesis	?	?
Nitrogen reduction	enhancement	?
Protein synthesis	enhancement	retardation

IV. A Hypothetical Model for the Role of Hormones in Plant Adaptation to Water Stress

Some tentative generalizations can be made on plant response to moderate and transient water stress. (1) Water stress causes changes in levels of two growth regulators, namely, an increase in ABA and a reduction in CK levels. These modifications seem to occur simultaneously and the two hormones possibly have a reciprocal effect on the levels of each others. (2) The change in hormone levels may conceivably modify cell membrane characteristics. On the other hand, the possibility of the stress affecting membranes and so determining hormone level cannot be precluded. (3) Modification in membrane characteristics may cause stomatal closure, changes in the size of metabolic pools, and in the activity of membrane-bound enzymes, so that the degradation processes are enhanced and synthetic processes retarded. (4) Certain effects of water stress last some time after cessation of the stress, hence implying a "memory". (5) The "overshoot" which is evident in a number of processes on relief of stress, is an integral part of the response. (6) Stress may cause some developmental changes which by their nature are irreversible: as induction of flowering and senescence. The above generalizations enable the presentation of a hypothetical model of a chain of events taking place during stress and subsequent recovery (Fig. 8).

A similar chain of events can be found as a response to other environmental stresses (e. g. high and low temperatures and mineral deprivation). This similarity raises the question of whether the response to stress, which is not merely a "water-conserving" response, but one of growth retardation, is advantageous to the stressed plant. It is very possible that this conduct enables tolerance and survival during stress. This premise is supported by the knowledge that deceleration of life processes is one of the most common survival mechanisms of which the most extreme examples are dormant buds, seeds and spore resistance.

V. Conclusions

Hsiao (1973) postulates that water stress primarily causes turgor reduction which in turn retards growth. This primary effect is responsible for secondary effects caused by metabolite accumulation which then results in all the known water-stress responses including changes in hormone levels. This hypothesis is questionable as it separates the known response to water stress from the additional information which can be deduced from the known response to other stresses, such as heat, salinity, mineral deprivation, or boron toxicity, where the turgor remains unchanged (Itai and Benzioni, 1974; Bernstein, 1963; Slatyer, 1961; Bussiba, 1975). All these stresses are characterized by hormone changes and other symptoms similar to those of water stress. It is likely, therefore, that an alternative hypothesis based on the hormonal balance as the primary effect and the other responses as secondary ones, is more appropriate. This hypothesis suggests that plants evolved to cope with water stress by an inbuilt "flow-sheet" regulated by the hormonal balance of CK/ABA. However, further

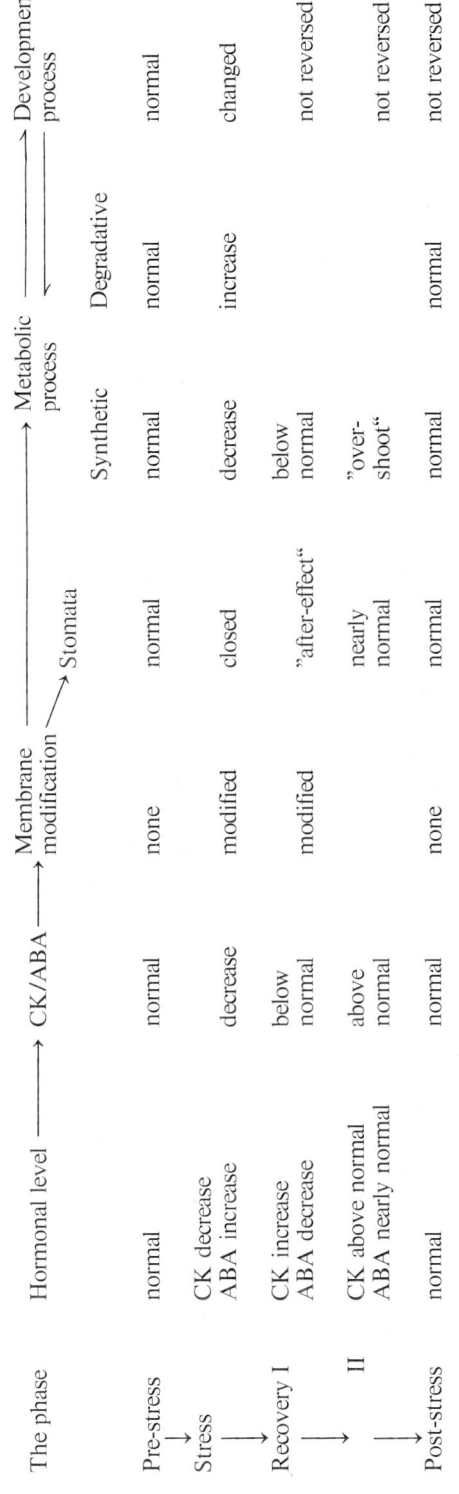

Fig. 8. A hypothetical model for the role of hormones in plant adaptation to water stress

information is required here on certain cardinal points, some of which are the following: (1) CK and ABA interaction with other hormones such as GA, IAA and ethylene in the response to water stress. (2) The effect of endogenous hormone levels on membrane characteristics. (3) Information on how stress causes hormonal shift. (4) A kinetic study of membrane changes during the phases of stress and recovery. (5) The mode of action of membrane change in affecting metabolic processes.

Acknowledgements. Supported in part by a grant from the Bernard and Louis M. Bloomfield Negev Research Institute for Silviculture and Applied Ecology. The help of Cynthia Bellon in editing the manuscript is gratefully acknowledged.

References

Arad, S., Mizrahi, Y., Richmond, A. E.: Leaf water content and hormone effects on RNAse activity. Plant Physiol. **52**, 510–512 (1973).

Back, A., Bittner, S., Richmond, A. E.: The effect of abscisic acid on the metabolism of kinetin in detached leaves of *Rumex pulcher*. J. Exp. Botany **23**, 744–750 (1972).

Ben-Yehoshua, S., Aloni, B.: Effects of water stress on ethylene production by detached leaves of Valencia orange *Citrus sinensis osbeck*. Plant Physiol. **53**, 863–865 (1974).

Benzioni, A., Itai, C., Vaadia, Y.: Water and salt stresses, kinetin and protein synthesis in tobacco leaves. Plant Physiol. **42**, 361–365 (1967).

Benzioni, A., Mizrahi, Y., Richmond, A. E.: Effect of kinetin on plant-response to salinity. New Phytologist **73**, 315–319 (1974).

Bernstein, L.: Osmotic adjustment of plants to saline media. II. Dynamic phase. Am. J. Botany **50**, 360–390 (1963).

Browning, G.: Flower bud dormancy in *Coffea arabica* L. II. Relation of cytokinins in xylem sap and flower buds to dormancy-release. J. Hort. Sci. **48**, 297–310 (1973).

Burrows, W. J., Carr, D. J.: Effects of flooding the root system of sun-flower plants on the cytokinin content in the xylem sap. Physiol. Plantarum **22**, 1105–1112 (1969).

Bussiba, S.: Involvement of ABA in the interrelationship between the response to various stresses and recovery from them. M. Sc. Thesis Ben-Gurion University, Beer Sheva (Hebrew) (1975).

Collins, J. C., Kerrigan, A. P.: Effect of kinetin and abscisic acid on water and ion transport in isolated maize roots. New Phytologist **73**, 309–314 (1974).

Cooper, M. J., Digby, J., Cooper, P. J.: Effects of plant hormones on stomata of barley. A study of the interaction between ABA and kinetin. Planta (Berl.) **105**, 43–49 (1972).

Cram, J. W., Pitman, M. A.: The action of abscisic acid on ion uptake and water flow in plant roots. Australian J. Biol. Sci. **25**, 1125–1132 (1972).

Darbyshire, B.: Changes in IAA oxidase activity associated with plant water potentials. Physiol. Plantarum **25**, 82–87 (1971).

El-Beltagy, A. S., Hall, M. A.: Effect of water stress upon endogenous C_2H_4 levels in *Vicia faba*. New Phytologist **73**, 47–53 (1974).

Feng, K. A.: Effects of kinetin on the permeability of *Allium cepa* cells. Plant Physiol. **51**, 868–870 (1973).

Feng, K. A., Unger, J. W.: Influence of kinetin on the membrane permeability of *Allium cepa* epidermal cells. Experientia **28**, 1310–1311 (1972).

Fischer, R. A.: After-effect of water stress on stomatal opening potential. II. Possible causes. J. Exp. Botany **21**, 386–404 (1970).

Giles, K. L., Beardsell, M. F., Cohen, D.: Cellular and ultrastructural changes in mesophyll and bundle sheath cells of maize in response to water stress. Plant Physiol. **54**, 208–212 (1974).

Glinka, Z., Reinhold, L.: ABA raises the permeability of plant cells to water. Plant Physiol. **48**, 103–105 (1971).

Horton, F. R.: Stomatal opening, the role of abscisic acid. Can. J. Botany **49**, 583–587 (1971).
Horton, F. R., Moran, L.: ABA inhibition of K influx into stomatal guard cells. Z. Pflanzenphysiol. **66**, 193–196 (1972).
Hsiao, T. C.: Plant responses to water stress. Ann. Rev. Plant Physiol. **24**, 519–570 (1973).
Ilan, I.: Evidence for hormonal regulation of the selectivity of ion uptake by plant cells. Physiol. Plantarum **25**, 230–233 (1971).
Imber, D., Tal, M.: Phenotypic reversion of flacca, a wilty mutant of tomato, by abscisic acid. Science **169**, 592–593 (1970).
Itai, C., Benzioni, A.: Regulation of plant response to high temperature treatments. In: Mechanisms of regulation of plant growth (eds. R. L. Bieleski, A. R. Ferguson, M. M. Cresswell). Bull. **12**, 477–482, Roy. Soc. New Zealand 1974.
Itai, C., Richmond, A. E., Vaadia, Y.: The role of root cytokinins during water and salinity stress. Israel J. Botany **17**, 187–195 (1968).
Itai, C., Vaadia, Y.: Kinetin-like activity in root exudate of water stressed sunflower plants. Physiol. Plantarum **18**, 941–945 (1965).
Itai, C., Vaadia, Y.: Cytokinin activity in water-stressed shoots. Plant Physiol. **47**, 87–90 (1971).
Jacoby, B., Dagan, J.: Effects of 6-N-Benzyladenine on primary leaves of intact bean plants on their sodium absorption capacity. Physiol. Plantarum **23**, 397–403 (1970).
Kahane, I., Poljakoff-Mayber, A.: Effects of substrate salinity on the ability for protein synthesis in pea roots. Plant Physiol. **41**, 1115–1119 (1968).
Kaufmann, M. R., Ross, K. J.: Water potential temperature and kinetin effects on seed germination in soil and solute systems. Am. J. Botany **57**, 413–418 (1970).
Kirkham, M. B., Gardner, W. R., Gerloff, G. E.: Internal water status of kinetin treated salt stressed plants. Plant Physiol. **53**, 241–243 (1974).
Kriedemann, P. E., Loveys, B. R., Fuller, G. L., Leopold, A. C.: ABA and stomatal regulation. Plant Physiol. **48**, 842–847 (1972).
Kulaeva, O. N.: The Effect of roots on leaf metabolism in relation to the action of kinetin on leaves. Soviet Plant Physiol. (Eng. Trans.) **9**, 182–189 (1962).
Larque-Saavedra, A., Wain, R. L.: ABA levels in relation to drought tolerance in varieties of *Zea mays* L. Nature **251**, 716–717 (1974).
Lejohn, H. B., Roselynn, M., Stevenson, K.: Cytokinins and magnesium ions may control the flow of metabolites and calcium ions through fungal cell membranes. Biochem. Biophys. Res. Commun. **34**, 1061–1066 (1973).
Levitt, J.: The mechanism of stomatal movement—once more. Protoplasma **82**, 1–17 (1974).
Livine, A., Vaadia, Y.: Water deficits and hormone relations. In: Water deficits and plant growth. Vol. 3 (ed. T. T. Kozlowski), pp. 255–271. London, New York: Academic Press 1972.
Loveys, B. R., Kriedemann, P. E.: Rapid change in abscisic acid-like inhibitors following erasions in vine leaf water potential. Physiol. Plant. **28**, 476–479 (1973).
Loveys, B. R., Kriedemann, P. E.: Internal control of stomatal physiology and photosynthesis. I. Stomatal regulation and associated changes in endogenous levels of abscisic and phaseic acids. Australian J. Plant Physiol. **1**, 407–415 (1974).
Loveys, B. R., Kriedemann, P. E., Torokfalvy, E.: Is abscisic acid involved in stomatal response to carbon dioxide? Plant Sci. Letters **1**, 335–338 (1973).
McMichael, B. L., Jordan, W. R., Powell, R. D.: Effect of water stress on ethylene production by intact cotton petioles. Plant Physiol. **49**, 658–662 (1972).
Meyer, R. E., Gingrich, J. R.: Osmotic stress effects of its application to a portion of wheat root system. Science **144**, 1463–1464 (1964).
Milborrow, B. V., Noddle, R. C.: Conversion of 5-(1,2-epoxy-2,6,6-trimethylcyclohexyl)-3-methylpenta-cis-2-trans-4-dienoic acid into abscisic acid in plants. Biochem. J. **119**, 727–734 (1970).
Miller, L. N., Kramer, P. J.: Effects of water stress on the growth of pine seedlings. Plant Physiol. **40**, suppl. XXIV (1965).
Mills, V. M., Todd, G. W.: Effects of water stress on the IAA oxidase activity in wheat leaves. Plant Physiol. **51**, 1145–1149 (1973).

Mizrahi, Y., Blumenfeld, A., Bittner, S., Richmond, A. E.: Abscisic acid and cytokinin contents of leaves in relation to salinity and relative humidity. Plant Physiol. **48**, 752–755 (1971).

Mizrahi, Y., Blumenfeld, A., Richmond, A. E.: The role of abscisic acid and salination in the adaptive response of plants to reduced root aeration. Plant Cell Physiol. (Tokyo) **13**, 15–21 (1972).

Mizrahi, Y., Scherings, S. G., Malis-Arad, S., Richmond, A. E.: Aspects of the effect of ABA on the water status of barley and wheat seedlings. Physiol. Plantarum **31**, 44–50 (1974).

Prisco, J. T., O'Leary, J. W.: The effects of humidity on growth and water relations of salt stressed bean plants. Plant Soil **39**, 263–276 (1973).

Raschke, K.: Interaction between stomatal response to CO_2 and ABA "optimize" stomatal moderation of water loss in leaves of *Xanthium strumarium*. Int. Assoc. Plant Physiol. 1st Meet., Würzburg, Abstracts, 83 (1974).

Reed, M. M., Bonner, B. A.: The effect of abscisic acid on the uptake of potassium and chloride into *Avena* coleoptile sections. Planta (Berl.) **116**, 173–185 (1974).

Shah, C. B., Loomis, R. S.: RNA and protein metabolism in sugar beet during drought. Physiol. Plantarum **18**, 240–254 (1967).

Skene, K. G. M.: Cytokinins in the xylem sap of a grape vine canes: changes in activity during cold-storage. Planta (Berl.) **104**, 89–92 (1972).

Skene, K. G. M., Kerridge, G. H.: Effect of root temperature on cytokinin activity in root exudate of *Vitis vinifera* L. Plant Physiol. **42**, 1131–1139 (1967).

Slatyer, R. O.: Effects of several osmotic substrates on the water relationship of tomato. Australian J. Biol. Sci. **14**, 518–540 (1961).

Stevenick, R. F. M. van: Abscisic acid stimulation of ion transport and alteration in K^+/Na^+ selectivity. Z. Pflanzenphysiol. **67**, 282–286 (1972).

Tal, M., Imber, D.: Abnormal stomatal behavior and hormonal balance in flacca, a wilty mutant of tomato. II. Auxin and ABA-like activity. Plant Physiol. **46**, 373–376 (1970).

Tal, M., Imber, O.: Abnormal stomatal behavior and hormonal imbalance in flacca, a wilty mutant of tomato. III. Hormonal effects on the water status in the plant. Plant Physiol. **47**, 840–850 (1971).

Tal, M., Nevo, Y.: Abnormal stomatal behavior and root resistance and hormonal imbalance in three wilty mutants of tomato. Biochem. Genet. **8**, 291–300 (1973).

Zabadal, T. J.: A water potential threshold for the increase of abscisic acid in leaves. Plant Physiol. **53**, 125–127 (1974).

C. Carbon and Nitrogen Metabolism Under Water Stress

M. KLUGE

I. Introduction

Water is an essential constituent of all living cells. In plant cells which consist mainly of osmotic systems, it is even more important. Hence, it is not surprising that one of the central events during the life of a plant is the effort to maintain its water status near the optimum. However, mainly in response to disadvantageous environmental conditions, plants can suffer from more or less severe deviations from the optimal water status. Such deviations can tend in both directions from the optimum; i.e. plants can suffer either from an excess of water (as during flooding of the root system; see Crawford and Thyler, 1969), or from deficiency of water (for example during drought periods). In the strict sense of the word, both of these situations represent water stress (see Levitt, 1972). Following the convention, in the discussion presented below the term "water stress" will be used in the sense of "water deficit".

During recent years, interest has been increasingly focused on biochemical aspects of water stress (see the reviews of Levitt, 1972; Hsiao, 1973; Kozlowski, 1972). Because plants and their life processes represent steady-state systems of biochemical reactions, all relationships between plants and water deficits have necessarily to some extent biochemical aspects (but see Oertli, this volume Part 1:B). However, in some cases certain biochemical processes can now be brought into such close causal relationships with water stress, or they provide ecological advantages under water stress conditions so obviously, that it might be possible to speak about "biochemistry of water stress". In the following, some selected and, it is believed, typical problems of this topic will be considered. This selection has to be limited and thus, necessarily, subjective. It will refer mainly to questions of carbon and nitrogen metabolism, because water deficits, on the one hand, have clearly negative effects on the productivity and final yield of biomass in plants. On the other hand, the main constituents of this biomass are carbon and nitrogen. Hence, ist seems reasonable to assume that the metabolism of carbon and nitrogen (see also Naylor, 1972) might be directly affected by water stress.

II. Carbon Metabolism Under Water Stress

1. Photosynthesis and Photorespiration

The flow of carbon in a photosynthesizing higher plant from CO_2 in the atmosphere to the photosynthetic end-products is limited, in a rough approximation, by two main resistances. One of them is determined by the movements of the stomata ("stomatal resistance"). The other one ("mesophyll resistance") integrates several components including diffusion of CO_2 from the intercellular spaces into the mesophyll cells and then to the sites of photosynthesis. It includes also the biochemical processes of carbon assimilation itself. There is overwhelming evidence in the literature that many (if not most) of the negative effects of water stress on carbon gain are caused by closure of stomata, thus increasing the stomatal resistance (see Hsiao, 1973). At the moment, there is nothing definitely known of any specific biochemical responses in the epidermal region[1] which could account for the behavior of stomata under water stress. Considering the fact that certain biochemical and metabolic processes are part of the mechanism of stomatal movements (e.g. nonautotrophic CO_2 fixation, malate synthesis due to cation uptake, see Allaway, 1973; Pearson, 1973) it seems to be most promising to continue investigations in this field.

Apart from the clearly dominating effects of stomatal closure with higher plants, negative non-stomatal effects of water stress on photosynthesis have also been observed. Some of these effects are due to events occurring on the subcellular level. For example, in chloroplasts isolated from water-stressed sunflower or cotton leaves, inhibition of the Hill reaction and of cyclic photophosphorylation occurs (for detailed discussion see Hsiao, 1973; cf. also Vieira da Silva, this volume Part 4: A).

Since the carboxylating step of photosynthesis is one of the major constituents of the mesophyll resistance, one would expect that water stress, when increasing the mesophyll resistance (Troughton, 1969; Troughton and Slatyer, 1969; Redshaw and Meidner, 1972; see Hsiao, 1973), acts directly on the activity of the carboxylating enzymes. However, low or even higher water deficits seem not to affect the activity of RuDP carboxylase (Huffaker et al., 1970). On the other hand, PEP carboxylase, the enzyme catalyzing the primary carboxylation step in C_4-plants, seems to be negatively influenced by water stress (Huffaker et al., 1970).

Further negative effects on the photosynthetic carbon gain can be expected from the increase in photorespiration which was observed with higher plants under water stress (Fischer, 1968; Redshaw and Meidner, 1972). This increase can be explained as follows: it is now well established that the production of CO_2 by photorespiration; i.e. carbon loss by the plant, is an inevitable consequence of RuDP carboxylase-catalyzed CO_2 fixation in the presence of oxygen (Lorimer and Andrews, 1973). Furthermore, it was shown by Ogren (1975) that at low CO_2 concentrations (with the O_2 concentration being high), the CO_2 production via photorespiration should be favored. This will become clear from Fig. 1. It is shown in this figure, that RuDP-carboxylase has a

---

[1] The effects of plant hormones are beyond the scope of this chapter. See Livne and Vaadia (1972).

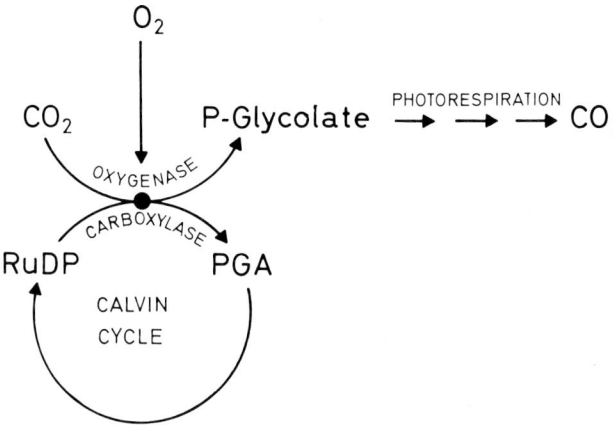

Fig. 1. The dual nature of RuDP-carboxylase as carboxylating and oxygenating enzyme and its possible role in photorespiration. *PGA* 3-phosphoglyceric acid; *P-Glycolate* phosphoglycolate; *RuDP* ribulose-1,5-di-phosphate. (After a scheme presented by Ögren during the Symposium Environ. and Biol. Control of Photosynthesis, August 1974, Hasselt/Belgium)

double function (for abbreviations see Fig. 2); i.e. it acts as a carboxylating and as well as an oxygenating enzyme. When CO_2 reacts exclusively, only 3-PGA will be formed. When CO_2 reacts together with O_2, P-glycolate will result in addition to 3-PGA. After dephosphorylation, the latter provides glycolate, which is an intermediate of photorespiration. Hence, because CO_2 and O_2 compete for RuDP (see Fig. 1), relatively low CO_2 concentrations must favor the oxygenation reaction, and consequently also the photorespiration, by enhanced production of its substrate. As indicated above, this situation can be expected to occur under water stress because of the low internal CO_2 concentration in the leaves due to the increased stomatal resistance to CO_2.

Interestingly enough, among those higher plants which occupy arid habitats, where frequent water stress is most likely to occur, specialists could be detected which are able to avoid the undesirable carbon loss due to photorespiration. These specialists are represented by the C_4 plants (for the problems of C_4 photosynthesis consult e.g. Black, 1973, and in this volume Huber and Sankhla, Part 5:C and Ludlow, Part 5:D). In C_4 plants, the CO_2 fixation via RuDP carboxylase and the concommitant Calvin cycle operate exclusively in the bundle sheath cells. The decarboxylation of the C_4 dicarboxylic acids (malate and aspartate) occur also only in the bundle sheath, thus locally providing a rather high CO_2 concentration at the sites of RuDP carboxylase activity. As discussed above, this high local CO_2 concentration can be expected to suppress the oxygenation of RuDP, which will decrease the supply of glycolate, and thus reduce photorespiration. Any CO_2 produced by photorespiration in spite of this mechanism will be recycled into malate via PEP carboxylase localized in the mesophyll (see Fig. 2).

2. Osmoregulation in Algae and Higher Plants

It is very obvious that certain plant species show significantly higher resistance to water stress than others. Plants can resist water stress because they have

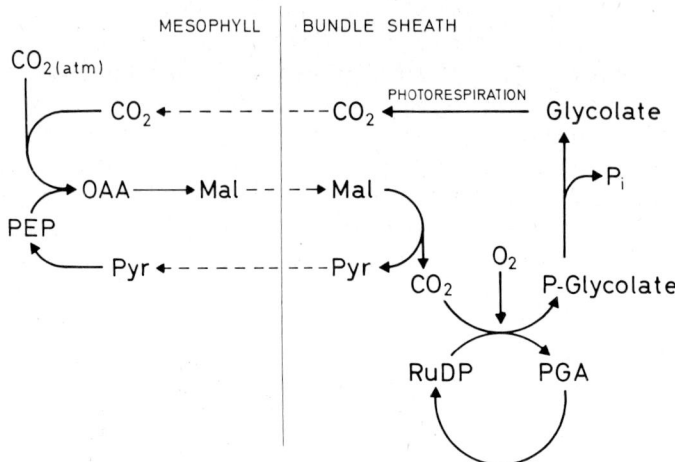

Fig. 2. The role of photorespiration in C_4-photosynthesis. *Mal* malate, *OAA* oxaloacetate, *PGA* 3-phosphoglyceric acid, *PEP* phospho-enol-pyruvate, *Pyr* pyruvate, *RuDP* ribulose-1,5-diphosphate. (After Ogren, as in Fig. 1)

developed mechanisms which either allow them to avoid the stress, and/or to tolerate it (see Levitt, 1972). Among the plurality of mechanisms which plants can use to avoid or to reverse water deficits, osmoregulation is one of the most striking and thus well-investigated features.

The term "osmoregulation" characterizes the effort of a plant to minimize differences in the water potential between the plant body and its environment by means of adjusting a suitable internal osmotic potential. For example: if the water potential of the environment decreases, i.e. it becomes more negative (for example by high water-vapor saturation deficit of the atmosphere, by decreasing water content of the soil or by increasing solute concentration in an aquatic environment), plants exhibiting osmoregulation can lower (make more negative) their own internal water potential by synthesis and accumulation of osmotically active substances, and vice versa. By this mechanism, plants are able not only to prevent loss of water to an environment having a large negative water potential, but also to take up water from such an environment, a fact that enables them to balance the water deficits of the cells.

Osmoregulation seems to be widely distributed in the plant kingdom. There are reports that in higher plants the sucrose content of the tissues increases under drought conditions, the starch content being simultaneously decreased (see Mothes, 1956). This effect has been explained in terms of enhanced activities of starch-degrading enzymes which may occur under water stress (Takaoki, 1968). On the other hand, results obtained by Kluge and Heininger (1973), who have studied osmotically stressed leaf slices of succulents, suggest that with water stress the photosynthetic flow of carbon can be shifted directly to low molecular weight (i.e. osmotically active) end products such as sucrose rather than to osmotically inactive polymers such as starch.

Osmoregulation is classically featured by those organisms which live in water of quickly and drastically changing osmotic potential. Water pools in the coastal tidal zones or brackish water in the mouths of rivers provide such environments. Kauss and his coworkers have studied the osmoregulatory behavior of *Ochromonas malhamensis* extensively (Kauss, 1967; Kauss, 1974; Schobert et al., 1972). In this organism, osmoregulation proceeds by raising the cellular concentration of α-galactosyl glycerides (fluoridoside or isofloridoside) at the expense of a polysaccharide (chrysolaminarin). The synthesis of isofloridoside is due to galactosyl transferase. There is evidence that this enzyme is regulated by osmotic stress, i.e. the enzyme activity increases under increasing stress. This enhancement of the enzyme activity is not caused by enhanced de novo synthesis of enzyme molecules but by allosteric regulation. However, the details of this regulation mechanism are as yet unknown.

In *Dunaliella parva*, a green halophilic alga, glycerol is synthesized and accumulated to balance the low water potentials of the environment (Wegmann, 1971; Ben-Amotz and Avron, 1973; Ben-Amotz, 1974). The formation and degradation of glycerol are not directly dependent on light. However, in *Dunaliella*, glycerol is also a photosynthetic product. Ben-Amotz (1974) assumes that two pathways are involved in glycerol synthesis, one of them using directly intermediates of photosynthetic carbon flow, the other one using products of starch degradation.

Platymonas subcordiformis, a further example of an alga living in brackish-water, makes use of mannitol, with starch as the osmotically inactive antagonist, to perform osmoregulation (Kirst, 1975). As in the case of *Dunaliella*, it can be assumed that the osmotically active substance is synthesized in the light via photosynthesis. In the dark, however, synthesis of mannitol proceeds via degradation of starch.

In the marine diatomee *Cyclotella meneghiniana*, osmoregulation is assumed to be due to the accumulation and disappearence of proline (Schobert, 1974). Proline accumulation is also a common feature in higher plants subjected to water stress (see below).

It has been outlined by Kauss (1974), that the occurrence of osmoregulation implies that the cells must have a mechanism to sense the osmotic pressure and convert this information into regulation of metabolism. As has been shown above, the general phenomena of osmoregulation and, in some cases, also metabolic pathways involved in osmoregulation (viz. *Ochromonas*) are well understood. However, the key questions as to the nature of the basic sensor of osmotic pressure and the localization of this sensor are still open and provide exciting problems for investigations in the future.

III. Nitrogen Metabolism Under Water Stress

1. Pools of Free Amino Acids Under Water Stress

As indicated above, the content of free proline may increase in plants suffering from water deficits (Kemble and MacPherson, 1954; Barnett and Naylor, 1966; Thompson et al., 1966; Steward et al., 1966; Steward, 1972a, b; 1973; Singh et al., 1973a, b). Treichel (1975) showed accumulation of proline in halophytes when the plants were irrigated with NaCl solutions. There was a close correlation between the osmotic potential of the irrigating solution and the extent of proline accumulation. This suggests that the increase of the proline pool in the halophytes is a water stress effect rather than a salt effect. This view is further supported by the finding that proline begins to accumulate in the leaves shortly after beginning irrigation with NaCl solutions, even before the applied ions have entered the leaves (Treichel, 1975).

The proline accumulation in water stressed plants is readily reversed if the stress situation is eliminated by rewatering the plants (Singh, 1973a, c). There is evidence that the accumulation of proline is due to de novo synthesis (Barnett and Naylor, 1966; Thompson et al., 1966) rather than to breakdown of proteins. Glutamate is believed to act as the precursor in proline synthesis, and Steward et al. (1966) suggest carbohydrates to be the ultimate source of the carbon stored in the accumulated proline. This could explain the observation that proline accumulation occurs only as long as stored carbohydrates are available (Steward et al., 1966).

It is tempting to speculate that proline accumulation might be involved in osmoregulation (see above). However, it must be doubted if the increase of the proline level can account for a substantial increase in the osmotic potential of the cells of higher plants. Hence, it has been argued that the major importance of proline accumulation is a hardening effect rather than osmoregulation (Singh et al., 1973b, c). There are also suggestions that proline may provide a convenient source of energy and nitrogen during immediate post-stress metabolism (Steward, 1973).

2. Protein Metabolism

Because the metabolism of proteins is one of the fundamental events of the cells, direct effects of water stress at this point would influence a great number of life processes featured by a plant. For example, if protein synthesis is negatively affected, one can expect that enzymes having short half-live periods will be depressed. This, of course, would directly influence the operation of metabolic pathways.

There have been numerous attempts to study the relationship between cellular water deficits and protein synthesis (Hsiao, 1973). The most elegant way to gain information in this field is to study populations of ribosomes in water stressed cells with respect to the ratio of single ribosomes (synthetically inactive) to polyribosomes (synthetically active). By this method, Hsiao (1970) has provided clear evidence that even mild water stress can shift the ribosomal profiles in favor of single ribosomes, indicating inhibition of protein synthesis. However, the moss *Tortula ruralis*, which is subjected to frequent desiccation and rehydration in its natural habitat, can conserve components of the protein-synthesizing machinery even during complete desiccation (Bewley, 1972; 1973). Thus, in this species, protein synthesis recovers immediately after rehydration. It seems to be reasonable to interpret this effect as an adaptive mechanism.

IV. Biochemical Aspects of Desiccation Resistance

The most simple and most effective protection against injuries due to extreme water loss is the ability to tolerate almost complete desiccation. This ability is very common among cryptogamic plants such as mosses (cf. the example of *Tortula* cited above), lichens and algae. It is also known to be exhibited by specialists among higher plants (for example: the ferns *Asplenium ruta-muraria, Ceterach officinarum, Notholaena maranthae,* and *Ramonda nathaliae* (Gesneriaceae), *Chamaegigas intrepidus* (Scrophulariaceae), *Myrothamnus flabellifolia* (Myrothamnaceae)).

The physiology of some of these poikilohydric plants has been studied by Ziegler and Vieweg (1970), Gündel (1972), Ziegler (1974). These authors showed that twigs of *M. flabellifolia* can be kept for months and years, detached from the plant, in an almost completely desiccated state. If these samples are rehydrated, respiratory CO_2 production can be observed after one hour and photosynthetic CO_2 fixation after some further hours. Gündel (1972) has shown that numerous important enzymes of respiration and photosynthesis, once synthesized, can survive desiccation of the leaves. For example, aldolase, $NAD^{\oplus}$-specific triosephosphate dehydrogenase, RuDP carboxylase, ribose-5-isomerase, phosphoribulokinase, $NADP^{\oplus}$-specific triosephosphatedehydrogenase have been found to remain potentially fully active in the desiccated tissues. It should be noted that all these enzymes are not bound to membranes in the cell but are soluble enzymes. Thus, Ziegler (1974) concludes that the delay in the onset of respiration and especially in photosynthesis is not due to the lack of soluble enzymes but to the lack of intact membrane structures involved in electron transport processes. These membrane structures seem to be destroyed by severe dehydration (Nir and Poljakoff-Mayber, 1967). After rehydration, the membranes have at first to be restored to allow electron transport and thus respiration and photosynthesis.

It should be mentioned also that freezing injuries occurring in plant cells can be explained in terms of negative effects on membrane structures (Heber, 1968; Santarius, 1969). Also in these cases the membrane structure might finally be damaged by local desiccation due to crystallization of ice which would cause local increases in ion concentrations up to levels high enough to inactivate the membrane systems (cf. Heber and Santarius in this volume Part 4:D). However, the effects described above do not as yet answer the question why certain specialists can tolerate desiccation of their cells, and other plants can not. As Ziegler (1974) suggests, the phenomenon of desiccation resistance in cormophytes is a structural rather than a biochemical problem. This author has outlined that poikilohydric plants show the ability to avoid mechanical strain of the cells during desiccation. This seems to be achieved mainly by the special quality of the vacuolar content which solidifies (Iljin, 1930; Rouschal, 1938) rather than disappears during drying out. This mechanism seems to be suitable especially for the protection of those structures which establish the osmotic system in the plant cells.

V. Conclusions

The selected examples discussed above show that water stress can clearly affect the metabolism of nitrogen and carbon in plants. In spite of the extensive knowledge of numerous phenomena and of insights into parts of causal chains involved in water-stress response, in all cases a fundamental question remains to be answered, i.e. how water stress (in particular mild or moderate water stress) is sensored and finally transformed into metabolic responses by the organism.

Hsiao (1973) proposes with certain reservations the following processes as possibly responsible for the postulated sensor mechanism: (1) Decrease of the turgor pressure during water stress (see also Levitt and Ben Zaken, 1975); (2) Increase of molecular or ionic concentrations due to water loss (i.e. decrease of volume) of the cells; (3) Alterations of the spatial properties of membrane systems and compartments in the cells; (4) Direct effects on the structure of macromolecules due to removal of hydration water.

It is tempting to speculate that stress dependent alterations of molecular and ionic concentrations [see (2)], or spatial properties of compartments and membranes [see (3)] might represent effective regulators of stress metabolism, since living cells in principle can make use of molecular concentrations or compartmentalization for controlling their metabolic pathways. However, all models of regulation of stress metabolism still remain working hypotheses and offer a wide field for exciting investigations in the future.

References

Allaway, W. G.: Accumulation of malate in guard cells of *Vicia faba* during stomatal opening. Planta (Berl.) **110**, 63–70 (1973).

Barnett, N. M., Naylor, A. W.: Amino acid and protein metabolism in Bermuda grass during water stress. Plant Physiol. **41**, 1222–1230 (1966).

Ben-Amotz, A.: Osmoregulation mechanism in the halophilic alga *Dunaliella parva*. In: Membrane transport in plants (eds. U. Zimmermann, J. Dainty), pp. 95–100. Berlin-Heidelberg-New York: Springer 1974.

Ben-Amotz, A., Avron, M.: The role of glycerol in the osmotic regulation of the halophilic alga *Dunaliella parva*. Plant Physiol. **51**, 875–878 (1973).

Bewley, J. D.: The conservation of polyribosomes in the moss *Tortula ruralis* during total desiccation. J. Exp. Botany **23**, 692–698 (1972).

Bewley, J. D.: Desiccation and protein synthesis in the moss *Tortula ruralis*. Can. J. Botany **51**, 203–206 (1973).

Black, C. C.: Photosynthetik carbon fixation in relation to net CO_2 uptake. Ann. Rev. Plant Physiol. **24**, 253–286 (1973).

Crawford, R. M. V., Thyler, P. D.: Organic acid metabolism in relation to flooding tolerance in roots. J. Ecol. **57**, 235–244 (1969).

Fischer, R. A.: Resistance to water loss in the mesophyll of leek *(Allium porrum)*. J. Exp. Botany **19**, 135–145 (1968).

Gündel, R.: Anatomische, cytologische und physiologische Untersuchungen an poikilohydren Kormophyten. Dissertation Darmstadt 1972.

Heber, U.: Freezing injury in relation to loss of enzyme activities and protection against freezing. Cryobiology **5**, 188–201 (1968).

Hsiao, T. C.: Rapid changes in levels of polyribosomes in *Zea mays* in response to water stress. Plant Physiol. **46**, 281–285 (1970).
Hsiao, T. C.: Plant responses to water stress. Ann. Rev. Plant Physiol. **24**, 519–570 (1973).
Huffaker, R. C., Radin, T., Kleinkopf, G. E., Cox, E. L.: Effects of mild water stress on enzymes of nitrate assimilation and of the carboxylative phase of photosynthesis in barley. Crop Sci. **10**, 471–476 (1970).
Iljin, W.: Über die Ursachen der Resistenz von Pflanzenzellen gegen Austrocknen. Protoplasma **10**, 379–414 (1930).
Kauss, H.: Isofloridosid und Osmoregulation bei *Ochromonas malhamensis*. Z. Pflanzenphysiol. **56**, 453–465 (1967).
Kauss, H.: Osmoregulation in *Ochromonas*. In: Membrane transport in plants (eds. U. Zimmermann, J. Dainty), pp. 90–94. Berlin-Heidelberg-New York: Springer, 1974.
Kemble, A. R., MacPherson, H. T.: Liberation of amino acids in perennial ryegrass during wilting. Biochem. J. **58**, 46–50 (1954).
Kirst, G. O.: Beziehungen zwischen Mannitkonzentration und osmotischer Belastung bei der Brackwasseralge *Platymonas subcordiformis* Hazen. Z. Pflanzenphysiol. **76**, 316–325 (1975).
Kluge, M., Heininger, B.: Untersuchungen über den Efflux von Malat aus den Vakuolen der assimilierenden Zellen von *Bryophyllum* und mögliche Einflüsse dieses Vorgangs auf den CAM. Planta (Berl.) **113**, 333–343 (1973).
Kozlowski, T. T. (ed.): Water deficits and plant growth. III. Plant responses and control of water balance. New York, London: Academic Press 1972.
Levitt, J.: Responses of plants to environmental stresses. New York, London: Academic Press 1972.
Levitt, J., Ben Zaken, R.: Effects of small water stresses on cell turgor and intercellular spaces. Physiol. Plantarum **34**, 273–279 (1975).
Livne, A., Vaadia, Y.: Water deficits and hormone relations. In: Water deficits and plant growth, Vol. 3 (ed. T. T. Kozlowski), pp. 255–274. New York, London: Academic Press 1972.
Lorimer, G. H., Andrews, T. J.: Plant photorespiration—an inevitable consequence of the existence of atmospheric oxygen. Nature **243**, 359 (1973).
Mothes, K.: Der Einfluß des Wasserzustandes auf Fermentprozesse und Stoffumsatz. In: Handbuch der Pflanzenphysiologie, Vol. 3 (ed. W. Ruhland), pp. 656–664. Berlin-Göttingen-Heidelberg: Springer 1956.
Naylor, A. W.: Water deficits and nitrogen metabolism. In: Water deficits and plant growth, Vol. 3 (ed. T. T. Kozlowski), pp. 241–251. New York, London: Academic Press 1972.
Nir, I., Poljakoff-Mayber, A.: Effect of water stress on the photochemical activity of chloroplasts. Nature **213**, 418–419 (1967).
Ogren, W. L.: Control of photorespiration in soybean and maize. In: Environmental and biological control of photosynthesis (ed. R. Marcell), pp. 45–52. The Hague: W. Junk 1975.
Pearson, C. J.: Daily changes in stomatal aperture and in carbohydrates and malate epidermis and mesophyll of leaves of *Commelina apanea* and *Vicia faba*. Australian J. Biol. Sci. **26**, 1035–1044 (1973).
Redshaw, A. J., Meidner, H.: Effects of water stress on the resistance to uptake of carbon dioxide in tobacco. J. Exp. Botany **23**, 229–240 (1972).
Rouschal, E.: Eine physiologische Studie an *Ceterach officinarum*. Flora **132**, 305–318 (1938).
Santarius, K. A.: Der Einfluß von Elektrolyten auf Chloroplasten beim Gefrieren und Trocknen. Planta (Berl.) **89**, 23–46 (1969).
Schobert, B.: The influence of water stress on the metabolism of diatoms. I. Osmotic resistance and proline accumulation in *Cyclotella meneghiniana*. Z. Pflanzenphysiol. **74**, 106–120 (1974).
Schobert, B., Untner, E., Kauss, H.: Isofloridosid und die Osmoregulation bei *Ochromonas malhamensis*. Z. Pflanzenphysiol. **67**, 385–398 (1972).
Singh, T. N., Aspinall, D., Paleg, L. G., Boggess, S. F.: Stress metabolism. II. Changes in proline concentration in excised plant tissue. Australian J. Biol. Sci. **26**, 57–63 (1973b).

Singh, T. N., Paleg, L. G., Aspinall, D.: Stress metabolism. I. Nitrogen metabolism and growth in the barley plant during water stress. Australian J. Biol. Sci. **26,** 45–56 (1973a).

Singh, T. N., Paleg, L. G., Aspinall, D.: Stress metabolism. III. Variations in response to water deficit in the barley plant. Australian J. Biol. Sci. **26,** 64–76 (1973c).

Steward, C. R.: Effects of proline and carbohydrates on the metabolism of exogenous proline by excised bean leaves in the dark. Plant Physiol. **50,** 551–555 (1972a).

Steward, C. R.: Proline content and metabolism during rehydration of wilted excised leaves in the dark. Plant Physiol. **50,** 679–681 (1972b).

Steward, C. R.: The effect of wilting on proline metabolism in excised bean leaves in the dark. Plant Physiol. **51,** 508–511 (1973).

Steward, C. R., Morris, C. J., Thompson, J. F.: Changes in amino acid content of excised leaves during incubation. II. Role of sugar in the accumulation of proline in wilted leaves. Plant Physiol. **41,** 1585–1590 (1966).

Takaoki, T.: Relation between drought tolerance and aging in higher plants. II. Some enzyme activities. Botan. Mag. (Tokyo) **81,** 297 (1968).

Thompson, J. F., Steward, C. R., Morris, C. J.: Changes in amino acid content of excised leaves during incubation. I. The effect of water content of leaves and atmospheric oxygen level. Plant Physiol. **41,** 1578–1584 (1966).

Throughton, J. H.: Plant water status and carbon dioxide exchange of cotton leaves. Australian J. Biol. Sci. **22,** 289–302 (1969).

Throughton, J. H., Slatyer, R. O.: Plant water status, leaf temperature, and the calculated mesophyll resistance to carbon dioxide of cotton leaves. Australian J. Biol. Sci. **22,** 815–827 (1969).

Treichel, S.: Der Einfluß von NaCl auf die Prolinkonzentration verschiedener Halophyten. Z. Pflanzenphysiol. **76,** 56–68 (1975).

Wegmann, K.: Osmotic regulation of photosynthetic glycerol production in *Dunaliella*. Biochim. Biophys. Acta **234,** 317–323 (1971).

Ziegler, H.: Biochemische Anpassungen der Pflanzen an extreme Standortbedingungen. Biol. Rundschau **12,** 81–95 (1974).

Ziegler, H., Vieweg, G. H.: Poikilohydre Pteridophyta und Spermatophyta. In: Die Hydratation und Hydratur des Protoplasmas der Pflanzen und ihre ökophysiologische Bedeutung. Protoplasmatologica II C (eds. H. Walter, K. Kreeb), pp. 88–108. Vienna-New York: Springer 1970.

D. Water Stress During Freezing

U. Heber and K. A. Santarius

I. Introduction

When the temperature drops below the freezing point, water becomes first supercooled and then is converted into ice. Plants, plant organs or cells exposed to subzero temperatures may or may not be damaged when freezing of intracellular water occurs. This depends both on the mode of freezing and on the nature or physiological state of the plant material. When the temperature decreases slowly, ice formation is initiated extracellularly and progresses outside the cells producing cell dehydration. Depending on the extent of cellular resistance, this is tolerated or harmful. Only when the rate of freezing is too fast to permit transfer of intracellular water to extracellular ice loci does intracellular freezing occur. It is lethal, owing to mechanical damage produced by growing ice crystals, except when freezing is so rapid as to produce "vitrification" of cells. We will consider only the effects of slow physiological freezing, which causes cell dehydration, and therefore represents water stress to plants, and also discuss briefly mechanisms permitting cells to tolerate such water stress. The field has been reviewed in recent years by different investigators (Meryman, 1966; Mazur, 1969, 1970; Weiser, 1970; Alden and Hermann, 1971; Levitt, 1972; Heber and Santarius, 1973). It is not our purpose to assess again merits and disadvantages of the different hypotheses put forward to explain frost damage and frost resistance, and the reader is urged to consult earlier reviews for more detailed and complementary information.

II. Frost Injury

1. Frost-sensitive Sites in Plant Cells

There are questions as to the nature of the events causing frost injury and the cell components which are susceptible to freezing damage. Cells contain and are composed of different classes of compounds. In view of their molecular stability there is no reason to assume that the multitude of low-molecular-weight cell constituents suffer damage during freezing. High molecular building blocks of plant cells are polymer carbohydrates and proteins. Carbohydrates are insensitive to freezing. Most soluble proteins also remain unaffected by freezing as is evidenced by the resistance of simple enzymes to conditions of freezing which would damage cells (Heber and Santarius, 1964; Santarius, 1969). The same holds true for nucleic acids (Mazur, 1966). In contrast, complex membrane-bound enzyme systems appeared to be highly sensitive to freezing. Especially chloroplast and mitochondrial membranes that are involved in cellular energy conservation became inactivated by freezing (Heber and Santarius, 1964). This inactivation was irreversible and affected in particular both photophoshorylation of thylakoid membranes and oxidative phosphorylation of mitochondria, whereas electron transport reactions were much more resistant toward dehydration. Freezing actually resulted in the uncoupling of phosphorylation from electron transport. Even if this constituted the only effect of freezing, it is obvious that after thawing, lack of ATP would lead to the breakdown of metabolism of cells, which would than have a damaged energy conservation system. Inactivation of energy-conserving membranes is therefore a sufficient, although certainly not the only, cause of cell death after freezing.

Under more extreme conditions, freezing also affected electron transport in chloroplast membranes (Heber and Ernst, 1967; Heber et al., 1973). The water-splitting site in chloroplast electron transport became inactivated after photophosphorylation was damaged, causing inactivation of electron transport through photosystem II. In contrast, electron transport through photosystem I was much less frost-sensitive and actually was considerably stimulated when photophosphorylation was lost. However, even this stimulation appears to reflect damage to the membranes.

From these results it may be generalized that inactivation of protoplasmic membranes is the primary cause of frost damage in plant cells.

2. Mechanisms of Freezing Injury

Membrane inactivation during freezing may be mechanical owing to the accumulation of ice crystals, and may result from the loss of stabilizing water from the membranes or be caused by the accumulation of solutes, which inevitably accompanies the conversion of cellular water into ice.

Earlier work has suggested that formation of ice crystals during freezing leads to mechanical damage of cell structures (Iljin, 1933). However, in vitro experiments have later established that, under normal conditions of freezing, membrane damage is rarely caused by mechanical effects of growing ice crystals. For instance, dehydration of chloroplast membranes in the absence of ice formation at 0°C results in the same damage as freezing (Santarius and Heber, 1967). Damage resembling that caused by freezing is also produced when biological membranes are exposed to concentrated solutions of various electrolytes at 0°C (Lovelock, 1953a; Meryman, 1968; Santarius, 1969; Santarius and Heber, 1970). Only eutectic freezing, which does not normally occur during freezing of cells, produces extensive mechanical damage. It always leads to complete inactivation of photophosphorylation of thylakoid membranes (Santarius, 1973a).

Simple removal of water from the membranes, which might have destabilizing effects, cannot be made responsible for the damage. The extent of water loss

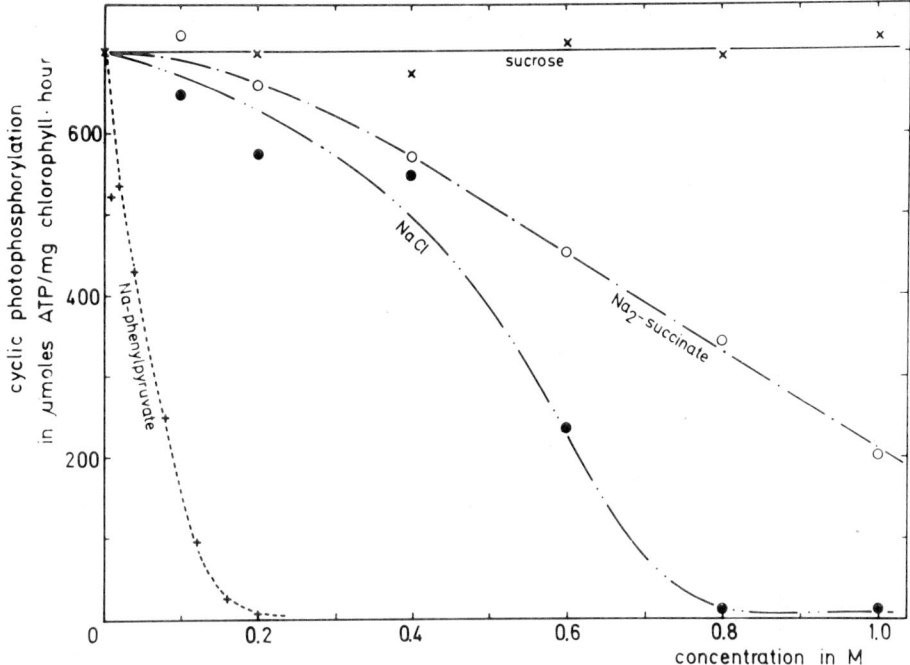

Fig. 1. Inactivation of photophosphorylation of isolated chloroplast membranes after exposure to various concentrations of sucrose, sodium succinate, NaCl and sodium phenylpyruvate for 3 h at 0° C. It should be noted that solute inactivation of biomembranes becomes particularly pronounced during freezing. Owing to the accumulation during freezing, much lower initial concentrations are then sufficient for damage. (From Santarius and Heber, 1972)

from a system is clearly a function of the freezing temperature. The lower it is, the more extensive is dehydration. At a given temperature, dehydration is determined solely by the vapor pressure over ice and is independent of the composition of the system. If inactivation was caused by membrane dehydration, it should also be independent of the composition of the system and should always increase with decreasing temperature. As a matter of fact, it is drastically influenced by the presence of very different solutes. In vitro there are conditions under which it is less extensive at lower than at higher temperatures (Santarius and Heber, 1970). For these reasons it is thought that neither mechanical damage nor membrane dehydration are decisive or even significant factors in causing injury to cells under most natural freezing conditions.

When cellular water is converted into extracellular ice, the concentration of intracellular solutes increases dramatically. In vitro experiments with isolated thylakoid membranes have shown that various cell constituents are toxic and cause membrane inactivation when present at high concentrations (Santarius, 1969, 1971; Heber et al., 1971). This is particularly true for various inorganic salts, but a number of organic compounds also are effective inactivators (Fig. 1). It was shown that the rate of membrane inactivation during freezing was strongly dependent (1) on the nature of the solutes present, (2) on the freezing temperature, (3) on the length of time during which freezing was exerted on the cell membranes, (4) on the rate of cooling and thawing, and (5) on the solute concentration before freezing (Santarius and Heber, 1970). Of the soluble cell constituents, which are toxic to cell membranes during freezing in vitro, semi-polar compounds such as phenylalanine, valine, leucine, isoleucine and others are especially effective. They damage membranes even at relatively low concentrations compared with other potentially toxic solutes (Fig. 1). Comparison between profiles of damage obtained by the freezing of isolated thylakoids in vitro and of intact leaves showed that frost damage which occurred in vivo is similar to that produced by semi-polar compounds during freezing in vitro (Heber et al., 1973).

From these observations it can be concluded that the accumulation of potentially membrane-toxic inorganic and organic cell constituents in the surroundings of frost-sensitive membranes, which is a consequence of the dehydration brought about by freezing, is a primary cause of the inactivation of membranes during freezing.

There is also the question as to the nature of membrane damage. Membrane-bound energy conservation, which is lost during injurious freezing, requires intact and osmotically active membranes. Such membranes have a low permeability even toward small ions such as protons. In fact, active proton pumping across the membranes, driven by electron transport, appears to be a prerequisite of energy conservation (Mitchell, 1966). It results in the formation of proton gradients which represent an electrochemical potential that can be used for phosphorylation. After freezing damage, such proton gradients can no longer be established across the membranes by electron transport, suggesting alterations in the permeability of the membranes (Heber, 1967). In contrast to intact membrane vesicles, such systems neither shrink in hypertonic nor swell in hypotonic solutions of normally non-penetrating compounds. Obviously, the permeability of the membranes is drastically increased after freezing and has led to membrane damage. Permeability changes actually appear to form the basis of freezing injury to cells. It is perfectly clear that the life functions of a cell depend on the maintenance of its internal composition, which is facilitated by highly selective membrane permeability. If this selectivity breaks down as a consequence of injurious freezing, cell death is an inevitable result.

At present the mechanisms, which during freezing cause membrane permeability alterations under the influence of the accumulation of inorganic membrane-toxic compounds, are not fully understood. The stabilization of biomembranes appears to be achieved in a complex fashion by hydrogen bonds and ionic and van der Waals interactions between the different membrane constituents. During freezing in the presence of electrolytes, the ionic strength in the surroundings of the membranes increases dramatically. Since ionic interactions between membrane components are non-directional, the increase in external ionic strength should be expected to suppress internal ionic interactions. In fact, recent investigations have shown that membrane damage during freezing is accompanied by the release of membrane proteins (Volger, unpublished). Presumably, semi-polar compounds disturb, during freezing, lipid/lipid interactions within the membranes, thereby destabilizing the membrane structure and leading to damage.

According to Levitt (1962, 1972) frost injury is due to an oxidation of SH groups of structural proteins to SS bridges which is irreversible and leads to the denaturation of membrane proteins. However, freezing of isolated thylakoids in the presence of an excess of cysteine and glutathione, or under nitrogen, which should be expected to prevent or reduce linkage of different proteins by SS bridges, did not prevent inactivation of photophosphorylation (Heber and Santarius, 1964). In addition, the membrane-bound ATPase which is responsible for the terminal energy-conservation step in photophosphorylation was not altered during freezing, which was sufficient to inactivate photophosphorylation (Heber, 1967).

Meryman (1970, 1971) suggested that the increase in the concentration of soluble compounds during freezing would increase concentration gradients of impermeable solutes across the membranes. Hypertonic stress would produce shrinkage of the membranes; below a minimum tolerable volume this shrinkage would cause irreversible membrane damage. This hypothesis does not explain why different solutes cause the same membrane damage at very different osmolar concentrations, and also, that membrane inactivation is dependent on the length of exposure time to membrane-toxic compounds and on the rate of cooling.

Even though there is disagreement as to the detailed mechanism, it appears established that injury to cells by freezing consists in the alteration of the permeability properties of sensitive biomembranes.

III. Frost Resistance

It is well known that frost resistance is acquired by a number of plant species in the fall, under what are called hardening conditions, and is usually lost by dehardening when dormancy is replaced by active growth in the spring. Possible mechanisms of membrane protection are shown in Fig. 2:

1. Cells might decrease their content of potentially membrane-toxic solutes in relation to other solutes thereby reducing the extent of accumulation of toxic solutes during freezing.

2. Cells might increase their concentration of non-toxic solutes in relation to that of potentially toxic solutes. Both (1) and (2) would produce the same result in decreasing the actual concentration of potentially toxic solutes to which membranes are exposed under freezing conditions (Fig. 2A).

3. Membranes may be shielded from the influence of potentially membrane-toxic solutes by special protective compounds (Fig. 2B).

4. Possible changes in the membrane structure may alter and in fast decrease the sensitivity of membranes toward inactivation by the accumulation of toxic solutes during freezing (Fig. 2C).

In the following we will briefly discuss whether these possible mechanisms are verified in, and employed by, nature to protect cells against freezing injury.

1. Protection by a Decrease in the Concentration of Membrane-toxic Solutes

Although the literature on changes in plant composition accompanying changes in hardiness is extensive, no evidence appears to exist on a decrease

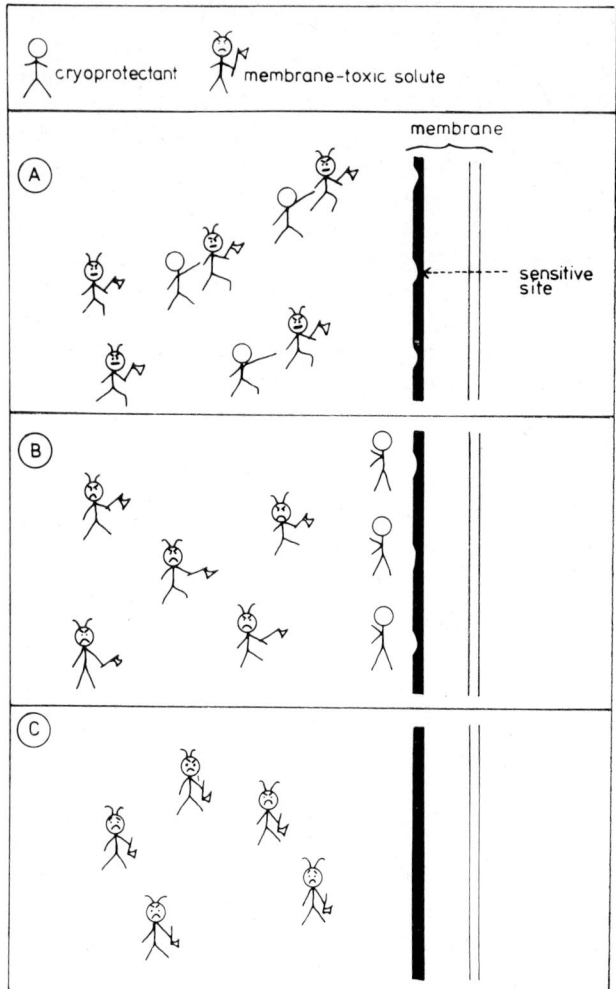

Fig. 2 A–C. Possible modes of membrane protection during freezing. (A) Unspecific colligative "dilution" of cryo-toxic compounds by membrane-compatible solutes. (B) Specific protection by solutes, which interact with membranes shielding them from the action of cryo-toxic compounds. (C) Increase of membrane resistance by changes in membrane structure. For more information see text

during hardening and an increase during dehardening in the concentration of solutes which have been recognized as damaging to biomembranes during freezing. Kappen and Ullrich (1970) reported that chloride concentration in nonaqueously isolated chloroplasts from halophytes did not alter much during changes in hardiness. At high concentrations, chlorides cause inactivation of thylakoid membranes.

2. Colligative Protection

Especially well-known is the fact that soluble sugars accumulate during hardening in many plant species. The significance of these observations has been questioned on the grounds that high concentrations of sugars may also occur in plant species incapable of hardening or when frost resistance is lost (Levitt, 1956). These objections do not take into account the intracellular compartmentation of cell constituents. In the absence of resistance, sugars may be distributed in cells in such a way that protection of sensitive biomembranes is not permitted.

Soluble sugars are tolerated by thylakoids up to high concentrations and are thus not membrane-toxic (Fig. 1). In fact, thylakoids or mitochondria frozen in the presence of sugars are protected against inactivation during freezing (Jagendorf and Avron, 1958; Duane and Krogmann, 1963; Heber and Santarius, 1964). The extent of protection is a function of the concentration of membrane-toxic solutes also present during freezing. Protection of membranes by sugars against freezing may be brought about unspecifically, simply by decreasing the concentration of toxic solutes to non-toxic levels colligatively (Fig. 2A) or specifically by shielding the membranes against toxic solutes (Fig. 2B). A clear-cut distinction between these possibilities does not appear to be possible and both modes of protection may well play a role simultaneously. Thus sugars protect thylakoids for instance against inactivation by high concentrations of membrane-toxic solutes to some extent even at 0 °C in the absence of freezing (Santarius, 1971). Under these conditions the presence of sucrose does not significantly change the concentration of membrane-toxic solutes. The observed effect therefore appears to reflect a specific membrane-stabilizing property of sucrose (Fig. 2B). Other cryoprotective agents such as sodium succinate show similar behavior. On the other hand, comparison of the effectiveness of sugars with that of other chemically unrelated cryoprotectants leads to the conclusion that another major component of the cryoprotective action of sugars is unspecific, colligative dilution of membrane-toxic solutes. This is suggested by the experiments of Figs. 3 and 4, which show protection of thylakoids against freezing damage by cryoprotectants chemically as different as sucrose, proline and sodium succinate. As a cryotoxic solute, sodium chloride was also present. Without cryoprotectant, even low concentrations of NaCl caused extensive membrane damage during freezing. Increasing the concentration of cryoprotectant in relation to that of NaCl increasingly protected the membranes. The molecular ratio of cryoprotectant to cryotoxic solute was the most important factor in determining the extent of protection ultimately achieved. In the case of sucrose and proline, all ratios higher than certain threshold values were protective. When succinate was used as cryoprotectant, both low and high ratios led to damage. Only in an intermediary range of ratios was protection observed. Serine, glycine, α-alanine, glutamate, citrate, malate and others were similar to succinate as far as cryoprotective action is concerned, while protection by different sugars, sugar alcohols, threonine, β- and γ-amino acids and others resembled that achieved by sucrose or proline (Santarius, 1971, 1973b; Heber et al., 1971; Santarius and Heber, 1972; Tyankova, 1972). The molar ratio of cryoprotectant to cryotoxic solute necessary for effective protection differed for different cryoprotectants.

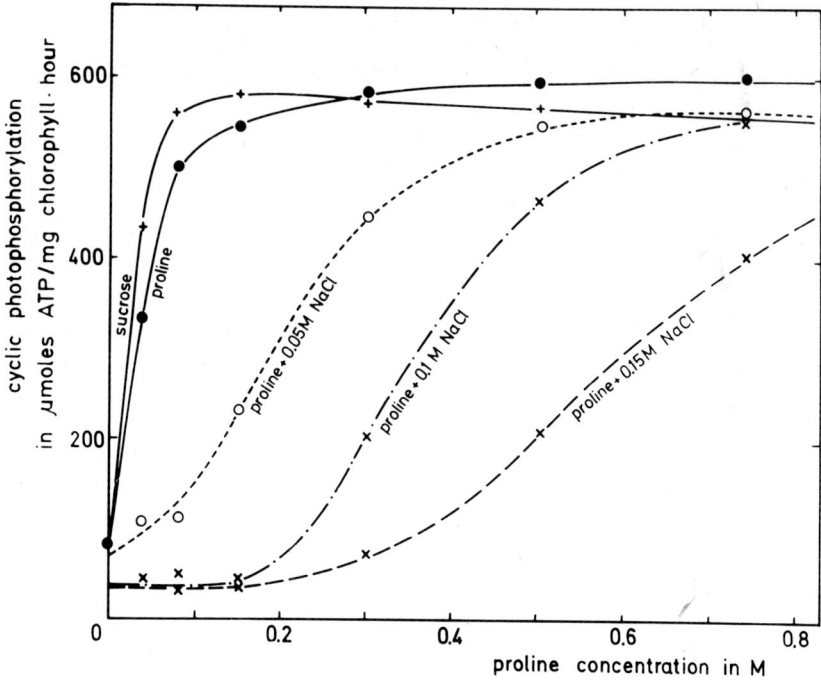

Fig. 3. Protection of thylakoids against freezing damage by proline and counteraction by salt. For comparison, protection by sucrose was also measured. As in the case of proline, increasing ratios of salt to sucrose shift protection curves to the right (not shown). Freezing for 3 h to −25° C. (From Tyankova, 1970)

Lovelock (1953b) has invoked a colligative mechanism of protection for red blood cells which when suspended in physiological saline were protected against freezing damage by glycerol. Such a mechanism is also capable of explaining many aspects of the protection of thylakoids and mitochondria by chemically very different low-molecular-weight solutes, although it does not account for observed individual differences in cryoprotective effectiveness and for protection in the absence of freezing.

A simple fact necessary for understanding colligative protection is that the total osmolar concentration of any solution coexisting and in equilibrium with ice is solely a function of the freezing temperature. A necessary assumption is that a potentially membrane-toxic solute is not damaging below, but leads to injury above, a certain threshold concentration. At any one freezing temperature, accumulation of a mixture of solutes in the unfrozen part of a membrane suspension or of cells will lead to the same total osmolarity, irrespective of the initial concentration or composition of the unfrozen system. If the composition of the membrane suspension or of the cells is such that potentially membrane-toxic solutes make up a high proportion of the total solutes, accumulation during freezing will lead to levels of toxic solutes sufficient for membrane damage. If on the other hand, membrane-neutral solutes predominate, solute accumulation

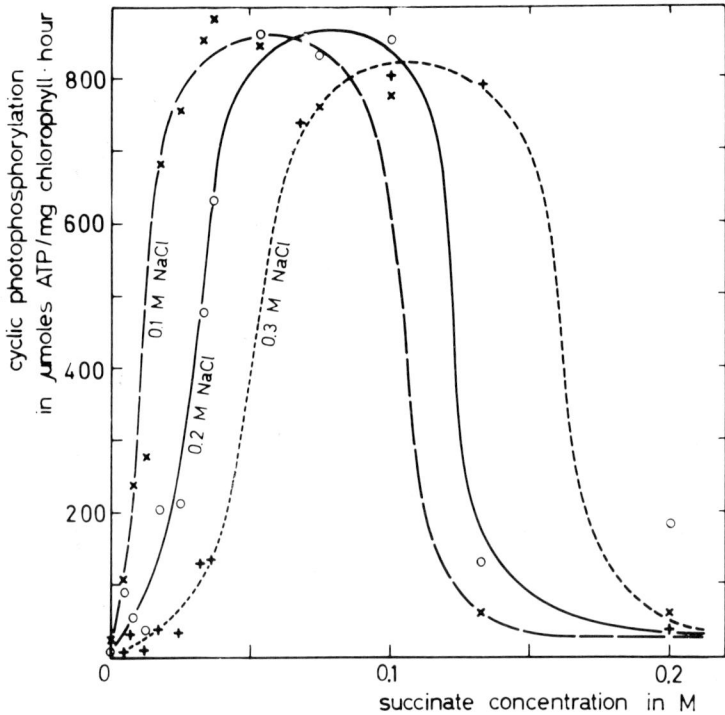

Fig. 4. The effect of various concentrations of NaCl on photophosphorylation of isolated chloroplast membranes during freezing for 3 h at $-25°C$ as a function of the succinate concentration. (From Santarius, 1971)

to the same total osmolarity as before will no longer build up a damaging concentration of toxic solutes and protection will be observed simply because most of the total osmolarity of the unfrozen solution coexisting with ice is caused, and represented by, the neutral solute. The latter "dilutes" membrane-toxic solutes.

Under some circumstances, even potentially membrane-toxic solutes may contribute to protection as seen in Fig. 4 for sodium succinate. In the absence of succinate, damage by freezing is caused by the accumulation of NaCl to damaging levels. Addition of succinate decreases the NaCl accumulation under freezing and is therefore protective. However, high concentrations of succinate in relation to chloride again result in freezing damage, presumably because freezing now accumulates succinate to levels, which are membrane-toxic. Indeed, high concentrations of succinate cause membrane inactivation even at 0°C in the absence of freezing (Fig. 1; cf. also Santarius, 1969). Thus at low ratios of succinate to chloride freezing causes chloride damage and at very high ratios succinate damage. Intermediary ratios are protective because neither chloride nor succinate reach damaging levels. Membrane protection by two potentially membrane-toxic solutes, which are simultaneously present, can obviously be observed only if the mode of interaction of the two solutes with the membranes is not identical.

Colligative protection by a decrease in the effective concentration of membrane-toxic solutes is a powerful concept to explain lacking specificity of protection. For this type of protection, the ratio of different solutes to one another, in particular that of neutral solutes to potentially membrane-toxic solutes, is of decisive influence, while the nature of neutral solutes is of little importance. In the light of physiological observations of an accumulation of sugars and other neutral solutes during hardening of plants, the well-documented protection of sensitive biomembranes against freezing damage by such compounds leaves little room for doubt that the accumulated solutes act as cryoprotectants. For physical reasons, a considerable part of protection must be colligative protection.

3. Protection by Proteins

Colligative protection should be independent of molecular properties and comparative effects should be exerted by the same osmolar concentrations of different neutral solutes. Any particularly high effectiveness of a compound as a cryoprotectant would foster the suspicion that it does not, or not mainly, act colligatively but rather via specific interactions with sensitive membranes (Fig. 2 B). It has already been mentioned that concentrations of different low-molecular-weight cryoprotectants, which are necessary for protection during freezing, differ very considerably. For instance, sodium succinate is, in the presence of 0.1 M NaCl, better by a factor of about 5 than sucrose (Santarius, 1971). The latter compound is better than glucose, but inferior to raffinose in its cryoprotective properties (Santarius, 1973b). A number of proteins isolated from frost-hardy plant material were found to be more effective even on a unit weight basis than low-molecular-weight cryoprotectants, such as sucrose, in protecting thylakoids against inactivation during freezing (Fig. 5; cf. Heber and Ernst, 1967). Protection was specific in that many other plant proteins were not, or were only slightly protective. On a molar basis, effectiveness was higher by almost three orders of magnitude than that of low-molecular-weight cryoprotectants. At extremely low concentrations (0.01 % or less), which produced little protection in the absence of other solutes, the cryoprotective proteins still considerably increased protection by other cryoprotectants such as sucrose. These observations are strongly suggestive of a specific mechanism of protection which is different from that outlined above. A specific role of these proteins in frost resistance was also indicated by the failure to isolate them from frost-sensitive plant material. The protective proteins had rather unusual properties in being heat stable. They were easily water-soluble and required high concentrations of ammonium sulphate for precipitation. Gel electrophoretic separation of a cryoprotective protein fraction yielded several active bands. The molecular weights of different active proteins as determined by sedimentation analysis (Heber and Kempfle, 1970) or by sodium dodecyl sulfate electrophoresis (Volger and Heber, 1975) ranged between 10,000 and 20,000 Daltons. The amino acid composition of two of the active proteins has been determined (Volger and Heber, 1975). Qualitatively it is not very different from that of many other proteins. Remarkable

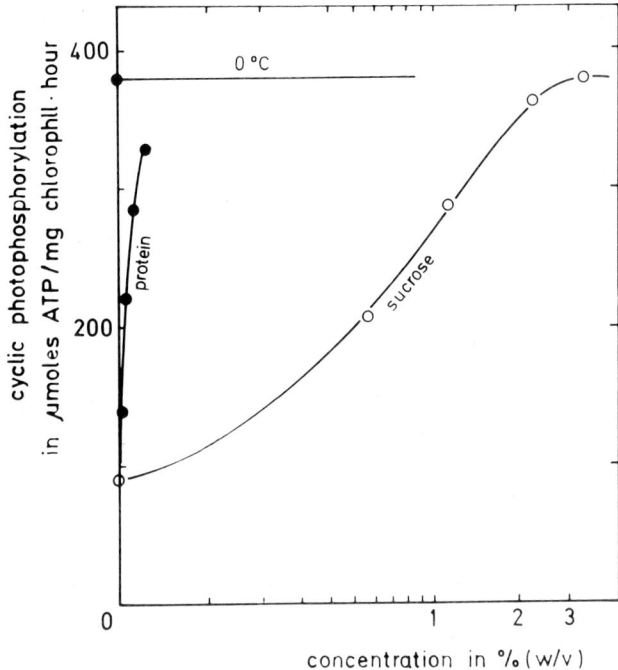

Fig. 5. Protection of thylakoids against freezing damage by a protein fraction from chloroplasts of hardy spinach leaves. Protection by sucrose is also shown on a unit weight basis for comparison. Freezing for 3 h at $-25\,°C$. Note non-linear scale on abscissa. (From Heber, 1970)

are only the unusually high percentage of polar amino acids and the low percentage of amino acids with non-polar side chains. Measurements of circular dichroism failed to indicate helical conformations.

Attempts to demonstrate binding of the proteins in dilute solution to the membranes were unsuccessful. Still, in view of the high effectiveness of these compounds it is thought that accumulation of the proteins during freezing leads to interactions with the membranes which render them insensitive to freezing.

4. Protection by Changes in Membrane Structure

While differences in the freezing sensitivity of thylakoid membranes from leaf material of different origin or physiological state are commonly observed, attempts have failed to correlate frost hardiness of the parent leaves to the sensitivity of photophosphorylation of thylakoids, which were carefully washed to free them from natural soluble cryoprotectants. Only recently have seasonal differences in the sensitivity of electron transport reactions of thylakoids become apparent (Santarius, 1974). Such differences might indicate structural differences in membranes from frost-hardy and frost-sensitive plant material. Indeed, model experiments suggest that structural membrane alterations can lead to increased membrane resistance to freezing (Fig. 2C).

It has been mentioned that semi-polar compounds cause membrane inactivation at elevated concentrations. They are able to overcome, in a fashion similar to that of inorganic salts, the protection against freezing provided by sucrose or other cryoprotectants. However, in contrast to salts, at very low concentrations compounds such as sodium phenylpyruvate, sodium dodecyl succinate or isoleucine, which contain non-polar side chains and a polar group can also cause protection. This is, within a very limited range of concentrations, expecially apparent, when membranes are frozen in solutions of an inorganic salt and a cryoprotectant such as sucrose. These solutions need to be balanced in order to prevent complete inactivation during freezing (Fig. 6). Addition of sodium phenylpyruvate or a similar compound at concentrations as low as 1 mM decreases membrane damage during freezing. Under the conditions of the experiment shown in Fig. 6, optimal protection occurred at 5 mM sodium phenylpyruvate. As should be expected from the membrane-toxic properties of this compound, increased concentrations then decreased protection and finally led to complete membrane inactivation during freezing.

There is the possibility that protection by phenylpyruvate is caused by colligative action of this compound at concentrations which are not yet membrane-toxic. However, available evidence suggests that this is an insufficient explanation. Adding sucrose or another neutral compound to the system at the same osmolar concentrations, which produced good protection by phenylpyruvate, did not increase protection by very much. Actually, phenylpyruvate binds to the mem-

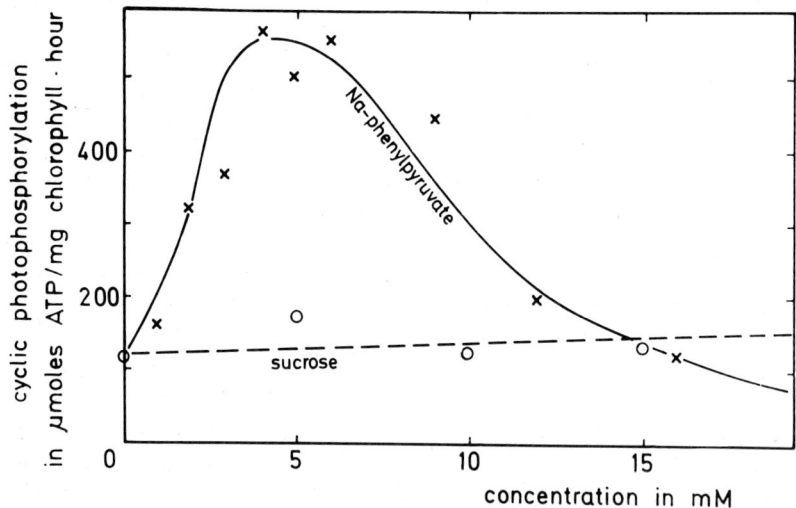

Fig. 6. Protection of thylakoids against freezing damage by low concentrations of sodium phenylpyruvate and inactivation by higher concentrations. Sucrose (150 mM) and sodium chloride (100 mM) were also present in the membrane suspension. *Dashed line:* extent of protection when sucrose was added instead of phenylpyruvate on top of the sucrose/NaCl mixture forming the solute basis of the membrane suspension. Freezing for 3 h to $-25°$C.
(From Overbeck, unpublished)

branes and changes the membrane structure as shown by slow changes of light scattering by the membranes on addition of low concentrations of phenylpyruvate. No such changes in light scattering occurred, when sucrose or other neutral solutes were added to the membranes. It is therefore proposed that increased membrane resistance to freezing caused by phenylpyruvate is a consequence of alterations in membrane structure.

IV. Conclusions

During slow freezing unprotected cells from potentially hardy organisms succumb to the water stress posed by the conversion of intracellular water to extracellular ice rather than to mechanical effects of ice or to temperature stress. Death occurs because sensitive cellular membranes suffer loss of semipermeability when certain intracellular solutes accumulate to toxic levels.

Hardening of the cells prevents damage and leads to frost resistance. Of main importance during hardening is the accumulation and proper intracellular distribution of cryoprotectants. These comprise very different and chemically unrelated compounds, which act through different mechanisms. Of common occurrence in plants is the accumulation of neutral solutes such as soluble carbohydrates during hardening. An important aspect of protection by sugars and their derivatives is their membrane compatibility. They do not exert harmful effects on biomembranes even at high concentrations. Also, they do not readily crystallize. Indeed, eutectic freezing, which would lead to mechanical damage, rarely occurs in biological systems. Simply by their presence, sugars reduce during freezing colligatively the concentration of potentially membrane-toxic compounds. In addition they appear to be membrane-stabilizing. The two effects together form the basis of protection. Even potentially membrane-toxic compounds such as succinate, glycine, serine and others will contribute to colligative protection, provided that their concentration remains, during freezing, below damaging levels. In contrast to protection by low-molecular-weight cryoprotectants, that by specific hydrophilic proteins appears to be dominated by direct interactions with the membranes as judged from their unusually high effectiveness all of these cryoprotectants act from outside on the membranes. Their molecular properties make it appear unlikely that they dissolve in the hydrophobic membrane phase. In contrast, semi-polar compounds exert profound effects on the membrane structure. At very low concentrations they increase protection, at higher concentrations they cause membrane inactivation presumably by altering the membrane structure. It is not yet clear whether protection by changes in the membrane structure is of physiological significance and represents, in addition to colligative and specific protection by hydrophilic solutes, a third mechanism of protection.

Acknowledgements. Investigations of our laboratory reported and reviewed in this contribution were supported by the Deutsche Forschungsgemeinschaft. We are grateful to M. R. Kirk for reading the manuscript.

References

Alden, J., Hermann, R. K.: Aspects of the cold-hardiness mechanism in plants. Botan. Rev. **37**, 37–142 (1971).
Duane, W. C., Krogmann, D. W.: Chloroplast storage with retention of photosynthetic activities. Biochim. Biophys. Acta **71**, 195–196 (1963).
Heber, U.: Freezing injury and uncoupling of phosphorylation from electron transport in chloroplasts. Plant Physiol. **42**, 1343–1350 (1967).
Heber, U.: Proteins capable of protecting chloroplast membranes against freezing. In: Ciba Found. Symp. on The Frozen Cell (eds. G. E. W. Wolstenholme, M. O'Connor), pp. 175–188. London: Churchill Ltd. 1970.
Heber, U., Ernst, R.: A biochemical approach to the problem of frost injury and frost hardiness. In: Cellular injury and resistance in freezing organisms (ed. E. Asahina). Proc. Intern. Conf. Low Temp. Sci., Vol. II, pp. 63–77. Sapporo (Japan): Bunyeido Printing Co. 1967.
Heber, U., Kempfle, M.: Proteine als Schutzstoffe gegenüber dem Gefriertod der Zelle. Z. Naturforsch. **25b**, 834–842 (1970).
Heber, U., Santarius, K. A.: Loss of adenosine triphosphate synthesis caused by freezing and its relationship to frost hardiness problems. Plant Physiol. **39**, 712–719 (1964).
Heber, U., Santarius, K. A.: Cell death by cold and heat and resistance to extreme temperatures. Mechanisms of hardening and dehardening. In: Temperature and life (eds. H. Precht, J. Christophersen, H. Hensel, W. Larcher), pp. 232–263. Berlin-Heidelberg-New York: Springer 1973.
Heber, U., Tyankova, L., Santarius, K. A.: Stabilization and inactivation of biological membranes during freezing in the presence of amino acids. Biochim. Biophys. Acta **241**, 578–592 (1971).
Heber, U., Tyankova, L., Santarius, K. A.: Effects of freezing on biological membranes in vivo and in vitro. Biochim. Biophys. Acta **291**, 23–37 (1973).
Iljin, W. S.: Über den Kältetod der Pflanzen und seine Ursachen. Protoplasma **20**, 105–124 (1933).
Jagendorf, A. T., Avron, M.: Cofactors and rates of photosynthetic photophosphorylation by spinach chloroplasts. J. Biol. Chem. **231**, 277–290 (1958).
Kappen, L., Ullrich, W. R.: Verteilung von Chlorid und Zuckern in Blattzellen halophiler Pflanzen bei verschieden hoher Frostresistenz. Ber. Deut. Botan. Ges. **83**, 265–275 (1970).
Levitt, J.: The hardiness of plants. New York, London: Academic Press 1956.
Levitt, J.: A sulfhydryl-disulfide hypothesis of frost injury and resistance in plants. J. Theoret. Biol. **3**, 355–391 (1962).
Levitt, J.: Responses of plants to environmental stresses. New York, London: Academic Press 1972.
Lovelock, J. E.: The haemolysis of human red blood-cells by freezing and thawing. Biochim. Biophys. Acta **10**, 414–426 (1953a).
Lovelock, J. E.: The mechanism of protective action of glycerol against haemolysis by freezing and thawing. Biochim. Biophys. Acta **11**, 28–36 (1953b).
Mazur, P.: Physical and chemical basis of injury in single-celled micro-organisms subjected to freezing and thawing. In: Cryobiology (ed. H. T. Meryman), pp. 213–315, New York, London: Academic Press 1966.
Mazur, P.: Freezing injury in plants. Ann. Rev. Plant Physiol. **20**, 419–448 (1969).
Mazur, P.: Cryobiology: The freezing of biological systems. Science **168**, 939–949 (1970).
Meryman, H. T. (ed.): Cryobiology. New York, London: Academic Press 1966.
Meryman, H. T.: Modified model for the mechanism of freezing injury in erythrocytes. Nature **218**, 333–336 (1968).
Meryman, H. T.: The exceeding of a minimum tolerable cell volume in hypertonic suspension as a cause of freezing injury. In: Ciba Found. Symp. on the Frozen Cell (eds. G. E. W. Wolstenholme, M. O'Connor), pp. 51–67. London: Churchill Ltd. 1970.
Meryman, H. T.: Osmotic stress as a mechanism of freezing injury. Cryobiology **8**, 489–500 (1971).

Mitchell, P.: Chemiosmotic coupling in oxidative and photosynthetic phosphorylation. Biol. Rev. **41**, 445–502 (1966).
Santarius, K. A.: Der Einfluß von Elektrolyten auf Chloroplasten beim Gefrieren und Trocknen. Planta (Berl.) **89**, 23–46 (1969).
Santarius, K. A.: The effect of freezing on thylakoid membranes in the presence of organic acids. Plant Physiol. **48**, 156–162 (1971).
Santarius, K. A.: Freezing: the effect of eutectic crystallization on biological membranes. Biochim. Biophys. Acta **291**, 38–50 (1973a).
Santarius, K. A.: The protective effect of sugars on chloroplast membranes during temperature and water stress and its relationship to frost, desiccation and heat resistance. Planta (Berl.) **113**, 105–114 (1973b).
Santarius, K. A.: Seasonal changes in plant membrane stability as evidenced by the heat sensitivity of chloroplast membrane reactions. Z. Pflanzenphysiol. **73**, 448–451 (1974).
Santarius, K. A., Heber, U.: Das Verhalten von Hillreaktion und Photophosphorylierung isolierter Chloroplasten in Abhängigkeit vom Wassergehalt. II. Wasserentzug über $CaCl_2$. Planta (Berl.) **73**, 109–137 (1967).
Santarius, K. A., Heber, U.: The kinetics of the inactivation of thylakoid membranes by freezing and high concentrations of electrolytes. Cryobiology **7**, 71–78 (1970).
Santarius, K. A., Heber, U.: Physiological and biochemical aspects of frost damage and winter hardiness in higher plants. In: Proc. of a colloqu. on the winter hardiness of cereals (ed. S. Rajki), pp. 7–29. Martonvásár: Agr. Res. Inst., Hungarian Acad. Sci. 1972.
Tyankova, L.: Stabilität von Thylakoidmembranen in Gegenwart von Aminosäuren bei Gefrieren. Ber. Deut. Botan. Ges. **83**, 491–497 (1970).
Tyankova, L.: The effect of amino acids on thylakoid membranes during freezing as influenced by side chain and position on the amino group. Biochim. Biophys. Acta **274**, 75–82 (1972).
Volger, H., Heber, U.: Cryoprotective leaf proteins. Biochim. Biophys. Acta **412**, 335–349 (1975).
Weiser, C. J.: Cold resistance and injury in woody plants. Science **169**, 1269–1278 (1970).

E. Cell Permeability and Water Stress

O. Y. LEE-STADELMANN and E. J. STADELMANN

I. Introduction

Deficit of water, which all higher land plants in nature may experience recurrently during their life span, often causes significant changes in their growth, development, morphology, physiology and biochemistry (cf. Hsiao, 1973). These alterations depend ultimately on the effect of water shortage on the living protoplasm. However, the effect of a given water shortage on protoplasm will vary from species to species and, in fact, will be less severe in plants possessing morphological and anatomical features which enable them to reduce water loss or increase water uptake (drought avoidance, cf. Levitt, 1972).

Although protoplasm has been investigated for many decades, our understanding of how protoplasm and its elements are involved in the basic mechanisms of drought sensitivity or drought resistance of a plant cell and of the entire plant is still very fragmentary. The reasons for this fragmentary understanding are based, to some extent, on the technical difficulties encountered in working with living cells; protoplasm in the fully developed higher plant cell represents only a thin layer of living matter adjacent to the cell wall (4 to 5% of the total cell volume). Furthermore, experimental work on living protoplasm requires tedious and laborious procedures, especially for quantitative measurements, because the methods which are applicable permit only limited use of automated devices or instruments.

One property of the living protoplasm, which can be measured quantitatively with relative ease and fairly accurately, is the permeability of the protoplasmic layer to water and to relatively harmless nonelectrolytes. Such data may be especially valuable for obtaining insight into changes occurring in the cell membrane (=plasmalemma) and tonoplast.

Data from the literature on the influence of water deficit on protoplasmic factors have sometimes led to confusion and contradictable interpretation. The major reasons, however, for this may not be inaccurate results, but rather too broad a generalization of the findings; data from experiments with different kinds of plant cells and therefore different types of protoplasm can be compared only after sufficient testing has been carried out, because a given water deficit does not yield the same effects on every type of protoplasm. Consequently, comparison of the effect of a single external factor on the living protoplasm, under a given water stress, will not be reliable when measurements were made in otherwise uncontrolled environments or with inappropriate controls; living protoplasm is far too sensitive to permit meaningful interpretation of its changes in such experiments, even when cells of the same type of tissue are compared. Furthermore, the adjustment of the protoplasm to water shortage varies not only with the magnitude of the water deficit but will also depend to a large extent on the history of the plant and its stage of growth and development.

II. Principles of Cell Permeability

The term permeability, as understood here, refers to that property of a membrane which determines its ability to allow passage of substances (mostly in liquid phase or dissolved in a liquid). Generally, the permeability of a membrane is demonstrable when the membrane separates two compartments and a migration (permeation) of a substance (permeator) from one compartment to the other one can be detected. Permeation occurs when a "driving force" exists between the two compartments.

Preference is given here to the term "membrane permeability" rather than "membrane transport," since the first expression better reflects the control function which the membrane has for the (passive and active) migration of a substance through it.

Two kinds of "driving force" are most common in biological short range (inter- and intracellular) transport processes: (1) an electrochemical or a pressure gradient between the two compartments (e. g., concentration or hydrostatic pressure difference; passive permeability), (2) conformational changes of carrier molecules or their migration inside the membrane, requiring energy supply (active permeability, or active transport).

The permeability of a membrane can be detected only indirectly by observing that permeation takes place. Permeation is usually demonstrated by measuring changes in the amount or concentration of the permeator (solute) in one or both compartments. Here it has to be assumed that inside each compartment the diffusion resistance for the permeator is negligible, compared to the permeation resistance of the membranes. If the rate of concentration change, the magnitude of the driving force and the membrane area are known a measure of membrane permeability can be derived (see p. 271).

A plant cell exhibits a high degree of compartmentalization. Different kinds of membranes (cytoplasmic membranes = cytomembranes, cf. Bennet, 1964) can be distinguished and are known to provide sites for enzymatic activity. These cytomembranes might also exhibit permeability to solute surrounding them. Many driving forces (e. g., concentration differences) may be acting simultaneously and a great number of transport processes may take place together with enzymatic reactions, the combined effects of which result in maintenance of cellular organization and metabolism. For these complex processes, no quantitative data on the permeability of the cytomembranes involved can be given yet. To obtain such numerical values, several parameters are needed (see below), the measurement of which, however, is only possible for the plant cell membrane (and the tonoplast) and limited to materials meeting specific requirements and suitable methods.

The interaction between structure and composition of a membrane and a specific permeator determines the magnitude of the membrane permeability for the permeating substance. An accurate value for passive permeability for solutes and water is of special interest for three reasons:

1. Measurements of permeability (and its changes) for a series of substances may lead to clues for membrane structure and composition (and its variability).

2. The values obtained experimentally must be met by any theory which attempts to derive permeability quantitatively from specific qualities of membrane and permeator.

3. A list of permeability values which reflect their variability with cell type and conditions may be helpful in recognizing differences in cytoplasmic or cell membrane properties which control the permeation process in cells.

In numerous experiments cell permeability (i.e. the permeability of the cell membrane and, when the permeator migrates into the vacuole, of the tonoplast in series; cf. Url, 1971) has been tested for a great number of substances (mostly harmless nonelectrolytes and water) during the recent decades. Most of these substances, except water, either do not occur at all or occur at much lower concentrations in the cell and its environment under natural conditions. Changes in water permeability are often observed with variations of cell conditions. Frequently the assumption is made that these changes might alter the availability of water for the protoplasm; only recently did it become clear that water permeability for the protoplasm is more than sufficiently high to have no restraining effect on metabolic and osmotic processes in the cell (with the probable exception of cellular freezing; cf. Huber and Höfler, 1930; see Oertli, this volume Part 1:C). Water permeability, however, is a sensitive indicator for changes in the cell membrane and/or tonoplast and detailed analysis of permeability changes yield important information about structure and function of biological membranes in situ.

Cytomembranes consist of phospholipids and lipids forming bilayers. Proteins seem to coat at least partially both membrane surfaces and some of the proteins reach through the bilayer (cf. Singer and Nicolson, 1972; Hölzl-Wallach, 1972; Singer, 1974). No pores seem to exist in the phospholipid bilayer (cf. Gutknecht, 1968; Klocke et al., 1972) and earlier results which suggested pores now must be interpreted otherwise (cf. Wartiovaara, 1950; Cass and Finkelstein, 1967).

Passive permeation, also of water and ions, is essentially a solution and diffusion process of the permeator in and across the phospholipid bilayer. The concentration difference of a permeator between both surfaces inside the phospholipid bilayer is the driving force. The permeation resistance of the hydrocarbon chain region of the phospholipid bilayer determines the passive permeability of such membrane (cf. Diamond and Wright, 1969).

III. Quantitative Determination of Permeability

One of the most frequently used plant materials for permeability research is the fully developed cell of higher plants. It was recognized from the beginning of physiological experiments with plant cells that the protoplasmic layer of such cells is the main barrier for an exchange of material between vacuole and an external solution. Furthermore, the protoplasmic layer acts as a differentially permeable membrane, which presents little resistance to water passage but exhibits considerably higher permeation resistance to relatively harmless substances used experimentally as permeators. In fact, this differential permeability of the protoplasm layer causes the cells to exhibit properties of an osmometer and thereby provides a convenient means of measuring internal concentration and concentration changes. For instance, permeability of the protoplasmic layer for water can be calculated quantitatively by using a nonpermeating solute as hypertonic solution and by measuring changes of the protoplast length of a cylindrical cell (see Fig. 1) after a rapid change to a second (higher or lower) concentration of the same plasmolyticum. When the second concentration is

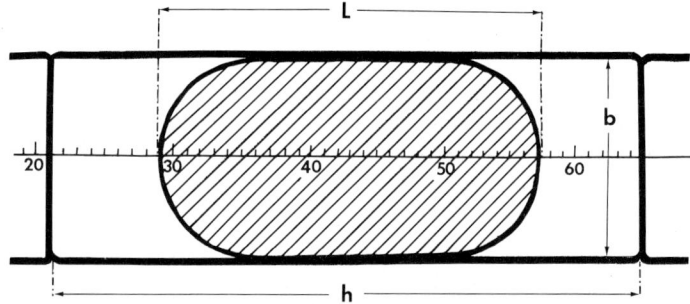

Fig. 1. Diagram of a cylindrical cell in longitudinal section with super-imposed micrometer scale in the middle of the cell. h inner cell length; b inner cell width; L length of the protoplast. (After Höfler, 1918)

also hypertonic but still strong enough so that protoplast ends do not yet touch the transversal cell walls, the water permeability constant (K_{wo}) can be determined by the equation (Stadelmann, 1963; 1966):

$$K_{wo} = 32.0 \cdot \frac{b}{C \cdot L_o \cdot (t_2 - t_1)} \cdot \left[\left(L_o - \frac{b}{3} \right) \cdot \lg \frac{L_1 - L_o}{L_2 - L_o} + \frac{b}{3} \cdot \lg \frac{L_2}{L_1} \right] \cdot F.$$

Calculation of K_{wo} is simplified when the second concentration is hypotonic (hypotonic deplasmolysis method). Here only the deplasmolysis time T is needed to obtain a good approximation of K_{wo}. In this case the following substitutions have to be made in above formula (Lee, 1975):

$$L_1 = L_B; \quad L_2 = h + \frac{b}{3}; \quad L_o = \left(L_B - \frac{b}{3} \right) \cdot \frac{C_p}{C} + \frac{b}{3}; \quad t_2 - t_1 = T.$$

Whereby K_{wo} is water permeability constant in cm/s; C is concentration (i.e. second concentration) of the nonpermeating deplasmolyticum (=osmoticum) in osmol kg^{-1}; L_o is protoplast length when osmotic equilibrium is reached in the concentration C, in (eye piece) micrometer units (MU); b is inner width of the cell in MU; t_1 and t_2 are times of the measurement of L_1 and L_2, during deplasmolysis, in s; L_1 and L_2 are protoplast length at times t_1 and t_2 in MU; L_B is protoplast length at the final equilibrium in the plasmolyzing concentration (first concentration) in MU; h is inner cell length in MU; C_p is first concentration of the osmoticum in osmol kg^{-1}; 32.0 is numerical factor with the unit osmol kg^{-1}; T is deplasmolysis time (time elapsed from the contact of the cell with the deplasmolyticum until incipient plasmolysis is reached) in sec; F is conversion factor of MU in cm, in cm MU^{-1}.

Since the cytomembranes are lacking pores, no interaction between solute and solvent transport takes place inside the membrane. Therefore, the solute permeability constant can be derived in a manner similar to the derivation of the water permeability constant from a simple equalization law (cf. Stadelmann, 1951).

The plasmolytic determination of cell permeability yields reliable values when appropriate precautions are taken (cf. Stadelmann, 1966). Changes of water permeability of the cell membrane between turgid and plasmolyzed state of the cell (assumed by Dainty, 1969; and Zimmermann and Steudle, 1974) do not occur or were found to be relatively small (cf. Steudle et al., 1975 Palta and Stadelmann, 1976). Also water permeability of artificial membranes is independent of the concentration difference or the concentration applied (Hanai and Haydon, 1966). Further difficulties mentioned by Dainty (1969) can be avoided by using the plasmometric method (accurate determination of protoplast volume) and fully differentiated cells of higher plants (negligible contribution of the protoplasm to the protoplast volume). Also errors by unstirred layers (Dainty, 1969) will be negligible due to cell size and magnitude of K_{wo} (cf. House, 1974).

The values for the permeability constant indicate the permeability of plasmalemma and tonoplast in series.

IV. Alterations of Cell Permeability by the Plant Water Deficit

Effects of water stress on cell permeability have been investigated for about the last 40 years. The early measurements, however, could not lead to a coherent description of the drought effects on cell permeability, because the experiments concerned a variety of different plant material and cell types. Furthermore, growth and drought regimes were not rigorously defined and different measures to determine permeability were used.

The availability of growth chambers and the derivation of formulas which allow comparison of permeability values opened the way to a more comprehensive investigation of drought effects on protoplasmic qualities.

The only work so far known which takes advantage of this development and considers also other cytological factors was done with subepidermal stem cells of *Pisum sativum* (Lee, 1975). This work makes possible in-depth analysis of permeability changes. Therefore, it seems justified to review the results and conclusions in more detail.

Seedlings of *P. sativum* were potted and kept in growth chambers under controlled conditions of temperature, light, and soil water content (Lee, 1975). Two drought regimes were tested: for early stress, watering was discontinued on the 6th day, and for late stress on the 21st (or 27th) day after planting. Permeability of cells from control plants and water-stressed plants was measured for water and several nonelectrolytes every fourth to fifth day over the entire growth period.

Water permeability of cells from control plants nearly doubled with plant growth and maturation (Fig. 2). When plants came close to full height (about 3 weeks after planting) change of water permeability became insignificant although a slight tendency remained towards a further increase as the plant continues to mature. Only in cells from older plants with senescing lower leaves did the water permeability decrease. Permeability for apolar nonelectrolytes (methyl urea and ethyl urea) showed a clear decrease after full growth, but increased in cells from plants which were prematurely senescing (Figs. 3 and 4).

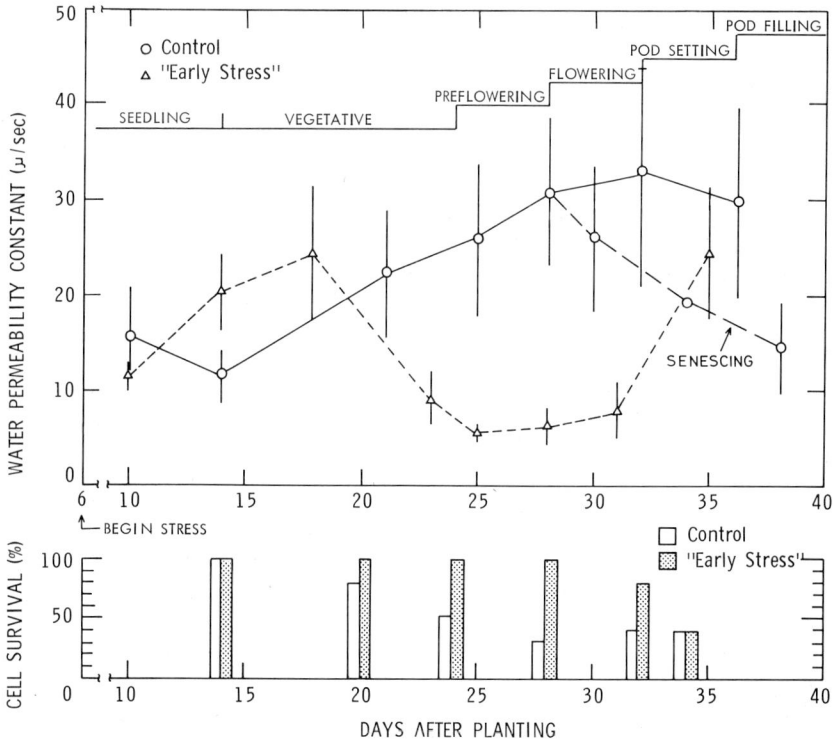

Fig. 2. Permeability constants for water (graph) and cell resistance (lower diagram). Material: subepidermal cells of the stem basis of *Pisum sativum* from well-watered control plants and plants subjected to early stress regime throughout the growth period. Stress began on the 6th day after planting. Permeability constants were obtained with the hypotonic deplasmolysis method, in 2nd or 3rd deplasmolysis. Plasmolyticum and deplasmolyticum: mannitol or glucose. Individual values are means of 21 to 26 cells from 4 to 5 experiments.

Lower diagram: Cell survival in per cent after 5th plasmolysis–deplasmolysis cycle

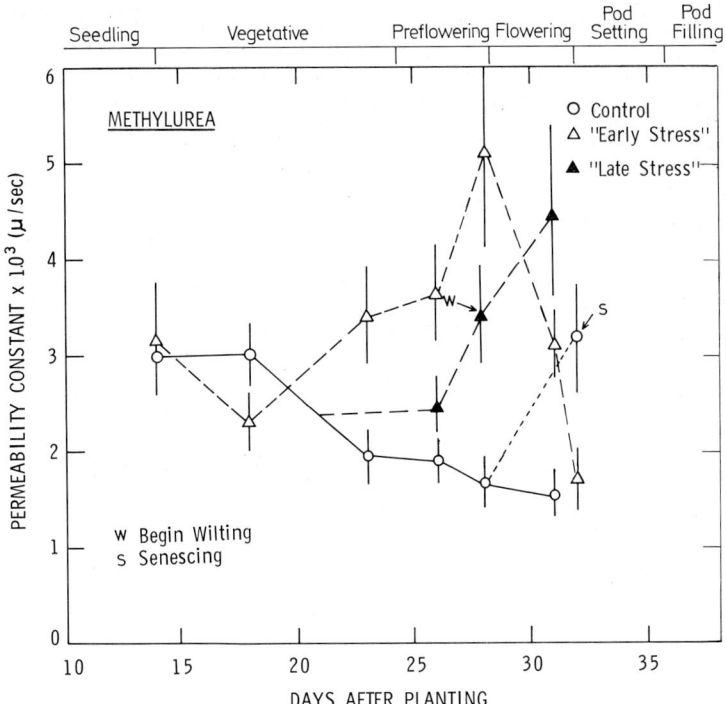

Fig. 3. Permeability constants for methyl urea for cells from early stress, late stress, and control plants. Early stress began on the 6th day after planting, and late stress on the 21st day after planting. Individual values are means of 18 to 27 cells from 3 to 5 experiments (early stress), 22 to 27 cells from 3 to 6 experiments (late stress) and 20 to 28 cells from 3 to 8 experiments (control)

Permeability changes of cells from water-stressed plants were considerably different from those of cells from control plants. Water permeability under early stress conditions changed in a triphasic mode: during the first 2 weeks of stress, the water permeability increased but further stress led to drastic decrease of water permeability (minimum after about 3 weeks of stress). When stress became extreme, however, water permeability increased again.

Changes in permeability for nonelectrolytes showed an inverse trend to those for water permeability: after a transient decrease (for ethyl urea see Fig. 4) permeability increased but subsequently decreased in materials from plants subjected to water stress for over 3 weeks.

By comparison, permeability changes in cells from late stress plants were biphasic: during the first days of stress, water and nonelectrolyte permeability were not greatly different in control and late stress cells. After 7 days of stress, however, when the first wilting of the lower part of the plant was observed, the water permeability began to decrease (Fig. 5). Permeability for nonelectrolytes changed, inversely again.

Such decrease in water permeability was similar to the one observed in subepidermal stem cells from senescing control plants (Fig. 2). These cells may

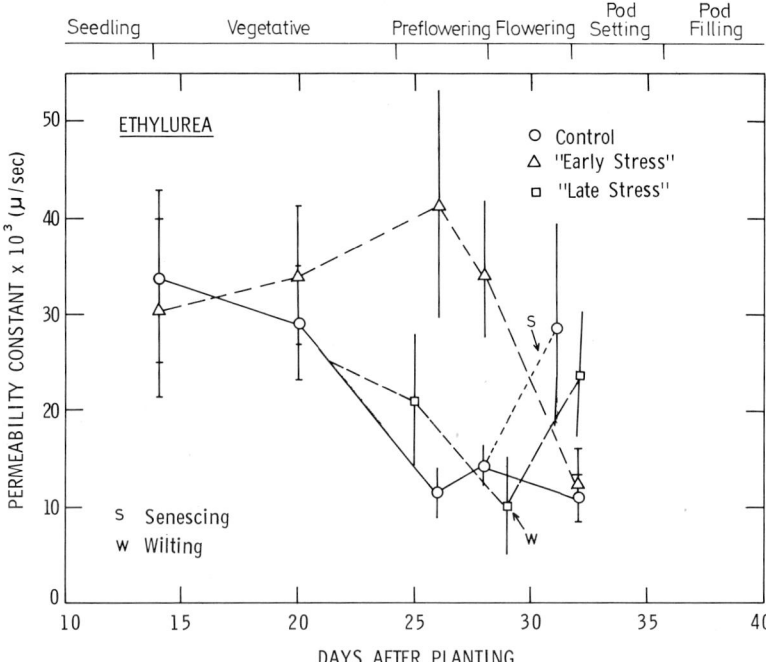

Fig. 4. Permeability constants for ethyl urea for cells from early stress, late stress, and control plants. Early stress began on the 6th day after planting and late stress on the 21st day after planting. Individual values are means of 18 to 26 cells from 3 to 8 experiments (early stress), 19 to 26 cells from 4 to 5 experiments (late stress), and 18 to 23 cells from 4 to 7 experiments (control)

also have been under water stress caused by internal water translocation to the upper parts of the plant at that stage of plant development (pod stage). Similarities of the colloidal changes of the cell (lowering water-holding capacity and reduced ability to swell) under water stress during aging were already proposed by Ratner (1944; cf. Henckel, 1964).

Effects of late stress on the cell were significantly different from those which resulted from prolonged early stress when the water deficit became extreme. Water permeability never increased in cells under extreme late stress as was observed in early stress cells. In fact, the drought-injured cells showed extremely low permeability for water. Cells with low water permeability, in both stress conditions, exhibited high cell resistance when tested by plasmolysis-deplasmolysis cycles (see Fig. 2 for early stress) or by drought resistance (Lee, 1975).

Mechanisms of drought injury, with respect to permeability changes, during early and late stress conditions seem to be significantly different.

The peculiar protoplasmic resistance observed in cells from wilting plants under late stress may be attributed to a stabilizing effect of some substances (sugar?) which are produced as a result of drought. The cell membrane remained semipermeable while the chloroplasts became discolored and appeared nonfunctional.

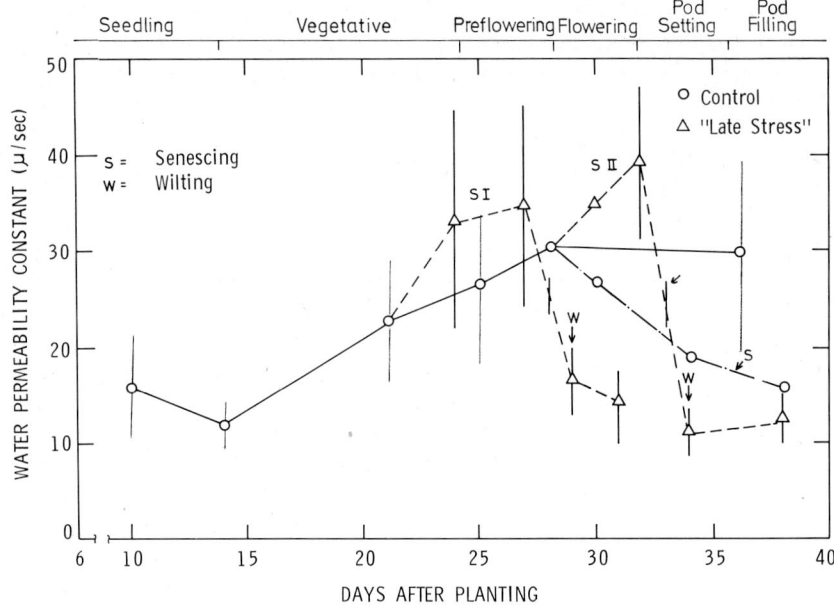

Fig. 5. Water permeability constants for subepidermal cells of the stem basis of *Pisum sativum* from well-watered control plants and plants subjected to late stress. Permeability constants were obtained with the hypotonic deplasmolysis method, in 2nd or 3rd deplasmolysis. Plasmolyticum and deplasmolyticum: mannitol or glucose. Individual values are means of 18 to 27 cells from 4 to 5 experiments. S I Stress began on the 21st day after planting. S II Stress began on the 28th day after planting

High protoplasmic or cell resistance in cells from early stress was observed throughout almost the whole stress period (see Fig. 2); stress resistance, therefore, developed during early stages of water stress. Appearance of chloroplasts did not change (even during the extreme stress period).

The increase of permeability for nonelectrolytes described above in cells from plants exposed to extended water stress agreed with the results of Levitt and Scarth (1936), and of Whiteside (1941). In all these experiments water stress developed gradually. Changes of permeability seem to be different, however, when plants were exposed to a sudden water stress resulting in a decrease of permeability for nonelectrolytes (Schmidt, 1939; Schmidt et al., 1940; Ross, 1961).

Changes of membrane permeability by water deficit greatly depend upon the stage of growth at which the water stress was applied. Furthermore, the permeability differences between control and stress cells were greatly due to changes of permeability of control cells.

The study of Lee (1975) allows some first insight into the complex sequence of protoplasmic alterations by drought; two important conclusions result from her observations and have to be fully considered by any hypothesis for the molecular basis of protoplasmic changes by drought: (1) permeability for water and for lipophilic nonelectrolytes change nearly always in the opposite direction;

(2) low water permeability and high permeability for lipophilic nonelectrolytes (high lipophily of the membrane) is related to a high cell resistance to damage and tolerance to drought in *P. sativum* subepidermal stem cells.

V. Possible Mechanisms for Changes in Cell Permeability by Plant Water Stress

Since cell permeability is regulated by plasmalemma and/or tonoplast, changes in permeability essentially reflect alterations of these membranes.

a) Membrane Composition. Changes in membrane composition are most widely assumed to be the cause for all observed permeability changes (cf. Kuiper, 1972). Exchange of membrane molecules may need a considerable time to become effective and to produce noticeable changes in permeability of cytomembranes.

b) Conformational Alterations of the Membrane. Changes of permeability may also result from conformational alterations of the membrane. This could involve changes in spacing of the membrane molecules caused by water stress. Since conformational changes might be induced by external factors (such as pretreatment of a tissue in salt or sugar solutions) and might occur relatively rapidly (within a few hours), it is conceivable that these are the first reactions of membranes to water and other stress conditions prior to an alteration in their composition. The well-known protective effect of sugars against stress damage may also result from a membrane stabilization (cf. Webb, 1965; Parker, 1972). Changes in the configuration of membrane protein (cf. sulfhydryl hypothesis, Levitt, 1962, 1972) could also lead to different spacing of the phospholipids because of the polar bonds between both types of molecules.

c) Protection and Restoration. Protection and restoration of cytomembranes is closely related to the mechanisms for membrane changes. Two specific external factors acquired special importance.

Calcium-bridge Hypothesis. Calcium may play an important role in stabilizing cell membranes and making them more resistant to stress conditions to which the cell is exposed (Stadelmann and Lee, 1974).

Table 1 shows that the Ca^{++} concentration of a pretreatment solution alters the membrane permeability of the cells from well watered plants (control). This

Table 1. Permeability constants for dimethyl urea, after 3 h pretreatment in spring water or in salt mixture solution. Plant age: 28 ± 1 days after planting (flowering stage). Numbers in parenthesis indicate the number of cells measured. Values are average of 3 to 4 experiments

Pretreatment	Permeability constants $\times 10^2$ in $\mu\ s^{-1}$	
	Control	Early stress
Salt mixture[a]	3.44 ± 0.731 (22)	3.09 ± 0.957 (21)
Spring water[b]	1.63 ± 0.329 (21)	4.00 ± 1.03 (26)

[a] 10 mM KCl + 0.2 mM $CaCl_2$ in distilled water.
[b] 1.8 mM Ca^{++}.

might indicate an effect of the Ca^{++} ions on the spacing of the phospholipid molecules. The association of Ca^{++} with the polar ends of two phospholipid molecules could lead to a change in the average distance between them, and thus affect permeability (phosphate groups, Dawson and Hauser, 1970; Stadelmann and Lee, 1974; cf. Umrath, 1956). The calcium-bridge hypothesis, however, cannot yet explain all the changes of permeability observed in stress and control cells after variation of the external Ca^{++} concentration. Nevertheless, this hypothesis might serve as a preliminary concept.

Effect of Sugars on Water Permeability and Cytoplasm. The protective effect of nonpermeating sugar solutions on cells or tissues against damage by drought and frost has been a focal point of interest (cf. Maximov, 1929; Tumanov and Trunova, 1957; Parker, 1972; Heber and Santarius, this volume Part 4:D).

Two steps are probably involved in the protective action of sugars: (1) Water can be removed (by osmosis) from cellular elements covered by a partially permeable membrane. This osmotic protection is most efficient at hypertonic concentrations and is nonspecific—produced by different kinds of sugars. (2) Water within the cytomembranes might be partially replaced by sugar; such sugar molecules may loosely associate with the polar head groups of the membrane phospholipids because of their molecular shape (Webb, 1965; Parker, 1972). In this way the cell membrane becomes "tighter", i.e., the hydrocarbon chains of the lipids are packed more densely, which results in lower water permeability (Lee, 1975). The association of the membranes with sugar might stabilize the membrane in the same way as the binding of Ca^{++} ions at low (e.g., 0.2 mM) Ca^{++} concentrations.

VI. Conclusions

Effect of water stress on passive cell permeability for water and nonelectrolytes has long been recognized but only recently has closer analysis been possible. The permeability changes of a cell during plant development indicate that a complex sequence of factors is involved. Since passive permeability in higher plant cells depends on the plasmalemma and the tonoplast, the permeability changes indicate alterations in the lipid part of these membranes which become more lipophilic by prolonged water stress. Some of these alterations could result from differences in the packing density of the phospholipid molecules. Conformational changes of the membrane proteins might cause such alterations in molecular packing. The water stress tolerance of the protoplasm and of the membranes increase simultaneously.

The data discussed above might provide a basis for a more direct testing of drought effects on membranes in the living cell. Intact cellular organization is an essential condition for studying and understanding the interaction of factors involved in drought effects on the cells and in drought tolerance. Increased drought tolerance of the protoplasm of economic plants is of greatest practical importance and this approach might contribute significantly to reach that goal. The usual biochemical methods often leading to destruction of life yield important but limited knowledge of individual factors. Permeability determinations (and other physical methods), however, make it possible to recognize that part of drought injury and drought tolerance which must be attributed to the organizational level characteristic of the cell as the smallest unit of life.

Further work along this line may concern probing of the membrane by using agents reacting with a specific site of the membrane in order to test the function of such a site under drought conditions. Such an approach would enhance the insight into stress-induced membrane alterations at the molecular level, about which little is presently known.

Paper No. 9324, Scientific Journal Series, Minnesota Agricultural Experiment Station, St. Paul, Minnesota, USA.

Acknowledgements. This paper was prepared during the tenure of a Humboldt-Award by Ed. Stadelmann for which thanks are given to the Alexander von Humboldt Foundation (West Germany). The authors are greatly indebted to Prof. Kreeb, Fachgebiet Ökophysiologie und Vegetationskunde, Institut für Landeskultur und Pflanzenökologie, Universität Hohenheim, Stuttgart, for his invitation and hospitality, and to Prof. G. Fritz, University of Florida, Gainsville for partial revision of English and style.

References

Bennet, H. S.: Introductory remarks. In: Intracellular membranous structure (eds. S. Seno, E. V. Cowdry), pp. 7–13. Okayama (Japan): Japan Soc. Cell. Biol. 1964.
Cass, A., Finkelstein, A.: Water permeability of thin lipid membranes. J. Gen. Physiol. **50**, 1765–1784 (1967).
Dainty, J.: The water relation of plants. In: Physiology of plant growth and development (ed. M. B. Wilkins), pp. 421–452. New York: McGraw-Hill 1969.
Dawson, R. M. C., Hauser, H.: Binding of calcium to phospholipids. In: Calcium and cellular function (ed. A. W. Cuthbert), pp. 17–41. New York: St. Martin's Press 1970.
Diamond, J. M., Wright, E. M.: Molecular force governing nonelectrolyte permeation through cell membranes. Proc. Roy. Soc. **172 B**, 273–316 (1969).
Gutknecht, J.: Permeability of *Valonia* to water and solutes: apparent absence of aqueous membrane pores. Biochim. Biophys. Acta **163**, 20–29 (1968).
Hanai, T., Haydon, D. A.: The permeability to water of bimolecular lipid membranes. J. Theoret. Biol. **11**, 370–382 (1966).
Henckel, P. A.: Physiology of plants under drought. Ann. Rev. Plant Physiol. **15**, 363–386 (1964).
Höfler, K.: Eine plasmolytisch-volumetrische Methode zur Bestimmung des osmotischen Wertes von Pflanzenzellen. Denkschr. Oesterr. Akad. Wien, Wiss. Math.-Naturwiss. Kl. **95**, 99–170 (1918).
Hölzl-Wallach, D. F.: The disposition of proteins in the plasma membranes of animal cells. Biochim. Biophys. Acta **255**, 61–83 (1972).
House, C. R.: Water transport in cells and tissues. London: E. Arnold Ltd. 1974.
Hsiao, T. C.: Plant responses to water stress. Ann. Rev. Plant Physiol. **24**, 519–570 (1973).
Huber, B., Höfler, K.: Die Wasserpermeabilität des Protoplasmas. Jb. Wiss. Bot. **73**, 351–511 (1930).
Klocke, R. A., Andersson, K. K., Rotman, H. H., Forster, R. E.: Permeability of human erythrocytes to ammonia and weak acids. Am. J. Physiol. **222**, 1004–1013 (1972).
Kuiper, P. J. C.: Water transport across membranes. Ann. Rev. Plant Physiol. **23**, 157–172 (1972).
Lee, O. Y.: Studies on the effect of water stress on the protoplasm of *Pisum sativum* subepidermal stem cells. Ph. D. Thesis. St. Paul, Minnesota, U.S.A. 1975.
Levitt, J.: A sulfhydryl-disulfide hypothesis of frost injury and resistance in plants. J. Theoret. Biol. **3**, 355–391 (1962).

Levitt, J.: Responses of plants to environmental stresses. New York, London: Academic Press 1972.
Levitt, J., Scarth, G. W.: Frost-hardening studies with living cells II. Permeability in relation to frost resistance and the seasonal cycle. Can. J. Res. **14C**, 286–305 (1936).
Maximov, N. A.: The plant in relation to water. A study of the physiological basis of drought resistance (ed. R. H. Yapp). London: Allen and Unwin 1929.
Palta, J., Stadelmann, E. J.: The effect of turgor pressure on water permeability of *Allium cepa* epidermis cell membranes. Plant Physiol. **57**, Ann. Meeting Suppl. p. 79 (1976).
Parker, J.: Protoplasmic resistance to water deficits. In: Water deficits and plant growth, Vol. III (ed. T. T. Kozlowski), pp. 125–176. New York, London: Academic Press 1972.
Ratner, E. I.: Interaction between roots and soil colloids as one of the problems of the physiology of mineral nutrition of plants. III. Age variation in the desorbing ability of the plants. Influence of wilting in the case of moisture deficiency. Compt. Rend. (Doklady) Acad. Sci. URSS **44**, 37–40 (1944).
Ross, H.: Viscosität und Permeabilität des Plasmas von *Lamium maculatum* bei Dürre-, Temperatur- und Schütteleffekten. Planta **56**, 125–149 (1961).
Schmidt, H.: Plasmazustand und Wasserhaushalt bei *Lamium maculatum*. Protoplasma **33**, 25–43 (1939).
Schmidt, H., Diwald, D., Stocker, O.: Plasmatische Untersuchungen an dürreempfindlichen und dürreresistenten Sorten landwirtschaftlicher Kulturpflanzen. Planta (Berl.) **31**, 559–596 (1940).
Singer, S. J.: Lipid-protein interactions in membranes. In: Comparative biochemistry and physiology of transport (eds. K. Bloch, L. Bolis, S. E. Luria, F. Lynen), pp. 95–101. Amsterdam: North Holland Publ. Co. 1974.
Singer, S. J., Nicolson, G. L.: The fluid mosaic model of the structure of cell membranes. Science **175**, 720–731 (1972).
Stadelmann, E. J.: Zur Messung der Stoffpermeabilität pflanzlicher Protoplasten. I. Die mathematische Ableitung eines Permeabilitätsmasses für Anelektrolyte. Sitzgsber. Oesterr. Akad. Wiss. Mathem.-naturwiss. Kl., Abt. I, **160**, 761–787 (1951).
Stadelmann, E. J.: Vergleich und Umrechnung von Permeabilitätskonstanten für Wasser. Protoplasma **57**, 660–718 (1963).
Stadelmann, E. J.: Evaluation of turgidity, plasmolysis, and deplasmolysis of plant cells. In: Methods in cell physiology (ed. D. M. Prescott), Vol. II, pp. 143–216. New York, London: Academic Press 1966.
Stadelmann, E. J., Lee, O. Y.: Inverse changes of water and non-electrolyte permeability. In: Comparative biochemistry and physiology of transport (eds. K. Bloch, L. Bolis, S. E. Luria, F. Lynen), pp. 434–441. Amsterdam: North Holland Publ. Co. 1974.
Steudle, E., Lüttge, U., Zimmermann, U.: Water relations of the epidermal bladder cells of the halophytic species *Mesembryanthemum crystallinum*: Direct measurements of hydrostatic pressure and hydraulic conductivity. Planta **126**, 229–246 (1975).
Tumanov, I. I., Trunova, T. I.: Hardening tissues of winter plants with sugar absorbed from the external solution. Sov. Plant Physiol. **4**, 379–388 (1957).
Umrath, K.: Über Plasmalemmabildung nach plasmatischen Versuchen. Protoplasma **46**, 762–767 (1956).
Url, W.: The site of penetration resistance to water in plant protoplasts. Protoplasma **72**, 427–447 (1971).
Wartiovaara, V.: Zur Erklärung der Ultrafilterwirkung der Plasmahaut. Physiol. Plant. **3**, 462–478 (1950).
Webb, S. J.: Bound water in biological integrity. Springfield, Ill.: Thomas 1965.
Whiteside, A. G. O.: Effect of soil drought on wheat plants. Sci. Agr. **21**, 320–334 (1941).
Zimmermann, U., Steudle, E.: Hydraulic conductivity and volumetric elastic modulus in giant algal cells: Pressure- and volume-dependence. In: Membrane transport in plants (eds. U. Zimmermann, J. Dainty), pp. 64–71. Berlin-Heidelberg-New York: Springer, 1974.

F. Water Stress and Dynamics of Growth and Yield of Crop Plants

T. C. HSIAO, E. FERERES, E. ACEVEDO, and D. W. HENDERSON

I. Introduction

Crop performance and yield are the results of genotypic expression as modulated by continuous interactions with the environment. Among the environmental factors, one of the most widely limiting for crop production on a global basis is water. A better understanding of how long-term growth and yield are affected by water stress should aid in improving irrigation efficiency and practices, in modifying plants for more efficient water use, and in developing effective dry-land agriculture. What we now know of such relationships is almost exclusively empirical, learned from thousands of irrigation trials conducted over many decades, the data of which have been summarized (Salter and Goode, 1967; Stanhill, 1973). A sounder basis should come from knowledge of the underlying physiology and crop ecology. Numerous short-term studies in the last decade have detailed a variety of physiological and metabolic changes in the plant caused by water stress (Hsiao, 1973; Boyer, 1975; Vieira da Silva, this volume Part 4:A). Many of these changes, such as the suppression of photosynthesis and protein synthesis, and the dramatic accumulation of the growth regulator abscisic acid (ABA), could have important consequences in terms of final crop yields. Yet very few studies have attempted to link the short-term changes caused by water stress at the subcellular and cellular level to crop performance in the field integrated over time and the whole organism or community. Bridging the gap between biochemical and physiological changes effected by stress at different times and the final harvested yield involves difficulties that are enormous and in many cases insurmountable for the immediate future. One major difficulty is our ignorance of how crop performance in general manifests and cumulates from the interactions of myriad and ever-changing physiological processes. Other difficulties are more specifically related to water stress: the extremely dynamic nature of plant water status; the dependence of water stress effects on the severity, duration, and timing of the stress during the plant ontogeny; and the complex interplay between water stress and other environmental variables. These difficulties notwithstanding, a substantial start should now be made to bridge the gap. Several reviews have been published (Slatyer, 1973; Fischer and Hagan, 1965) which analyze with insight the effects of water stress on yield in terms of known detailed physiological changes. What is needed is a combination of both theoretical and experimental approaches, emphasizing experimental tests of integrative hypotheses derived from what is known at the short-term and sub-organ level. In this regard, modeling and simulation have an important role.

The first part of this paper attempts an overall view and summary of how crop yields may be affected by water status in terms of the interacting underlying parameters and processes. Emphasis is placed on the gaps in our knowledge and need for integration. The remainder of the paper amplifies selected aspects of the overall view by examining certain plant behavior patterns in the field with accompanying variations in water status. These examinations illustrate the complexity and dynamic nature of plant behavior as related to water, and, in addition, focus attention on some important plant adjustment and adaptation processes in the field.

II. Overview of Growth and Yield as Affected by Water

1. General View of CO_2 Assimilation and Assimilate Partition

Yields often consist of dry matter. Regardless of whether the yield of interest is the fruit, the storage root, or the whole aerial part of the plant, dry-matter production and distribution among plant parts are of prime concern in considering the effects of water stress on productivity. In other words, the central issue is the cumulative net assimilation of CO_2 and the partition of assimilates. With this rather simplistic approach, Fig. 1 summarizes broadly selected effects of water stress on plants, their probable relations to yield, and the dependence of the relations on the ontogenetic stage at which water stress occurs. The figure depicts cases involving a distinct reproductive growth, with harvest taken during or after such growth and including the reproductive organ. Stress effects whose relation to yield is either not at all clear or too complicated are necessarily ignored. Also, broad parameters, such as sink size, are used in place of more specific ones to gloss over details so as to keep the concepts general and the scheme uncomplicated. The three central parameters for determining cumulative assimilation are total leaf area or source size, net assimilation rate per unit leaf area or source intensity, and longevity of the leaves or source duration. This somewhat arbitrary division between source size and source duration separates the period when total leaf area is being defined by vegetative growth from the period when it is largely affected by leaf senescence. For harvestable yield, there is the additional parameter of the size of the sink for utilizing or storing assimilates, such as the number of fruits on the plant. Closely linked to sink size (and only implicit in Fig. 1) is the partition of assimilates among the plant parts, which determines how much of the total dry matter actually ends up as yield. A reduction in sink size leads to an accumulation of assimilates, which in some cases is reported (Neales and Incoll, 1968; Thorne and Koller, 1974) as having a negative feedback effect on photosynthesis and thus on source intensity (line 5).

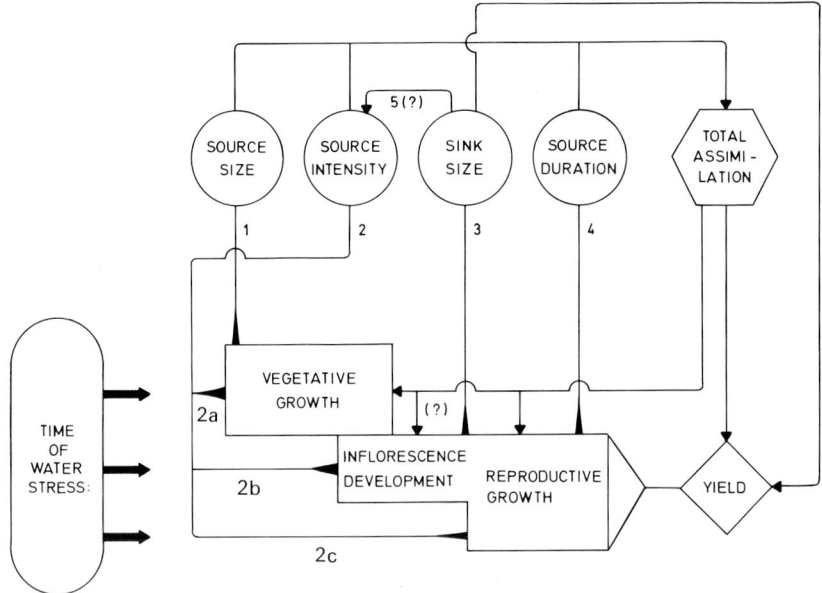

Fig. 1. General effects of water stress on yield as viewed in the context of temporal variations in CO_2 assimilation, source–sink relationships for assimilates, and plant ontogeny. The effects are considered to vary with time, and the cumulative effects would be the integrals over time. Arrows (⟶): negative effects. For example, stress during the vegetative growth stages can reduce source intensity (line 2a), which in turn can lead to a reduction in total assimilation and yield. Question marks (?): effects which are not well established. For example, it is not certain that a reduction in total assimilation would generally inhibit or reduce inflorescence development

Even this over-simplified scheme makes evident the multiplicity of the interactions among water stress, growth, development, ontogenetic stages, and yield. For example, if water stress occurs during flowering, it could diminish source intensity by causing stomatal closure and inhibiting the biochemical steps of CO_2 assimilation (line 2b). Sink size could be reduced (line 3) by effects such as impaired pollination. In addition, the reduced sink may have its separate negative effect on source intensity (line 5), resulting in a still lower rate of assimilation. Final yield would then be reduced by the diminished assimilation as well as an inadequate number of fruits (smaller sink size). If water stress occurs during fruit maturation, again source intensity is reduced (line 2c). Combined with this may be an acceleration of leaf senescence, thus shortening source duration (line 4). The resultant reduction in total assimilation would most likely lead to an underfilling of the fruits and a reduction in yield. In the more complicated cases, stress occurs at not only one but several growth stages; thus the reduction in yield would depend on cumulative interactions of sets of parameters that vary with time. These interactions, in addition, are modulated by genetic variation in the ability to adapt to water stress.

For the broad overview, the severity of water stress is not specifically considered. In the following sections, selected aspects in Fig. 1 are amplified and attention is given to stress severity and the need to differentiate plant processes with respect to their sensitivity to stress.

2. Expansive Growth, Leaf Area Development, and CO_2 Assimilation

From the viewpoint of dry-matter production, expansive growth deserves special attention, since it is the means of developing leaf area for intercepting light and carrying out photosynthesis (source size in Fig. 1). With many crops, the process most sensitive to water stress appears to be expansive growth (Hsiao, 1973). Apparently this is the consequence of the critical role of turgor in the growth process (Hsiao et al., 1975). Figure 2A shows that stresses too mild to close stomata and inhibit photosynthesis can readily reduce leaf area development (Relation I). If leaf area is low and intercepts only a fraction of the incident radiation, it would be a factor limiting CO_2 assimilation by the crop stand (Watson, 1947; Loomis et al., 1971) (line 1), though there is no direct stress effect on photosynthesis (Relation II). Most of the reduction in leaf area appears to be the consequence of slowed cell expansion. Prolonged suppression of cell expansion, however, could have a postulated (Hsiao, 1973) negative feedback effect on cell division (line 2) and hence restrict the potential size of a leaf. In addition, suppression of cell division would lead to a slowing of the rate of leaf initiation (line 3). Another possible effect of reduced cell expansion is, however, in a sense counteractive to the aforementioned effects. When cell expansion is restricted, solutes may build up in cells if solute uptake or internal production remains relatively unaffected. The build-up in solutes and the consequent lowering of tissue solute potential (also referred to as osmotic potential) Ψ_s (line 4) and total water potential (Ψ) lead to accelerated water uptake and a restoration of turgor (Ψ_p) and hence growth. The maintenance of turgor through solute build-up, often referred to as osmotic adjustment, is elaborated on hereafter in an example of field behavior. Another conteractive effect depicted in Fig. 2A is the probable decrease in total respiration associated

Fig. 2 A and B. Probable interactions among water stress, leaf expansive growth, CO_2 assimilation, and related parameters. The diagrams have been presented in a dynamic simulation format (Forrester, 1968) for ease of comprehension. The presentation is only conceptual, however and not to be taken as a simulation model. Material flow is represented by (⇢), and information flow by (--→). Rates are denoted by symbolic valves (⊳), which, when placed on material flow lines, indicate that the flows are regulated by the rates. Levels or quantities are represented by (☐), whereas variables affecting rates and levels are signified by (○). Variables which are held constant for a given situation are symbolized by (—○—). (A) A case when plant water status is low enough to inhibit expansive growth but not CO_2 assimilation directly. (B) Adds more details on photosynthesis and focuses on a case of water stress sufficient to inhibit stomatal opening and CO_2 assimilation. Relation I is based on Hsiao et al. (1975); and Relations II, III, and IV, on general literature. Double arrows (←→) in Relations II and III indicate that the position and shape of the curves are affected by the variables which are the sources of information flow (lines 8 and 9). Relation IV is probably also influenced by leaf history and age but is not so shown because of a lack of data

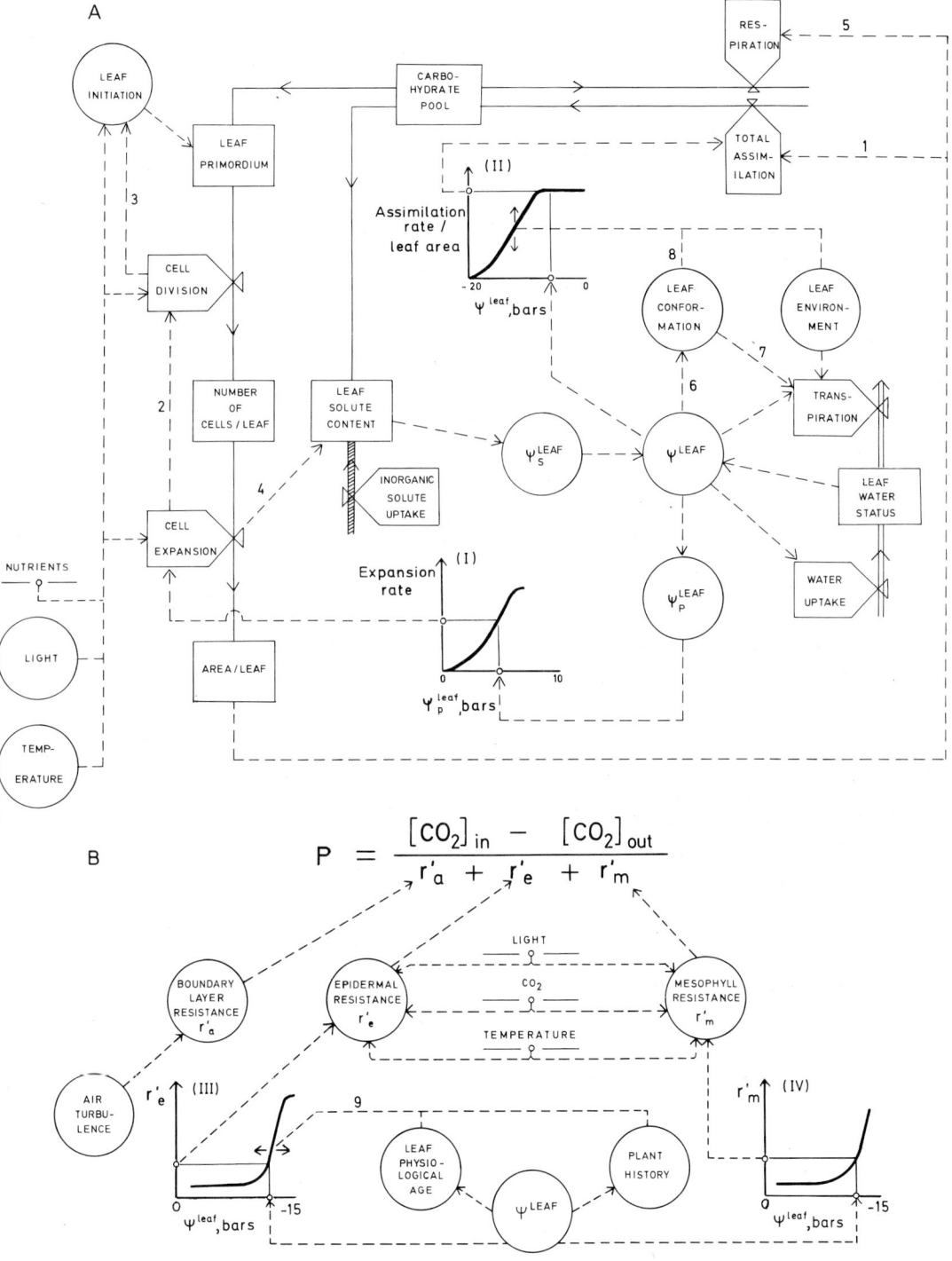

Fig. 2 A and B

with a smaller leaf area (line 5), which would lessen the competition with growth for the available carbohydrates or energy substrates.

One other interaction is also worth noting. A more substantial reduction in leaf water status than that depicted in Relations I and II of Fig. 2A may affect leaf orientation and conformation (line 6). The usual consequences are leaf rolling in grass species and drooping in some dicots. These changes commonly reduce radiation load on the leaf (Luxmoore et al., 1971). Although that promotes a more favorable water balance for the leaf (line 7), it also reduces the light absorbed for photosynthesis (line 8).

Some other interactions, largely poorly understood, are not included in Fig. 2A, though additional aspects, relating to root and to reproductive growth, are discussed further in subsequent sections.

Figure 2A depicts the situation when leaf water stress is sufficient to suppress growth but not low enough to affect CO_2 assimilation directly. With more severe water stress, Ψ drops below the threshold level for stomatal closure, leading to reduced CO_2 assimilation. That situation is shown in Fig. 2B. Partial or total stomatal closure causes increases in epidermal resistance of the leaf to the inward passage of CO_2 (r'_e) (Relation III). The literature (Hsiao, 1973) also indicates that there is frequently a hand-in-hand increase in leaf "mesophyll" resistance (r'_m) (Relation IV). The latter increase represents inhibitory effects of stress on the non-stomatal (and presumably mostly biochemical) components of photosynthesis. An increase in r'_e and r'_m would reduce CO_2 assimilation rate (P), though the magnitude of the reduction will depend also on the boundary layer resistance (r'_a), as is seen from the expression for P (Fig. 2B). Found elsewhere are more detailed discussions of stomatal behavior as related to environmental parameters, especially water (Raschke, 1975; see this volume Part 3).

Leaf Ψ is given as the indicator of water stress in Relations II, III, and IV of Fig. 2A and B. Recent evidence suggests, however, that a more relevant indicator might possibly be Ψ_p, or even relative water content (RWC) (Hsiao et al., 1976). It can be speculated that changes in sensitivity to water stress of r'_e with leaf age and history (line 9, Fig. 2B) may be only apparent and would become minimal if Ψ_p were used instead of Ψ as the water status indicator (Hsiao et al., 1976).

Figures 2A and B necessarily omit many known interactions, most of which are still poorly defined in quantitative terms. Nonetheless, the interlocking nature of the various components in the diagram is clear and emphatically points to the importance of considering the system as a whole and the danger of over-concentration on any single process, especially when concerned with water stress and plant performances spanning substantial time intervals.

3. Development, Reproduction, and Partition of Assimilates

When water stress occurs during the reproductive phase and the yield consists of reproductive organs, the plant reactions that must be considered are still more complex. Implicit in Fig. 1 is the fact that fruit or grain yield can be reduced while total dry-matter production (total assimilation) remains relatively constant. The key consideration here is the partitioning of assimilates among the various plant parts (Milthorpe and Moorby, 1974). Though the underlying processes and regulatory mechanisms are undoubtedly very complex and mostly not elucidated, a perspective can be taken in terms of the size and strength of the fruit sink for assimilates. Thus inflorescence development, fertilization, abortion of fertilized ovaries, and fruit abscission are prime factors; and effects of water stress on them deserve special attention. Unfortunately, available information is largely qualitative and descriptive. In the absence of a better understanding and quantitative relationships, we have merely outlined, as a flow chart in Fig. 3, many of the relevant factors to consider, and the key question to be answered. Most of the points made in Fig. 3 are self-evident. Notable again is the dynamic nature of plant responses to water stress, highlighting the importance of the timing of water stress. In addition, the complexity of the interactions and the large number of factors which should be taken into account is once more illustrated. It is also useful to point out that two pivoting points in Fig. 3 relate to flowering habit and harvest time, aspects far removed from photosynthesis or assimilation. Currently there is a tendency, in ecophysiology in general and plant-water relations in particular, to focus too much attention on the processes of CO_2 assimilation. For many crop plants it is becoming more evident that yields consisting of reproductive or storage organs may not be directly related to CO_2 assimilation rates (Evans, 1975). Therefore we must not lose sight of other parts of the system, especially those associated with growth patterns, development, and assimilate partition, which may be equally and perhaps even more significant and also are amenable to genetic modification.

Given the importance of assimilate partition among plant parts in determining yield, a pertinent question is how the translocation of assimilates may be affected by water stress. Several reviewers (Wardlaw, 1968; Slatyer, 1969) have concluded that water stress which is not too severe probably affects translocation for the most part indirectly, via altered source-sink relationships for assimilates. For example, reduced leaf cell expansion would reduce sink strength for vegetative growth and lessen the competition with fruit growth for assimilates. Because translocation and the partition of assimilates have a crucial role in determining fruit growth, more definite study of water stress in this context is sorely needed.

Roots also constitute an important sink for assimilates. In many cases of mild to moderate water stress, the weight ratio of root to shoot increased (Pearson, 1966; El Nadi et al., 1969; Hoffman et al., 1971). If not severe, water deficit apparently favors root growth relative to that of shoots in some manner, increasing the relative strength of the root sink. The greater relative root growth occurs despite the double effect of soil drying: not only does the water potential decrease, but the soil's mechanical strength also increases with consequently greater resistance to root growth (Taylor and Gardner, 1963). A possible explanation of this paradoxical behavior lies in the ability to adjust osmotically. A review of very limited evidence (Hsiao, 1973) suggests that roots may be superior to shoots in ability to adjust osmotically to water shortage. Both the postulated adjustment and greater relative growth require translocation of more assimilates to the root. Preferential root growth would enable a plant to obtain water from a soil more effectively. The importance of root growth in water supply to the plant is discussed further below.

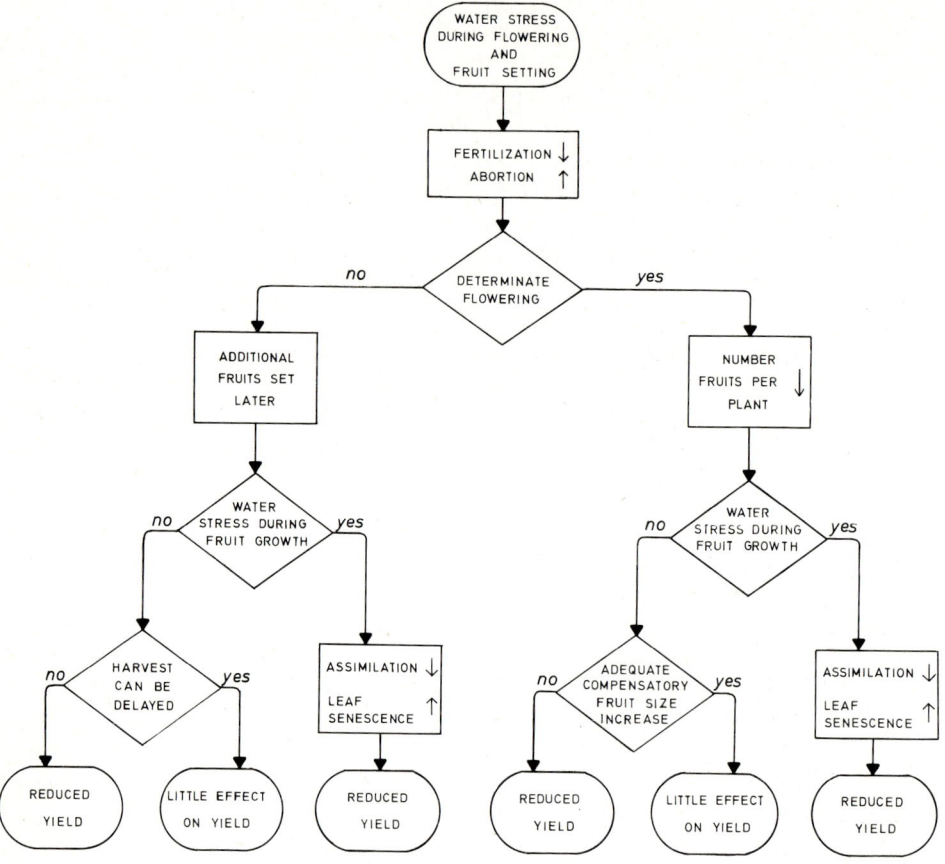

Fig. 3. Fruit (grain) yield as affected by water stress during reproductive growth. The diagram, in the format of a computer flow chart, is based on data accumulated in the literature. Vertical arrows next to variables within the rectangles indicate either increases (↑) or decreases (↓) in the variable. Many of the effects of stress depicted merely represent empirical observations, and the underlying mechanisms remain obscure. The general assumptions underlying the diagrams are: that the crop has achieved a full canopy prior to flowering; that the water stress does not last too long regardless of the time of occurrence; and that the stress is severe enough to cause the stated changes but is not overly severe. Omitted for brevity are two well known effects of stress: abscission of the younger fruits when stress is in the early reproductive phase of growth (Hsiao, 1973, p. 543); and enhancement of fruit quality by a mild to moderate stress in the late maturation phase (Fischer and Hagan, 1965). These effects could easily be added by the principles outlined in Figs. 1 and 3

Figure 4 depicts possible interactions among mild water stress, assimilation shoot or leaf growth, and root growth. Though not explicit in the diagram, assimilate partition among plant parts and its regulation obviously play an important role.

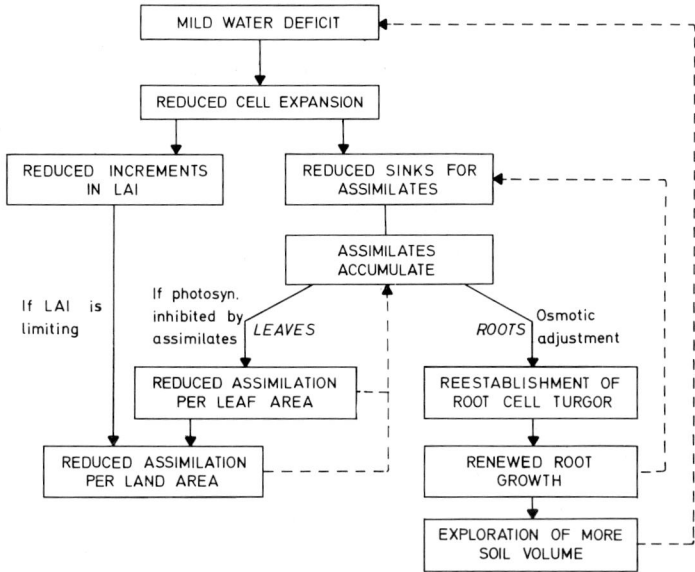

Fig. 4. Possible interactions and effects on leaf and root growth of water stresses not severe enough to affect stomata and photosynthesis directly. Dashed lines (------): negative feedback effects. Leaf area index is abbreviated LAI. For clarity and simplicity, osmotic adjustment in leaves is assumed to be small and hence negligible. See, however, subsequent sections for leaf osmotic adjustment in the field. (After Hsiao and Acevedo, 1974)

III. Some Behavior Observed in the Field

Some specific examples of crop behavior in the field as related to water are given here to illustrate further the complexity of interactions among crop and environmental factors and to identify some adaptive and adjustment processes which enable a crop to produce reasonable yields under conditions of water shortage. In dry-land agriculture, or when irrigation and rainfall are infrequent, plants must explore water stored in the soil. The uptake of soil water as related to root growth is chosen as the first illustration of dynamic crop behavior in the field[1].

1. Dynamics of Root Growth and Water Supply

Because they are normally hidden from view, roots have been neglected in most field studies of plant growth; instead, they have concentrated on the aerial part. Realization of the importance of roots and their activities in supplying water and inorganic nutrients to the shoots and in being a substantial sink

---

[1] Sections III.1 and III.2 of this chapter are based on the Ph. D. Thesis of Acevedo (1975) being prepared for publication elsewhere.

for photosynthetic assimilation, however, has recently generated a notable increase in research on root behavior in the field. Knowledge on roots should consequently be greatly expanded in the next few years as the result. This section examines the dynamic interaction between root growth, soil water status, and water supply of maize in the field. General aspects of water transport in the root and of root water uptake are discussed in detail in other chapters (see this volume Part 2).

The water-holding capacity of most soils is not large relative to the transpiration rate for plants with root systems that are not unusually extensive growing in environments of high evaporative demand. In the absence of frequent rainfall or irrigation, plants must be dependent on the water stored in the soil, the utilization of which is determined by the rate of water transport to the soil layer immediately adjacent to the root surface as well as the rate of root moving into moist soil. Thus, root growth, hydraulic conductivities, and water potentials of the soil and the plant interact in determining the water supply to a crop. These complex interactions may be appreciated better when one attempts to answer questions such as: how extensively must a soil volume be permeated by roots for the effective use of the stored water? Or, how and by what are the limits set for the distance water can be transported through the soil to the roots at a rate sufficient to prevent plant water stress?

a) Soil and Root Resistances to Water Transport as Related to Root Density. A conceptual answer to the questions raised above requires a quantitative treatment of the interactions among the soil and plant water parameters. Starting with the appropriate transport equation and the assumptions used by Gardner (1960), an equation can be derived (Acevedo, 1975) in which the rate of water transport to the root surface q is related to the gradient of water potential $(\Psi^r - \Psi^0)$, the soil hydraulic conductivity K, and the root length density L_v

$$q = \frac{4\pi L_v \Delta z K}{\ln(\pi L_v) + 2\ln r}(\Psi^r - \Psi^0). \tag{1}$$

Here q is given in cm (of water depth) s^{-1}, i.e., cm^3 of water transported to the roots per cm^2 of land area per second for a soil depth interval Δz. L_v is in cm (of root length) cm^{-3} (or soil), and K in cm^2 s^{-1} bar^{-1}. The mean root radius, in cm, is represented by r and Δz is in cm. The water potentials at the root surface, Ψ^r, and in the soil midway between roots, Ψ^0, are expressed in bars. The main assumptions (Gardner, 1960) are: that roots act as uniform and long sinks for water and are evenly and parallelly spaced in the soil for a given L_v; that water moves only radially from the soil to the root surface; that the soil K is constant in the radial path; and that transport is steady state.

Equation (1), though derived with all those simplifying assumptions and therefore containing considerable uncertainties, makes clear that in addition to the obvious parameters of soil hydraulic conductivity and Ψ gradient, water transport to the roots depends also on the root length density in the soil.

To make this point more emphatic, we describe water transport with a more familiar equation (see this volume Part 2).

$$q = \frac{1}{R_s}(\Psi^r - \Psi^0) \tag{2}$$

where R_s is the soil resistance to water transport. Comparing Eqs. (1) and (2) (Feddes and Rijtema, 1972), it becomes obvious that under the conditions specified for Eq. (1),

$$R_s = \left[\frac{\ln(\pi L_v) + 2\ln r}{4\pi L_v \Delta z}\right]\frac{1}{K}. \tag{3}$$

Thus, root geometry, as expressed by the terms enclosed within the square brackets, is as crucial in determining soil resistance as is soil hydraulic conductivity. Increases in L_v through root growth should decrease R_s at any given soil K. It is important to point out, however, that K increases approximately logarithmically with soil water content (see Caldwell, this volume Part 2:A). The depletion of soil water can cause such fast declines in K as to obviate any influence due to changes in L_v.

Aside from soil resistance, root growth also reduces root resistance R_r. With roots of most species, increases in total root length arise largely from branching and very little from the sustained elongation of the main roots. Hence, increases in L_v, enlarging the root surface area for water absorption and reducing the radial resistance of the root, are accompanied also by an increase in paths connected in parallel and thus reducing the longitudinal resistance in the root system.

Root resistance is extremely difficult and tedious to assess in the field. Recent studies (e.g., Acevedo, 1975) have applied Eq. (2) to a soil depth interval Δz. For steady-state transport,

$$q = \frac{1}{R_s + R_r}(\Psi^b - \Psi^0) \tag{4}$$

where Ψ^b is the water potential at the root-stem junction (base of the stem). The rate of root water uptake q from each depth interval is measured, as is Ψ^0 for that interval. R_s is calculated from estimated values of L_v, r, and K using Eq. (3). Ψ at the base of the stem is extrapolated from the vertical Ψ profile of leaves at various horizontal layers of the canopy. R_r is then computed from Eq. (4). A plot of computed R_r vs. L_v (Fig. 5), taken from field data of Acevedo (1975), shows a strong dependence of R_r on L_v, as expected from the arguments presented above.

With resistances of the root and the soil both being reduced by increases in L_v, water absorption from soil should be closely dependent on the location and density of the roots. In fact, many studies (e.g., Veihmeyer and Hendrickson,

1938) have relied on soil water depletion in the absence of rain or irrigation to indicate the depth of root penetration and proliferation. To estimate more accurately the amount of water removed by root absorption, it is necessary to take into account movements within the soil in response to gradients of soil water potential. In principle the procedure is simple. To carry it out in field studies, however, requires numerous measurements of soil water contents and Ψ, as well as a detailed knowledge of spatial and temporal variations in soil hydraulic conductivities. These difficulties notwithstanding, several recent studies (Arya et al., 1975; Allmaras et al., 1975) have succeeded in obtaining density (see also Greacen et al., this volume Part 2:B). Figure 6 is given as an example which shows that absorption from different horizontal wet soil layers is strongly dependent on root length density.

b) **Root Growth and Soil Water Extraction.** With root density being such a crucial factor in water uptake and transport, root growth, death, and permeability to water deserve special attention in studying plant utilization of soil water. Detailed and painstaking observations (e. g., Browning et al., 1975) have established that root systems are much more dynamic than is generally realized (see also Caldwell, this volume Part 2:A). In some species, the smaller and branch roots appear to remain alive for only a few weeks, being constantly replaced by new root networks growing into the same soil space. Old roots, apparently suberized and low in permeability to water, may suddenly send out numerous young laterals when the soil environment is made favorable. In a heterogeneous

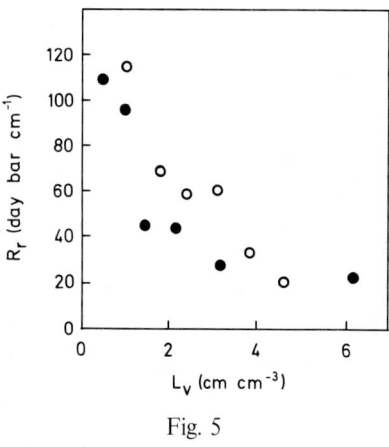

Fig. 5

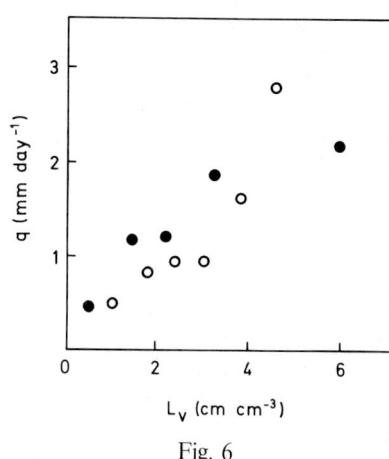

Fig. 6

Fig. 5. Calculated root resistance R_r as related to measured root length density L_v for a maize crop growing in Yolo clay loam. Values are means of various 30-cm soil depth intervals for the time period of 62–74 days after planting and represent two treatments: Frequently irrigated (●●) and irrigated weekly beginning 55 days after planting (○○). L_v was measured by the line intercept method of Newman (1966). See text for details on R_r. (After Acevedo, 1975)

Fig. 6. Computed water absorption (q) by maize roots as related to root length density L_v in a moist Yolo clay loam. Same crop, soil depth interval, time, and treatment symbols as Fig. 5. (From Acevedo, 1975)

soil, root density and functionality presumably vary with small-scale soil conditions. Therefore, spatial variation in water uptake can be extremely complex.

For now, however, the importance of root growth to water uptake can be clearly seen only in a less complex setting. In the uniform and deep Yolo clay loam of high water-holding capacity (field capacity approximately 35% v/v) at Davis, California, some annual crops can be grown to maturity without any rain or irrigation after planting. The transpiration requirement of the crop must be met by the water stored in the soil. Recent detail data of Acevedo (1975) dramatically illustrate the importance of root growth and development in that soil in providing stored soil water for a maize crop (see Figs. 7 A and B). Figure 7 A shows that root proliferation deepened as the season progressed, reaching a depth of almost 3 m at harvest. The zone of maximum water absorption is also displaced downward with season (Fig. 7 B), in dynamic correspondence with the downward progression of roots. The peak of water absorption is apparently shaped jointly by the drying of the upper soil layers and by the growth of roots into the deeper moist soil. Comparing Fig. 7 A and B, it is seen that absorption maxima are generally approached after root length density reaches 1 to 2 cm cm^{-3}. It is worth repeating that in this case the growth of the roots into moist soil is essentially the only means by which new water is made available to the crop. A striking demonstration of this point is when root growth into deeper layers is impeded by compacting the soil down to 120 cm. In this example, a maize crop, also grown at Davis, suffered seriously from water deficiency and wilted badly. Only when the roots eventually managed to penetrate the compacted layer to reach new water did the plants recover from stress and resume growth (Das, 1973).

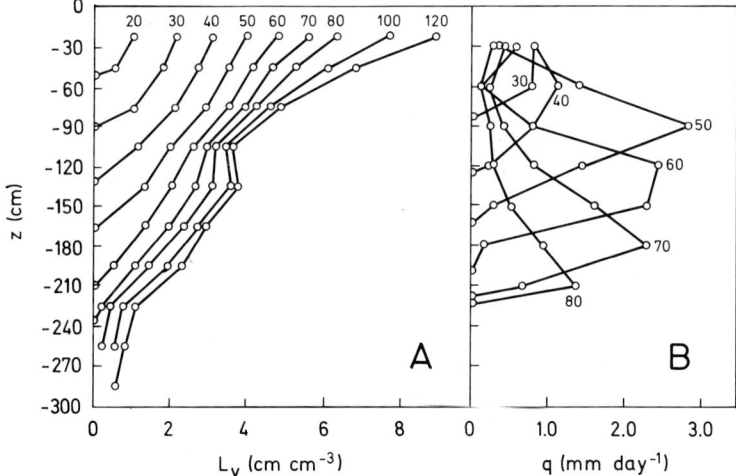

Fig. 7 A and B. Estimated root length density L_v (A) and computed root water absorption q (B) as dependent on soil depth z of an unirrigated maize crop growing in Yolo clay loam. Numbers on the curves denote number of days from planting. The soil was fully wetted to a depth of 3 m before planting and received only about 1 cm of rain during the growing season. (From Acevedo, 1975)

While root growth has a profound effect on the use of soil water, soil water status in turn affects root growth. As mentioned briefly in Sect. II of this chapter, however, roots might be able to adjust osmotically so as to maintain a substantial excess of turgor pressure over soil back pressure, to permit root growth either as root Ψ is lowered or as the soil mechanical impedance is increased. This might be the explanation for the increase in root density in the dry soil layers evident in Fig. 7A. In spite of this ability, however, root growth is slowed or stopped if the soil becomes dry enough. In many such cases, the suppression of root growth was attributed mostly to the indirect effect of drying on increased soil mechanical strength (Taylor and Gardner, 1963).

2. Diurnal Growth as Related to Plant Water Status and Temperature

Because of the daily variation in the evaporative demand of the atmosphere, Ψ of the plant oscillates diurnally. Midday depression of Ψ^{leaf} is most likely experienced by all exposed plants on sunny days (e.g., Klepper, 1968). The daytime lowering in Ψ^{plant} could inhibit key processes underlying productivity, such as expansive growth and stomatal opening (see Figs. 1 and 2). Indeed, midday slowing or stoppage of growth attributable to a low plant water status was clearly delineated as early as the first part of this century (Coster, 1927)—in the humid tropics! The phenomenon is apparently common also in the temperate region (e.g., Loomis, 1934). Boyer (1968) reasoned from the relationship between Ψ^{leaf} and leaf enlargement of sunflower that growth should be confined mainly to the night period, when plant water status is most favorable, and confirmed that conclusion with measurements of growth in the greenhouse. Hsiao (1973) and Hsiao and Acevedo (1974), from data then existing, speculated on how and under what conditions the elimination of midday depression in Ψ^{leaf} and of the associated growth reduction might enhance long-term dry matter production.

Generally, stomata opening is less sensitive to water stress than growth. However, low Ψ^{plant} around noon is also thought to be one cause of midday stomatal closure (Kramer, 1937; Slatyer, 1967), which in turn would cause a depression in CO_2 assimilation at that time.

The diurnal oscillation in the environment is even larger in amplitude in most arid areas than in the temperate zones, and the evaporative demand at midday is higher (cf. Hall et al., this volume Part 3:D). Does this mean that crops would generally make little or no expansive growth during the day in those areas? And further, with the large midday depression in water potential, do most crops close their stomata at that time in arid areas?

At Davis, California, summer days are mostly hot and completely sunny. The high evaporative demand during the day (e.g., a saturation deficit of 20 mm of mercury) leads to pronounced diurnal fluctuations in Ψ^{leaf}, as recently found in maize and sorghum by Acevedo et al. (1976). These data (Fig. 8) show that Ψ^{leaf} was minimum around noon and that the amplitude of oscillation was as much as 8 to 10 bars. Previous work with maize in growth chambers demonstrated that lowering Ψ^{leaf} by 4 bars can stop growth completely (Fig. 9), and that lowering Ψ^{leaf} by 8 to 10 bars (to $\Psi = -10$ to -12 bars) can

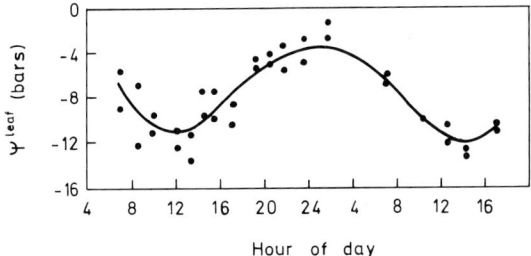

Fig. 8. Diurnal pattern of Ψ in leaves of an unirrigated field maize crop at Davis, California. Data points represent individual measurements of expanding leaves at positions 12 and 13 (counting upward) 41–42 days after planting. The curve represents a least square fitting of fourth degree polynomial of the data. The crop is the same as that depicted in Fig. 7. (From Acevedo et al., 1976; cf. footnote on p. 289)

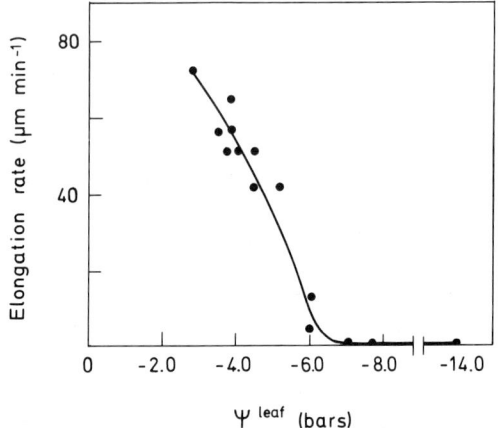

Fig. 9. Rate of elongation of the youngest leaf (third from bottom) as related to Ψ of the next older (second) leaf in maize seedling (10 days old) growing in a growth chamber (approx. 1,100 ft. c.). Ψ of the second and third leaves were within approximately 1 bar of each other. (Unpublished data from Acevedo, Hsiao, and Henderson, 1971)

cause marked stomatal closure (Beadle et al., 1973). Those results would imply, at first glance, not only that growth would be stopped at midday in the field in Davis, but also that CO_2 assimilation would be markedly reduced. Instead and surprisingly, the measurements depicted in Fig. 10 show that growth was substantial at midday in spite of the low Ψ^{leaf}. In addition, growth rate increased as the morning progressed although Ψ^{leaf} was falling. With stomatal opening and CO_2 assimilation generally being less sensitive than growth to water stress (see Sect. II, 2 of this chapter), the expectation is that under these conditions there would also be no substantial stomatal closure. That was confirmed with measurements taken on the same crop but at another time. Figure 11 reproduces data on the diurnal pattern of stomata 15 days earlier. On that day Ψ^{leaf} of the unirrigated treatment dropped to -11 to -13 bars near noontime and that of the irrigated control was perhaps 2 or 3 bars higher. Neither treatment showed any sign of midday stomatal closure.

What accounted for the lack of substantial effect on growth and the total absence of effect on stomata of the rather large midday depression in Ψ^{leaf}? The answer appears to lie in the diurnal oscillation in tissue solute content as it affects Ψ_s^{leaf} and hence Ψ_p^{leaf}. The data (Fig. 12) show that as the morning progressed, Ψ_s^{leaf} declined apparently from the accumulation of osmotically active solute in leaves, such that Ψ_p^{leaf} rose in spite of the decline in Ψ^{leaf}. Additionally, the daily minimum in Ψ_s^{leaf} appeared to lag behind the minimum in Ψ^{leaf} by about 2 h, so that Ψ_p^{leaf} recovered earlier during the day than would be permitted by the afternoon recovery in Ψ^{leaf} alone. The maintenance or enhancement of Ψ_p^{leaf} is presumably partly responsible for the acceleration or sustension of growth during the late morning and early afternoon.

Plant water status is not, however, the only important parameter that varies diurnally. Temperature oscillates similarly, though in the opposite direction. Parts of the growth pattern (cf. Fig. 10) appear to be dictated by temperature, as might be expected also from results of other studies (Friend, 1966; Watts, 1972). The slow growth at night and in the very early morning was probably due to low temperature, since it was found that growth and temperature are linearly related during those periods. More information on this and temperature and water interactions as related to growth can be found in the paper of Acevedo et al. (1976) and in work of Watts (1974).

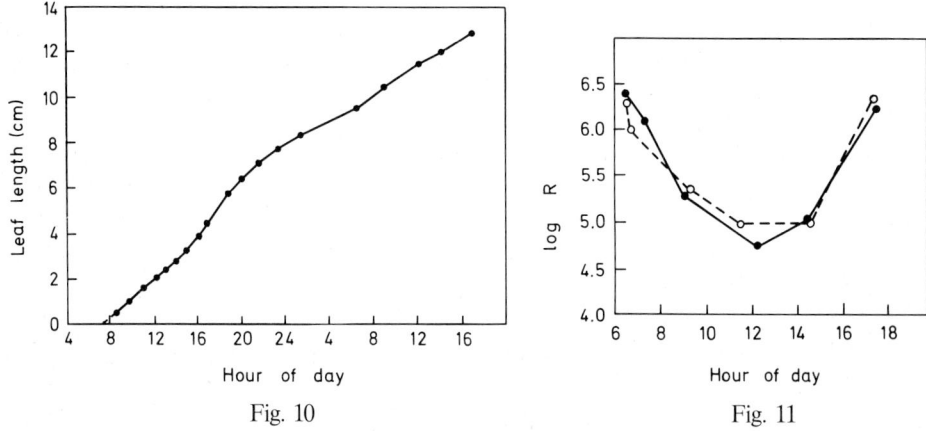

Fig. 10   Fig. 11

Fig. 10. Diurnal course of leaf elongation of unirrigated maize at Davis, California. Data are means for 10 plants of leaves at positions 12 and 13 and were obtained concurrently and from the same plot as the Ψ^{leaf} measurements depicted in Fig. 8. (From Acevedo et al., 1976)

Fig. 11. Daily trend in stomatal opening as indicated by leaf resistance to the mass-flow of air (R) in the upper layer of the canopy of irrigated (●—●) and unirrigated (○—○) maize. Measurements were taken with a Gregory-Pearse porometer. The quantity log R (R in g cm^{-2} s^{-1}) correlates closely with stomatal aperture (Hsiao and Fischer, 1975); a decrease indicates stomatal opening. The crop is the same as that depicted in Fig. 8 but measurements were made on day 26 after planting. (From Acevedo, 1975)

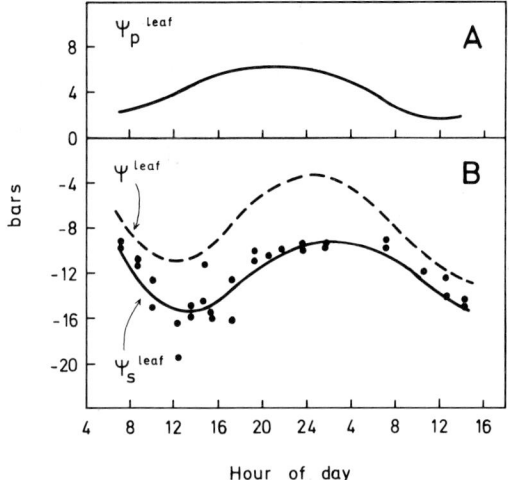

Fig. 12 A and B. Diurnal patterns of (A) pressure potential Ψ_p and (B) water potential Ψ and solute potential Ψ_s, in expanding leaves of an unirrigated maize crop at Davis, California. Data points represent individual Ψ_s^{leaf} measurements taken on the same samples used to obtain data in Fig. 8. For comparison, the Ψ^{leaf} curve in Fig. 8 is reproduced (---). The Ψ_s^{leaf} curve (—) represents polynomial regression of the points. The Ψ_p^{leaf} curve was computed as the difference between the Ψ^{leaf} and Ψ_s^{leaf} curves. (From Acevedo et al., 1976; cf. footnote on p. 289)

3. Contrasting Yield Behavior of Maize and Sorghum Under Water Stress

The ability of sorghum to sustain yield under water stress has been known for many years; and comparisons have frequently been made with the behavior of maize, which suffers marked yield reductions under the same conditions (Miller, 1916; Martin, 1930; Sullivan and Blum, 1970). Table 1 gives an example of the contrasting yields that were obtained with and without irrigation in Davis, California. Sorghum yield was only slightly reduced by a lack of applied water, whereas maize yield was about halved. Sorghum, however, yielded less than maize under frequent irrigation, at least under the commercial management practices used in that study.

Table 1. Grain yields of maize and sorghum in Davis, California, with and without irrigation, 1974[a]

	Frequently irrigated[b]		Unirrigated[b]	
	Yield (kg ha^{-1})	Harvest index	Yield (kg ha^{-1})	Harvest index
Maize[c]	11 300	0.47	5 200	0.43
Sorghum[c]	8 780	0.48	8 000	0.55

[a] Unpublished data of Fereres and Acevedo.
[b] The crops were planted in a deep Yolo clay loam fully wetted to a depth of 3 m and were either frequently (nearly weekly) irrigated or left unirrigated.
[c] Cultivar of maize: Dekalb XL 22; sorghum: Pioneer 846.

Published comparisons between maize and sorghum commonly sought to identify only one or two factors as the cause for the contrasting behavior. A careful examination of the literature and our own experience (Fereres, 1976), however, indicate that the cause is multifaceted and encompasses physiological, developmental, and morphological parameters. This is elaborated on in some detail here to serve as an example of the concerted effect of several diverse plant phenomena that enables one species to out-perform another under a water-limiting environment. The aspects specifically discussed are: rooting characteristics, stomatal characteristics, CO_2 assimilation, and developmental morphology as related to source and sink for assimilates.

a) **Rooting Characteristics.** It is sometimes thought that the yield advantage of sorghum over maize when water is in short supply might have a principal cause in the root systems of the two crops (Sullivan and Blum, 1970). Miller (1916) reported that although both root systems were about the same in extensiveness, the ratio of secondary to primary roots was twice as great for sorghum as for maize at several growth stages in the field. He invoked this difference as a major cause for the greater drought tolerance of sorghum but did not explain how a higher proportion of secondary roots should confer an advantage. In a recent detailed comparison of the two crops, which were growing on stored water without irrigation in a deep Yolo loam at Davis, Vega (1972) found that root length density and rooting depth were similar for both crops for most of the season. At the very early seedling stage, however, root density was greater for the sorghum crop, apparently because sorghum was planted, following commercial practices, at a population density about three times that of maize. Water use by the two crops was about the same for the season (Vega, 1972). Comparisons were also made between frequently irrigated and unirrigated treatments. Vega found that total root length per unit land area was less in the irrigated treatment in maize but unaffected by irrigation in sorghum. With both crops, irrigation altered the vertical distribution of roots in the soil, increasing root density at the lower depth while reducing it near the surface. Results similar to those of Vega were obtained subsequently on the same soil by Acevedo and Fereres (unpublished). In a study where rooting activities were observed indirectly by measuring ^{32}P uptake, Lavy and Eastin (1969) reported similar patterns for maize and sorghum at the end of the season, although some differences were observed at early stages of growth.

From this evidence, one can conclude that differences between the root systems of the two crops are minor and insufficient to explain the differing yield behavior under short water supply.

b) **Stomata and CO_2 Assimilation.** Several studies have compared maize and sorghum for stomatal behavior under water stress. Whether the plants were growing in the field (Glover, 1959; Turner, 1974) or in an artificial environment (Sanchez-Diaz and Kramer, 1971; Beadle et al., 1973; Beardsell and Cohen, 1975), stomata began to close earlier in maize than in sorghum leaves as water stress developed. The threshold Ψ^{leaf} for closure (see Fig. 2B) was usually lower (more negative) in sorghum than in maize. Generally, the threshold Ψ^{leaf} for closure in different species varied from study to study. It also varies with leaf position (insertion level) on the stem (Jordan et al., 1975). There is indication that the threshold is at the point where bulk tissue Ψ_p drops to nearly zero (Hall et al., this volume Part 3:D). These variations might be associated with variations in tissue Ψ_s, which would result in different Ψ_p at the same tissue Ψ (Hsiao et al., 1975; see also Sect. II.2 of this chapter).

Although the threshold Ψ^{leaf} values for stomatal closure are apparently influenced by several factors, the difference in values between maize and sorghum appears to be consistent. This implies that sorghum should be able to maintain a normal rate of photosynthesis for a longer time than maize when water shortage develops. Indeed, Beadle et al. (1973), in a growth-chamber study, found that the decline in net CO_2 assimilation started at a lower Ψ in sorghum than in maize as Ψ^{leaf} was reduced. With sorghum, CO_2 assimilation remained substantial when Ψ^{leaf} fell to -12 bars, whereas it approached zero in maize leaves at the same Ψ^{leaf}. There appear to be differences among cultivars in sensitivity of assimilation and stomata to reductions in tissue Ψ, among maize as well as among sorghum (Sullivan and Blum, 1970; Blum and Sullivan, 1972). The general trend, however, is that sorghum is less sensitive. This difference between the two species, delineated mostly in the laboratory or growth chamber, has yet to be verified systematically in the field on a long-term basis.

A difference in ability to adjust osmotically, if present, could account for the difference between maize and sorghum in stomatal and photosynthetic behavior. As discussed in Sect. III.2 of this chapter and elaborated on elsewhere (Hsiao et al., 1975), osmotic adjustment enables a plant to maintain turgor or water content while tissue Ψ is reduced. Curves of Ψ^{leaf} vs. RWC (Sanchez-Diaz and Kramer, 1971; Beardsell and Cohen, 1975), although obtained in artificial environments, suggest that sorghum is either more capable of osmotic adjustment or generally contains more solutes than maize leaves. Field data of Fereres et al. (1976), demonstrate full osmotic adjustment under mild but prolonged water stress. Osmotic adjustment in maize under the same conditions appeared to be not as complete in the companion study by Acevedo (1975).

c) **Developmental Morphology and Source-Sink Relations.** Yields in both crops often consist of grain. As pointed out in Sect. II of this chapter, water status affects developmental events and thus has a role in determining sink size or the numbers of grains (fruits) to be filled. Further, the filling of the grains is determined by source size and intensity as well as by factors influencing the partition of assimilates among competing sinks. The yield advantage of sorghum over maize in water-limiting situations is attributable partly to a greater ability to adjust or maintain grain number according to conditions, and to differences in assimilate partition between the two crops.

Current sorghum varieties are much more "plastic" in reproductive development than maize. The plasticity lies in the ability to form tillers and branch flower heads freely. Should water stress reduce the number of young fruits in sorghum via either inhibited pollination or abortion and death of young fruits, tillering or branch head formation can make up for most of the reduction in grain number. The "second flush" of grain, of course, would have a later maturation date, requiring harvest postponement. A good example of this behavior is seen in Table 2. Sorghum was subjected to severe water stress at different stages of development in the field. Stress around heading time reduced yield substantially, whereas the reduction was slight or nil with stress occurring either much earlier or later. Most interesting is the stress treatment at the beginning of head emergence, resulting in severe "head blasting." Head emergence was stopped during stress; and after the release of stress, heads emerged but parts of the heads had died. Branch heads formed later in response, from the flag-leaf node and the node below, accounting for the increase in number of heads per plant over the control treatment (Table 2). In spite of the head blast, the final yield was still 78% of the control, reflecting the compensation effect of the branch heads. Maize varieties, being determinate in reproductive growth after the ear reaches a certain size, cannot compensate for stress effects on the number of young fruits by forming new ears.

Table 2. Responses in yield, number of heads, and grain weight of sorghum (cv. RS 610) to water stress at various growth stages[a]

	Control	Early stress	Stress at head emergence Moderate	Stress at head emergence Severe	Late stress
Yield (kg ha^{-1})	6080	5960	5180	4760	5780
Heads per plant	1.16	1.09	1.28	1.93	1.09
1000-grain weight (g)	22.0	24.4	25.4	26.7	20.8
Grains per head[b]	950	890	630	370	1010

[a] Unpublished data of D. W. Henderson, obtained in 1960 at Davis, California.
[b] Calculated from grain weight and head weight.

Early inflorescence development in both maize (Denmead and Shaw, 1960) and sorghum (Whiteman and Wilson, 1965) appears to be tolerant of short periods of fairly severe water stress, though no careful comparisons have yet been made of the two crops. Both crops are known to be sensitive to water stress around the time of flowering (for maize: Robins and Domingo, 1953; Denmead and Shaw, 1960; for sorghum: Musick and Grimes, 1961; Lewis et al., 1974). The major effect is a reduction in the number of grains per ear or panicle. Again, the advantage of sorghum is its ability to compensate for this reduction. Grain size can be altered by water stress in both crops.

Commercial cultivars of field maize, for all practical purposes, do not tiller. In contrast, common cultivars of sorghum tiller readily and apparently adjust the number of head-bearing shoots per plant according to planting density, favorableness of the environment, and nutrient and water supply. Head formation on tillers lags behind that on the main shoot, thus ensuring that at least some heads will not be at the most sensitive stage should a short period of water stress occur at heading and flowering, thereby providing an adaptive advantage. Probably more important, however, is another advantage offered by tillering. When planted at a given population density in dry-land agriculture, sorghum would reduce its number of heads per hectare if water starts to be limited during the early vegetative stage. This reduces the leaf area and vegetative material and in essence concentrates the available water and assimilates to form and mature a reduced number of heads. Maize, on the other hand, may invest all the resources to form the number of ears dictated by the planting density but completely fail to fill the young fruits if rain should continue to be scarce later in the season.

Important in addition to the difference in developmental flexibility might be differences in source–sink relations. In both crops most of the dry matter in the grain comes from material assimilated after flowering (Fisher and Wilson, 1971a; Tanaka and Yamaguchi, 1972). With sorghum, most of the assimilates are supplied by the head and the upper few leaves (Fisher and Wilson, 1971b), especially in the case of commercial plantings, because of the favorable light environment of the upper part of the canopy (Eastin, 1972). In contrast, in maize the source of assimilates for the grain is more diffused, consisting of the five or six leaves above and at the ear node (Tanaka and Yamaguchi, 1972). Even leaves below the ear may translocate some assimilates to the ear (Palmer et al., 1973). Probably contributing to this translocation pattern is the canopy architecture of maize, which permits more light penetration to the lower leaves than with sorghum at the same leaf area index (LAI). Water stress accelerates the senescence of old leaves (e.g., Fischer, 1973), which progresses from the base of the stem upward (Martin, 1930). Hence, stress would likely have more of a restrictive effect in maize than in sorghum on source size and intensity.

The preceding discussion emphasized the differences between maize and sorghum. Often ignored was the fact that there are differences in resistance to water stress among cultivars of the same species, especially in sorghum (Blum, 1973). It is possible that if cultivars in the upper extremes of maize and the lower extremes of sorghum in stress resistance were compared, little or no difference would be found between the species. One study (Sullivan and Blum, 1970) of Ψ values for stomatal closure in sorghum and in a resistant maize cultivar points to this possibility.

A good example of differences between sorghum cultivars is seen in Fig. 13. Cultivar DD 38 maintained its grain yield regardless of water stress, whereas cultivar Ryer 15 suffered yield loss with each increment in stress. The maintenance of yield in DD 38 apparently resulted largely from altered partition of assimilates, as indicated by the change in grain-to-stover ratio from 0.82 to 1.29. In contrast, grain yield in Ryer 15 roughly paralleled vegetative dry matter at different levels of water stress, and the grain-to-stover ratio varied only slightly (from 1.43 to 1.30). In this respect maize under stress (Acevedo, 1975) behaved similarly to Ryer 15.

IV. Concluding Remarks

Crop yields respond to water deficits in dynamic and complex ways. This chapter, in focusing on the processes underlying the responses, has made all too clear the inadequacy of our understanding of the physiology of water stress and of plant adaptation. It is evident that information is scant on yield-determining plant parameters and their interactions under water-limiting conditions. The prospect of basing yield predictions on a reasonably comprehensive understanding of the component processes is therefore very distant.

The discussions in this chapter revealed many gaps in our knowledge in stress physiology. These include ignorance of: the parameter best indicating plant water status in terms of plant function; the processes mediating osmotic adjustment; stress effects on developmental and fertilization processes; quantitative relations between stress and translocation; and interactions between stress and senescence. All these aspects have important ramifications in determining yields and adaptation to water stress. Much of the ignorance reflects neglects and imbalance in research on the physiology and ecology of water stress rather than limitations in techniques or basic theories. For example, while photosynthesis receives much current attention at the subcellular and single-leaf level, leaf senescence is seldom dealt with. While many workers study changes in growth regulators effected by water stress, especially ABA, no sustained effort appears to be directed at relating those changes to stress-induced fruit abortion and abscission (cf. Fig. 3), reduction in expansive growth (cf. Fig. 2A), or senescence.

More broadly speaking, much needs to be done to link long-term crop performance and water use in the field to detailed physiology and metabolism as related to water. True, the task is made exceedingly difficult by the dynamic nature of the system, as exemplified in Sect. III.1 and III.2 of this chapter, and the highly integrative behavior of the plant. The results, however, should be refreshingly rewarding, illuminating facets hitherto unrevealed in traditional

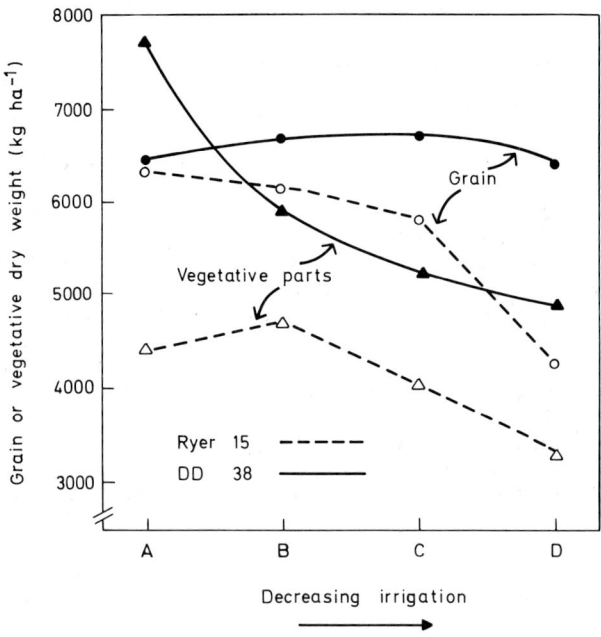

Fig. 13. Contrasting effects of irrigation on the production of vegetative dry matter and grain by sorghum cultivars Ryer 15 and DD (double dwarf) 38. The crops were planted in a fully moist Yolo clay loam. Treatment A was irrigated every 10–12 days with about 9 cm of water while treatments B and C were irrigated respectively twice and once with about 18 cm of water during the growing season. Treatment D received no irrigation. (Unpublished data from D. W. Henderson, 1957)

agronomical and basic physiological studies, not to mention the practical benefits that can arise.

References

Acevedo, E.: The growth of maize (*Zea mays* L.) under field conditions as affected by its water relations. Ph. D. Thesis, Davis, Calif. 1975.
Acevedo, E., Fereres, E., Henderson, D. W., Hsiao, T. C.: Diurnal fluctuation of the water potential components in field maize (*Zea mays* L.) leaves and its significance on leaf extension (in preparation, 1976).
Allmaras, R. R., Nelson, W. W., Voorhies, W. B.: Soybean and corn rooting in Southwestern Minnesota: II. Root distributions and related water inflow. Soil Sci. Soc. Am. Proc. **39**, 771–777 (1975).
Arya, L. M., Blake, G. R., Farrell, D. A.: A field study of soil water depletion patterns in presence of growing soybean roots: III. Rooting characteristics and root extraction of soil water. Soil Sci. Soc. Am. Proc. **39**, 437–444 (1975).
Beadle, C. L., Stevenson, K. R., Neumann, H. H., Thurtell, G. W., King, K. M.: Diffusive resistance, transpiration and photosynthesis in single leaves of corn and sorghum in relation to leaf water potential. Can. J. Plant Sci. **53**, 537–544 (1973).
Beardsell, M. F., Cohen, D.: Relationships between leaf water status, abscisic acid levels, and stomatal resistance in maize and sorghum. Plant Physiol. **56**, 207–212 (1975).
Blum, A.: Component analysis of yield responses to drought of sorghum hybrids. Exp. Agric. **9**, 159–167 (1973).
Blum, A., Sullivan, C. L.: A laboratory method for monitoring net photosynthesis in leaf segments under controlled water stress. Experiments with sorghum. Photosynthetica **6**, 18–23 (1972).
Boyer, J. S.: Relationships of water potential to growth of leaves. Plant Physiol. **43**, 1056–1062 (1968).
Boyer, J. S.: Photosynthesis of low water potentials. Phil. Trans. Roy. Soc. Lond. B. **273**, 501–512 (1976).
Browning, V. D., Taylor, H. M., Huck, M. G., Klepper, B.: Water relations of cotton: A rhizotron study. Auburn, Alabama: Agric. Exp. Stn. Auburn Univ. 1975.
Coster, C.: Die täglichen Schwankungen des Längenzuwachses in den Tropen. Rec. Trav. Bot. Neér. **24**, 257–305 (1927).
Das, K. C.: Dynamics of corn root growth as affected by compact subsoil and its influence on crop response to irrigation. Ph. D. Thesis, Davis, Calif. 1973.
Denmead, O. T., Shaw, R. H.: The effects of soil moisture stress at different stages of growth on the development and yield of corn. Agron. J. **52**, 272–274 (1960).
Eastin, J. D.: Photosynthesis and translocation in relation to plant development. In: Sorghum in the seventies (eds. N. G. P. Rao, L. R. House), pp. 214–246. New Delhi: Oxford and IBH Publ. Co. 1972.
El Nadi, A. H., Brouwer, R., Locker, J.: Some responses of the root and the shoot of *Vicia faba* plants to water stress. Neth. J. Agr. Sci. **17**, 133–142 (1969).
Evans, L. T.: The physiological basis of crop yield. In: Crop physiology (ed. L. T. Evans), pp. 327–355. Cambridge: Univ. Press 1975.
Feddes, R. A., Rijtema, P. E.: Water withdrawal by plant roots. J. Hydrology **17**, 33–59 (1972).
Fereres, E.: Growth, development and yield of sorghum in the field under variable water supply. Ph. D. Thesis, Davis, Calif. 1976.
Fereres, E., Acevedo, E., Henderson, D. W., Hsiao, T. C.: Seasonal changes in water potential and turgor maintenance in sorghum under water stress (in preparation, 1976).
Fischer, R. A.: The effect of water stress at various stages of development of yield processes in wheat. In: Plant responses to climatic factors. Proc. Uppsala Symp. (ed. R. O. Slatyer), pp. 233–240. Paris: UNESCO 1973.
Fischer, R. A., Hagan, R. M.: Plant water relations, irrigation management and crop yield. Exp. Agr. **1**, 161–177 (1965).

Fisher, K. S., Wilson, G. L.: Studies of grain production in *Sorghum vulgare*. I. The contribution of preflowering photosynthesis to grain yield. Australian J. Agr. Research **22**, 33–37 (1971 a).

Fisher, K. S., Wilson, G. L.: Studies of grain production in *Sorghum vulgare*. II. Sites responsible for grain matter production during the post-anthesis period. Australian J. Agr. Res. **22**, 39–47 (1971 b).

Forrester, J. W.: Principles of systems. Cambridge: Wright-Allen Press Inc. 1968.

Friend, D. J. C.: The effects of light and temperature on the growth of cereals. In: The growth of cereals and grains (eds. F. L. M. Thorpe, J. D. Ivins), pp. 181–200. London: Butterworths 1966.

Gardner, W. R.: Dynamic aspects of water availability to plants. Soil Sci. **89**, 63–73 (1960).

Glover, J.: The apparent behavior of maize and sorghum stomata during and after drought. J. Agr. Sci. **153**, 412–416 (1959).

Hoffman, G. J., Rawlins, S. L., Garber, M. J., Cullen, E. M.: Water relations and growth of cotton as influenced by salinity and relative humidity. Agron. J. **63**, 822–826 (1971).

Hsiao, T. C.: Plant responses to water stress. Ann. Rev. Plant Physiol. **24**, 519–570 (1973).

Hsiao, T. C., Acevedo, E.: Plant responses to water deficits, water use efficiency, and drought resistance. Ag. Meteorol. **14**, 59–84 (1974).

Hsiao, T. C., Acevedo, E., Fereres, E., Henderson, D. W.: Water stress, growth, and osmotic adjustment. Phil. Trans. Roy. Soc. Lond. B. **273**, 479–500 (1976).

Hsiao, T. C., Fischer, R. A.: Mass flow porometers. In: Measurement of stomatal aperture and diffusive resistance (ed. E. T. Kanemasu), pp. 5–11. Bulletin 809. Washington State Univ.: College of Agr. Res. Cent. 1975.

Jordan, W. R., Brown, K. W., Thomas, J. C.: Leaf age as a determinant in stomatal control of water loss from cotton during water stress. Plant Physiol. **56**, 595–599 (1975).

Klepper, B.: Diurnal pattern of water potential in woody plants. Plant Physiol. **43**, 1931–1934 (1968).

Kramer, P. J.: The relation between rate of transpiration and rate of absorption of water in plants. Am. J. Botany **24**, 10–15 (1937).

Lavy, T. L., Eastin, J. D.: Effect of soil depth and plant age on ^{32}phosphorus uptake by corn and sorghum. Agron. J. **61**, 677–680 (1969).

Lewis, R. G., Hiler, E. A., Jordan, W. R.: Susceptibility of grain sorghum to water deficit at three growth stages. Agron. J. **66**, 589–591 (1974).

Loomis, R. S., Williams, W. A., Hall, A. E.: Agricultural productivity. Ann. Rev. Plant Physiol. **22**, 431–468 (1971).

Loomis, W. E.: Daily growth of maize. Am. J. Botany **21**, 1–6 (1934).

Luxmoore, R. J., Millington, R. J., Marcellos, H.: Soybean canopy structure and some radiant energy relations. Agron. J. **63**, 111–114 (1971).

Martin, J. H.: The comparative drought resistance of sorghum and corn. Agron. J. **22**, 993–1003 (1930).

Miller, E. C.: Comparative studies of the root system and leaf area of corn and the sorghums. J. Agr. Res. **6**, 311–332 (1916).

Milthorpe, F. L., Moorby, J.: An introduction to crop physiology. London: Cambridge Univ. Press 1974.

Musick, J. T., Grimes, D. W.: Water management and consumptive use by irrigated grain sorghum in western Kansas. Kansas Tech. Bull. 113 (1961).

Neales, T. F., Incoll, L. D.: The control of leaf photosynthesis rate by the level of assimilate concentration in the leaf: A review of a hypothesis. Botan. Rev. **34**, 107–125 (1968).

Newman, E. I.: A method of estimating the total length of root in a sample. J. Appl. Ecol. **3**, 139–145 (1966).

Palmer, A. F. E., Heichel, G. H., Musgrave, R. B.: Patterns of translocation, respiration loss and redistribution of ^{14}C in maize labeled after flowering. Crop Sci. **13**, 371–376 (1973).

Pearson, R. W.: Soil environment and root development. In: Plant environment and efficient water use (eds. W. H. Peirre, D. Kirkham, J. Pesek, R. Shaw), pp. 95–126. Madison: Am. Soc. Agron. and Soil Sci. Soc. Am., 1966.

Raschke, K.: Stomatal action. Ann. Rev. Plant Physiol. **26**, 309–340 (1975).

Robins, J. S., Domingo, C. E.: Some effects of severe soil moisture deficits at specific stages in corn. Agron. J. **45**, 612–621 (1953).
Salter, P. J., Goode, J. E.: Crop responses to water at different stages of growth. Commonwealth (G. B.) Bur. Hort. Plant. Crops 1967.
Sanchez-Diaz, M. F., Kramer, P. J.: Behavior of corn and sorghum under water stress and during recovery. Plant Physiol. **48**, 613–616 (1971).
Slatyer, R. O.: Plant-water relationships. New York, London: Academic Press 1967.
Slatyer, R. O.: Physiological significance of internal water relations to crop yield. In: Physiological aspects of crop yield (eds. J. D. Eastin, F. A. Haskins, C. Y. Sullivan, C. H. M. van Bavel), pp. 53–83. Madison, Wisc.: Am. Soc. Agron. and Crop Sci. Soc. Am. 1969.
Slatyer, R. O.: The effect of internal water status on plant growth, development and yield. In: Plant responses to climatic factors. Proc. Uppsala Symp. (ed. R. O. Slatyer), pp. 177–188. Paris: UNESCO 1973.
Stanhill, G.: Simplified agroclimatic procedures for the effect of water supply. In: Plant responses to climatic factors. Proc. Uppsala Symp. (ed. R. O. Slatyer), pp. 461–474. Paris: UNESCO 1973.
Sullivan, C. Y., Blum, A.: Drought and heat resistance of sorghum and corn. In: 25th Ann. Rep. Corn and Sorghum Res. Conf., pp. 55–66. Chicago, Ill.: Am. Seed Trace Assoc. 1970.
Tanaka, A., Yamaguchi, J.: Dry matter production, yield components and grain yield of the maize plant. J. Facul. Agro., Hokkaido Univ., Sapporo, Japan **57**, 71–131 (1972).
Taylor, H. M., Gardner, H. R.: Penetration of cotton seedling taproots as influenced by bulk density, moisture content, and strength of soil. Soil Sci. **96**, 153–156 (1963).
Thorne, J. H., Koller, R. H.: Influences of assimilate demand on photosynthesis, diffusive resistances, translocation, and carbohydrate levels of soybean leaves. Plant Physiol. **54**, 201–207 (1974).
Turner, N. C.: Stomatal behavior and water status of maize, sorghum, and tobacco under field conditions. II. At low soil water potential. Plant Physiol. **53**, 360–365 (1974).
Vega, J. D. C.: Comparative dynamics of root growth and subsoil water availability in unirrigated corn and sorghum. Ph. D. Thesis, Davis, Calif. 1972.
Veihmeyer, F. J., Hendrickson, A. H.: Soil moisture as an indication of root distribution in deciduous orchards. Plant Physiol. **13**, 169–177 (1938).
Wardlaw, I. F.: The control and pattern of movement of carbohydrates in plants. Botan. Rev. **34**, 79–105 (1968).
Watson, D. J.: Comparative physiological studies on the growth of field crops. I. Variations in net assimilation rate and leaf area between species and varieties and within and between years. Ann. Botany N. S. **11**, 41–76 (1947).
Watts, W. R.: Leaf extension in *Zea mays*. II. Leaf extension in response to independent variation of the temperature of the apical meristem of the air around the leaf and of the root zone. J. Exp. Botany **23**, 713–721 (1972).
Watts, W. R.: Leaf extension in *Zea mays*. III. Field measurements of leaf extension in response to temperature and leaf water potential. J. Exp. Botany **25**, 1085–1096 (1974).
Whiteman, P. C., Wilson, G. L.: Effect of water stress in the reproductive development of *Sorghum vulgare*. Univ. Queensland Botan. Papers **4**, 233–239 (1965).

Part 5
Water Relations and CO_2 Fixation Types

Preface

For the purpose of a comprehensive approach to a water stress concept it is necessary to discuss not only physiological tolerance to water stress but also all "constitutional" properties based on the morphogenotype by which the plant can avoid water losses. Such "constitutional" properties refer to life-form and structure of the plant, as e.g., the morphology and anatomy of the assimilating organs (Kummerow, 1973; cf. also Stålfelt, 1956: sclerophylly) or structure of the stomata (Pisek et al., 1970), but they refer also to physiological properties that can indirectly avoid or at least retard water stress. It has been known for a long time (cf. Haberlandt, 1896; Volkens, 1887) that succulents, on the one hand, and plants with a so-called "Kranz" type of leaf anatomy, on the other, constitute a conspicuously high percentage of the vegetation of hot arid and semi-arid regions. Today we know that these morphological and anatomical properties are, at the same time, the marks of a special metabolic organization, and that the succulence has more functions than water storage alone. Findings have increased of plants that fix CO_2 not only by the well-known enzyme ribulose-1,5-diphosphate carboxylase, but also by the detour of primarily producing C_4 products via β-carboxylation of phosphoenol-pyruvate. It seems rewarding to analyze such plant types more thoroughly with respect to their water balance, and also to pay more attention to the biochemical and physiological principles of the C_4 "syndrome" and of Crassulacean acid metabolism (CAM).

It is interesting to observe that, initially, an exact distinction between CAM, C_4, and C_3 species was made. This is shown clearly by the genetic differentiations that were made between C_3 and C_4 *Atriplex* species (Björkman et al., 1971). The genetically fixed properties of many plant types capable of C_4 metabolism, however, do not appear under all test conditions and thus this capacity is often overlooked. In view of more refined methods, it is small wonder that presently the capacity of C_4 metabolism or CAM is being found in more and more plant families, genera and species, and it appears that this will possibly prove true of a considerable percentage of all species of the higher plants (cf. Sankhla et al., 1975; Downton, 1975).

The distinction between CAM and C_4 plants seems to be clear at first sight, since the former exhibit a temporal, and the latter a spatial separation of the primary fixation of CO_2 from the Calvin cycle. This definite distinction applies, without doubt, to many plant types, but begins to blur where transitions and intermediates between C_3 and C_4 or C_3 and CAM types are found (Winter and Lüttge, Part 5:B; Huber and Sankhla, Part 5:C).

According to the carbon fixation pathway, one can understand CAM as an intermediate type between the C_3 and the C_4 types (Huber and Sankhla, Part 5:C) and thus the different plant species can exhibit, depending for example on the prevailing ecological conditions, "option" (Osmond, 1975) for one or the other type. Also the findings are remarkable that the C_4-type carboxylation is present in certain organs of C_3 plants. Even freshwater green algae, some marine algae and tissue cultures were regarded as representatives of C_4 metabolism (cf. Part 5:C). However (cf. Raven, 1974; Tolbert, 1974), it must be proved that the formation of malate and aspartate resulting from a C_4 dicarboxylic acid cycle serves as a CO_2 source for the Calvin cycle. With aquatic plants the fact must also be considered that fixation of CO_2 from bicarbonate increases ^{13}C uptake. Thus according to their isotope discrimination (cf. Huber and Sankhla, Part 5:C) such plants are shifted into the range of C_4 plants but have no C_4 metabolism (cf. Deuser and Degens, 1967; Ziegler, 1976, unpubl.). On the other hand, evidence of the C_4 dicarboxylic acid cycle in lower plants shows that in the course of plant evolution at least the general biochemical pathway of the C_4 syndrome and CAM did not appear as late as with the phanerogams and it might be surprising then that CAM was found in no other gymnosperms (cf. Troughton, 1971) than *Welwitschia mirabilis* (Dittrich and Huber, 1974). It means also that the morphological and anatomical constituents of plants with these CO_2 fixation types might be not necessarily indispensable prerequisites for the respective type of metabolism. Meanwhile, C_4 plants have been found without "Kranz" type of leaf anatomy (cf. Huber and Sankhla, Part 5:C) and also CAM plants among non-succulent Bromeliaceae (Kluge et al., 1973). Further evidence is, however, necessary to find out whether we can speak about a general basic mechanism which has probably evolved polyphyletically in different physiological and morphological plant types increasing their CO_2 gain or at least the water-use efficiency of photosynthesis (cf. Part 3).

However, unambiguously the constitutional properties of the CAM and C_4 plants may indicate a special adaptation of the plants to life in arid regions, it is still difficult to prove the ecophysiological evidence of the direct connection with water stress by laboratory or field experiments. This is shown in the comprehensive discussions, especially by Ludlow, in Part 5:D. Certainly, other habitat factors, such as high temperatures, photoperiod, radiation intensity, and salinity, can also cause such types of adaptation of the C-metabolism. A key aspect of this problem is presented by those plant types that can shift between C_3 and C_4 or C_3 and CAM type, of which the study of Winter and Lüttge (Part 5:B) serves as an example. They question whether the salinity-induced shift from C_3 to CAM of *Mesembryanthemum crystallinum* originates only from an osmotic effect (i.e. water stress), or specifically from the NaCl. It could be that the shift to CAM in these plants is the consequence of a chain of events which can come about in different ways. Von Willert (1975a, b), for instance, assumes that the production of malate is a direct result of the increase of inorganic phosphate (P_i) which he believes compensates feed-back inhibition of phosphoenol-pyruvate-carboxylase caused by malate. The P_i increase may be induced, among other reasons, by salinity or osmotic stress. In semi-arid habitats, the water-use efficiency of most CAM and also C_4 plants is definitely greater than that of most C_3 plants. Kluge (Part 5:A) explains how the water balance of the CAM type can be regulated largely by the C-metabolism, and also how succulents come to take up less and less external CO_2 with continuing aridity until, finally, their stomata remain closed even during the dark period. Thus, CAM displays its optimal effect not in extremely arid habitats but rather where the climate consists of alternating dry and rainy periods. The high water-use efficiency of C_4 grasses seems to originate less from an increasing reduction of transpiration than from CO_2 uptake higher than that of C_3 plants under comparably arid conditions (Ludlow, Part 5:D). This, among other reasons, would cause the C_4 grasses in an arid habitat to flower and ripen faster and thus improve their chances of survival (cf. also Hsiao, Part 4:F).

A further essential approach to the problem of the ecological evidence of the adaptation of C_4 or CAM type to hot, dry habitats can be achieved by examining correlations between the geographical distribution and frequency of these plant types and climatic parameters as precipitation and levels of day and night temperatures. It would be worthwhile to devote an entire chapter of this volume to this question, but for now we can only refer to a few studies on this topic. For instance with *Sempervivum* species, according to Osmond et al. (1975), a correlation exists between the drought stress of the habitat and a prevalent tendency of the plants to dark CO_2 fixation. This becomes even more obvious with the Bromeliaceae: those species that live in arid regions (or as epiphytes) show, almost without exception, typical CAM characteristics (Medina and Troughton, 1974). A study of Teeri and Stowe (1976) tries to answer the question raised above by means of a mathematical analysis of the correlation between the geographical distribution of C_4 grasses and climatic parameters in North America. Their finding is that, in general, aridity as well as soil salinity are related to the distribution of C_4 grasses. A significant correlation, however, could be found only with the heat factor, i.e., the minimum night temperatures during the growing season. A mainly climate-related distribution of C_4 and CAM plants seems to be the case also in the Namib desert of South-West Africa (Schulze et al., 1976; Schulze and Schulze, 1976). In the coastal regions of the Namib, the proportion of the investigated C_4 and CAM plants is shifted in favor of the CAM types. In the warmer inland zone, the C_4 grasses are predominant and form the vast, partly periodic and partly perennial grassland. One of the main reasons for their existence is surely the effect of the more moderate drop in night temperatures there. The behavior of *Welwitschia mirabilis* is flexible: the closer its habitat is to the coast, the stronger is its tendency to CAM. This reveals that dark fixation takes place preferably in regions with cool nights, which, because of the lower-than-dewpoint-temperatures, are also damp and foggy.

Obviously, one cannot generally expect a predominance of C_4 type-carboxylating plants even in hot deserts. In the Colorado hot desert of California, for instance, only four species with a C_4 syndrome were found among the 33 species investigated (Philpott and Troughton, 1974). The investigations of Mooney and Troughton (1974) show that in Baja California and in Chile the proportion of the CAM plants increases with increasing aridity, but dominates only in narrow confines over the C_3 species. At approximately 100 mm of annual precipitation, CAM plants constitute only 25% of the desert flora and are, according to the authors' interpretation, less effective in the competition with deciduous C_3 plants. C_4 plants are rare and are found only in saline habitats.

In this section we have only made initial attempts and it cannot be decided clearly whether temperature factor or water conditions are decisive for the distribution of CAM and C_4 plants. Still outstanding are, above all, analyses of microclimates and CO_2 exchange of plants in such habitats. Thus it is hoped that future and current research will supply more material for a comprehensive chapter on the possible causal relations between climate, water conditions, and geographical distribution of certain CO_2 fixation types among plants. For now, we take advantage of the opportunity to obtain from the following contributions a survey of the recent knowledge of the CAM and C_4 syndromes in consideration of their ecological relevancy, especially for the water balance of the plants.

References

Björkman, O., Nobs, M., Pearcy, R., Boynton, J., Berry, J.: Characteristics of hybrids between C_3 and C_4 species of *Atriplex*. In: Photosynthesis and photorespiration (eds. M. D. Hatch, C. B. Osmond, R. C. Slatyer), pp. 105–119. New York: Wiley Interscience 1971.

Deuser, W. G., Degens, E. T.: Carbon isotope fractionation in the system CO_2 (gas)—CO_2 (aqueous)—HCO_3^- (aqueous). Nature **215**, 1033–1035 (1967).

Dittrich, P., Huber, W.: Carbon dioxide metabolism in members of the Chlamydospermae. In: Proceedings of the third international congress on photosynthesis (ed. M. Avron), pp. 1573–1578. Amsterdam: Elsevier Publ. Co. 1974.

Downton, W. J. S.: The occurrence of C_4 photosynthesis among plants. Photosynthetica **9**, 96–105 (1975).

Haberlandt, G.: Physiologische Pflanzenanatomie, 2nd ed. Leipzig: W. Engelmann 1896.

Kluge, M., Lange, O. L., Eichmann, M. v., Schmid, R.: Diurnaler Säurerhythmus bei *Tillandsia usneoides*: Untersuchungen über den Weg des Kohlenstoffes sowie die Abhängigkeit des CO_2-Gaswechsels von Lichtintensität, Temperatur und Wassergehalt der Pflanze. Planta (Berl.) **112**, 357–372 (1973).

Kummerow, J.: Comparative anatomy of sclerophylls of Mediterranean climatic areas. In: Mediterranean type ecosystems, origin and structure (eds. F. di Castro, H. A. Mooney), Ecological Studies Vol. 7, pp. 157–170. Berlin-Heidelberg-New York: Springer 1973.

Medina, E., Troughton, J. H.: Photosynthetic patterns in the Bromeliaceae. Carnegie Inst. Year Book **73**, 805–809. Stanford, California (1974).

Mooney, H., Troughton, J. H., Berry, J. A.: Arid climatic and photosynthetic systems. Carnegie Inst. Year Book **73**, 793–805. Stanford, California (1974).

Osmond, C. B.: Environmental control of photosynthetic options in Crassulacean plants. In: Environmental and biological control of photosynthesis (ed. R. Marcelle), pp. 311–321. The Hague: W. Junk 1975.

Osmond, C. B., Ziegler, H., Stichler, W., Trimborn, P.: Carbon isotope discrimination in alpine succulent plants supposed to be capable of Crassulacean acid metabolism (CAM). Oecologia (Berl.) **18**, 209–217 (1975).

Philpott, J., Troughton, J. H.: Photosynthetic mechanisms and leaf anatomy of hot desert plants. Carnegie Inst. Year Book **73**, 790–793. Stanford, California (1974).

Pisek, A., Knapp, H., Ditterstorfer, J.: Maximale Öffnungsweite und Bau der Stomata, mit Angaben über ihre Größe und Zahl. Flora **159**, 459–479 (1970).

Raven, J. A.: Carbon dioxide fixation. In: Botanical Monographs Vol. 10, Algal physiology and biochemistry (ed. W. D. P. Stewart), pp. 434–455. Oxford, London, Edinbourgh, Melbourne: Blackwell Scientific Publ. 1974.

Sankhla, N., Ziegler, H., Vyas, O. P., Stichler, W., Trimborn, P.: Ecophysiological studies on Indian arid zone plants. V. A screening of some species for the C_4-pathway of photosynthetic CO_2-fixation. Oecologia (Berl.) **21**, 123–130 (1975).

Schulze, E.-D., Schulze, I.: Distribution and control of photosynthetic pathway in plants growing in the Namib Desert, with special regard to *Welwitschia mirabilis* Hook. fil. Madoqua, series II, 5, in press (1976).

Schulze, E.-D., Ziegler, H., Stichler, W.: Environmental control of Crassulacean acid metabolism in *Welwitschia mirabilis* Hook. fil. in its range of natural distribution in the Namib Desert. Oecologia (Berl.) **24**, 323–334 (1976).

Stålfelt, M. G.: Morphologie und Anatomie des Blattes als Transpirationsorgan. Handb. d. Pflanzenphysiologie, Vol. 3 (ed. W. Ruhland), pp. 324–341. Berlin-Göttingen-Heidelberg: Springer 1956.

Teeri, J. A., Stowe, L. G.: Climatic patterns and the distribution of C_4 grasses in North America. Oecologia (Berl.) **23**, 1–12 (1976).

Tolbert, N. E.: Photorespiration. In: Botanical Monographs Vol. 10, Algal physiology and biochemistry (ed. W. D. P. Stewart), pp. 474–504. Oxford, London, Edinbourgh, Melbourne: Blackwell Scientific, Publ. 1974.

Troughton, J. H.: Aspects of the evolution of the photosynthetic carboxylation reaction in the plants. In: Photosynthesis and photorespiration (eds. M. D. Hatch, C. B. Osmond, R. C. Slatyer), pp. 124–129. New York: Wiley Interscience 1971.

Volkens, G.: Die Flora der ägyptisch-arabischen Wüste, auf Grundlage anatomisch-physiologischer Forschungen dargestellt. Berlin: Gebr. Bornträger 1887.

Willert, D. J. v.: Die Bedeutung des anorganischen Phosphats für die Regulation der Phosphoenolpyruvat Carboxylase von *Mesembryanthemum crystallinum*. Planta (Berl.) **122**, 273–280 (1975a).

Willert, D. J. v.: Stomatal control, osmotic potential and the role of inorganic phosphate in the regulation of the Crassulacean acid metabolism in *Mesembryanthemum crystallinum*. Plant Sci. Letters **4**, 225–229 (1975b).

A. Crassulacean Acid Metabolism (CAM): CO_2 and Water Economy

M. KLUGE

I. Introduction

The term Crassulacean acid metabolism characterizes a type of carbon metabolism occurring almost exclusively in succulent plants, and originally only observed in species of the Crassulaceae. Plant species exhibiting CAM[1] feature essentially the following activity:

1. Diurnal fluctuation of the malic acid (not malate) content of the photosynthesizing tissue, with malic acid accumulation during the night and malic acid consumption during the day.

2. Diurnal fluctuation of the starch content inverse to the oscillation of malic acid, that is, the starch content decreases during the night and increases during the day.

3. Net consumption of atmospheric CO_2 by the plant during the night and depression of net CO_2 uptake during the day.

The interest in CAM has increased considerably during the last decade because of striking biochemical similarities between this metabolic pathway and the C_4 pathway of photosynthesis, and because CAM is now increasingly understood as an adaptive mechanism that enables succulent plants to maintain a positive carbon balance under arid environmental conditions.

The literature on CAM has been extensively reviewed (see, e.g.: Ranson and Thomas, 1960; Ting, 1971, 1972; Sutton, 1974). For this reason, it is attempted to provide here only a short and rather general discussion of CAM that will contribute to the understanding of physiological adaptations of plants to arid environments, and which will serve as an introduction to the problems discussed below by Winter and Lüttge (see this volume Part 5:B).

II. Carbon Metabolism of CAM Plants

The problems discussed in the section are summarized in the scheme given in Fig. 1.

This scheme includes a model of the carbon flow in CAM as well as presently-known information about the compartmentalization of its biochemical sequences in the cell.

---

[1] For explanation of abbreviations see Notation, p. 320.

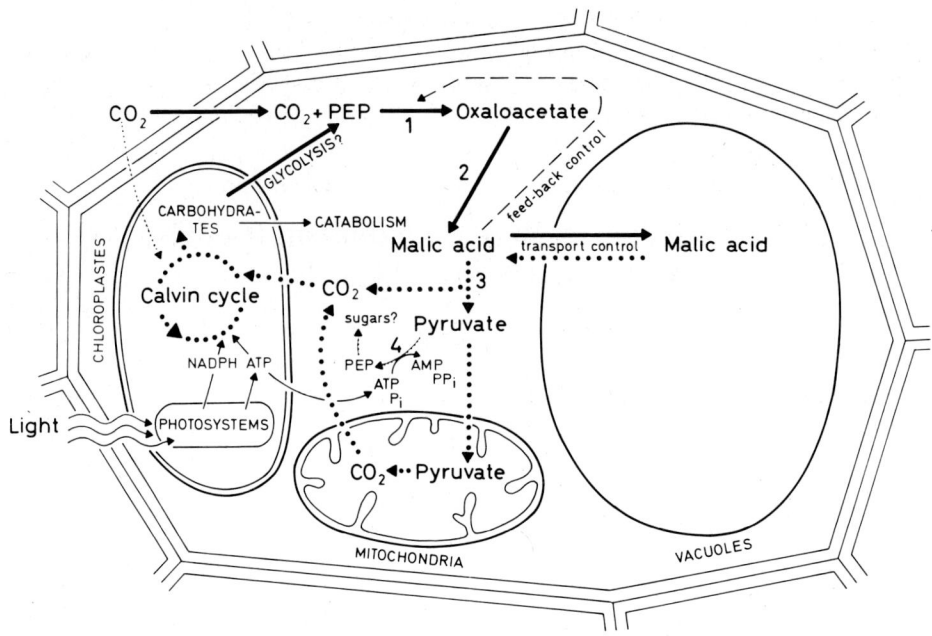

Fig. 1. Model of the carbon flow in CAM including its compartimentalization in the cell as far as is presently known. Reactions dominating during night (———); reactions dominating during the day (......); regulatory properties (-----). (After Kluge, 1972)

1. Malic Acid Synthesis

It is now generally accepted that the malic acid accumulation occurring at night in CAM plants is due to a dark fixation of atmospheric CO_2 and of CO_2 that is endogenously supplied by respiration. This CO_2 dark fixation proceeds via β-carboxylation of phosphoenol-pyruvate, catalyzed by PEP carboxylase:

$$PEP + CO_2 + H_2O \rightarrow OAA + P_i. \tag{1}$$

OAA, the product of this step, is then reduced to malic acid by malate dehydrogenase:

$$OAA + NADH_2 \rightarrow \text{malic acid} + NAD^{\oplus}. \tag{2}$$

Bradbeer et al. (1958) suggested that the carboxylation of PEP might be preceded by a further reaction of CO_2 fixation, that is, by the carboxylation of RuDP. This concept of a double CO_2 dark fixation as mechanism of the nocturnal malic acid synthesis in CAM is expressed in Eq. (3) (for arguments in favor of this scheme, see Bradbeer et al., 1958).

$$CO_2 + RuDP \rightarrow 2\,PGA \rightarrow 2\,PEP \xrightarrow{2CO_2} 2\text{ Malate}. \tag{3}$$

However, recent findings (Sutton and Osmond, 1972; Kluge et al., 1974; Cockburn and McAulay, 1975) tend more to the view of a single CO_2 fixation step, with PEP carboxylase being the only CO_2 fixing enzyme, than to Bradbeer's double fixation hypothesis.

2. Conversion of Malic Acid into Carbohydrates

The malic acid consumption occurring during the day is initiated in most of the CAM plants by malic enzyme [Eq. (4)].

$$\text{malic acid} + NADP^{\oplus} \rightarrow CO_2 + \text{Pyruvate} + NADPH_2. \tag{4}$$

In some species of CAM plants, PEP carboxykinase seems to be the decarboxylating enzyme [Eq. (5)] (Dittrich et al., 1973).

$$\text{malic acid} \xrightarrow[NAD^{\oplus} \quad NADH_2]{MDH} OAA \xrightarrow[ATP \quad ADP]{\text{PEP carboxykinase}} PEP + CO_2. \tag{5}$$

It should be mentioned that in both cases the depletion of malic acid during the day is due to a decarboxylation, that is, to the production of CO_2. There is no doubt that this CO_2 is fed directly into photosynthesis, thus providing an endogenous source of substrate for the Calvin cycle. As a consequence of this, photosynthesis is essentially involved in the conversion of malic acid into carbohydrates.

There is evidence that if malic enzyme is the decarboxylating enzyme, some of the pyruvate derived from this reaction can be oxidized completely to CO_2, thus providing further substrate for photosynthesis (see reviews cited above). However, Kluge and Osmond (1972) showed pyruvate-P_i-dikinase to be active in some CAM species. Therefore, it can be assumed that pyruvate might be converted to PEP, thus allowing carbohydrate synthesis via glucoseogenesis.

Hence, by the photosynthetic refixation of CO_2 produced from malic acid, the carbon flow in CAM proceeds from malic acid into end products of photosynthesis such as starch and other glucans (Sutton, 1974). This pathway is nearly identical with the C_4 pathway of photosynthesis (Hatch and Slack, 1970; Black, 1973).

3. Regulation

As can be seen from the scheme given in Fig. 1, problems of regulation results from the fact that free CO_2 is an intermediate in the flow of carbon from malic acid to carbohydrates. In the light, potentially two CO_2 fixing systems can operate.

On one hand, there is the PEP carboxylase/malate dehydrogenase system with malic acid as the end product, and, on the other hand, the Calvin cycle, producing carbohydrates as end products. The question then arises of how CO_2 produced by malic acid decarboxylation can be directed, nearly quantitatively, into carbohydrate synthesis via the Calvin cycle, rather than being trapped and recycled via PEP-carboxylase into the malic acid pool.

Considering that PEP-C has a substantially higher affinity for CO_2 than RuDP-C, and that in CAM plants the PEP-C is much more active than RuDP-C (see Kluge and Osmond, 1972), the problem discussed above becomes even more striking. In the C_4 plants, interference of the two CO_2 fixing pathways is prevented by spatial separation, that is, by localization of the PEP-C system in the mesophyll, and by the malate decarboxylation with concomitant refixation to the CO_2 via RuDP-C in the bundle sheath (see Hatch and Slack, 1970; Black, 1973). In CAM plants, however, the two CO_2 fixing pathways are separated by time. This is achieved by biochemical and physiological mechanisms of regulation.

At the moment, mainly two models are being discussed to explain mechanisms by which CAM might be regulated.

a) Endogenously Oscillating Enzyme Activities. This model is advocated by Queiroz and his coworkers (see Queiroz, 1974). These authors assume PEP-C activity to be low during the day in response to a basic diurnal oscillator yet unknown. According to this concept, PEP-C is not able to compete with RuDP-C for CO_2 during the day because of its temporally low activity due to the endogenous rhythm.

b) Feed-back Regulation of PEP-Carboxylase. This model (Kluge, 1969, 1971) is based on the finding that malate is a potent inhibitor of PEP-C (Ting, 1968; Kluge and Osmond, 1972). Thus, during the early hours of the light period, the PEP-C activity will be low due to malate-induced feed-back inhibition. This will last as long as the malic acid level of the tissue stays high (i.e., as long as CO_2 production from malic acid continues).

A more recent version of this model includes the malic acid fluxes across the tonoplast (Kluge and Heininger, 1973; Kluge and Lüttge, 1974; Lüttge et al., 1975). It has been postulated that the tonoplast fluxes control the cytoplasmic malic acid level which is finally responsible for the enzyme regulation. There is reasonable evidence that the tonoplast fluxes of malate might be controlled by a turgor mechanism (Kluge and Lüttge, 1974; Lüttge et al., 1975; Lüttge and Ball, 1974).

III. Gas Exchange of CAM Plants

The most peculiar feature of the gas exchange of CAM plants is the net fixation of atmospheric CO_2 during the night (see Fig. 2). In contrast to this, the CO_2 exchange occurring during the day is characterized by an extended period of depression where the curve approaches the compensation point. Hence, CAM plants usually fix more CO_2 from the atmosphere during the night than during the day, and in extreme cases, mainly in response to environmental conditions such as drought, CAM plants may take up CO_2 exclusively in the dark (see Fig. 3; cf. Neales, 1975).

This unusual pattern of CO_2 exchange is accompanied by an equally unusual pattern of stomatal movements. In contrast to the majority of higher plants, CAM plants open the stomata during the night and close them for a considerable period during the day (Nishida, 1963; Kluge and Fischer, 1967; see also Ting

Crassulacean Acid Metabolism (CAM): CO_2 and Water Economy 317

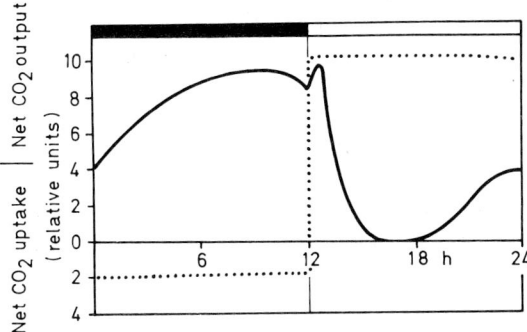

Fig. 2. Scheme of CO_2 exchange in CAM-plants (——) and non-CAM plants (........) in a 12:12 h dark:light rhythm

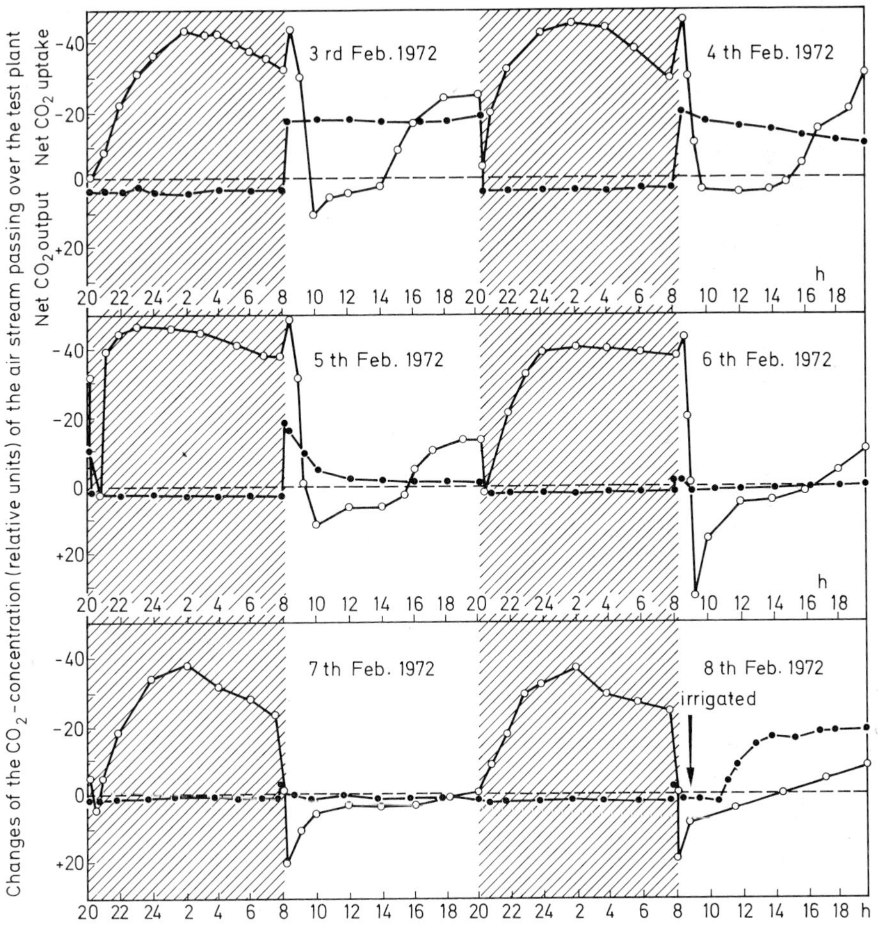

Fig. 3. CO_2 exchange of *Kalanchoë daigremontiana* a CAM-plant (—o—o—) and *Coleus sp.*, a non-CAM-plant (—●—●—) under experimental drought conditions. Last irrigation: Feb. 2, 8:00 p.m., beginning with the dark period (shaded area). Dark:light 12:12 h; 20,000 lux. (After Kluge, 1972)

et al., 1972). The interdependence of CO_2 exchange and the movement of the stomata is clearly demonstrated by the experiments of Nishida (1963); Kluge and Fischer (1967). These authors show that CO_2 exchange follows exactly the course of stomatal opening. However, in contrast to what one would automatically conclude from these results, Kluge and Fischer (1967) found that the basic pattern of CAM-type gas exchange is not determined by the stomatal behavior.

On the contrary, in CAM plants the movements of the stomata are controlled by the CO_2 exchange of the mesophyll cells, and again the CO_2 exchange of these cells is a function of the biochemical events of CAM. Hence, taking into account that stomata close when the partial pressure of CO_2 in the subepidermal spaces is high, and that they open when it is low (Raschke, 1966), the relationship between the CO_2 exchange of the mesophyll, on one hand, and the stomatal movements, on the other, can be explained as follows (see also Raschke, 1966; Kluge and Fischer, 1967; Levitt, this volume Part 3:C):

During the night, CO_2 is removed from the subepidermal spaces by the very active CO_2 dark fixation. Thus, the stomata will open as long as CO_2 is fixed and malic acid is synthesized. During the day, the stomata close because CO_2 is produced by malic acid decarboxylation, and with the rate of the photosynthetical refixation being a limiting factor, the CO_2 will increase in the subepidermal spaces. This hypothesis is evidenced by the finding that the extension of the depression in CO_2 uptake during the day is very closely correlated with the period of malic acid consumption, that is, with the production of internal CO_2 (Kluge, 1968).

Since on the one hand the stomatal movements are regulated by the carbon metabolism of CAM, and on the other the stomatal movements control the transpiratory water loss of the leaves, the water balance is directly affected and controlled by the carbon metabolism in CAM plants.

IV. Ecological Aspects of CAM

The typical habitats of CAM plants are arid environments. Those higher plants which occupy biotops where water is deficient are confronted with the dilemma that either the water balance is endangered if the stomata are opened during the day to facilitate photosynthesis, or the carbon balance becomes negative if the plant closes the stomata during day in order to prevent transpiratory water loss. However, by fixing CO_2 from the atmosphere and opening the stomata by night while closing them by day, CAM plants avoid this difficulty. It is obvious that the opening of stomata during the night, when evaporation is low, reduces the transpiratory water loss significantly, compared to the transpiration that would occur during the day. Ting et al. (1972) have calculated that the water loss for a given epidermal diffusion resistance would be increased three-fold in desert environments of Arizona if the plants opened the stomata during the day rather than during the night. The fact that CAM plants have a higher rate of carbon gain from the atmosphere per amount of water loss than plants without CAM can be ascribed to their inverted pattern of gas exchange (Ting et al., 1972). The higher "water use efficiency" of CAM plants is thus expressed by the transpiration/net photosynthesis (T/P) ratio (Ting et al., 1972; Szarek and Ting, 1975). Some typical T/P ratios of CAM and non-CAM plants are given in Table 1 (see also Winter and Lüttge, this volume Part 5:B).

Table 1. Transpiration/net photosynthesis (T/P) of CAM-, C_4- and C_3-species, $T/P = H_2O$ loss $(g \cdot dm^{-2} \cdot h^{-1})/CO_2$ uptake $(g \cdot dm^{-2} \cdot h^{-1})$. (Data from Szarek and Ting, 1975)

Type of Species	Day	Night
CAM	150–600	25–150
C_4	250–350	—
C_3	450–600	—

The ecological advantage of the CAM type of gas exchange becomes most striking if CAM plants are experimentally subjected to drought periods (Fig. 3). It can be seen that under water stress the daytime CO_2 uptake with open stomata is increasingly reduced. In contrast to this, the nocturnal CO_2 uptake is clearly less affected by drought (i.e., by low water content of the soil), and after some period of water stress, carbon dioxide is taken up exclusively during the night (see also Ting et al., 1972; Neales, 1975; Bartholomew, 1973). Szarek et al. (1973) showed that in its natural habitat *Opuntia basilaris* finally reduces its nocturnal CO_2 uptake completely if the drought period continues for weeks and months. Therefore, during extended drought, there is neither important carbon loss nor carbon gain. Respiratory CO_2 produced during the night is retained in the plant and recycled via CAM. On the other hand, *O. basilaris* can respond to precipitation very quickly. After one heavy rainfall the opening of stomata at night is initiated immediately, thus reestablishing the normal CAM pattern of gas exchange (Szarek and Ting, 1974).

From all these results it can be deduced that the major advantage of CAM is to maintain a positive carbon balance or at least to prevent a negative balance even during longer periods of drought. In contrast, it can be seen from Fig. 3, in which the gas exchange of a CAM plant and a plant without CAM are directly compared during experimental water stress, that the mesophytic plants without CAM exhibit a negative carbon balance after relatively short drought treatment.

As outlined by Ting (1972), CAM plants generally show slow growth rates, which means low primary production in natural habitat. This makes it obvious that CAM is not a mechanism for providing high rates of photosynthesis (as is true for the metabolism of C_4 plants), but rather for maintaining, as pointed out earlier, a most advantageous carbon balance during seasonal periods when precipitation is rare or non-existent. This hypothesis is verified by the fact that typical habitats of CAM plants are found in climatic zones where periods of drought alternate regularly with periods of precipitation. This is true, for example, in the summer dry deserts of California, where the ecology of CAM plants has been intensively studied (Szarek et al., 1973; Szarek and Ting, 1974). In hot deserts where precipitation is low and irregular, CAM plants are less abundant and more restricted to specific habitats. Lange et al. (1975) studied one of these rare exceptions *(Caralluma)* in the Negev deserts. They showed that the carbon gain of this species is minimal under the desert conditions of the Negev.

V. Conclusions

There is now sufficient evidence to allow CAM to be interpreted as an adaptive variant of the photosynthetic pathway of carbon fixation, as is also true for the C_4 pathway of photosynthesis. The example of the CAM plants shows that plants conquer arid environments and establish themselves ecologically not only by means of adaptive anatomical structures such as specialized leaf anatomy and morphology (Haberlandt, 1904), but also by specialized metabolic pathways (biochemical adaptations) which guarantee a maximum possible carbon gain combined with a minimum loss of water. The CAM of the succulent plants represents such a mechanism.

Notation

CAM: Crassulacean acid metabolism
MDH: Malate dehydrogenase
OAA: Oxaloacetate
PEP: Phosphoenol-pyruvate
PEP-C: Phosphoenol-pyruvate carboxylase
3-PGA: 3-Phosphoglyceric acid
RuDP: Ribulose-1,5-diphosphate
RuDP-C: Ribulose-1,5-diphosphate carboxylase

References

Bartholomew, B.: Drought response in the gas exchange of *Dudleya farinosa* (Crassulaceae) grown under natural conditions. Photosynthetica **7**, 114–120 (1973).
Black, C. C.: Photosynthetic carbon fixation in relation to net CO_2 uptake. Ann. Rev. Plant Physiol. **24**, 253–282 (1973).
Bradbeer, J. W., Ranson, S. L., Stiller, M. C.: Malate synthesis in Crassulacean leaves. I. The distribution of C^{14} in malate of leaves exposed to $C^{14}CO_2$ in the dark. Plant Physiol. **33**, 66–70 (1958).
Cockburn, W., McAulay, A.: The pathway of carbon dioxide fixation in Crassulacean plants. Plant Physiol. **55**, 87–89 (1975).
Dittrich, P., Campbell, W. H., Black, C. C.: Phosphoenolpyruvate carboxykinase in plants exhibiting Crassulacean acid metabolism. Plant Physiol. **52**, 357–361 (1973).
Haberlandt, G.: Physiologische Pflanzenanatomie, 3rd ed. Leipzig: Verlag W. Engelmann 1904.
Hatch, M. D., Slack, C. R.: Photosynthetic CO_2-fixation pathways. Ann. Rev. Plant Physiol. **21**, 141–162 (1970).

Kluge, M.: Untersuchungen über den Gaswechsel von *Bryophyllum* während der Lichtperiode. II. Beziehungen zwischen dem Malatgehalt des Blattgewebes und der CO_2-Aufnahme. Planta (Berl.) **80**, 359–377 (1968).

Kluge, M.: Veränderliche Markierungsmuster bei $^{14}CO_2$-Fütterung von *Bryophyllum tubiflorum* zu verschiedenen Zeitpunkten der Hell-Dunkelperiode. I. Die $^{14}CO_2$-Fixierung unter Belichtung. Planta (Berl.) **88**, 142–150 (1969).

Kluge, M.: Studies on CO_2 fixation by succulent plants in the light. In: Photosynthesis and photorespiration (eds. M. D. Hatch, C. B. Osmond, R. C. Slatyer), pp. 283–287. New York: Wiley Interscience 1971.

Kluge, M.: Die Sukkulenten: Spezialisten im CO_2-Gaswechsel. Biologie in unserer Zeit **4**, 120–129 (1972).

Kluge, M., Fischer, K.: Über Zusammenhänge zwischen dem CO_2-Austausch und der Abgabe von Wasserdampf durch *Bryophyllum daigremontianum* Berg. Planta (Berl.) **77**, 212–223 (1967).

Kluge, M., Heininger, B.: Untersuchungen an *Bryophyllum* über den Efflux von Malat aus den Vacuolen der assimilierenden Zellen und mögliche Einflüsse dieses Vorgangs auf den CAM. Planta (Berl.) **113**, 333–343 (1973).

Kluge, M., Kriebitsch, Ch., Willert, D. v.: Dark fixation of CO_2 in Crassulacean acid metabolism. Are two carboxylation steps involved? Z. Pflanzenphysiol. **72**, 460–463 (1974).

Kluge, M., Lüttge, U.: The role of malic acid fluxes in regulation of Crassulacean acid metabolism. In: Membrane transport in plants. (eds. M. Zimmermann, J. Dainty), p. 101. Berlin-Heidelberg-New York: Springer 1974.

Kluge, M., Osmond, C. B.: Pyruvate P_i dikinase in Crassulacean acid metabolism. Naturwissenschaften **58**, 414–415 (1971).

Kluge, M., Osmond, C. B.: Studies on phosphoenolpyruvate carboxylase and other enzymes of Crassulacean acid metabolism of *Bryophyllum tubiflorum* and *Sedum praealtum*. Z. Pflanzenphysiol. **66**, 97–105 (1972).

Lange, O. L., Schulze, E.-D., Kappen, L., Evenari, M., Buschbom, U.: CO_2-exchange pattern under natural conditions of *Caralluma negevensis*, a CAM plant of the Negev desert. Photosynthetica **9**, 318–326 (1975).

Lüttge, U., Ball, E.: Proton and malate fluxes in cells of *Bryophyllum daigremontianum* leaf slices in relation to potential osmotic pressure of the medium. Z. Pflanzenphysiol. **73**, 326–338 (1974).

Lüttge, U., Kluge, M., Ball, E.: Osmoregulation of vacuolar malic acid storage: a basic principle of the oscillatory behaviour of Crassulacean acid metabolism. Plant Physiol. (submitted, 1975).

Neales, T. F.: Effect of ambient CO_2 concentration on rate of transpiration of *Agave americana* in the dark. Nature (London) **228**, 880–882 (1970).

Neales, T. F.: The gas exchange patterns of CAM plants. In: Environmental and biological control of photosynthesis (ed. E. Marcelle). Proc. Conf. Limburgs Univ. Cent., Diepenbeck, Belgium Aug. 26–30, 1974, pp. 299–310. The Hague: W. Junk 1975.

Nishida, K.: Studies on stomatal movement of Crassulacean plants in relation to acid metabolism. Physiol. Plantarum **16**, 281 (1963).

Queiroz, O.: Circadian rhythms and metabolic patterns. Ann. Rev. Plant Physiol. **25**, 115–134 (1974).

Ranson, S. C., Thomas, M.: Crassulacean acid metabolism. Ann. Rev. Plant Physiol. **11**, 81–110 (1960).

Raschke, K.: Die Reaktionen des CO_2-Regelsystems in den Schließzellen von *Zea mays* auf weißes Licht. Planta (Berl.) **68**, 111–140 (1966).

Sutton, B. G.: Regulation of carbohydrate metabolism in succulent plants. Thesis Australian Nat. Univ., Canberra 1974.

Sutton, B. G., Osmond, C. B.: Dark fixation of CO_2 by Crassulacean plants. Evidence for a single carboxylation step. Plant Physiol. **50**, 360–365 (1972).

Szarek, S. R., Johnson, H. B., Ting, I. P.: Drought adaptations in *Opuntia basilaris*. Plant Physiol. **52**, 539 (1973).

Szarek, S. R., Ting, J. P.: Seasonal patterns of acid metabolism and gas exchange in *Opuntia basilaris*. Plant Physiol. **54**, 76–81 (1974).

Szarek, S. R., Ting, J. P.: Photosynthetic efficiency of CAM plants in relation to C_3 and C_4 plants. In: Environmental and biological control of photosynthesis (ed. E. Marcelle). Proc. Conf. Limburgs Univ. Cent., Diepenbeck, Belgium Aug. 26–30, 1974, pp. 289–297. The Hague: W. Junk 1975.

Ting, I.: CO_2 metabolism in corn roots. III. Inhibition of P-enolpyruvate carboxylase by L-malate. Plant Physiol. **43**, 1919–1924 (1968).

Ting, I.: Non autotrophic CO_2 fixation and Crassulacean acid metabolism. In: Photosynthesis and photorespiration (eds. M. D. Hatch, C. B. Osmond, R. O. Slatyer), pp. 169–185. New York-London-Sydney-Toronto: Wiley Interscience 1971.

Ting, I., Johnson, H. B., Szarek, S. R.: Net CO_2 fixation in Crassulacean acid metabolism plants. Proc. Symp. South. Sec. Am. Soc. Plant Physiol. (ed. C. C. Black), pp. 22–53. Mobiles, North Carolina: Univ. S. Alabama 1972.

B. Balance Between C_3 and CAM Pathway of Photosynthesis

K. WINTER and U. LÜTTGE

I. Introduction

Apart from primary photosynthetic CO_2 fixation by RuDP carboxylase ("C_3 photosynthesis"), photosynthetic mechanisms are observed among higher plants by which primary CO_2 fixation is catalyzed by PEP carboxylase with formation of oxalo-acetic acid (OAA). OAA is reduced to malate and, subsequently, malate is decarboxylated, and the resulting CO_2 is refixed by RuDP carboxylase. Malate synthesis and malate decarboxylation with subsequent refixation of CO_2 may be separated spatially in particularly differentiated tissues of a leaf ("C_4 photosynthesis") or they may be separated by time, that is, they occur in a given cell during different phases of a rhythm (Crassulacean acid metabolism, "CAM") (see Kluge, this volume Part 5:A).

The question has been asked whether these modes of photosynthesis are determined genetically and whether shifting between these mechanisms occurs in nature or can be triggered experimentally. It has been attempted to affect the balance between primary CO_2 fixation by RuDP carboxylase and by PEP carboxylase by treating young seedlings of C_3- and C_4-plants with various hormones (see Huber and Sankhla, this volume Part 5:C). Crossing and breeding experiments have revealed that C_4 photosynthesis is associated with a syndrome of genetically determined structural, physiological and biochemical properties (Björkman et al., 1971). This suggests that genetically intermediate forms between C_3- and C_4-photosynthesis are unlikely to occur in different species or different varieties of a species. Hence, it is even more difficult to envisage that in a given species or variety net gain of organic matter can be attained alternatively by C_3- and C_4-photosynthesis.

A shift in the prevailing mode of photosynthesis due to environmental factors is, a priori, more likely to occur between CAM and C_3. In CAM, phases with predominant operation of PEP carboxylase alternate with those of predominant operation of RuDP carboxylase. In the latter, not only CO_2 from malate decarboxylation but also CO_2 from the atmosphere can be fixed, depending on the physiological conditions (Osmond and Allaway, 1974). It is well known from the literature that environmental conditions may interact with the short-term regulation of this rhythm. In addition, environmental factors appear to be highly important in long-term regulation of the balance between C_3 photosynthesis and CAM. This suggests a high degree of ecological adaptability of CAM plants (for a review and literature survey, see Kluge, 1974). A convincing example is that in *Kalanchoë blossfeldiana* the establishment of CAM rhythmicity is under phytochrome-dependent photoperiodic control. This plant displays C_3 photosynthesis under long days and CAM under short days (Queiroz and Morel, 1974). By an apparently quite different mechanism, the halophyte *Mesembryanthemum crystallinum* can be shifted from C_3 to CAM when grown under highly saline conditions (Winter and von Willert, 1972). This interesting system will be assessed in the present review.

II. Adaptation to Salinity

The Aizoaceae *Mesembryanthemum crystallinum* L. was originally common in South Africa but has spread throughout areas having a "mediterranean climate". Figure 1 shows experimental plants that were grown for 2 weeks in various

Fig. 1. 7-weeks-old *M. crystallinum* plants treated for 2 weeks with 0, 100, 200, 400 and 600 mM NaCl respectively (from left to right). Note that the best growth is obtained at 100 mM NaCl. (From Winter, 1975)

culture solutions containing up to 600 mM NaCl (the salt was added in steps of 50 mM per day). As a halophyte, *M. crystallinum* not only tolerates high levels of NaCl but actually requires about 100 mM NaCl for maximal growth under the environmental conditions used (12 h light, 16,000 lux Xenon, 25°C, 50% relative humidity; 12 h dark, 15°C, 65% relative humidity). Shoot fresh weight is 2.4 times, and dry weight 1.9 times greater during growth in 100 mM NaCl than in the absence of added NaCl. Accumulation of Na^+ and Cl^- from the nutrient solution lowers the osmotic potential of the cells, so maintaining a gradient in water potential between the plant and the growth medium (Winter, 1975).

III. Environmental Control of Photosynthetic Pathways

1. Salinity Effects

a) **Gas Exchange.** Different levels of NaCl in the culture solution greatly affect the CO_2 gas exchange and the transpiration patterns of *M. crystallinum* during the daily light-dark cycle (Fig. 2; Winter, 1975). Net CO_2 gas exchange of the plants grown without added NaCl or with 100 mM NaCl in the growth medium is characterized by a more or less constant net CO_2 uptake during the light period and a constant net CO_2 production during the dark period. By contrast, net CO_2 gas exchange patterns of the plants grown with 200, 400 and 600 mM NaCl indicate typical features of CAM with a pronounced depression of net CO_2 uptake during the day and a net CO_2 uptake during the night. The ratio of night-time to day-time net CO_2 fixation reaches the highest values at 600 mM NaCl. Figure 2, depicting these patterns of CO_2 gas exchange for individual plants, shows a wide range of absolute rates of CO_2 exchange per plant which is due to differences in shoot development under various conditions of salinity (cf. Fig. 1). Figure 3 shows a more quantitative comparison. If CO_2 assimilation is expressed on the basis of fresh weight, dry weight or leaf area,

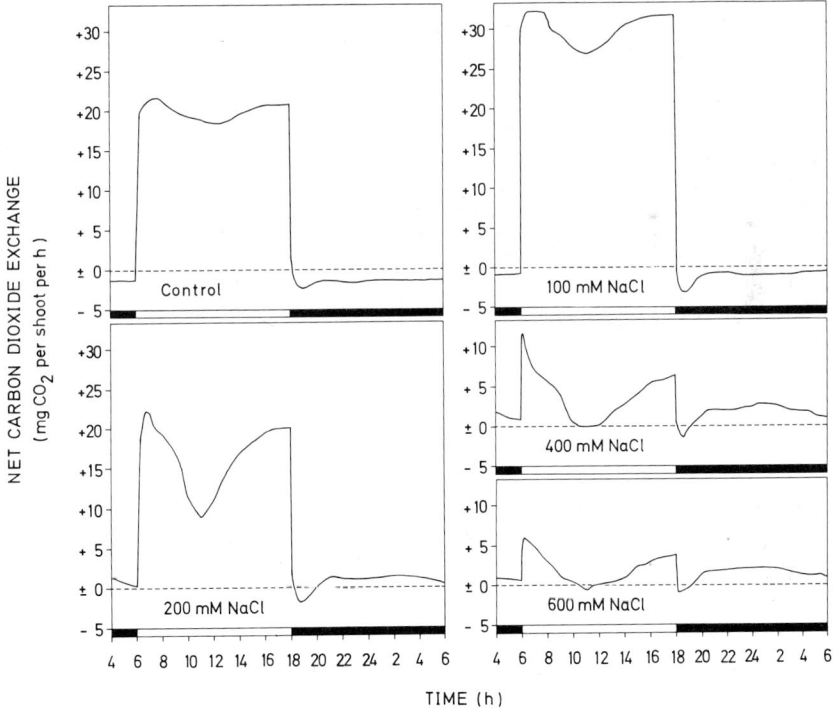

Fig. 2. Net CO_2 gas exchange of the plants shown in Fig. 1. Positive values refer to net CO_2 uptake, negative values represent net CO_2 loss from the plants. (From Winter, 1975)

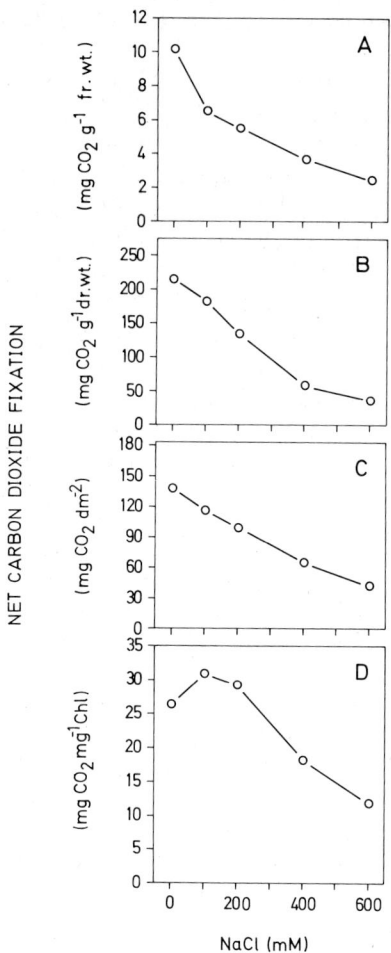

Fig. 3 A–D. Net CO_2 fixation over 24 h (12 h light plus 12 h dark) of the *M. crystallinum* shoots shown in Fig. 1. Net CO_2 assimilation, calculated by integrating the curves in Fig. 2, is expressed on a fresh weight (A), dry weight (B), leaf area (C) and chlorophyll (D) basis. (From Winter, 1975)

then net carbon gain over 24 h decreases with increased salinity, notwithstanding the fact that the plants grow best at 100 mM NaCl. If net carbon gain over 24 h is expressed on a chlorophyll basis, however, then there is a correlation with plant productivity (Winter, 1975).

The loss of water vapor (transpiration) responds to saline conditions in the culture solution in a way similar to CO_2 uptake (Winter, 1975). This confirms that NaCl-treated *M. crystallinum* plants display the inverted stomatal rhythm characteristic for CAM. An ecologically important consequence of this gas-exchange pattern is an approximately two-fold increase in water use efficiency, with the assimilation-transpiration quotient (mg CO_2 g H_2O^{-1}, calculated over a 24h period) rising from <10 in the absence of added NaCl to ~17 at high NaCl levels.

b) Metabolism. As a result of increased night-time CO_2 fixation, large diurnal fluctuations in the levels of malate are observed in the mesophyll of plants treated with high salt concentrations. The malate levels range from ~5 μM per g fresh wt, at the end of the light phase to ~30 μM per g fresh wt, at the end of the dark phase (Winter, 1973a, 1973b). Similar diurnal malate fluctuations were observed in plants grown under natural conditions at the seashore of Tel Aviv (Israel) (Winter, 1975). By contrast, in glycophytically grown control plants, the malate level is constant at about 1 μM per g fr. wt. throughout the day. During 15 min exposure to $^{14}CO_2$ in the dark the leaves of salt-treated plants (400 mM NaCl) incorporate about 10 times more ^{14}C than the control plants (Winter et al., 1974). In the control plants, only 32% of the total ^{14}C recovered is detectable in malate, whereas malate is the major labeled compound (75%) in the salt-treated plants. A pulse-chase experiment shows that the malate labeled in the salt-treated plants in the dark does not appreciably equilibrate with acids of the TCA cycle or with other metabolic pools (Fig. 4A). In the light, the ^{14}C is transferred from malate into carbohydrates (Fig. 4B).

This behavior suggests an involvement of PEP carboxylase in salt induction of net CO_2 dark fixation in *M. crystallinum*. Treichel et al. (1974) detected a considerable increase in specific activity of PEP carboxylase in crude extracts of plants grown under highly saline conditions, as compared with crude extracts obtained from control plants.

During CO_2 fixation, PEP carboxylase discriminates against the C isotope ^{13}C to a lesser extent than RuDP carboxylase (Whelan et al., 1973). Control plants of *M. crystallinum* and plants treated with 100 mM NaCl show $\delta^{13}C$ values of −28 to −30‰, irrespective of the thermoperiod applied together with the photoperiod (i.e., 12h light, 25°C: 12h dark, 30°C vs 12h light 25°C: 12h dark, 5°C). Plants treated with 400 mM NaCl in a photo-thermo-period of 12h light, 25°C: 12h dark, 5°C display net CO_2 fixation during the dark

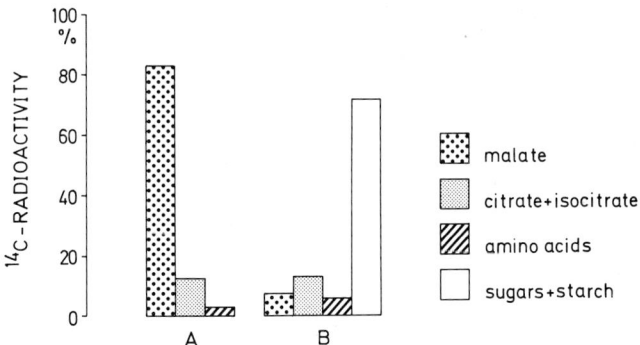

Fig. 4. Percentage distribution of ^{14}C radioactivity in leaves of a salt-treated (400 mM NaCl) *M. crystallinum* plant after exposure to a pulse of $^{14}CO_2$ in the dark for 15 min, followed by a 4h chase in ^{14}C free atmosphere in the dark *A*, or by a chase in ^{14}C-free atmosphere for 4h in the dark plus 12h in the light *B*. (From Winter, 1975)

phase and have a $\delta^{13}C$ of $\sim -24‰$, which is less negative than that of the controls and of the 100 mM NaCl plants. This is further evidence in favor of an increased contribution of primary CO_2 fixation via PEP carboxylase to total carbon gain of salt-treated *M. crystallinum* plants. No net CO_2 fixation during the dark phase is observed when the 400 mM NaCl plants are kept in a rhythm of 12 h light, 25° C/12 h dark, 30° C, stressing the role of temperature in short-term regulation of CAM (Kluge, 1974). The $\delta^{13}C$ of these plants is $\sim -29‰$, as in the controls (Osmond et al., 1973; K. Winter, J. H. Troughton, and U. Lüttge, unpublished results). In general, the $\delta^{13}C$ values reported in the literature for CAM plants are less negative than the values given above for *M. crystallinum* performing CAM. However, in contrast to constitutive CAM plants, the *M. crystallinum* plants that were used in determining the above values were induced to perform CAM when the experimental leaves were almost mature and also contained carbon that had been previously fixed, predominantly by RuDP-carboxylase. Nevertheless, the differences in $\delta^{13}C$ values of *M. crystallinum* plants exhibiting CAM or C_3 photosynthesis are significant. It is expected that a still more pronounced correlation between photosynthetic pathways and $\delta^{13}C$ values would be obtained in this system by using aqueous extracts rather than the whole leaves for $\delta^{13}C$ determinations (Lerman et al., 1974).

2. Water Stress Effects

High Na_2SO_4, KCl and K_2SO_4 levels in the culture medium have effects on net CO_2 gas exchange in *M. crystallinum* similar to that of high NaCl salinity (Winter, 1973c). This suggests that the induction of CAM in *M. crystallinum* is not a Na^+ or Cl^- specific effect. Salt may act indirectly by changing the water relations of the plants. To test this possibility, *M. crystallinum* plants grown in the absence of added NaCl were subjected to water stress by keeping them for some time in a cooled nutrient solution (10° C) or in a nutrient solution of low oxygen content (i.e., without aeration). Both treatments are well known to reduce the water uptake of many plants (Kramer, 1969). They greatly diminish growth of *M. crystallinum* (Winter, 1974a, 1974b).

Net CO_2 gas exchange of plants grown in a cooled nutrient medium exhibits typical features of CAM with net CO_2 uptake during the night (Winter, 1974a). At the same time, specific activity of PEP carboxylase increases considerably (Winter, 1974b). That *M. crystallinum* plants perform CAM when kept under conditions of reduced water supply is also borne out by measurements of malate levels, which oscillate in a diurnal fashion in plants having cooled or low oxygen treated roots, but remain constant in the controls (Fig. 5A). Pulse-chase experiments on $^{14}CO_2$ fixation (Fig. 5B) reveal a pattern similar to that shown in Fig. 4A, and thus further confirm that *M. crystallinum* responds similarly to salinity and other conditions that diminish water supply.

IV. Regulation of the Balance Between C_3 and CAM

The phenomenon of CAM induction in *M. crystallinum* confronts us with the problem of what the mechanism of the regulation of the balance between

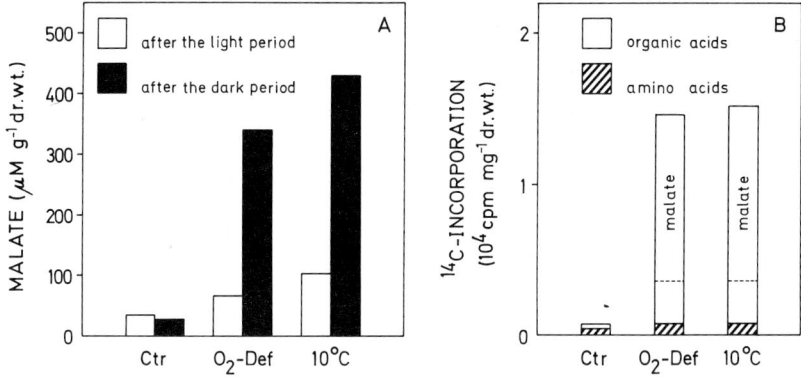

Fig. 5 A and B. Effects of water stress caused by cooling or by O_2-deficiency of the nutrient solution of *M. crystallinum* plants. (A) Diurnal variation of malate levels in the leaves. (B) $^{14}CO_2$ dark fixation, where the leaves were exposed to $^{14}CO_2$ in a 30 min pulse, followed by a 4 h chase in ^{14}C-free atmosphere. (From Winter, 1975)

C_3 and CAM is. Since no definite conclusions are available regarding the mechanism of CAM regulation per se (Queiroz and Morel, 1974; Lüttge et al., 1975b), it is certainly too early to present a model explaining the shift between C_3 and CAM. However, a few possibilities can be evaluated.

1. Balance of Electrical Charge

Malate synthesis by PEP carboxylase is known to play an important role in avoiding imbalance of electric charge within plant cells caused by ion transport and metabolism of mineral anions (e.g., NO_3^-, PO_4^{3-}, SO_4^{2-}) (recent reviews: Osmond, 1975; Smith and Raven, 1975). However, the diurnal fluctuations of malate in salt-treated *M. crystallinum* are not paralleled by diurnal fluctuations of Na^+, K^+ and Cl^-. Furthermore, titratable protons are found to fluctuate in a stoichiometric relation to malate ($2H^+ : 1$ $malate^{2-}$) (Lüttge et al., 1975a). Thus, in salt-induced CAM of *M. crystallinum*, charge balance is maintained by simultaneous fluctuations of H^+ and $malate^{2-}$ (malic acid fluctuations). Therefore, it is unlikely that CAM is induced as a consequence of charge balance requirements.

2. Succulence

In halophytes, increased succulence has often been reported as a means to avoid high salt concentrations within cells (Jennings, 1968). Succulence of the assimilatory organs seems to be a prerequisite for CAM, because succulence is thought to be equivalent to a high capacity of the vacuoles for accumulation of malate during the dark. In this way, an overflow of malate in the cytoplasm, which might result in a premature feed-back inhibition of PEP carboxylase, appears to be prevented (Kluge and Osmond, 1972; but see Kluge et al., 1973).

In *M. crystallinum*, the induction of CAM by salinity cannot be correlated unequivocally with increased leaf succulence. A gradual increase in succulence, as expressed by the ratio surface area/water content, is apparent when salinity is increased (Winter, 1973a). However, this correlation is not seen when succulence is expressed as the ratio of fresh weight to dry weight or H_2O/chlorophyll; these ratios attain their highest values during optimum growth in 100 mM NaCl, when the plants do not perform CAM. Furthermore, the response of CAM in 400 mM NaCl-treated *M. crystallinum* plants to changes of the environmental conditions is often inversely related to water content or "succulence". If such plants are subjected to additional water stress by removal of the roots from the nutrient solution, leaf water content decreases, but night-time net CO_2 fixation increases considerably (Winter, 1974c). Conversely, re-transfer to a nutrient solution free of added NaCl may lead to an appreciable increase of leaf water content with a concomitant decrease and gradual disappearance of night-time net CO_2 fixation within a few days (Winter, 1974c). The large bladder cells on the upper and lower leaf surfaces make an evaluation of leaf succulence in *M. crystallinum* in relation to CAM still more difficult. The bladder cells are not involved in the diurnal malate fluctuations, but they make a considerable contribution to the total leaf water content (Fig. 6).

3. Water Stress and Phytohormones

It has been shown above that water stress appears to be the most important factor regulating the balance between C_3 and CAM in *M. crystallinum*. The response of plants to water stress in general seems to include phytohormonal effects, especially via the balance between cytokinins and abscisic acid (ABA), which are also involved in senescence (see Itai and Benzioni,

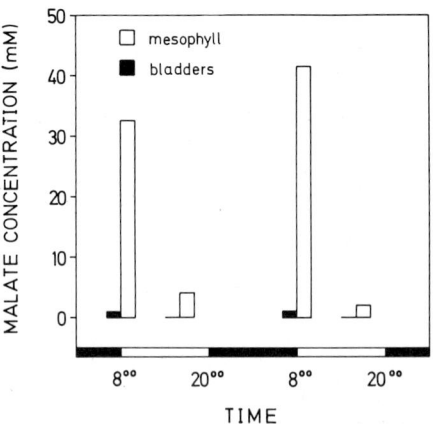

Fig. 6. Diurnal fluctuations of malate levels (mM/l) in leaf tissue and in bladder cells of 400 mM NaCl-treated *M. crystallinum* plants. The small amount of malate in the bladder cells at 8:00 a.m. is presumably contamination with sap from the mesophyll

this volume Part 4:B). This hormonal balance may play a role in regulation of the balance between PEP carboxylase activity and RuDP carboxylase activity in seedlings (see Huber and Sankhla, this volume Part 5:C). It also may be involved in the well-known phenomenon that, in typical CAM plants, CAM develops with leaf age (Wolf, 1932; Avadhani et al., 1971; Jones, 1975). In *M. crystallinum*, salt-induced CAM appears first in mature leaves and only later in the younger leaves, as they develop (Winter, 1973a). Nothing is known about the role of phytohormones in CAM, and surprisingly little information is available on the hormonal balance in halophytes (Waisel, 1972). Thus far, all attempts to induce features of CAM in *M. crystallinum* by spraying the leaves with ABA or by adding ABA to the culture solution or the use of leaf slices to investigate possible effects of ABA on $^{14}CO_2$ dark fixation have not given convincing evidence (Winter, 1975).

4. Osmoregulation

Leaf slices of salt-treated *M. crystallinum* plants (von Willert, 1973) and of the CAM plant *Kalanchoë daigremontiana* (Lüttge et al., 1975b) were floated in external solutions of varied mannitol or sorbitol concentration to study effects of osmotic pressure difference between the cells and the medium. These investigations suggest that the diurnal fluctuations of malic acid levels in CAM leaves are under osmoregulatory control and that turgor pressure may affect malic acid fluxes at the tonoplast (Lüttge et al., 1975b).

The observation that CAM can be induced in *M. crystallinum* in vivo by conditions of water stress appears to agree with this hypothesis, and hence

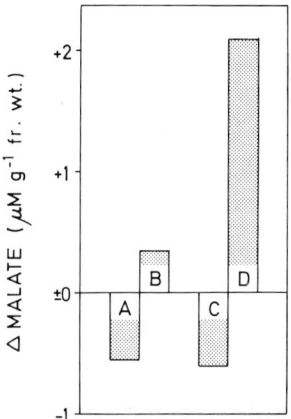

Fig. 7. Changes in malate levels (Δ malate) in leaf slices of glycophytically-grown *M. crystallinum* plants after incubation in media at 15°C of high (200 mM mannitol) and low (0 mM mannitol) osmotic pressure, respectively. Incubation medium: 0.1 mM $CaSO_4$ + 5 mM tris-HCl (pH 7.5); treatments: *A* 12 h dark (0 mM mannitol), *B* 12 h dark (200 mM mannitol), *C* B + 10 h light (0 mM mannitol), *D* B + 10 h light (200 mM mannitol)

this may not only be relevant for short-term regulation but also for long-term induction of CAM in a system like that of *M. crystallinum*. In experiments with leaf slices of glycophytically grown *M. crystallinum*, exhibiting no CAM-like diurnal fluctuations in vivo, the leaf slices synthesize and accumulate malate when the leaf water content is reduced through an increase of the osmotic pressure of the incubation medium (von Willert, 1974; Fig. 7). Surprisingly, this malate accumulation is much higher in the light than in the dark, and malate accumulation in the dark is not followed by a decrease of malate level in the light (Fig. 7; Winter, 1975). Therefore, this rapid response of malate metabolism in leaf slices to water stress is not directly comparable with malate metabolism during CAM. Blue light interactions with PEP carboxylase and organic acid metabolism (Kamiya and Miyachi, 1974), and/or the increased synthesis of PEP as a CO_2 acceptor via RuDP carboxylase and phosphoglyceric acid (Osmond and Allaway, 1974), may be involved in the yet unexplained effect of light in the experiment of Fig. 7.

This demonstrates that we are only beginning to understand the balance between C_3 and CAM in *M. crystallinum*. A crucial question in this context is, how alterations in plant water status (resulting in a changed cell water content, cell volume and turgor pressure etc.) affect the enzymatic machinery.

V. Ecological Aspects

A shift from C_3 to CAM could occur during the life cycle of *M. crystallinum* growing in its natural habitats. This annual species usually germinates in the humid time of the year and dies some months later in the subsequent drought period. It may be that under those conditions, *M. crystallinum* fixes CO_2 mainly by the C_3 pathway during the first weeks after germination, i.e. in the humid period, and changes to CAM when the drought season starts.

On the basis of our results with *M. crystallinum* it could be supposed that CAM is a predominant mechanism by which plants adapt to high soil salinity. However this is not so. A recent survey on coastal and desert plants of Israel and the Sinai shows that, among halophytes, the capacity to perform CAM is mainly restricted to members of the Aizoaceae (Winter, 1975).

References

Avadhani, P. N., Osmond, C. B., Tan, K. K.: Crassulacean acid metabolism and the C_4 pathway of photosynthesis in succulent plants. In: Photosynthesis and photorespiration (eds. M. D. Hatch, C. B. Osmond, R. O. Slatyer), pp. 288–293. New York-London-Sydney-Toronto: Wiley Interscience 1971.

Björkman, O., Nobs, M., Pearcy, R., Boynton, J., Berry, J.: Characteristics of hybrids between C_3 and C_4 species of *Atriplex*. In: Photosynthesis and photorespiration (eds. M. D. Hatch, C. B. Osmond, R. O. Slatyer), pp. 105–119. New York-London-Sydney-Toronto: Wiley Interscience 1971.

Jennings, D. H.: Halophytes, succulence and sodium in plants—a unified theory. New Phytologist **67**, 899–911 (1968).

Jones, M. B.: The effect of leaf age on leaf resistance and CO_2 exchange of the CAM plant *Bryophyllum fedtschenkoi*. Planta (Berl.) **123**, 91–96 (1975).

Kamiya, A., Miyachi, S.: Effects of blue light on respiration and carbon dioxide fixation in colorless *Chlorella* mutant cells. Plant Cell Physiol. (Tokyo) **15**, 927–937 (1974).

Kluge, M.: Metabolism of carbohydrates and organic acids. In: Progress in Botany (eds. H. Ellenberg, K. Esser, H. Merxmüller, E. Schnepf, H. Ziegler) Vol. 36, pp. 90–98. Berlin-Heidelberg-New York: Springer 1974.

Kluge, M., Lange, O. L., Eichmann, M. von, Schmid, R.: Diurnaler Säurerhythmus bei *Tillandsia usneoides*: Untersuchungen über den Weg des Kohlenstoffs sowie die Abhängigkeit des CO_2-Gaswechsels von Lichtintensität, Temperatur und Wassergehalt der Pflanze. Planta (Berl.) **112**, 357–372 (1973).

Kluge, M., Osmond, C. B.: Studies on phosphoenolpyruvate carboxylase and other enzymes of Crassulacean acid metabolism of *Bryophyllum tubiflorum* and *Sedum praealtum*. Z. Pflanzenphysiol. **66**, 97–105 (1972).

Kramer, P. J.: Plant and soil water relationship. New York: McGraw-Hill 1969.

Lerman, J. C., Deleens, E., Nato, N., Moyse, A.: Variation in the carbon isotope composition of a plant with Crassulacean acid metabolism. Plant Physiol. **53**, 581–584 (1974).

Lüttge, U., Ball, E., Tromballa, H. W.: K^+ independence of osmoregulated oscillations of malate^{2-} levels in the cells of CAM-leaves. Biochem. Physiol. Pflanzen **167**, 267–283 (1975a).

Lüttge, U., Kluge, M., Ball, E.: Osmoregulation of vacuolar malic acid storage: A basic principle in oscillatory behaviour of Crassulacean acid metabolism. Plant Physiol. **56**, 613–616 (1975b).

Osmond, C. B.: Ion accumulation and carbon metabolism in cells of higher plants. In: Encyclopedia of Plant Physiology New Series, Transport in plant cells and tissues (eds. U. Lüttge, M. G. Pitman) Chap. 19. Berlin-Heidelberg-New York: Springer 1975 (submitted).

Osmond, C. B., Allaway, W. G.: Pathways of CO_2 fixation in the CAM plant *Kalanchoë daigremontiana*. I. Patterns of $^{14}CO_2$ fixation in the light. Australian J. Plant Physiol. **1**, 503–511 (1974).

Osmond, C. B., Allaway, W. G., Sutton, B. G., Troughton, J. H., Queiroz, O., Lüttge, U., Winter, K.: Carbon isotope discrimination in photosynthesis of CAM plants. Nature **246**, 41–42 (1973).

Queiroz, O., Morel, C.: Photoperiodism and enzyme activity. Towards a model for the control of circadian metabolic rhythms in the Crassulacean acid metabolism. Plant Physiol. **53**, 596–602 (1974).

Smith, F. A., Raven, J. A.: H^+ transport and regulation of cell pH. In: Encyclopedia of Plant Physiology New Series, Transport in plant cells and tissues (eds. U. Lüttge, M. G. Pitman) Chap. 18. Berlin-Heidelberg-New York: Springer 1975 (submitted).

Treichel, S. P., Kirst, G. O., Willert, D. J. von: Veränderungen der Aktivität der Phosphoenolpyruvat-Carboxylase durch NaCl bei Halophyten verschiedener Biotope. Z. Pflanzenphysiol. **71**, 437–449 (1974).

Waisel, Y.: Biology of halophytes. New York-London: Academic Press 1972.

Whelan, T., Sackett, W. M., Benedict, C. R.: Enzymatic fractionation of carbon isotopes by phosphoenolpyruvate carboxylase from C_4 plants. Plant Physiol. **51**, 1051–1054 (1973).

Willert, D. J. von: Nächtliche Malatanhäufung in Blattstreifen von *Mesembryanthemum crystallinum* in wäßriger Lösung. I. Methode, Eigenschaften der Blattstreifen und ein Vergleich mit intakten Blättern. Ber. Deut. Botan. Ges. **86**, 477–483 (1973).

Willert, D. J. von: Der Säurestoffwechsel in Abhängigkeit von osmotischem Wert und NaCl-Belastung. Abstract: Tagung Deut. Botan. Ges., Würzburg, S. 103 (1974).

Winter, K.: CO_2-Fixierungsreaktionen bei der Salzpflanze *Mesembryanthemum crystallinum* unter variierten Außenbedingungen. Planta (Berl.) **114**, 75–85 (1973a).

Winter, K.: CO_2-Gaswechsel von an hohe Salinität adaptiertem *Mesembryanthemum crystallinum* bei Rückführung in glykisches Anzuchtmedium. Ber. Deut. Botan. Ges. **86**, 467–476 (1973b).

Winter, K.: Zum Problem der Ausbildung des Crassulaceensäurestoffwechsels bei *Mesembryanthemum crystallinum* unter NaCl-Einfluß. Planta (Berl.) **109**, 135–145 (1973c).

Winter, K.: Evidence for the significance of Crassulacean acid metabolism as an adaptive mechanism to water stress. Plant Sci. Letters **3**, 279–281 (1974a).

Winter, K.: Einfluß von Wasserstreß auf die Aktivität der Phosphoenolpyruvat-Carboxylase bei *Mesembryanthemum crystallinum*. Planta (Berl.) **121**, 147–153 (1974b).
Winter, K.: NaCl-induzierter Crassulaceen-Säurestoffwechsel bei der Salzpflanze *Mesembryanthemum crystallinum*. Abhängigkeit des CO_2-Gaswechsels von der Tag/Nacht-Temperatur und von der Wasserversorgung der Pflanzen. Oecologia **15**, 383–392 (1974c).
Winter, K.: Die Rolle des Crassulaceen-Säurestoffwechsels als biochemische Grundlage zur Anpassung von Halophyten an Standorte hoher Salinität. Dissertation Darmstadt, 1975.
Winter, K., Lüttge, U., Ball, E.: $^{14}CO_2$ dark fixation in the halophytic species *Mesembryanthemum crystallinum*. Biochim. Biophys. Acta **343**, 465–468 (1974).
Winter, K., Willert, D. J. von: NaCl induzierter Crassulaceensäurestoffwechsel bei *Mesembryanthemum crystallinum*. Z. Pflanzenphysiol. **67**, 166–170 (1972).
Wolf, J.: Beitrag zur Kenntnis des Säurestoffwechsels succulenter Crassulaceen. Planta (Berl.) **15**, 572–644 (1932).

C. C_4 Pathway and Regulation of the Balance Between C_4 and C_3 Metabolism

W. HUBER and N. SANKHLA

I. Introduction

Recent studies do not appear to support the validity of the concept of a single type of photosynthetic carbon assimilation occurring in all green plants. At least three pathways, namely, the C_3, C_4 and Crassulacean acid metabolism have been characterized (Hatch et al., 1971; Black, 1973; Laetsch, 1974). In addition, studies on epidermal tissue point towards operation of a photosynthetic process that has little to do with classical CO_2 reduction, and is primarily involved in the osmotic regulation of stomatal opening (Allaway, 1973; Willmer et al., 1973a, 1973b). Furthermore, there is evidence that, at least in some plants, a shift could occur from C_4 to C_3 or C_3 to CAM pathway (Kennedy and Laetsch, 1973; Khanna and Sinha, 1973; Laetsch, 1974; Sankhla and Huber, 1975). This shift seems to be regulated, at least partially, by factors such as light, temperature, leaf ontogeny, nutrition and substrate availability, and by the endogenous levels of bioregulants. Winter and Lüttge discuss the balance between the C_3 and CAM pathways of photosynthesis in Part 5:B of this volume. This report aims at elucidating the basic aspects of the physiology and biochemistry of the C_4 pathway, and at the same time evaluates the role of some environmental and internal factors in the regulation of balance between the C_3 pathway and C_4 pathway of carbon reduction. The possible selective advantages of C_4 photosynthesis for plant adaptation to environmental stress are also evaluated.

II. Carbon Metabolism of C_4 Plants

Recently, several reviews have been published on the current views on the biochemistry of the C_4 pathway of photosynthesis (Hatch et al., 1971; Coombs, 1971; Black, 1973; Björkman, 1973; Hatch and Boardman, 1973; Laetsch, 1974). The prominent features of the most widely accepted model of C_4 photosynthesis (Fig. 1) and the latest findings on the subject are briefly summarized in the following.

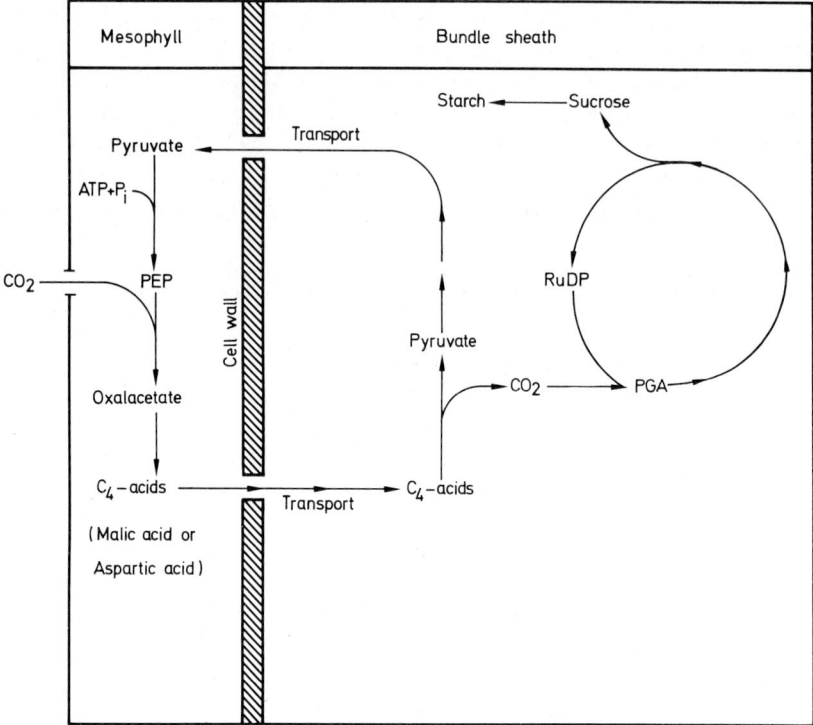

Fig. 1. The most widely accepted model of C_4 photosynthesis. (After Björkman, 1973)

It is well established (Kortschak et al., 1965) that when leaves of C_4 plants such as maize and sugarcane are fed with $^{14}CO_2$ in the light the radioactivity appears first in C_4-dicarboxylic acids, malate, and aspartate, and not in 3-PGA as in C_3 plants. This observation has been confirmed and extended to a number of other grasses and also to dicotyledonous plants by Hatch and Slack (1966) and coworkers. Comparative observations on radioactive labeling, on kinetics of intermediate products, and on enzyme activities revealed that the primary fixation of CO_2 occurs by carboxylation of PEP. CO_2 is trapped by PEP-carboxylase in the cytoplasm of mesophyll cells in order to form oxalacetate, which in most plants is reduced to malate in chloroplasts, or in some plants is converted to aspartate in the cytoplasm (Hatch et al., 1971; Hatch and Boardman, 1973; Björkman, 1973). Sometimes both malate and aspartate can be formed in one plant (Kennedy and Laetsch, 1973). The next step is the transportation of the products to the bundle-sheath cells. The decarboxylation of malate is accomplished in the chloroplasts of bundle-sheath cells, which are very rich in malic enzyme. This results in the formation of pyruvate, $NaDPH_2$ and CO_2. Recently, it has been hypothesized that the reducing power originating from this reaction overcomes the deficiency of photosystem II in bundle-sheath chloroplasts (Osmond, 1974). In contrast to the decarboxylation of malate, aspartate is first converted to oxalacetate (Hatch and Mau, 1973), which is either decarboxylated by PEP-carboxykinase (Black, 1973), or is reduced to malate, which in turn can be decarboxylated in mitochondria by malic enzyme (Hatch and Kagawa, 1974).

In a recent study (Gutierrez et al., 1974) on differences in activities of three decarboxylating enzymes and on cytological characteristics, C_4 plants are divided into three groups: NADP-malic enzyme species, NAD-malic enzyme species, and PEP-carboxykinase species (Fig. 2). However, the possibility that the same

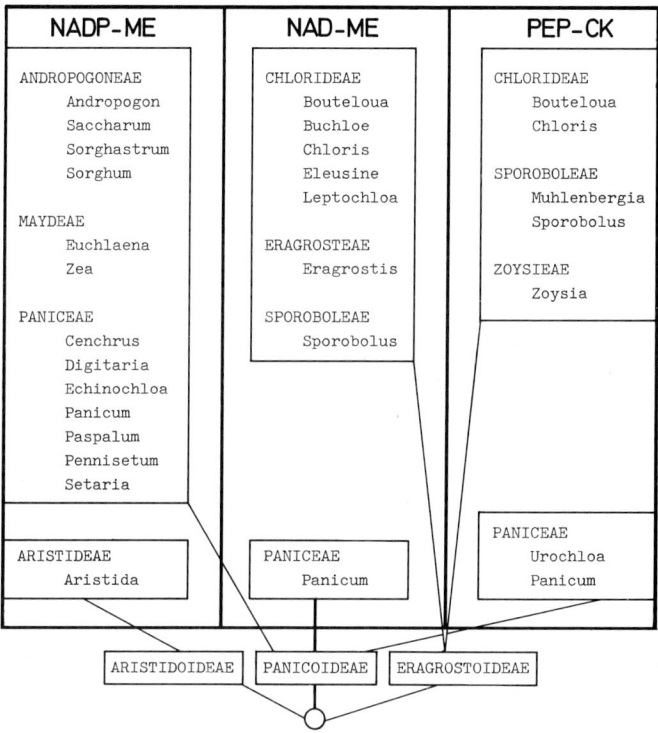

Fig. 2. Grouping of C_4 plants on the basis of the activities of decarboxylating enzymes and cytological characteristics. (After Gutierrez et al., 1974)

C_4 species may use more than one C_4-acid decarboxylation enzyme for carboxyl donation to the C_3 pathway cannot be excluded. In NADP-malic enzyme species (Fig. 3), the malate is decarboxylated by NADP-malic enzyme to form NADPH, whereby CO_2 is produced. The NADPH provides part of the reducing power for CO_2 fixation in the bundle-sheath cells through the C_3 pathway (Downton, 1970; Edwards and Black, 1971; Hatch, 1971; Huber et al., 1973). However, in these species most or all of the reducing power may be regenerated by the mesophyll chloroplasts. Most of the NAD-malic species (Fig. 3) transport aspartate to the bundle-sheath cells, where it is converted to oxalacetate. The latter produces malate, which is converted to pyruvate and CO_2 by aspartate

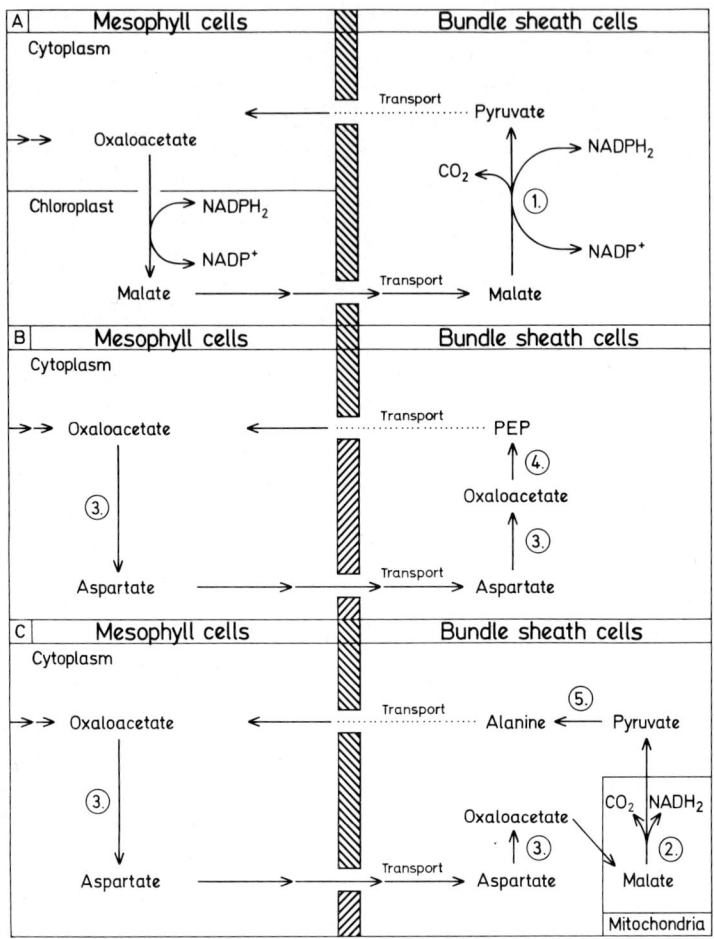

Fig. 3 A–C. Scheme of possible carboxylation reactions occurring in C_4 plants: (A) in NADP-malic enzyme species (NADP-ME), (B) in PEP-carboxykinase species (PEP-CK), and (C) in NAD-malic enzyme species (NAD-ME). *1* NADP-malic enzyme; *2* NAD-malic enzyme; *3* Aspartate aminotransferase; *4* PEP-carboxykinase; *5* Alanine aminotransferase. (Data from Gutierrez et al., 1974)

transaminase, malate dehydrogenase, and NAD-malic enzyme (Hatch and Kagawa, 1974). In PEP-carboxykinase species (Fig. 3), aspartate, after its translocation to the bundle-sheath cells, is converted by aspartate transaminase to oxalacetate. The latter is converted to PEP and CO_2 by PEP-carboxykinase (Edwards et al., 1971, 1974). These reactions produce CO_2, but no reducing power. Thus, in contrast to NADP-malic enzyme species (with agranal bundle-sheath chloroplasts), the granal bundle-sheath chloroplasts of NAD-malic enzyme and PEP-carboxykinase species may supply, through non-cyclic electron transport, the reducing power for fixation of CO_2 through the C_3 pathway. This interpretation appears to be in keeping with the finding that a large part of photosystem II is in bundle-sheath cells of NAD-malic enzyme and PEP-carboxykinase species (Ku et al., 1974; Gutierrez et al., 1974).

The CO_2 released by the carboxylations is refixed in the bundle-sheath cells by the RuDP-carboxylase of the Calvin-Benson cycle, while the pyruvate is returned to the mesophyll and phosphorylated to PEP by pyruvate-P_i-kinase (Hatch and Slack, 1968; Sugiyama, 1973). Thus, one can distinguish between two sequential carboxylations. The first is catalyzed by PEP-carboxylase, occurs in mesophyll cells, and utilizes atmospheric CO_2; the second is catalyzed by RuDP-carboxylase, takes place in bundle-sheath cells, and utilizes internally generated CO_2 (Björkman, 1973). In some species the carbon fixed in the bundle sheath is converted into starch in bundle-sheath chloroplasts (de Fekete and Vieweg, 1974), and in others it is transported back to mesophyll, where it is utilized in the synthesis of sucrose (Downton and Hawker, 1973).

The mesophyll cells generally lie next to the bundle-sheath cells, and the cells are interconnected by plasmodesmata (Crookston and Moss, 1973). This suggests that there is apparently no barrier impeding the intercellular transport that must take place in the C_4 pathway. However, some authors propose that PEP-carboxylase may be localized in non-chlorophyllous tissue and cytoplasm of mesophyll cells, and that RuDP-carboxylase may be present in the mesophyll chloroplasts (Baldry et al., 1971; Bucke and Long, 1971; Coombs, 1971; Coombs and Baldry, 1972; Coombs et al., 1973a, 1973b). According to these authors, the only function of the bundle-sheath chloroplasts seems to be the storage of starch. Chloroplasts capable of light-dependent CO_2 fixation by carboxylation of RuDP have been isolated even from young maize leaves (O'Neal et al., 1972). However, recent findings indicate that the C_4 pathway is localized in the mesophyll cells; and the carboxylation phase of the C_3 pathway, in bundle-sheath cells (Black, 1973; Hatch and Kagawa, 1974; Ku et al., 1974). Studies with isolated mesophyll and bundle-sheath cells have also proved very useful with respect to the location of enzyme activities (Chen et al., 1973; Kanai and Edwards, 1973; Ku et al., 1974), and photochemical behavior of these cells. Both mesophyll cells and mesophyll chloroplasts (Salin et al., 1973; Kagawa and Hatch, 1974) have been found to be capable of reducing oxalacetate to malate in a light-dependent reaction linked to O_2-evolution. It has been shown that bundle-sheath cells decarboxylate malate (Huber et al., 1973), and transfer the β-carboxyl of this malate into products of the C_3 pathway (Dittrich et al., 1973). In isolated bundle-sheath cells from maize, the major product of photosynthesis is 3-PGA, which is particularly due to the limited capacity of agranal chloroplasts to reduce newly-formed PGA (Farineau, 1975a). Malate is decarboxylated in bundle-sheath cells and causes the production of CO_2 and NADPH. However, 50% of the PGA molecules formed after malate addition cannot be reduced (because of the insufficient reducing power), and addition of malate cannot dramatically modify the pattern of ^{14}C distribution in the photosynthetic products (Farineau, 1975b). These results confirm the hypothesis that a fraction of PGA molecules produced during photosynthesis is reduced in mesophyll cells. Moreover, mesophyll chloroplasts are known to reduce PGA with a concomitant evolution of O_2 (Hatch and Kagawa, 1974).

For better understanding it seems to be necessary that the enzymes that participate in C_4 metabolism are characterized in more detail. It has been reported that a great number of enzymes involved in the C_4 pathway are activated by light (Graham et al., 1970; Johnson and Hatch, 1970; Hatch and Mau, 1973). The activity of PEP-carboxylase may be regulated by oxalacetate, glucose-6-phosphate, Mg^{2+}, and adenylates (Lowe and Slack, 1971; Coombs et al., 1973a, 1974). The enzyme isolated from a C_3 plant displays markedly different properties from that isolated from a closely-related C_4 plant (Hatch et al., 1972; Ting and Osmond, 1973a). This difference reflects the various effects of the same enzyme in different plants: in C_4 plants the malate resulting from the joint action of PEP-carboxylase and malate dehydrogenase is a photosynthetic intermediate, whereas in C_3 plants, it is not. Furthermore, the PEP-carboxylase in etiolated maize plants differs from the enzyme in light-grown maize both in ionic character and in saturation kinetics of PEP (Ting and Osmond, 1973b). Other than the green tissue, the etiolated tissue of sugarcane exhibits properties that more closely resemble the PEP-carboxylase of C_3 plants with low K_m PEP and $K_{0.5}Mg^{2+}$ values, and is markedly less sensitive to NaCl and glucose-6-phosphate (Goatly and Smith, 1974).

As is usually stated, the C_4 pathway requires the C_3 cycle only as an acceptor of CO_2 from the decarboxylation step in the bundle-sheath cells (Black, 1973). Thus, it is obvious that a C_4 plant must derive some PEP from PGA which is produced from RuDP carboxylation. The C_4 pathway cannot be an alternative for the C_3 pathway because it can not in itself regenerate more acceptor molecules than were available before carboxylation (Latzko and Kelly, 1974). The C_3 cycle, on the other hand, can achieve this. The reactions leading to regeneration of the CO_2-acceptor are evident in Fig. 4.

Another assumption is that in C_4 metabolism the mesophyll cells act as "CO_2-trapping antennas" and the bundle-sheath cells as "CO_2-reducing sinks" (Chen et al., 1974), and that the mutual dependence and coordination of these activities is indispensable for the smooth operation of the pathway (Björkman, 1973).

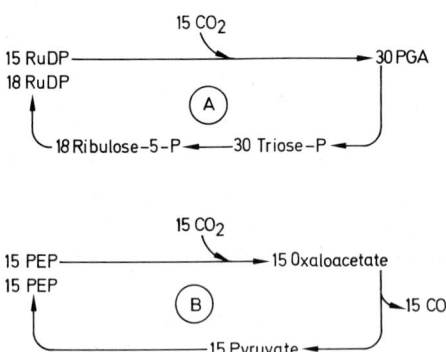

Fig. 4 A and B. Regeneration of acceptor molecules by (A) C_3 pathway and (B) C_4 pathway. (After Latzko and Kelly, 1974)

III. General Characteristics of C_4 Plants

a) The 'Kranz' or Wreath-Syndrome. The C_4 plants possess several diagnostic traits. These include a specialized leaf anatomy with separate mesophyll and bundle-sheath cell layers, low compensation point, absence of CO_2 release in the light, formation of C_4 acids as the first photosynthetic products, lack of photosynthetic CO_2 assimilation in response to O_2 concentration, high activities of the enzymes of the C_4 pathway, and low ^{13}C discrimination. This anatomical, ecophysiological and biochemical assortment of characteristics has been termed the 'Kranz' (wreath) syndrome. It is present in several members of monocotyledons and dicotyledons, and is associated with photosynthetic fixation of carbon. This intricate complex of biochemical and photochemical events involving division of labor between bundle-sheath cells and mesophyll cells has more recently been termed 'cooperative photosynthesis' (Karpilov, 1970). Plants possessing the C_4 pathway can easily be identified by analyzing and evaluating the structural and functional aspects of the syndrome.

b) Leaf Anatomy. The anatomy of the leaf offers one of the easiest ways to identify a C_4 plant. The conspicuous features of the leaf anatomy of C_4 plants have recently been reviewed (Laetsch, 1974). Principally, in a transverse section, the leaf of a C_4 plant shows a chlorenchymatous sheath of large, thick-walled cells around the vascular bundles, which, again, can be surrounded by one or more layers of loosely fitted mesophyll cells. This structural specialization of leaf, which has been termed 'Kranz'-type, is found in several plants belonging to both monocotyledons and dicotyledons (Figs. 5 and 6). There are several types of green chlorenchymatous sheaths found in plants, but only those composed of distinct, thick-walled cells with specialized plastids are associated with the C_4 pathway (Crookston and Moss, 1970). Several variations of this basic pattern are known (Laetsch, 1974). Leaves of *Arundinella hirta* present a special case: in addition to a sheath of large, bright green cells around the vascular bundles, there are strands of large parenchyma cells that appear to be identical with the bundle-sheath cells. They run parallel to the vascular bundles, but are not associated with any vascular tissue (Crookston and Moss, 1973).

The association of the C_4 pathway with the 'Kranz'-type leaf anatomy eventually led to the discovery of biochemical and chloroplast-ultrastructural differences between C_4 species. The differences concern the degree of grana development in the chloroplasts of bundle-sheath cells, the position of chloroplasts in these cells, and the activities of enzymes related to the pathway of photosynthesis (Laetsch, 1974; Gutierrez et al., 1974). On the basis of ultrastructural characteristics of bundle-sheath chloroplasts and their enzyme complement, thus far three groups of C_4 plants have been distinguished in the Gramineae. Species possessing NADP-malic enzyme have bundle-sheath chloroplasts in the centrifugal position and they lack well-developed grana; NAD-malic enzyme species have centripetally located bundle-sheath chloroplasts that contain grana; whereas, the phosphoenolpyruvate carboxykinase species have bundle-sheath chloroplasts in the centrifugal position and contain grana (Gutierrez et al., 1974). The C_4 dicotyledonous species so far examined are divided into two groups: those having high NADP-malic enzyme and bundle-sheath chloroplasts with poorly developed grana in centripetal

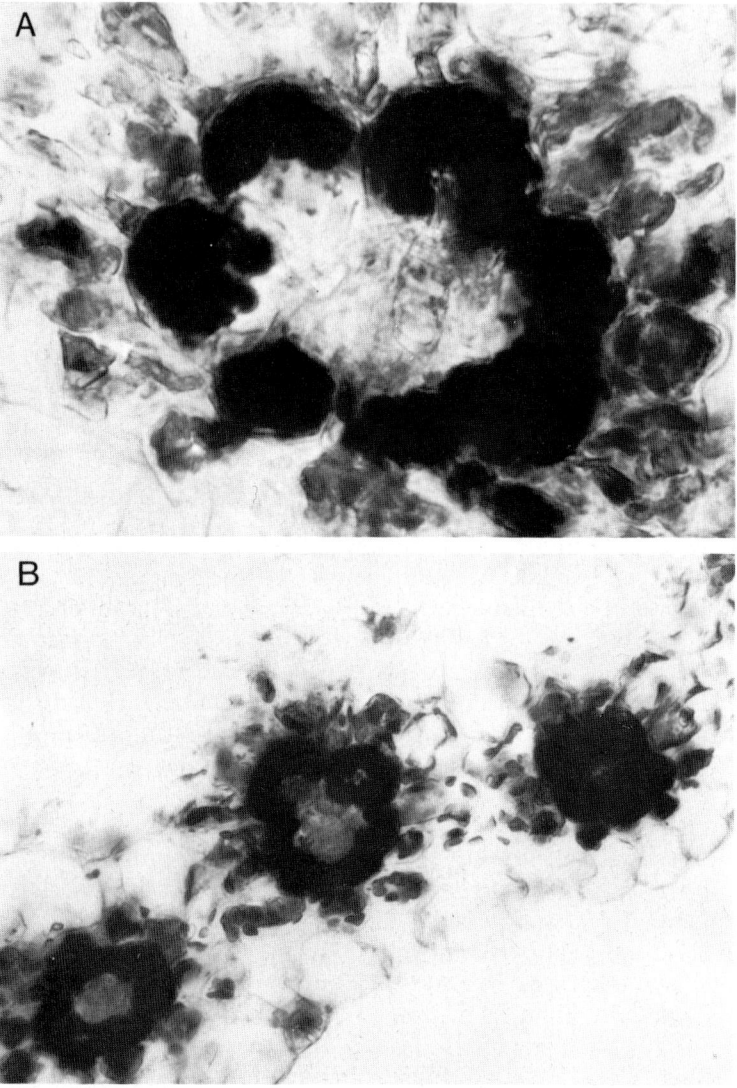

Fig. 5 A and B. Transverse sections of (A) *Amaranthus edulis* leaf and (B) *Pennisetum typhoides* leaf showing typical 'Kranz' type of leaf anatomy

position, and those having NAD-malic enzyme and bundle-sheath chloroplasts with well-developed grana in centripetal position (Gutierrez et al., 1974). This variation in the organelle ultrastructure sometimes forms the basis of specific variations in the pathway of carbon metabolism and electron transport (Black, 1973). However, the presence of high concentration of organelles, such as chloroplasts, mitochondria, and peroxisomes in all types, is noteworthy (Black and Mollenhauser, 1971). In addition, a peripheral reticulum seems to be particularly

C_4 Pathway and Regulation of the Balance Between C_4 and C_3 Metabolism 343

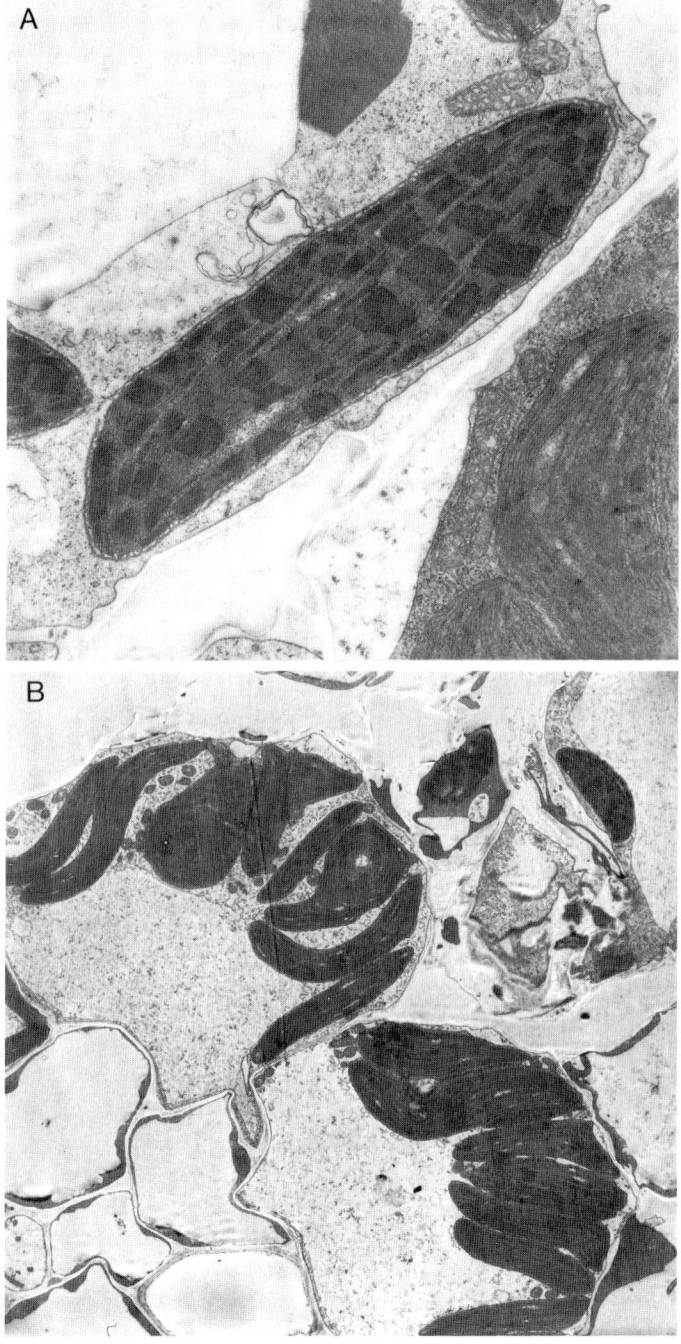

Fig. 6 A and B. Electromicrographs showing (A) mesophyll chloroplasts and (B) bundle-sheath chloroplasts in maize

well organized in chloroplasts of the C_4 plant group, although it is also found in C_3 and CAM plants (Laetsch, 1974).

c) **CO_2 Compensation Point.** C_4 plants are characterized by a very low CO_2 compensation point (Downton and Tregunna, 1968; Chen et al., 1970; Black, 1971). When actively photosynthesizing plants are kept in closed, illuminated chambers, the CO_2 concentration progressively decreases in the chamber till a time after which no further change in CO_2 concentration can be detected. At this time, photosynthetic CO_2 uptake equals respiratory CO_2 evolution. This CO_2 concentration at which the intake and output of CO_2 by the leaves reaches equilibrium is defined as the photosynthetic CO_2 compensation concentration (Jackson and Volk, 1970).

The C_3 plants have photosynthetic CO_2 compensation concentrations in the range of 30–70 ppm of CO_2, while the C_4 plants show a range of 0–10 ppm of CO_2 (Forrester et al., 1966a, b; Tregunna and Downton, 1967; Zelitch, 1967; Meidner, 1962; Chen et al., 1970; Crookston and Moss, 1970; Black, 1973). Generally, the low compensation point and the presence of a chlorenchymatous bundle-sheath with plastids specialized for starch formation are found coupled together. However, recently it has been demonstrated that under certain greenhouse conditions the leaves of *Amaranthus edulis*, which possess 'Kranz'-type leaf anatomy, occasionally yield high compensation point leaves (Lester and Goldsworthy, 1973). Their occurrence could not be correlated with leaf age, moisture stress, or the prevailing environmental conditions. It is quite possible that *A. edulis* may be a facultative C_3 plant when grown under suitable environmental conditions.

d) **Photorespiration.** C_3 plants have a characteristic respiratory CO_2 release in light (Decker, 1955). This is found to be coupled with glycolate formation and its subsequent metabolism (Jackson and Volk, 1970; Zelitch, 1971, 1973). C_4 plants, on the other hand, are characterized by their apparent lack of photorespiration (Forrester et al., 1966a, 1966b; Black, 1973). Thus, the presence or absence of detectable photorespiration serves as a very useful criterion in delimiting C_3 and C_4 plants. However, there is sufficient evidence to suggest that C_4 plants possess photorespiratory activity and that this activity is localized in the bundle-sheath cells (Chollet and Ogren, 1973; Chen et al., 1974). The metabolism of glycolate leading to the production of CO_2 has been shown to occur in bundle-sheath cells (Chollet and Ogren, 1972; Mühlbach and Wegmann, 1973) that contain a large number of the microbodies and enzymes involved in photorespiration, such as RuDP oxygenase, catalase, glycolate oxidase, hydroxypyruvate reductase (NAD^+), and phosphoglycolate phosphatase (Frederick and Newcomb, 1971; Huang and Beevers, 1972; Chen et al., 1974). However, RuDP oxygenase, phosphoglycolate phosphatase, and hydroxypyruvate reductase, which have a key role in the photorespiratory pathway (Tolbert and Yamazaki, 1969), were absent in the isolated mesophyll cells.

In any case, it appears that in C_4 plants, the photorespired CO_2 is not released, but is trapped in the mesophyll cells that have high levels of PEP-carboxylase (Latzko and Kelly, 1974). The magnitude and function of photorespiration in C_4 plants remains unknown. However, it is clear that the lack of apparent photorespiration may be responsible for the high photosynthetic capacity of these plants. The possibility of increasing plant productivity by controlling photorespiration has been recently demonstrated (Zelitch, 1973).

e) $^{14}CO_2$ **Fixation Products and Pulse Chase Studies.** In sharp contrast to the finding (Bassham and Calvin, 1957) that the earliest product of photosynthesis in green plants is phosphoglyceric acid, Kortschak et al. (1965) demonstrated that surgarcane leaves, when fed with $^{14}CO_2$ in light, fix 80% of $^{14}CO_2$ into malic and aspartic acid within a fraction of a second. These observations were confirmed by Hatch and Slack (1966), who demonstrated that oxalacetic acid was the initial product of CO_2 assimilation, and that it was rapidly converted into malic and aspartic acids. When the exposure to $^{14}CO_2$ is prolonged, the labeling of C_4-acids decreases, while it increases in 3-phosphoglyceric acid and other products.

Pulse chase experiments, on the other hand, give an estimate of the movement of radioactivity between intermediates under steady-state conditions, and provide more precise information relating to the nature and sequence of chemical events (Hatch and Slack, 1970; Black et al., 1973). In these experiments, the leaves are fed with a pulse of $^{14}CO_2$ for a short time, and are quickly transferred to $^{12}CO_2$ and exposed to light for subsequent chase of the products. It is observed that the label is finally transferred from C_4 acids to hexoses, which indicates that in light, C_4 plants can synthesize sugars from C_4-acids. However, if, after pulse with $^{14}CO_2$ in light, the chase is observed in darkness, the label from C_4-acids is almost quantitatively transferred to alanine as a result of transaminase reactions (Black et al., 1973). Many investigators prefer to pre-illuminate the leaves to ensure building-up of intermediate compounds or an energy source that might support CO_2 fixation in the subsequent dark period.

f) $^{13}C/^{12}C$ **Ratio.** A more reliable method of determining C_3 and C_4 photosynthetic pathways is the use of carbon isotope analysis (Smith and Brown, 1973; Osmond and Ziegler, 1975). Fractionation of the isotopes of carbon is due to preferential utilization of ^{12}C and partial exclusion of ^{13}C by plants during carbon assimilation (Smith, 1972). C_3 plants show a greater discrimination against the heavy isotopes of carbon than do C_4 plants. Thus, ^{13}C-values of -9 to $-16‰$ indicate the 'Kranz-syndrome', while ^{13}C-values of -23 to $-32‰$ are suggestive of C_3 plants (Bender, 1971; Smith and Epstein, 1971; Troughton, 1971).

The advantages of this method are that only small amounts of tissue are required, that the results are very accurate, and that even dried specimens from herbarium sheats may be used (Smith and Brown, 1973). CAM plants can also have a higher $^{13}C/^{12}C$ ratio, but this is subject to modulation depending on the growth conditions of the plants (Osmond and Ziegler, 1975). The lack of CAM plants in the Gramineae, however, ensures that $^{13}C/^{12}C$ ratios can be used as a diagnostic test for the C_4 pathway in this family, as well as in non-succulent plants belonging to other taxa (Smith and Brown, 1973).

The mechanism of this differential discrimination at the molecular level is not yet clear. However, it is possible that it stems from the differential carboxylation reactions involved in the C_3 and C_4 pathways (Osmond and Ziegler, 1975).

g) **Other Criteria.** Various other criteria may also prove useful in establishing, at least tentatively, the type of photosynthesis in the leaves of higher plants.

These include: temperature optimum for photosynthesis, response of photosynthesis to light intensity, lack of response by photosynthetic CO_2 assimilation to O_2 concentration, the chlorophyll a/b ratios in the leaves, and the characteristics of the mesophyll cells (Black, 1971, 1973; Chang and Troughton, 1972; Björkman, 1973; Holden, 1973; Chen et al., 1974). The optimum temperature for photosynthesis in C_3 plants is in the broad range of 10°–25°C, while for C_4 plants the range is 30°–45°C (see also Ludlow, this volume Part 5:D). The extremes are represented by *Caltha infraloba*, a C_3 plant of the alpine region with an optimum at 10°C (Phillips and McWilliam, 1971), and *Tidestromia oblongifolia*, a C_4 plant of hot deserts with an optimum between 45–50°C (Björkman, 1971). Another diagnostic characteristic of C_4 plants relates to their increasing capacity for CO_2 uptake with increasing light intensity, sometimes even till full sunlight (Björkman, 1973).

Photosynthesis in C_3 plants is generally inhibited at oxygen levels above atmospheric concentrations (21%), while C_4 plants are insensitive to oxygen concentrations up to 100% (Forrester et al., 1966a, b; Björkman et al., 1968; Downes and Hesketh, 1968). Recently it has been observed that even the chlorophyll a/b ratio in leaves of C_4 plants, both with and without grana in the bundle-sheath chloroplasts, was greater than in C_3 plants (Holden, 1973; Chang and Troughton, 1972). Even the characteristics of mesophyll cells in C_3 and C_4 plants may prove to be distinctive. Indeed, the mesophyll cell of a C_4 plant leaf appears to be unlike any other photosynthetic cell, and is characterized by (1) high PEP-carboxylase for CO_2 fixation, (2) highly active pyruvate-P_i-dikinase, (3) very low RuDP-carboxylase, (4) limited capacity for glucose synthesis, (5) low rate or complete absence of photorespiration, (6) a $^{14}CO_2$ fixation that is linked to the photosynthetic activity of the mesophyll chloroplasts in a stoichiometry in which fixation of 1 mole CO_2 corresponds to reduction of 1 mole malate and evolution of 1 atom of oxygen, (7) the light-dependent conversion, not associated with CO_2 fixation, of 3-PGA to hexose phosphates (Chen et al., 1974).

IV. Factors Affecting Shift

The adaptive strategies leading to ecological advantages in a plant may, in any ecosystem, be considered with respect to metabolic as well as environmental parameters. Recently, there has been some evidence that suggests that the balance between C_4 and C_3 metabolism within a plant, like any other metabolic process, may be regulated by a host of factors. The following discussion summarizes this evidence.

1. Ontogeny

Substantial quantitative and qualitative changes may occur in the primary photosynthetic products of a C_4 leaf during ontogeny (Kennedy and Laetsch, 1973; Khanna and Sinha, 1973). Mature *Portulaca oleracea* leaves fix $^{14}CO_2$, during an exposure period of 10 sec, primarily into organic and amino acids. In contrast, younger leaves can fix more than 60% of the total $^{14}CO_2$ into alanine (Kennedy and Laetsch, 1973; Huber et al., 1973a). Senescent leaves of *P. oleracea* conduct both C_3 and C_4 type photosynthesis during a very short exposure to $^{14}CO_2$, and display a quantitative shift of primary products towards the C_3 pathway with a concomitant reduction of the label in malate and aspartate. With increasing leaf age, this shift has a definite relationship with the leaf ontogeny and is not a function of nutrition, temperature, or the flowering status of the plant (Kennedy and Laetsch, 1973). Young seedlings of Sorghum and Pennisetum are known to have a predominantly C_4 pathway (Khanna and Sinha, 1973; Huber et al., 1973a). However, after flowering, RuDP-carboxylase was predominant in the leaves of both Sorghum and Pennisetum, and after $^{14}CO_2$ exposure, a higher 3-phospho-glycerate/malate ratio was obtained. The leaf anatomy and chlorophyll a/b ratio, however, remained unchanged (Khanna and Sinha, 1973). This change in predominance from C_4 to C_3 pathway also appears to be associated with the stage of development of the leaves.

2. Growth Conditions

The conditions under which an organism is grown appear to have a profound influence on CO_2 exchange, $^{14}CO_2$ fixation, and radioactive photosynthetic products, including enzyme activity and carboxylation reactions. For instance, *Anacystis nidulans*, a blue-green alga, can fix CO_2 by the C_3 pathway, C_4 pathway, or by reductive carboxylation of succinate to glutamate (Döhler, 1974a). It was observed that the CO_2 concentration during growth and during the measurement period plays a decisive role in influencing the type of CO_2 fixation pathway. In addition to the normal C_3 pathway, $^{14}CO_2$ is mainly incorporated into aspartate and glutamate when the plant is grown at 35°C in an atmosphere of 0.03 % or 3 % CO_2, and the measurements are made at low CO_2 concentration. The patterns of labeling were also very similar to those of the C_4 pathway of photosynthesis.

Graham and Whittingham (1968) studied the effect of different CO_2 concentrations, during growth and during measurement, on the ^{14}C-labeled photosynthetic products in *Chlorella pyrenoidosa*. This alga, under normal growth conditions, shows the C_3 pathway. However, when the alga was grown at a high CO_2 concentration (5 % volume) at 25°C, and photosynthesis was measured at low CO_2 concentration, a long stationary phase of photosynthetic induction was obtained that showed label predominantly in malate and aspartate, thus indicating the C_4 pathway. This may possibly be due to low activity of RuDP-carboxylase in *Chlorella* cells grown at 5 % CO_2 concentration (Döhler, 1974b).

In addition to CO_2-content, the temperature appears to play a very important role for photosynthesis. Becker (1972) studied the effect of growth temperature on the distribution of radiocarbon in products of photosynthesis in several algae. His studies clearly demonstrate that an alteration of temperature brings about profound alterations in metabolism. The temperature also affected the activities of aldolase, glycerine-3-phosphate dehydrogenase and glutamate-dehydrogenase in synchronous cultures of *Chlorella* (Berger and Pirson, 1967). The sensitivity of individual enzymes to growth temperature was very different. According to Bassham (1971), the regulation of the activity of key enzymes of the

Calvin cycle (RuDP-carboxylase, FDP-ase) plays a decisive role in controlling the rate of photosynthesis. Björkman and Gauhl (1969) observed that the activity of RuDP-carboxylase decreased significantly when the growth temperature of *Distichlis spicata* was raised from 16°C to 30°C or 40°C. Treharne and Cooper (1969) found different temperature-dependencies in the activity of RuDP-carboxylase and PEP-carboxylase in several grasses.

Döhler (1974b) studied the effect of growth conditions on the photosynthetic pathway of *Chlorella vulgaris*. The algae were grown in low (0.035 %) and high (3.5 % volume) CO_2 concentrations and at different temperatures (15°C, 25°C, 35°C). The exchange, $^{14}CO_2$ fixation, and ^{14}C incorporation into photosynthetic products were studied during the photosynthetic induction period at 10°C and 35°C. The autoradiographic studies of the kinetics of the appearance of labeled products at 10°C and 35°C indicated that at the given CO_2 concentrations, as well as at growth temperatures of 15°C and 35°C, the Calvin cycle is the main carboxylation pathway. However, at the beginning of the illumination period at 25°C, $^{14}CO_2$ was incorporated mainly into malate and aspartate, while at 10°C, into 3-phosphoglycerate. High CO_2 concentrations during the growth phase had only a small effect on the percentage distribution of ^{14}C-labeled products. Furthermore, a very active PEP-carboxylase could be found in extracts of *Chlorella* grown at 25°C in air and in 3.5 % CO_2. Higher activity of RuDP-carboxylase was found in air-grown *Chlorella* cells than in CO_2-grown cells. The results indicate that, during the growth of algae, temperature affects CO_2 fixation and enzyme activities more than does CO_2 concentration.

3. Nutrition

Until recently, little was known about the adaptation of C_4 plants to salinity. The studies of Shomer-Ilan and Waisel (1973) have indicated that low salt-status plants of the halophytic grass *Aeluropus litoralis*, in spite of possessing 'Kranz'-type bundle-sheaths, show CO_2 fixation by the C_3 pathway. It is interesting that the leaves indicate an enhanced activity of PEP-carboxylase, as well as a significant CO_2 fixation by the C_4 pathway, when these plants are exposed to a salt solution. This observation appears to be in contradiction to that of Downton and Törökfalvy (1975). However, Shomer-Ilan and Waisel (1973) also observed that the balance between the C_4 and C_3 pathways in maize and *Chloris* was affected similarly by salt treatment.

Sankhla and Huber (1974a) studied the effect of salt on the activities of photosynthetic enzymes and $^{14}CO_2$ assimilation in the young leaves of *Pennisetum typhoides*, a C_4 plant. Sodium chloride inhibited the rate of $^{14}CO_2$ fixation and the activities of PEP-carboxylase and RuDP-carboxylase, but increased the activities of both NAD- and NADP-specific malate dehydrogenase (Figs. 7, 8 and 9). At high concentration (1.7×10^{-2} M), NaCl also increased the incorporation of $^{14}CO_2$ in the fraction of organic acids (malate), but less radioactivity was recorded in the amino-acid fraction (alanine). In wheat, on the other hand, sodium chloride drastically inhibited the activity of RuDP-carboxylase, but the activity of PEP-carboxylase increased significantly (Table 1). A shift from C_3 to CAM, in response to salinity, has also been recorded by Winter and Lüttge

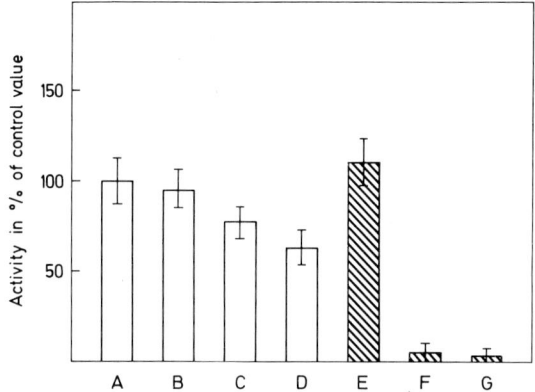

Fig. 7. Effect of abscisic acid and sodium chloride on the activity of RuDP-carboxylase in leaves of *Pennisetum typhoides*: A Control; B 3.8×10^{-6} M ABA; C 7.4×10^{-6} M ABA; D 1.5×10^{-5} M ABA; E 1.7×10^{-3} M NaCl; F 8.5×10^{-3} M NaCl; G 1.7×10^{-2} M NaCl

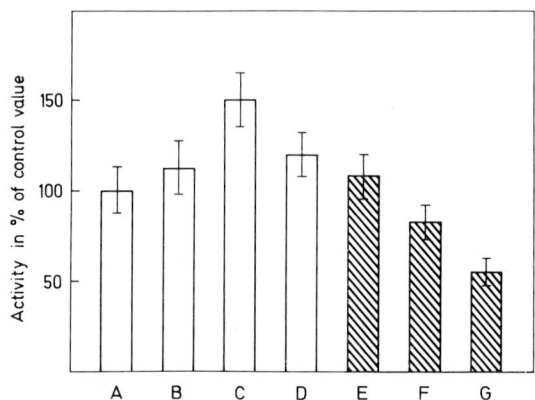

Fig. 8. Effect of abscisic acid and sodium chloride on the activity of PEP-carboxylase. Concentrations as in Fig. 7

Table 1. Effect of NaCl on the activity of photosynthetic enzymes in wheat and *Lemna minor*

Treatment (Molar)	Activity in % of control			
	Wheat		*Lemna*	
	PEP-carboxyl.	RuDP-carboxyl.	PEP-carboxyl.	RuDP-carboxyl.
Control	100	100	100	100
NaCl 1.7×10^{-3}	84	98	179[b]	102
NaCl 4.2×10^{-3}	—	—	216[b]	135[a]
NaCl 8.5×10^{-3}	128[a]	35[b]	247[b]	230[b]
NaCl 1.7×10^{-2}	149[b]	13[b]	—	—

[a, b] Values significantly different from control with P = 0.05 or 0.01.

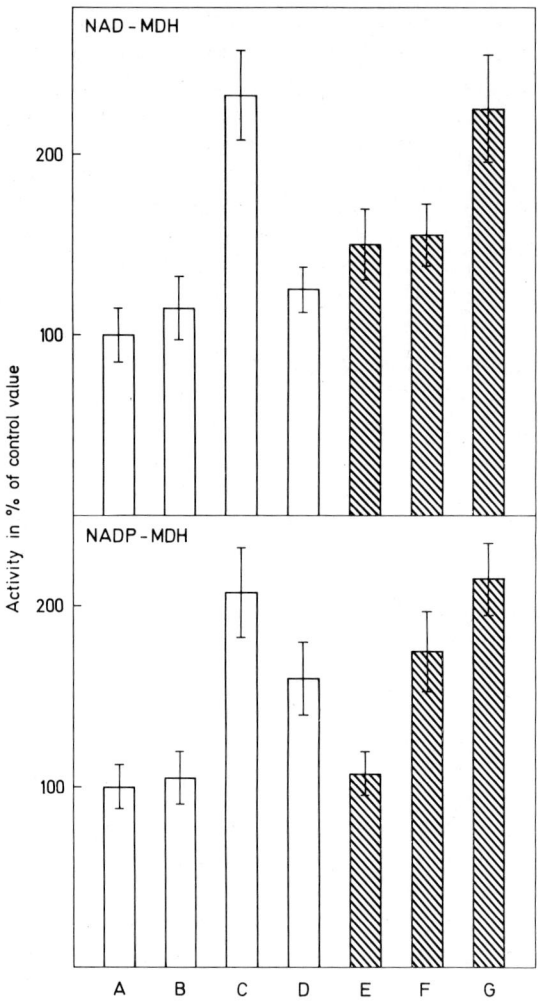

Fig. 9. Effect of abscisic acid and sodium chloride on the activities of NAD-malate dehydrogenase and NADP-malate dehydrogenase. Concentrations as in Fig. 7

(this volume Part 5:B). It is interesting to note that C_4-type plants have higher demands for sodium than C_3-type plants (Brownell and Crossland, 1972).

The metabolic conditions for the photosynthesis of marine plants are very different from those of freshwater plants and terrestrial glycophytes. Aspartate was reported to be a major product of short-term fixation in *Ulva lactuca* (Patil and Joshi, 1970). It was also detected in several brown algae of the west coast of India (Joshi et al., 1974) and in *Aegiceras majus*, a mangrove member of tropical sea shores. Besides aspartate, a significant incorporation was observed in the PEP-PGA fraction. These findings indicate that the C_4 pathway is operative in marine algae. It was also found that the enzymatic mechanism that is required for both the C_3 and C_4 pathways is operative in marine plants (Karkar and Joshi, 1973). In *Enteromorpha tubulosa*, more aspartate and less PGA are formed in short-term photosynthesis. It is possible that the simultaneous operation of both types of CO_2 fixation pathway may exist in the marine algae. This is in agreement with the studies of Smith and Epstein (1971) regarding the $^{13}C/^{12}C$ ratio, which suggests that algae form an intermediate group in the plant kingdom. Since CO_2 concentration in sea water is low, an efficient mechanism of photosynthesis must be postulated for marine algae. Similarly, CO_2 availability often becomes a limiting factor in mangroves due to sunken stomata in leaves. As PEP-carboxylase is a better acceptor of CO_2 than RuDP-carboxylase (Maruyama et al., 1966), the adaptation of the C_4 pathway under saline conditions appears to be an ecological necessity.

Grossman and Cresswell (1973) studied the influence of nitrogen supply in the nutrient media on the CO_2 compensation point of C_4 photosynthetic plants. In the case of maize and sugarcane, when the plants were transferred from nitrate nitrogen-receiving conditions to a medium containing ammonia nitrogen as the sole nitrogen source, the compensation point increased from zero to 14 ppm CO_2. This was possibly due to synthesis of higher levels of photorespiratory substrate in the presence of ammonia nitrogen. This study indicates an important involvement of nitrogen availability and nitrogen metabolism in photorespiration.

4. Bio-regulants

Bio-regulants play an important role in the growth and metabolism of plants. Recently, abscisic acid, a widely distributed hormone with a broad spectrum of activity, has been shown to have an important function in the control of several growth processes (Addicott, 1972; Sankhla, 1971; Sankhla and Sankhla, 1968; Milborrow, 1974).

It is well known that abscisic acid brings about closure of stomata, reduces the rate of photosynthesis, and inhibits the activity of RuDP-carboxylase (Mittelheuser and van Steveninck, 1971; Poskuta et al., 1972; Wellburn et al., 1973; see also Itai and Benzioni, this volume Part 4:B). There was, however, little knowledge of the mechanism of the action of hormones in the process of photosynthesis. The present authors have investigated the effect of single and combined hormones on the photosynthesis of C_4 and C_3 plants (Sankhla and Huber, 1974a, b, 1975; Huber and Sankhla, 1974a, 1974b). The results appear to provide new information on the role of bio-regulants in the process of photosynthesis. In leaves of *P. typhoides* seedlings, abscisic acid inhibited the rate of $^{14}CO_2$ fixation and the activity of RuDP-carboxylase (Fig. 7), but increased the activities of PEP-carboxylase (Fig. 8), malic enzyme, and both NADP- as well as NAD-specific (Fig. 9) malate dehydrogenase (Sankhla and Huber, 1974b; Huber and Sankhla, 1974b). The leaves of the seedlings grown in the presence of abscisic acid incorporated, in comparison to the control, more radio-activity in the fraction of organic acids and less in the amino-acid fraction. This indicates that abscisic acid probably favors the operation of the C_4 pathway. On the other hand, gibberellic acid was found to decrease the activities of photosynthetic enzymes, appeared to favor greater incorporation of radio-carbon in alanine, and reduced the same in malate (Huber and Sankhla, 1974a). Furthermore, as in growth, abscisic acid and gibberellic acid in combination, tended to antagonize each other in their effects on enzyme activity, as well as on incorporation of $^{14}CO_2$ into photosynthetic products (Sankhla and Huber, 1974b). Gibberellic acid, when added in combination with salt, also reversed some of the effects of salinity on growth and photosynthesis of *P. typhoides* (Sankhla and Huber, 1974a).

Once it was deduced that bio-regulants can greatly influence the flow of ^{14}C into individual photosynthetic products and the activities of photosynthetic enzymes, a series of experiments were planned using wheat, a conventional C_3 plant. Extracts of the leaves obtained from young wheat seedlings grown in the presence of ABA indicated, in comparison to the control, decreased activity of RuDP-carboxylase and a reduction in the rate of $^{14}CO_2$ fixation (Tables 2 and 3). However, the activity of PEP-carboxylase was greatly enhanced by ABA treatment. Concomitant with this effect, relatively more radioactivity was found to be incorporated in the fraction of malate, but less in phosphoglyceric acid (Sankhla and Huber, 1975). The balance between the C_4 and C_3 pathways depends primarily on the activity of PEP-carboxylase relative to RuDP-carboxylase, and on the availability of substrate (CO_2), which in turn depends on spatial arrangements and diffusion resistances in the leaf. It is conceivable that any factor, capable of affecting either the expression of one carboxylation reaction relative to the other (Kennedy and Laetsch, 1973) or the substrate level, might alter the type of photosynthetic carbon reduction. Since abscisic acid is probably present in all higher plants (Milborrow, 1974), it is possible that the hormone might represent one such factor. It is not clear why abscisic acid, although promoting the activity of PEP-carboxylase, inhibits the total $^{14}CO_2$ fixation. Besides its effect on several processes of growth and metabolism, abscisic acid is known to bring about closure of stomata (Cummins et al., 1971; Mittelheuser and van Steveninck, 1971; see Itai and Benzioni, this volume Part 4:B). Therefore it might be that by increasing the activity of PEP-carboxylase, the hormone tends to compensate for the closure of stomata. However, in *Lemna minor*, a plant with nonfunctioning but always open stomata (Wagner, unpublished[1], although the activity of PEP-carboxylase (Tables 1 and 2) was greatly enhanced by ABA and salt treatment, no significant difference could be found in the labeling pattern (Table 3) of the primary products of photosynthesis (Bauer et al., 1976). This appears to be in agreement with the above hypothesis. On the other hand, it is possible that the effect of abscisic acid on photosynthesis represents a secondary response emanating from the influence of the hormone on plastid ultrastructural morphogenesis or plastid/enzyme association (Wellburn et al., 1973).

V. Natural C_3-C_4 Intermediates

On the basis of leaf anatomy, cell ultrastructure, photorespiration, and primary photosynthetic products, Kennedy and Laetsch (1974) have shown that *Mollugo*

---

[1] Zulassungsarbeit, TU München, 1973.

Table 2. Effect of ABA on enzyme activity and $^{14}CO_2$ fixation in wheat and *Lemna minor*

Treatment (Molar)	% $^{14}CO_2$ fixed g^{-1} fr. wt.		Activity in % of control			
	Wheat	Lemna	Wheat		Lemna	
			RuDP-carboxyl.	PEP-carboxyl.	RuDP-carboxyl.	PEP-carboxyl.
Control	100	100	100	100	100	100
ABA 3.8×10^{-6}	79^a	40^b	16^b	113	125^a	201^b
ABA 7.4×10^{-6}	55^b	50^b	12^c	213^b	128^a	222^b
ABA 1.5×10^{-5}	—	49^b	9^{b,d}	162^b	144^a	245^c

$^{a-c}$ Values significantly different from control with P=0.05, 0.01 and 0.001.
^{d} Since ABA inhibited germination, the seedlings of wheat were grown in the presence of 7.4×10^{-6} M ABA for three days, and then transferred to this concentration.

Table 3. Effect of ABA on the distribution of ^{14}C in photosynthetic products after a pulse of 10 sec

Compound	% of total ^{14}C fixed					
	Wheat			Lemna		
	control	ABA$_1$	ABA$_2$	control	ABA$_1$	ABA$_2$
Malate	19.9	24.4	29.0	6.2	5.5	6.6
Alanine	11.0	8.8	17.0	6.3	5.5	8.9
Aspartate/Sucrose	9.7	9.5	17.3	2.9	2.0	3.7
PGA	34.7	28.6	10.1	23.4	27.0	24.1
Sugar-P-ester	22.4	28.6	26.6	39.2	44.9	47.4

ABA$_1 = 3.8 \times 10^{-6}$ M; ABA$_2 = 7.4 \times 10^{-6}$ M

verticillata, a member of Aizoaceae, possesses features that are intermediate between C_3 and C_4 plants. This is an important finding because, despite the polyphyletic evolution of C_4 photosynthesis, no plant species was known to have an assortment of characters intermediate between the C_3 and C_4 syndromes. CAM plants, however, can be considered as C_3-C_4 intermediates (Laetsch, 1974), at least with respect to their carbon fixation.

Recently some objections have been raised against the role of leaf anatomy in providing spatial separation in C_4 photosynthesis. It has been demonstrated that structurally homogenous tissue cultures of *Froelichia* can carry out C_4 photosynthesis (Laetsch and Kortschak, 1972), that tissues initially with C_3 characteristics can develop C_4 traits during ontogeny (Downton, 1971), that in *Sorghum* and *Pennisetum* an alteration in the predominance of C_4 to the C_3 pathway appears to be a function of developmental stage (Khanna and Sinha, 1973), and that in *P. oleracea* the senescent leaves can simultaneously carry out C_3 and C_4 photosynthesis (Kennedy and Laetsch, 1973).

Even certain CAM plants fix CO_2 into malate in light at short-term assimilation (Avadhani et al., 1971), although they have no 'Kranz' anatomy. These observations suggest that CO_2 fixation pathways and the leaf anatomy of C_4 plants might not be causally related. However, on the basis of his experiments on hybridization between C_3 and C_4 species of *Atriplex*, Björkman (1971, 1973) argues that C_4 photosynthesis can occur only if the component biochemical steps are coordinated, synchronized, and spatially compartmented. Especially interesting (Duffus and Rosie, 1973) is the case of barley pericarp, which contains chloroplasts and probably amyloplasts and starch-bearing chloroplasts. The PEP-carboxylase in the pericarp has been shown to be 100 times more active in CO_2 fixation than RuDP-carboxylase. In addition, the pericarp also possesses PEP-synthetase, pyrophosphatase, NAD- and NADP-dependent malate dehydrogenases, malic enzyme, and fructose-1,6-diphosphatase. This means that the pericarp is, at least in theory, capable of synthesizing, carboxylating and regenerating PEP. A fixation of CO_2 by this mechanism is possible, provided that RuDP-carboxylase is able to refix the CO_2 released by malic enzyme. This evidence suggests that barley, generally considered to be a C_3 plant, may possess C_4 metabolism in the green pericarp of the grain. Further studies on naturally occurring C_3-C_4 intermediates may yield valuable information on the relationship between structure and function in C_4 plants (Laetsch, 1974).

VI. Ecological Implications

In essence, the C_4 pathway represents a mechanism that is concerned with prefixing and concentrating CO_2 in the photosynthetic tissue (Evans, 1971). In certain ecosystems, the pathway may confer remarkable ecological adaptations, at least in certain plants, with respect to their behavior and survival under extreme climatic conditions. As shown by Björkman (1973), this pathway enables the plant to cope with the restrictions imposed on photosynthesis by high oxygen and low CO_2 concentration in the atmosphere. Therefore, the adaptive advantage of the C_4 pathway is particularly great under conditions of high irradiance, high thermal load, and low CO_2 concentration.

Haberlandt (1914) was probably the first who correlated 'Kranz'-type leaf anatomy with arid conditions: short and irregular periods of precipitation, periods of high temperature and high light intensity. In such regions, the advantage of C_4 plants is enormous because they have developed the ability to respond to such an environment. In contrast to the C_3 plants, which have the temperature optimum for net CO_2 uptake in the range of 10°–25°C, the C_4 plants exhibit the photosynthetic temperature optimum range of 30°–45°C. In addition, the illuminated leaves of C_4 plants possess the ability to reduce the CO_2 concentration in an enclosed space to almost zero, and usually require high light intensities to saturate photosynthesis (Björkman, 1973).

An extremely interesting case is that of *Tidestromia oblongifolia*, reported by Björkman et al. (1972). This herbaceous perennial, growing in Death Valley, Calif., carries out almost all of its photosynthesis and growth during the extremely hot summer. Under controlled conditions the optimum temperature for net photosynthesis was directly proportional to light intensity up to full noon sunlight. Even at these unusually high photosynthetic rates and resulting low intercellular space CO_2 concentrations, CO_2 was not a severely limiting factor for photosynthesis. This example illustrates how efficiently a C_4 plant is adapted to such extreme conditions.

Under such conditions, C_4 plants derive a substantial advantage from their more efficient internal trapping of CO_2, which occurs during photorespiration. An extreme example is *Triodia*, a grass of the hot Australian inlands, which survives by utilizing the C_4 pathway in leaves that are especially resistant to water loss (McWilliam and Mison, 1974).

Furthermore, C_4 plants are reported to be twice as efficient as C_3 plants in utilizing water. Thus they are expected to grow more in periods of moisture stress than C_3 plants. Recent studies on photosynthetic pathways in short-grass prairie species indicate that photosynthetic adaptation to temperature in C_3 and C_4 grasses may even allow many species to occupy the same site with minimal interspecific competition. This strategy for ecosystem utilization appears to be based on phenologies that allow the use of habitat factors in the growing season matched to physiological attributes (Williams III and Markley, 1973; Williams III, 1974).

Several agricultural ecosystems indicate that C_4 plants are highly competitive. For instance, even under cultural practices that favor C_3 crops, a variety of C_4 weeds grow luxuriantly, while in permanent pastures of C_4 grasses, there is an almost complete lack of weeds (Black, 1971). However, it should be borne in mind that the mere presence of the C_4 pathway does not reflect the only proof of successful adaptation for these plants. There may be several other means of adaptation that play a decisive role in their survival under extreme conditions.

An exciting finding in research relating to stress physiology concerns pronounced accumulation of ABA during water, osmotic, mineral, or salinity stress (see Itai and Benzioni, this volume Part 4:B). Exogenous application of ABA or salt also brings about profound changes in the photosynthesis of

C_3 and C_4 plants (Sankhla and Huber, 1974a, 1975). The activities of some photosynthetic enzymes is increased, and more label is incorporated into C_4 acids. Although the exact role of ABA in stress is not yet fully explained, these evidences point out that ABA may afford protection against water loss, and even facilitate the adaptation of plants to stress (see Itai and Benzioni, this volume Part 4:B; Hiron and Wright, 1973).

Similarly, salt is known to bring about a shift from C_3 to CAM or C_3 to C_4 pathway (see Winter and Lüttge, this volume Part 5:B; Sankhla and Huber, 1975). Marine algae, as well as mangroves, have also been shown to possess predominantly a C_4 pathway (Joshi et al., 1974). However in these plants an ability to perform C_3 photosynthesis has also been retained. For existence in a saline ecosystem, the C_4 pathway might have its own value in affording an additional advantage to the plants for their survival, and it appears to be an ecological necessity.

It seems that the C_4 pathway is also important for the adaptation of *Chlorella* and *Anacystis* (Döhler, 1974a, 1974b) to certain growth conditions. Studies of Willmer and Dittrich (1974) implicate a role by the C_4 pathway even in the metabolism of guard cells. However, additional information is necessary for correct evaluation of these findings.

VII. Conclusions

The bulk of evidence presented in the foregoing discussion indicates that it may be possible to regulate the balance between C_4 and C_3 pathways. The type of pathway may be greatly modulated by such factors as growth conditions, leaf ontogeny, mode of nutrition, and bio-regulants. The C_4 pathway can be experimentally induced in green algae, blue-green algae, and even C_3 plants. Under natural conditions, several plants living in desert regions, estuaries, and other extreme habitats have been shown to possess the C_4 pathway. Examples are also known where a shift from C_3 to Crassulacean acid metabolism was induced in response to stress. The C_4 pathway, therefore, may be looked upon as a complementary mechanism offering selective ecological advantage to the plant. To date, 13 families, 117 genera and 485 species of the angiosperms have been shown to possess the C_4 pathway of photosynthesis (Downton, 1975).

Examples of plants showing characteristics intermediate between the C_3 and C_4 pathway have also been recorded. Of the various criteria used to establish the identity of C_4 plants, the leaf anatomy has proved to be the most useful. However, the evidence presented in this chapter indicates, at least by implication, that spatial compartmentalization of leaf tissue need not be a prerequisite for the operation of C_4 photosynthesis. However, it has been argued that a perfect co-ordination between the biochemical steps in the spatial compartments of leaf tissue appears to be the key to an efficient C_4 photosynthesis. The mere presence of specialized anatomy or high PEP-carboxylase activity, or even the capacity to fix initially a high proportion of CO_2 into C_4 acids, is not sufficient for the characterization of the C_4 pathway (Björkman, 1973). To establish unequivocally the operation of the C_4 pathway in a species, the question must be solved, whether the produced C_4 acids represent the true intermediate products of photosynthesis, or whether they are merely a product of the activity of an isoenzyme of PEP-carboxylase that has no function in the C_4 photosynthesis. Further, the rapidity with which the transfer of label from C_4 acids to C_3 intermediates is accomplished by a shuttle between mesophyll and bundle-sheath cells, requires that the chase periods be accurately controlled. Further work will prove to be of great value in resolving these questions.

Acknowledgements. The authors are grateful to Prof. H. Ziegler for helpful suggestions and for very kindly providing the electron micrographs. The work was supported by grants from the Alexander von Humboldt Foundation and the Deutsche Forschungsgemeinschaft.

References

Addicott, F. T.: Biochemical aspects of abscisic acid. In: Plant growth substances. Proc. 7th Int. Conf. on Plant Growth Substances, Canberra 1970 (ed. D. J. Carr), pp. 272–280. Berlin-Heidelberg-New York: Springer 1972.

Allaway, W. G.: Accumulation of malate in guard cells of *Vicia faba* L. during stomatal opening. Planta (Berl.) **110**, 63–70 (1973).

Avadhani, P. N., Osmond, C. B., Tan, K. K.: Crassulacean acid metabolism and the C_4 pathway of photosynthesis in succulent plants. In: Photosynthesis and photorespiration (eds. M. D. Hatch, C. B. Osmond, R. O. Slatyer), pp. 288–293. New York: Wiley Interscience 1971.

Baldry, C. W., Bucke, C., Coombs, J.: Progressive release of carboxylating enzymes during mechanical grinding of sugarcane leaves. Planta (Berl.) **97**, 310–319 (1971).

Bassham, J. A.: The control of photosynthetic carbon metabolism. Science **172**, 526–534 (1971).

Bassham, J. A., Calvin, M.: The path of carbon in photosynthesis. Englewood Cliffs, N. J.: Prentice-Hall 1957.

Bauer, R., Huber, W., Sankhla, N.: Effect of abscisic acid on photosynthesis in *Lemna minor* L. Z. Pflanzenphysiol. **77**, 237–246 (1976).

Becker, W.: Physiologische Untersuchungen zur Photosynthese von Algen unter extremen Temperaturbedingungen. Dissertation Tübingen 1972.

Bender, M. M.: Variations in the $^{13}C/^{12}C$ ratios of plants in relation to the pathway of photosynthetic carbon dioxide fixation. Phytochemistry **10**, 1239–1244 (1971).

Berger, C., Pirson, A.: Aktivitätsänderungen einiger Enzyme synchronisierter *Chlorella*-Zellen im Licht-Dunkel-Wechsel II. Veränderungen bei gehemmtem Wachstum. Flora A **158**, 164–180 (1967).

Björkman, O.: Comparative photosynthetic CO_2 exchange in higher plants. In: Photosynthesis and photorespiration (eds. M. D. Hatch, C. B. Osmond, R. O. Slatyer), pp. 18–32. New York: Wiley Interscience 1971.

Björkman, O.: Comparative studies on photosynthesis in higher plants. In: Photophysiology (ed. A. C. Giese), Vol. 8, pp. 1–63. New York-London: Academic Press 1973.

Björkman, O., Gauhl, E.: Carboxydismutase activity in plants with and without β-carboxylation photosynthesis. Planta (Berl.) **88**, 197–203 (1969).

Björkman, O., Hiesey, W. M., Nobs, M., Nicholson, F., Hart, R. W.: Effect of oxygen concentration on dry matter production in higher plants. Carnegie Inst. Year Book, pp. 228–245. Washington, D.C.: Carnegie Inst. of Washington 1968.

Björkman, O., Pearcy, R. W., Harrison, A. T., Mooney, H.: Photosynthetic adaptation to high temperature: A field study in Death Valley, California. Science **175**, 786–789 (1972).

Black, C. C.: Ecological implications of dividing plants into groups with distinct photosynthetic production capacities. In: Advances in ecological research (ed. J. B. Cragg), Vol. 7, pp. 87–114. New York-London: Academic Press 1971.

Black, C. C.: Photosynthetic carbon fixation in relation to net CO_2 uptake. Ann. Rev. Plant Physiol. **24**, 253–286 (1973).

Black, C. C., Campbell, W. H., Chen, T. M., Dittrich, P.: The monocotyledons: their evolution and comparative biology. III. Pathways of carbon metabolism related to net carbon dioxide assimilation by monocotyledons. Quart. Rev. Biol. **48**, 299–313 (1973).

Black, C. C., Mollenhauser, H. H.: Structure and distribution of chloroplasts and other organelles in leaves with various rates of photosynthesis. Plant Physiol. **47**, 15–23 (1971).

Brownell, P. F., Crossland, C. J.: The requirement for sodium as a micronutrient by species having the C_4 dicarboxylic photosynthetic pathway. Plant Physiol. **49**, 794–797 (1972).

Bucke, C., Long, S. P.: Release of carboxylating enzymes from maize and sugar cane leaf tissue during progressive grinding. Planta (Berl.) **99**, 199–210 (1971).

Chang, F. H., Troughton, J. H.: Chlorophyll a/b ratios in C_3 and C_4 plants. Photosynthetica **6**, 57–65 (1972).

Chen, T. M., Brown, R. H., Black, C. C.: CO_2 compensation concentration, rate of photosynthesis, and carbonic anhydrase activity of plants. Weed Sci. **18**, 399–403 (1970).

Chen, T. M., Campbell, W. H., Dittrich, P., Black, C. C.: Distribution of carboxylation and decarboxylation enzymes in isolated mesophyll cells and bundle sheath strands of C_4 plants. Biochem. Biophys. Res. Commun. **51**, 461–467 (1973).

Chen, T. M., Dittrich, P., Campbell, W. H., Black, C. C.: Metabolism of epidermal tissues, mesophyll cells, and bundle sheath strands resolved from mature nutsedge leaves. Arch. Biochem. Biophys. **163**, 246–262 (1974).

Chollet, R., Ogren, W. L.: The Warburg effect in maize bundle sheath photosynthesis. Biochem. Biophys. Res. Commun. **48**, 684–688 (1972).

Chollet, R., Ogren, W. L.: Photosynthetic carbon metabolism in isolated maize bundle sheath strands. Plant Physiol. **51**, 787–792 (1973).

Coombs, J.: The potential of higher plants with the phosphopyruvic acid cycle. Proc. Roy. Soc. Ser. B **179**, 221–235 (1971).

Coombs, J., Baldry, C. W.: The C_4 pathway in *Pennisetum purpureum*. Nature (London) **238**, 268–270 (1972).

Coombs, J., Baldry, C. W., Bucke, C.: The C_4 pathway in *Pennisetum purpureum*. I. The allosteric nature of PEP-carboxylase. Planta (Berl.) **110**, 95–107 (1973a).

Coombs, J., Baldry, C. W., Bucke, C.: The C_4 pathway in *Pennisetum purpureum*. II. Malate dehydrogenase and malic enzyme. Planta (Berl.) **110**, 109–120 (1973b).

Coombs, J., Maw, S. L., Baldry, C. W.: Metabolic regulation in C_4 photosynthesis: PEP-carboxylase and energy charge. Planta (Berl.) **117**, 279–292 (1974).

Crookston, R. K., Moss, D. N.: The relation of carbon dioxide compensation and chlorenchymatous vascular bundle sheaths in leaves of dicots. Plant Physiol. **46**, 564–567 (1970).

Crookston, R. K., Moss, D. N.: A variation of C_4 leaf anatomy in *Arundinella hirta* (Gramineae). Plant Physiol. **52**, 397–402 (1973).

Cummins, W. R., Kende, H., Raschke, K.: Specificity and reversibility of the rapid stomatal response to abscisic acid. Planta (Berl.) **99**, 347–351 (1971).

Decker, J. P.: A rapid, post-illumination deceleration of respiration in green leaves. Plant Physiol. **30**, 82–84 (1955).

Dittrich, P., Salin, M. L., Black, C. C.: Conversion of carbon 4 of malate into products of the pentose cycle by isolated bundle sheath strands of *Digitaria sanguinalis* (L.) Scop. leaves. Biochem. Biophys. Res. Commun. **55**, 104–110 (1973).

Döhler, G.: C_4-Weg der Photosynthese in der Blaualge *Anacystis nidulans*. Planta (Berl.) **118**, 259–269 (1974a).

Döhler, G.: Einfluß der Kulturbedingungen auf die photosynthetischen Carboxylierungsreaktionen von *Chlorella vulgaris*. Z. Pflanzenphysiol. **71**, 144–153 (1974b).

Downes, R. W., Hesketh, J. D.: Enhanced photosynthesis at low oxygen concentrations: differential response of temperate and tropical grasses. Planta (Berl.) **78**, 79–84 (1968).

Downton, W. J. S.: Preferential C_4-dicarboxylic acid synthesis, the post-illumination CO_2 burst, carboxyl transfer step, and grana configuration in plants with C_4-photosynthesis. Can. J. Botany **48**, 1795–1800 (1970).

Downton, W. J. S.: Adaptive and evolutionary aspects of C_4 photosynthesis. In: Photosynthesis and photorespiration. (eds. M. D. Hatch, C. B. Osmond, R. O. Slatyer), pp. 3–17. New York: Wiley Interscience 1971.
Downton, W. J. S.: The occurrence of C_4 photosynthesis among plants. Photosynthetica **9**, 96–105 (1975).
Downton, W. J. S., Hawker, J. S.: Enzymes of starch and sucrose metabolism in *Zea mays* leaves. Phytochemistry **12**, 1551–1556 (1973).
Downton, W. J. S., Törökfalvy, E.: Effect of sodium chloride on the photosynthesis of *Aeluropus litoralis*, a halophytic grass. Z. Pflanzenphysiol. **75**, 143–150 (1975).
Downton, W. J. S., Tregunna, E. B.: Carbon dioxide compensation—its relation to photosynthetic carboxylation reactions, systematics of the Gramineae and leaf anatomy. Can. J. Botany **46**, 207–215 (1968).
Duffus, C. M., Rosie, R.: Some enzyme activities associated with the chlorophyll containing layers of the immature barley pericarp. Planta (Berl.) **114**, 219–226 (1973).
Edwards, G. E., Black, C. C.: Photosynthesis in mesophyll cells and bundle sheath cells isolated from *Digitaria sanguinalis* (L.) Scop. leaves. In: Photosynthesis and photorespiration. (eds. M. D. Hatch, C. B. Osmond, R. O. Slatyer), pp. 153–168. New York: Wiley Interscience 1971.
Edwards, G. E., Kanai, R., Black, C. C.: Phosphoenolpyruvate carboxykinase in leaves of certain plants which fix CO_2 by the C_4-dicarboxylic acid pathway of photosynthesis. Biochem. Biophys. Res. Commun. **45**, 278–285 (1971).
Edwards, G. E., Kanai, R., Ku, S. B., Gutierrez, M., Huber, S. C.: Compartmentation and coordination of CO_2 assimilation in isolated mesophyll protoplasts and bundle sheath cells of C_4 plants. In: Mechanism of regulation of plant growth. (eds. R. L. Bieleski, A. R. Ferguson, M. M. Cresswell), Bull. 12, pp. 203–211. Wellington: Roy. Soc. New Zealand 1974.
Evans, L. T.: Evolutionary, adaptive, and environmental aspects of the photosynthetic pathway: assessment. In: Photosynthesis and photorespiration. (eds. M. D. Hatch, C. B. Osmond, R. O. Slatyer), pp. 130–136. New York: Wiley Interscience 1971.
Farineau, J.: Photoassimilation of CO_2 by isolated bundle-sheath strands of *Zea mays* I. Stimulation of CO_2 assimilation by adding various intermediates of the photosynthetic cycle; evidence for a deficient photosystem II activity. Physiol. Plantarum **33**, 300–309 (1975a).
Farineau, J.: Photoassimilation of CO_2 by isolated bundle-sheath strands of *Zea mays* II. Role of malate as a source of CO_2 and reducing power for the photosynthetic activity of isolated bundle-sheath cells of *Zea mays*. Physiol. Plantarum **33**, 310–315 (1975b).
Fekete, M. A. R. de, Vieweg, G. H.: Über die Aktivität der Synthetase und Phosphorylase im Maisblatt bei verschiedenen Stärkegehalten. Planta (Berl.) **117**, 83–91 (1974).
Forrester, M. L., Krotkov, G., Nelson, C. D.: Effect of oxygen on photosynthesis, photorespiration and respiration in detached leaves I. Soybean. Plant Physiol. **41**, 422–427 (1966a).
Forrester, M. L., Krotkov, G., Nelson, C. D.: Effect of oxygen on photosynthesis, photorespiration and respiration in detached leaves II. Corn and other monocotyledons. Plant Physiol. **41**, 428–431 (1966b).
Frederick, S. E., Newcomb, E. H.: Ultrastructure and distribution of microbodies in leaves of grasses with and without CO_2-photorespiration. Planta (Berl.) **96**, 152–174 (1971).
Goatly, M. B., Smith, H.: Differential properties of phosphoenolpyruvate carboxylase from etiolated and green sugar cane. Planta (Berl.) **117**, 67–73 (1974).
Graham, D., Hatch, M. D., Slack, C. R., Smillie, R. M.: Light-induced formation of enzymes of the C_4 dicarboxylic acid pathway of photosynthesis in detached leaves. Phytochemistry **9**, 521–532 (1970).
Graham, D., Whittingham, C. P.: The path of carbon during photosynthesis in *Chlorella pyrenoidosa* at high and low carbon dioxide concentrations. Z. Pflanzenphysiol. **58**, 418–427 (1968).
Grossman, D., Cresswell, C. F.: Influence of nitrogen supply in nutrient media on carbon dioxide compensation point (T) of C_4 photosynthetic plants. South Afr. J. Sci. **69**, 244–246 (1973).

Gutierrez, M., Gracen, V. E., Edwards, G. E.: Biochemical and cytological relationships in C_4 plants. Planta (Berl.) **119**, 279–300 (1974).

Haberlandt, G.: Physiological plant anatomy. Translation of the 4th ed. London: McMillan 1914.

Hatch, M. D.: Mechanism and function of the C_4 pathway of photosynthesis. In: Photosynthesis and photorespiration. (eds. M. D. Hatch, C. B. Osmond, R. O. Slatyer), pp. 139–152. New York: Wiley Interscience 1971.

Hatch, M. D., Boardman, N. K.: Biochemistry of photosynthesis. In: Chemistry and biochemistry of herbage. (eds. G. W. Butler, R. W. Bailey), Vol. 2, pp. 25–55. New York-London: Academic Press 1973.

Hatch, M. D., Kagawa, T.: NAD-malic enzyme in leaves with C_4 photosynthesis and its role in C_4 acid decarboxylation. Arch. Biochem. Biophys. **160**, 346–349 (1974).

Hatch, M. D., Mau, S. L.: Activity, location and role of aspartate aminotransferase and alanine aminotransferase isoenzymes in leaves with C_4 pathway photosynthesis. Arch. Biochem. Biophys. **156**, 195–206 (1973).

Hatch, M. D., Osmond, C. B., Slatyer, R. O. (eds.): Photosynthesis and photorespiration. New York: Wiley Interscience 1971.

Hatch, M. D., Osmond, C. B., Troughton, J. H., Björkman, O.: Physiological and biochemical characteristics of C_3 and C_4 *Atriplex* species and hybrids in relation to the evolution of the C_4 pathway. Carnegie Inst. Wash. Year Book **71**, 135–141 (1972).

Hatch, M. D., Slack, C. R.: Photosynthesis by sugarcane leaves. A new carboxylation reaction and the pathway of sugar formation. Biochem. J. **101**, 103–111 (1966).

Hatch, M. D., Slack, C. R.: A new enzyme for the interconversion of pyruvate and phosphopyruvate and its role in the C_4-dicarboxylic acid pathway of photosynthesis. Biochem. J. **106**, 141–146 (1968).

Hatch, M. D., Slack, C. R.: The C_4-dicarboxylic pathway of photosynthesis. Progr. Phytochem. **2**, 35–106 (1970).

Hiron, R. W. P., Wright, S. T. C.: The role of endogenous abscisic acid in the response of plants to stress. J. Exp. Botany **24**, 769–781 (1973).

Holden, M.: Chloroplast pigments in plants with the C_4-dicarboxylic acid pathway of photosynthesis. Photosynthetica **7**, 41–49 (1973).

Huang, A. H. C., Beevers, H.: Microbody enzymes and carboxylases in sequential extracts from C_4 and C_3 leaves. Plant Physiol. **50**, 242–248 (1972).

Huber, S. C., Kanai, R., Edwards, G. E.: Decarboxylation of malate by isolated bundle sheath cells of certain plants having the C_4-dicarboxylic pathway of photosynthesis. Planta (Berl.) **113**, 53–66 (1973).

Huber, W., Sankhla, N.: Effect of gibberellic acid on the activities of photosynthetic enzymes and $^{14}CO_2$ fixation products in leaves of *Pennisetum typhoides* seedlings. Z. Pflanzenphysiol. **71**, 275–280 (1974a).

Huber, W., Sankhla, N.: Activity of malate dehydrogenase in leaves of abscisic and gibberellic acids treated seedlings of *Pennisetum typhoides*. Z. Pflanzenphysiol. **71**, 86–89 (1974b).

Huber, W., Sankhla, N., Ziegler, H.: Eco-physiological studies on Indian arid zone plants I. Photosynthetic characteristics of *Pennisetum typhoides* (Burm. f.) Stapf and Hubbard and *Lasiurus sindicus* Henr. Oecologia **13**, 65–71 (1973a).

Jackson, W. A., Volk, R. J.: Photorespiration. Ann. Rev. Plant Physiol. **21**, 385–432 (1970).

Johnson, H. S., Hatch, M. D.: Properties and regulation of leaf nicotinamide-adenine dinucleotide phosphate-malate dehydrogenase and 'malic' enzyme in plants with the C_4-dicarboxylic acid pathway of photosynthesis. Biochem. J. **119**, 273–280 (1970).

Joshi, G. V., Karkar, M. D., Gowda, C. A., Bhosale, L.: Photosynthetic carbon metabolism and carboxylating enzymes in algae and mangrove under saline conditions. Photosynthetica **8**, 51–52 (1974).

Kagawa, T., Hatch, M. D.: Light-dependent metabolism of carbon compounds by mesophyll chloroplasts from plants with the C_4-pathway of photosynthesis. Australian J. Plant Physiol. **1**, 51–64 (1974).

Kanai, R., Edwards, G. E.: Separation of mesophyll protoplasts and bundle sheath cells from maize leaves for photosynthetic studies. Plant Physiol. **51**, 1133–1137 (1973).

Karkar, M. D., Joshi, G. V.: Photosynthetic carbon metabolism in marine algae. Botan. Marina **16**, 216–220 (1973).

Karpilov, Y. S.: The photosynthesis of xerophytes. Kishinev: Academcy of Science Moldavian SSR 1970 (cit. by O. Björkman: Comparative studies on photosynthesis in higher plants. In: Photophysiology (ed. A. C. Giese), Vol. 8, pp. 1–63. New York-London: Academic Press 1973).

Kennedy, R. A., Laetsch, W. M.: Relationship between leaf development and primary photosynthetic products in the C_4 plant *Portulaca oleracea* L. Planta (Berl.) **115**, 113–124 (1973).

Kennedy, R. A., Laetsch, W. M.: Plant species intermediate for C_3, C_4 photosynthesis. Science **184**, 1087–1089 (1974).

Khanna, R., Sinha, S. K.: Change in predominance from C_4 to C_3 pathway following anthesis in *Sorghum*. Biochem. Biophys. Res. Commun. **52**, 121–124 (1973).

Kortschak, H. P., Hartt, C. E., Burr, G.: Carbon dioxide fixation in sugarcane leaves. Plant Physiol. **40**, 209–214 (1965).

Ku, S. B., Gutierrez, M., Kanai, R., Edwards, G. E.: Photosynthesis in mesophyll protoplasts and bundle sheath cells of various types of C_4 plants. II. Chlorophyll and Hill reaction studies. Z. Pflanzenphysiol. **72**, 320–337 (1974).

Laetsch, W. M.: The C_4 syndrome: a structural analysis. Ann. Rev. Plant Physiol. **25**, 27–52 (1974).

Laetsch, W. M., Kortschak, H. P.: Chloroplast structure and function in tissue cultures of a C_4 plant. Plant Physiol. **49**, 1021–1023 (1972).

Latzko, E., Kelly, G. J.: Carbon metabolism. In: Progress in Botany **36**, pp. 77–89. Berlin-Heidelberg-New York: Springer 1974.

Lester, J. N., Goldsworthy, A.: The occurrence of high CO_2-compensation points in *Amaranthus* species. J. Exp. Botany **24**, 1031–1034 (1973).

Lowe, J., Slack, C. R.: Inhibition of maize leaf phosphopyruvate carboxylase by oxaloacetate. Biochim. Biophys. Acta **235**, 207–209 (1971).

Maruyama, H., Eusterday, R. L., Chang, H. C., Lane, M. D.: The enzymatic carboxylation of phosphoenol pyruvate. I. Purification and properties of phosphoenol pyruvate carboxylase. J. Biol. Chem. **241**, 2405–2412 (1966).

McWilliam, J. R., Mison, K.: Significance of the C_4-pathway in *Triodia irritans* (Spinifex), a grass adapted to arid environments. Australian J. Plant Physiol. **1**, 171–175 (1974).

Meidner, H.: The minimum intercellular-space CO_2-concentration of maize leaves and its influence on stomatal movements. J. Exp. Botany **13**, 284–293 (1962).

Milborrow, B. V.: The chemistry and physiology of abscisic acid. Ann. Rev. Plant Physiol. **25**, 259–307 (1974).

Mittelheuser, C. J., van Steveninck, R. F. H.: Rapid action of abscisic acid on photosynthesis and stomatal resistance. Planta (Berl.) **97**, 83–86 (1971).

Mühlbach, H. P., Wegmann, K.: Photosynthetischer CO_2-Einbau und Photorespiration bei isolierten Zellen und Protoplasten aus Sonnenblumen und Mais. Hoppe-Seyler's Z. Physiol. Chem. **354**, 1226 (1973).

O'Neal, D., Hew, C. S., Latzko, E., Gibbs, M.: Photosynthetic carbon metabolism of isolated corn chloroplasts. Plant Physiol. **49**, 607–614 (1972).

Osmond, C. B.: Carbon reduction and photosystem II deficiency in leaves of C_4 plants. Australian J. Plant Physiol. **1**, 41–50 (1974).

Osmond, C. B., Ziegler, H.: Schwere Pflanzen und leichte Pflanzen: Stabile Isotope im Photosynthesestoffwechsel und in der biochemischen Ökologie. Naturwiss. Rundschau **28**, 323–328 (1975).

Patil, B. A., Joshi, G. V.: Photosynthetic studies in *Ulva lactuca*. Botan. Marina **13**, 111–115 (1970).

Phillips, P. J., McWilliam, J. R.: Thermal response of the primary carboxylating enzymes from C_3 and C_4 plants adapted to contrasting temperature environments. In: Photosynthesis and photorespiration. (eds. M. D. Hatch, C. B. Osmond, R. O. Slatyer), pp. 97–104. New York: Wiley Interscience 1971.

Poskuta, J., Antoszewski, R., Faltynowicz, M.: Photosynthesis, photorespiration and respiration of strawberry and maize leaves as influenced by abscisic acid. Photosynthetica **6**, 370–374 (1972).

Salin, M. L., Campbell, W. H., Black, C. C.: Oxaloacetate as the Hill oxidant in mesophyll cells of plants possessing the C_4-dicarboxylic acid cycle of leaf photosynthesis. Proc. Nat. Acad. Sci. **70**, 3730–3734 (1973).

Sankhla, N.: Morphological, physiological and biochemical studies on arid zone plants: A comparative study of the effect of bioregulants on growth of some desert as well as other plants. D. Sc. Thesis, Univ. Jodhpur, India 1971.

Sankhla, N., Huber, W.: Eco-physiological studies on Indian arid zone plants. IV. Effect of salinity and gibberellin on the activities of photosynthetic enzymes and $^{14}CO_2$-fixation products in leaves of *Pennisetum typhoides* seedlings. Biochem. Physiol. Pflanzen (BPP) **166**, 181–187 (1974a).

Sankhla, N., Huber, W.: Effect of abscisic acid on the activities of photosynthetic enzymes and $^{14}CO_2$-fixation products in leaves of *Pennisetum typhoides* seedlings. Physiol. Plantarum **30**, 291–294 (1974b).

Sankhla, N., Huber, W.: Regulation of balance between C_3 and C_4 pathway: Role of abscisic acid. Z. Pflanzenphysiol. **74**, 267–271 (1975).

Sankhla, N., Sankhla, D.: Reversal of ($\pm$)-abscisin II induced inhibition of lettuce seed germination and seedling growth by kinetin. Physiol. Plantarum **21**, 190–195 (1968).

Shomer-Ilan, A., Waisel, Y.: The effect of sodium chloride on the balance between the C_3 and C_4 carbon fixation pathways. Physiol. Plantarum **29**, 190–193 (1973).

Smith, B. N.: Natural abundance of the stable isotopes of carbon in biological systems. Bio. Science **22**, 226–231 (1972).

Smith, B. N., Brown, W. V.: The Kranz syndrome in the Gramineae as indicated by carbon isotopic ratios. Am. J. Botany **60**, 505–513 (1973).

Smith, B. N., Epstein, S.: Two categories of $^{13}C/^{12}C$ ratios for higher plants. Plant Physiol. **47**, 380–384 (1971).

Sugiyama, T.: Purification, molecular, and catalytic properties of pyruvate phosphate dikinase from the maize leaf. Biochemistry **12**, 2862–2868 (1973).

Ting, I. P., Osmond, C. B.: Photosynthetic phosphoenolpyruvate carboxylase: characteristics of alloenzymes from leaves of C_3 and C_4 plants. Plant Physiol. **51**, 439–447 (1973a).

Ting, I. P., Osmond, C. B.: Multiple forms of plant phosphoenolpyruvate associated with different metabolic pathways. Plant Physiol. **51**, 448–453 (1973b).

Tolbert, N. E., Yamazaki, R. K.: Leaf peroxisomes and their relation to photorespiration and photosynthesis. Ann. N. Y. Acad. Sci. **168**, 325–341 (1969).

Tregunna, E. B., Downton, J.: Carbon dioxide compensation in members of the *Amaranthaceae* and some related families. Can. J. Botany **45**, 2385–2387 (1967).

Treharne, K. J., Cooper, J. P.: Effect of temperature on the activity of carboxylases in tropical and temperature gramineae. J. Exp. Botany **20**, 170–175 (1969).

Troughton, J. H.: Aspects of the evolution of the photosynthetic carboxylation reaction in plants. In: Photosynthesis and photorespiration. (eds. M. D. Hatch, C. B. Osmond, R. O. Slatyer), pp. 124–129. New York: Wiley Interscience 1971.

Wellburn, F. A. M., Wellburn, A. R., Soddart, J. L., Treharne, K. J.: Influence of gibberellic and abscisic acids and the growth retardant, CCC, upon plastid development. Planta (Berl.) **111**, 337–346 (1973).

Williams III, G. J.: Photosynthetic adaptation to temperature in C_3 and C_4 grasses. A possible ecological role in the shortgrass prairie. Plant Physiol. **54**, 709–711 (1974).

Williams III, G. J., Markley, J. L.: The photosynthetic pathway type of North American shortgrass prairie species and some ecological implications. Photosynthetica **7**, 262–270 (1973).

Willmer, C. M., Dittrich, P.: Carbon dioxide fixation by epidermal and mesophyll tissues of *Tulipa* and *Commelina*. Planta (Berl.) **117**, 123–132 (1974).

Willmer, C. M., Kanai, R., Pallas, J. E., Black, C. C.: Detection of high levels of phosphoenolpyruvate carboxylase in leaf epidermal tissue and its significance in stomatal movements. Life Sci. **12**, 151–155 (1973a).

Willmer, C. M., Pallas, J. E., Black, C. C.: Carbon dioxide metabolism in leaf epidermal tissue. Plant Physiol. **52**, 448–452 (1973b).
Zelitch, I.: Water and CO_2 transport in the photosynthetic process. In: Harvesting the sun. (eds. A. San Pietro, F. A. Greer, T. J. Army), pp. 231–248. New York-London: Academic Press 1967.
Zelitch, I.: Photosynthesis, photorespiration and plant productivity. New York-London: Academic Press 1971.
Zelitch, I.: Plant productivity and the control of photorespiration. Proc. Nat. Acad. Sci. **70**, 579–584 (1973).

D. Ecophysiology of C_4 Grasses

M. M. LUDLOW

I. Introduction

In recent years the C_4 pathway of carbon dioxide fixation in plants has been associated with specific anatomical, physiological, and biochemical characteristics (Hatch et al., 1971; Black, 1973). Grasses exhibiting this pathway have become known as C_4 grasses and they are restricted to the sub-families Panicoideae and Eragrostoideae of the family Gramineae (Poaceae) (Smith and Brown, 1973). Although C_4 grasses occur in areas from the hot-wet tropics to the cool temperate their greatest concentration is in the semi-arid tropics and sub-tropics, which are characterised by high solar radiation, high daytime temperatures and evaporative demand, and frequent diurnal and seasonal water stress (Hartley, 1958a, b; Hartley and Slater, 1960; Coaldrake, 1964; Cooper, 1970).

It has been proposed that the C_4 syndrome evolved as an adaptation to hot, dry (and possibly saline; Lipschitz et al., 1974) conditions (Laetsch, 1969; Black, 1971; Björkman, 1973). To test this hypothesis the environmental conditions (solar radiation, temperature, and water stress) of areas where C_4 grasses occur naturally, or to which they have become adapted (with or without the assistance of man) will be described. In addition, the physiological responses of C_4 grasses to these environmental conditions will be discussed, and where appropriate, compared with C_3 plants to see if they have intrinsic characteristics which would give them an ecological advantage for growth and survival in such areas. The C_3 plants have been grouped for convenience as herbaceous (mainly mesophytic agricultural dicotyledons and monocotyledons), trees (and other woody species from temperate regions), and semi-arid and arid shrubs. There are many C_4 herbaceous and shrub plants, but as data are only available for *Atriplex* they have been included with C_3 semi-arid and arid shrubs, which appear to have similar characteristics.

Most emphasis is placed upon water stress because of its dominant influence on growth and survival, and because this aspect of the C_4 syndrome has not been reviewed previously. Unfortunately, there is a paucity of data on the physiological responses of C_4 grasses (especially to water stress), despite their importance as tropical crop (maize, sorghum, sugar cane, and millet) and pasture (Hutton, 1970; Ludlow and Wilson, 1968, 1971a) plants and as a component of the world's natural grasslands (Harlan, 1956). There is more physiological information for crop plants whose genetic composition has been altered by breeding and selection than for tropical pasture plants which have been mainly selected from wild populations. However, there is only limited information on the ecologically more interesting C_4 native grasses and "resurrection" plants (Gaff, 1971). Most information reviewed here is from plants grown under controlled conditions (growth cabinets, growth rooms, glasshouses, and phytotrons). However, it has been shown that the performance of field-grown plants differs quantitatively, if not qualitatively, from those grown under controlled conditions. Therefore in the case of response to water stress, data from potted plants grown under controlled conditions are separated from those of field-grown plants so that more meaningful comparisons can be made among species (see Jordan and Ritchie, 1971; McCree, 1974; Turner, 1974). Moreover, plants exhibit considerable adaptation to environmental conditions in both the field and controlled conditions (Caldwell, 1972; Björkman, 1973; Lange et al., 1974; McCree, 1974; McWilliam and Ferrar, 1974). Thus it is evident that great care must be exercised in extrapolating data from controlled environments to the field, and from experimental conditions to other situations.

II. Environmental Conditions

Annual short-wave radiation received in the tropics is high with comparatively little seasonal variation compared with the sub-tropics which have a marked seasonal variation and even higher annual receipts because of less cloudiness (Cooper, 1970; Monteith, 1972). Both regions have high midday illuminances and quantum fluxes of photosynthetically active radiation, with maximum values being received in summer when temperatures are the highest and the majority of the rainfall is received. C_4 grasses occur mainly in areas where average temperatures of the coldest month exceeds 5–$10°C$ and their spread into higher latitudes is limited by minimum winter temperatures (Hartley, 1958a, b; Hartley and Slater, 1960; Cooper, 1970). While foliage of C_4 grasses is killed by comparatively mild frosts (higher than $-5°C$; Hacker et al., 1974; Ivory, 1975; Ludlow, Taylor and Rowley, unpublished), some species survive and regrow from dormant crown buds even in areas where frosts are frequent and severe.

Some areas where C_4 grasses occur have a high uniformally-distributed rainfall (e.g. hot-wet tropics), but generally rainfall is seasonal with annual totals varying from high in monsoonal climates to low in semi-arid tropics and sub-tropics.

Seasonal rainfall has a predominantly summer incidence (Hartley, 1958a, b; Hartley and Slater, 1960), and the resulting water shortage during the remainder of the year is the most important climatic limitation to growth (Cooper, 1970). Under such conditions seasonal leaf water potentials down to -100 bars have been recorded (Peake et al., 1975; M. J. Fisher, personal communication). The most extreme deficits (-1500 to -2000 bars) are experienced by leaves of C_4 grasses with "resurrection" characteristics which remain air-dry for several months in the arid deserts of South-West Africa (Gaff, 1971). Temporary deficits (down to -20 bars) can occur during the day, and short-term deficits (down to -50 bars) for periods varying from days to weeks, when evaporative demand exceeds soil water supply in areas of high, uniform rainfall and during the rainy season in other areas.

III. Physiological Responses to Environmental Conditions

1. Solar Radiation

When net photosynthesis is not limited by environmental and physiological factors, the light response curve of C_4 grass leaves is not light-saturated at midday summer illuminances, whereas in C_3 plants saturated occurs between a third and a half of this value (Fig. 1; Ludlow and Wilson, 1971 a). The absence of saturation in C_4 grass leaves results from a continual decline of stomatal resistance with increasing illuminance, together with a low intracellular resistance (Gifford, 1971; Ludlow and Wilson, 1971 a; McPherson and Slatyer, 1973). However, the marked difference which exists between leaves of C_3 and C_4 species is gradually eroded as self and mutual shading increases in individual plants and in communities, such that net photosynthesis of communities of C_3 and C_4 species respond almost linearly to solar radiation, being steeper in C_4 than in C_3 crop species (Monteith, 1972; Gifford, 1974; Ludlow, unpublished).

Because of this strong dependence of net photosynthesis on solar radiation, the growth rate of C_4 grasses can be reduced by up to 40% as a result of seasonal variation of solar radiation in the sub-tropics (Monteith, 1972). Moreover, the growth of individual plants and communities of C_4 grasses is reduced by shading, despite increases in the relative size of the photosynthetic system (usually expressed as leaf area ratio) which partially compensate for the lower net assimilation rate. Although the growth of C_4 grasses is reduced more than that of

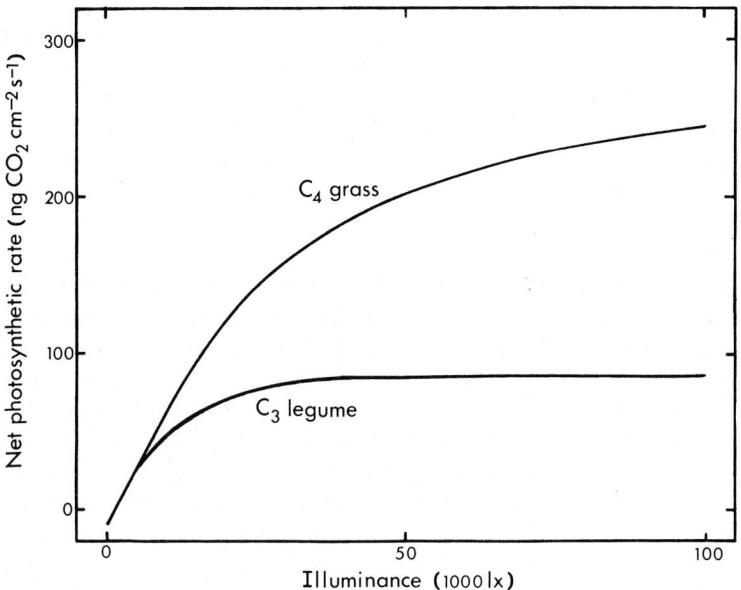

Fig. 1. Light response curves of net photosynthesis of a C_4 grass and a C_3 legume at optimum leaf temperatures and ambient carbon dioxide concentrations

C_3 legumes, neither appears to have a superior ability to grow under shaded conditions (Ludlow et al., 1974). Therefore, when they are grown together, success in competition for light probably arises when one gains preferential access to incident radiation.

When environmental (e.g. water stress; Fig. 4) or physiological (e.g. leaf age, Ludlow and Wilson, 1971 b) factors limit net photosynthesis of C_4 grasses light-saturation occurs. Under these conditions, small reductions in radiation will probably have few detrimental effects on the growth of C_4 plants. In fact, they may partially alleviate the effects of high ambient temperatures or high leaf temperatures associated with water stress or effects of mineral nutrient deficiencies, similar to some tropical plantation crops which are normally grown under shade trees.

2. Temperature

Cardinal temperatures for both growth and net photosynthesis are, in general, higher for C_4 grasses than for herbaceous C_3 plants (Cooper and Tainton, 1968; Cooper, 1970; Ludlow and Wilson, 1970, 1971 a; Black, 1973; Hirose, 1973; Doley and Trivett, 1974; Ivory, 1975; cf. also Huber and Sankhla, this volume Part 5:D):

	Minimum	Optimum (°C)	Maximum
Photosynthesis			
C_4 grasses	5–10	35–45	45–60
C_3 plants (herbaceous)	−10–0	15–30	35–45
Growth			
C_4 grasses	10–15	30–40	40–50
C_3 plants (herbaceous)	0–10	10–30	30–40

However, C_3 shrubs from arid and semi-arid areas have optimum (37°C) and maximum (52°C) temperatures for net photosynthesis similar to those of C_4 grasses (Hellmuth, 1971; see also Larcher, 1969). Net photosynthesis of both C_4 grasses and C_3 shrubs from semi-arid and arid areas respond more to temperature than does that of herbaceous C_3 plants, when only temperature limits photosynthesis (Fig. 2; Hellmuth, 1971; Ludlow and Wilson, 1971 a).

Foliage of most C_4 grasses is killed by mild frosts of −2 to −3°C, but some species of the genera *Paspalum*, *Setaria*, and *Chloris* can withstand frosts of −5° (Smith and Boyd, 1969; Cary and Mayland, 1970; Hacker et al., 1974; Ivory, 1975; Ludlow, Taylor and Rowley, unpublished). Although most C_4 grasses suffer chilling injury, some native (e.g. *Themeda australis*), "weed" and pasture (e.g. *Sorghum halepense*, *Cynodon dactylon*, *Digitaria sanguinalis* and *Echinochloa crus-galli*) grasses have acquired some chilling tolerance which has allowed them

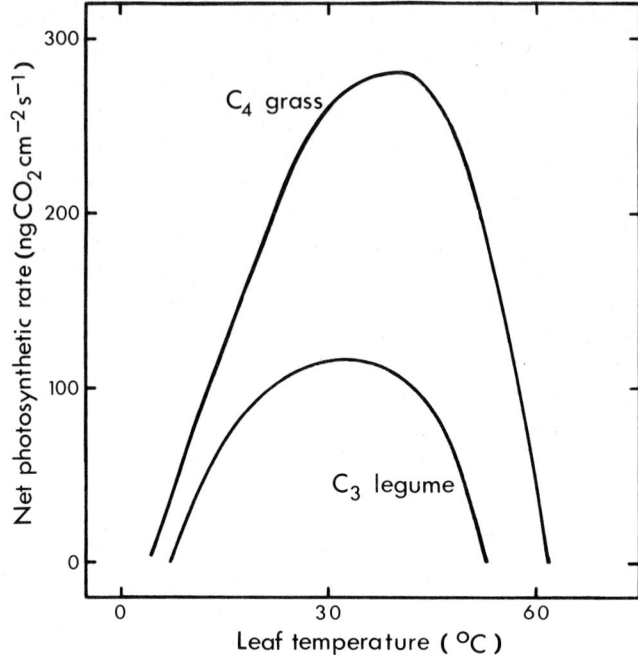

Fig. 2. Influence of leaf temperature on net photosynthesis of a C_4 grass *Cenchrus ciliaris* and a C_3 legume *Calopogonium mucunoides* at high illuminance and ambient carbon dioxide concentrations

to invade and survive in areas with a much more severe winter environment than is found in their natural habitats (McWilliam and Ferrar, 1974). This tolerance appears to be associated with greater membrane stability at chilling temperatures of about 12°C. C_4 grasses are also injured when leaves at or below chilling temperatures are exposed to relatively high illuminances (Taylor and Craig, 1971; Taylor and Rowley, 1971). On the other hand, the susceptibility of C_3 plants to chilling and frost injury varies from the highly sensitive tropical and sub-tropical plants to the highly resistant ones of high latitudes.

The thermal death point and heat tolerance of leaves of herbaceous C_3 plants are lower than those of woody C_3 species, which in turn are lower than those of C_4 grasses and C_3 shrubs from semi-arid and arid areas (Hellmuth, 1971; Levitt, 1972).

3. Water Stress

The physiological responses discussed are those associated with the carbon and water balance, and the expansion, death, and drought tolerance of leaves.

a) Net Photosynthesis. Under otherwise optimum conditions leaf net photosynthesis declines as leaf water potential decreases, either linearly (El-Sharkawy and Hesketh, 1964; Boyer, 1970a, b; Doley and Trivett, 1974) or in a reverse sigmoid (Fig. 3). Differences in the shape of the photosynthetic response may well be associated with effects of factors (such as growth conditions) other than those of water potential on net photosynthesis because both shapes can be found in the one species; for example, corn (Boyer, 1970a; cf. Beadle et al., 1973) and sorghum (El-Sharkawy and Hesketh, 1964; cf. Beadle et al., 1973). The leaf water potential at which net photosynthesis is zero varies as much among C_4 grasses as among all plants (Table 1). In controlled environments values for cultivated C_4 grasses are similar to herbaceous and woody C_3 species. However, the value for *Astrebla* which is a native of semi-arid Australia, but when grown under controlled conditions, approaches the low values of C_3 plants from semi-arid and arid areas of Australia and America grown in both the field and in controlled conditions (see also Caldwell, 1974). Although the average values for maize and *Sorghum* in Table 1 are similar when they were compared in the same experiment by Beadle et al. (1973), values for sorghum (-16 bars) were lower than those for maize (-11 bars). In summary, the values seem to be related to the environment to which plants have adapted rather than to the fact of whether plants are C_3 or C_4. There are insufficient data to draw any conclusions on the difference between plants grown under controlled conditions and in the field. However, circumstantial evidence based on stomatal behaviour (see later), and net photosynthesis of leaves from controlled environments and of communities in the field suggests that the leaf water potential at which photosynthesis is zero of field-grown plants is lower than that of controlled environment-grown plants (Fig. 4).

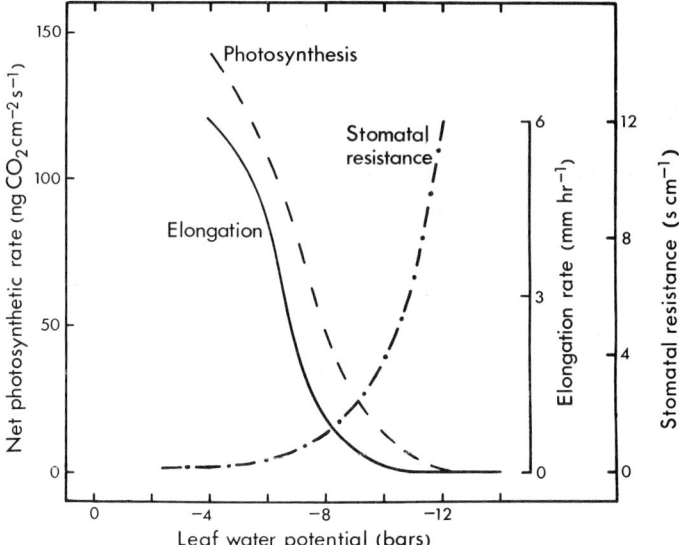

Fig. 3. Influence of leaf water potential on net photosynthetic rate, elongation rate and stomatal resistance of leaves of *Panicum maximum* at high quantum fluxes, optimum temperature, and ambient carbon dioxide concentrations. (From Ludlow and Ng, 1976)

Table 1. Leaf water potential (bars) at which net photosynthesis is zero. Mean values are given, and the range in parentheses. Equivalent relative water contents (RWC) are also given

	Controlled environments	Field	References[a]
C$_4$ grasses			
Zea mays	−15	−40	1, 2, 3, 4, 5, 24
Sorghum bicolor	−17		3, 6, 7
Panicum maximum	−12		8
Astrebla lappacea	−50		9
C$_3$ plants			
Herbaceous			
Atriplex hastata, cotton	−21		1, 2, 4, 7, 11,
Lolium temulentum, soybean, sugar	(−13 to −40)		12
beet, sunflower, tomato	[63% (55 to 72) RWC]		13
Phaseolus vulgaris		−12	14
Trees			
Abies, Pinus	−14		10, 15
Fagus, Olea, Quercus	(−12 to −16)		16, 17
Pyrus		−35 [83% (79–88) RWC]	
Semi-arid and arid shrubs			
Acacia harpophylla	−75	−75	18, 19
Atriplex, Chilipsis,		−61	20, 21, 22
Encelia, Eurotia, Larrea		(−35 to −80)	
Reaumuria hirtella		−340	23

[a] 1 Boyer (1970a); 2 Boyer (1970b); 3 Beadle et al. (1973); 4 Glinka and Katchansky (1970); 5 Wesselius and Brouwer (1972); 6 Pasternak (1971); 7 El-Sharkawy and Hesketh (1964); 8 Ludlow and Ng (1976); 9 Doley and Trivett (1974); 10, Brix (1962); 11 Slatyer (1970); 12 Wardlaw (1969); 13 Lawlor (1973); 14 Millar and Gardner (1972); 15 Hinckley and Ritchie (1972); 16 Kriedemann and Caterford (1971); 17 Larcher (1969); 18 van den Driessche et al. (1971); 19 Tunstall (1973); 20 Moore et al. (1972); 21 Odening et al. (1974); 22 Oechel et al. (1972a, b); 23 Whiteman and Koller (1964); 24 Heichel and Musgrave (1970).

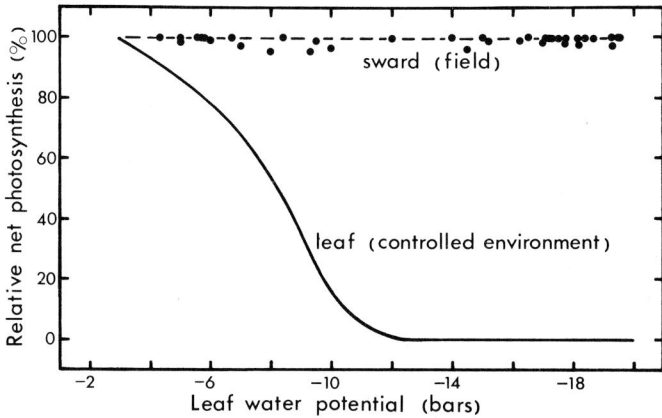

Fig. 4. Influence of leaf water potential on relative net photosynthesis of a *Panicum maximum* leaf from a plant grown in controlled environment rooms and of a mixed *Setaria anceps–Macroptilium atropurpureum* sward in the field. (From Ludlow and Davis, unpublished)

At moderate leaf water potentials (-6 to -8 bars) cultivated C_4 grasses such as *Panicum maximum* grown in controlled environments behave like C_3 plants with low photosynthetic rates and with light saturation at low quantum fluxes (Fig. 5). If field grown plants behave similarly, they may not always be able to express their photosynthetic potential because greater deficits than these develop even when soil water supply is adequate but atmospheric demand, is high, as well as when soil water supply is inadequate (Begg et al., 1964; Peake et al., 1975; Ludlow and Ibaraki, unpublished). However, young fully-expanded leaves of *P. maximum* can recover from leaf water potentials of -80 to -100 bars in both the field (D. C. I. Peake, personal communication; M. J. Fisher, personal communication) and controlled environments (Ludlow and Ng, 1974; Ludlow, 1975). In controlled environments, at least, this is accompanied by complete recovery of net photosynthesis. The generality of this behavior is not yet determined.

b) Respiration. Dark respiration rate of most plants, including the C_4 grasses *Astrebla* (D. Doley, personal communication) and *Panicum maximum* (Ludlow and Ng, 1976), declines progressively as leaf water potential decreases (Boyer, 1970a; Hsiao, 1973). *Sorghum* (Shearman et al., 1972) and loblolly pine (Brix, 1962) are exceptional in that dark respiration increases to a maximum at leaf water potentials of about -20 bars and then it declines.

Photorespiration of C_3 plants decreases as leaf water potential decreases (Boyer, 1970a; Hsiao, 1973). Photorespiration cannot be detected in unstressed leaves of C_4 grasses (Ludlow and Wilson, 1972), and there is no good evidence that it increases with increasing water stress. The fact that the carbon dioxide compensation point increases in stressed leaves (Meidner, 1967; Glinka and Katchansky, 1970) is not evidence of increased photorespiration unless photosynthesis is unaffected, because the compensation point is determined by the balance between photosynthesis and respiration (Ludlow and Jarvis, 1971).

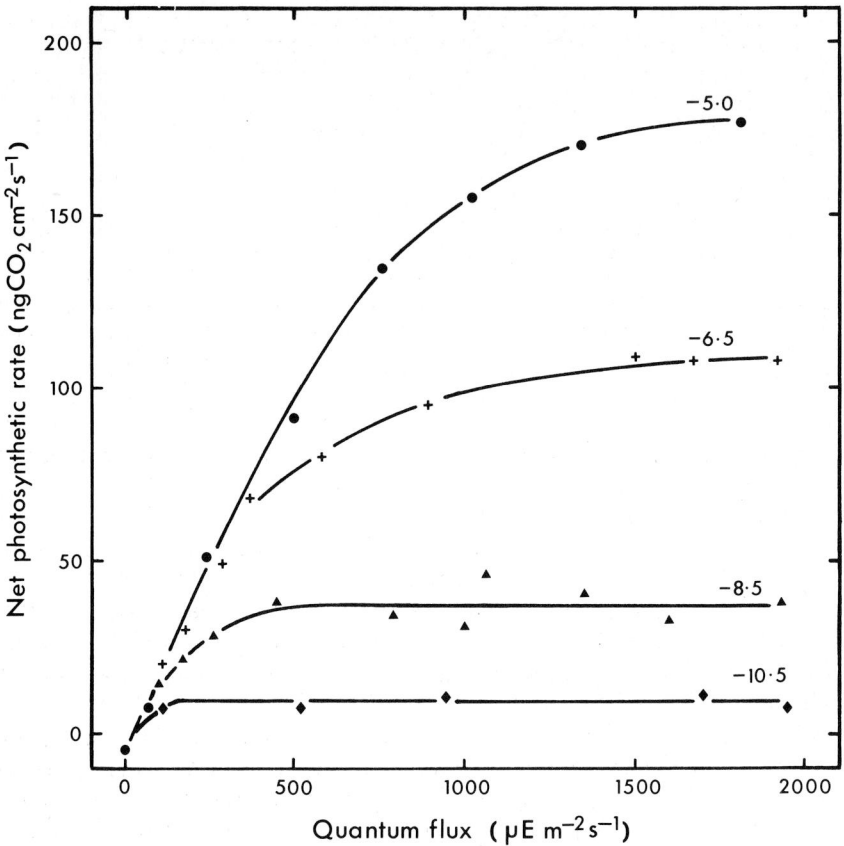

Fig. 5. Light response curves of leaves of *Panicum maximum* at various leaf water potentials (bars) measured at optimum temperature and ambient carbon dioxide concentrations. (From Ludlow and Ng, 1976)

c) Leaf Expansion. Leaf expansion is one of the most sensitive plant processes to water stress, more sensitive than net photosynthesis in a number of C_3 herbaceous plants and in *Zea mays* (Wardlaw, 1969; Boyer, 1970a; Lawlor and Milford, 1973). On the other hand, net photosynthesis and leaf expansion of *Panicum maximum* seem to behave similarly (Fig. 3). The shape of the response curve is either reverse sigmoid (e.g. *P. maximum*) or linear. The limited data available from controlled environment experiments indicate that neither the sensitivity of leaf expansion to water stress, nor the leaf water potential at which expansion reaches zero, differ between C_3 and C_4 plants (Table 2).

Although net photosynthesis may recover partially or almost completely after water stress is removed there is no evidence that it exceeds that of controls of the same physiological age (Ludlow, 1975). However in both *Z. mays* (Acevedo et al., 1971; Boyer, 1970a) and *P. maximum* (Ludlow and Ng, unpublished) elongation of leaves after mild stress is removed can exceed that of controls,

Table 2. Leaf water potential (bars) at which leaf expansion reaches zero of plants grown in controlled environments. Means and range (parentheses) of values are given

C_4 grasses		References[a]
Zea mays	−9	1, 2, 3, 4
Panicum maximum	−11	
C_3 plants (herbaceous)		
Bean, cotton, onion, potato, red clover, ryegrass, sugar beet, sunflower, soybean	−9 (−3 to −16)	2, 3, 5, 6, 7 8, 9, 10, 11

[a] 1 Acevedo et al. (1971); 2 Boyer (1970a); 3 Lawlor (1969); 4 Ludlow and Ng (1976); 5 Jordan (1970); 6 Lawlor and Milford (1973); 7 Lawlor (1972b); 8 Garwood and Gowman (1973); 9 Wardlaw (1969); 10 Millar et al. (1971); 11 Gandar and Tanner (1974).

and wholly or partially compensate for the reduced elongation rate during the stress.

If leaves of C_4 grasses in the field are as sensitive to water deficits as those from controlled environments, most leaf expansion would occur at night because of the deficits which develop during the day, even when the soil is at field capacity. However, there is a small amount of information which indicates that leaf expansion of Z. mays in the field is less sensitive than that of plants grown under controlled conditions, and that leaf expansion proceeds during the day (Cary and Wright, 1971; Watts, 1974).

d) **Leaf Death and Senescence.** It is well known that water loss from plants under stress is lowered by reduction of leaf area as leaves die progressively from the base of the plant (Oppenheimer, 1960); for example, the percentage of dead leaves on P. maximum plants increases as leaf water potential decreases until the last fully-expanded leaf is dead between −100 and −120 bars in the field (D. C. I. Peake, personal communication) and under controlled conditions (Ludlow, 1975). Ludlow (1975) proposed that death is due to direct killing by water stress rather than to accelerated senescence, and that it proceeds progressively up the plant because the dehydration avoidance aspect of drought tolerance (Levitt, 1972) increases with leaf number from the oldest to the youngest leaf (Ludlow, 1975; Ng et al., 1975). In P. maximum, the ontogenetic drift of net photosynthesis (Ludlow, 1975) and chemical composition (Wilson and Ng, 1975) of leaves, and net assimilation rate and relative growth rate of individual plants (Ng et al., 1975) appears to be suspended by water stress within the range −12 to −92 bars. It is not known whether this behavior is peculiar to C_4 grasses or only to P. maximum.

e) **Stomatal Resistance.** Stomatal resistance is the major physiological control for reducing water loss and preventing the development of deleterious water deficits. Furthermore, it exerts a predominant influence over net photosynthesis of C_4 grasses during water stress (Boyer, 1970b; Doley and Trivett, 1974; Ludlow and Ng, 1976). Therefore it is of interest to compare the response of stomata of C_4 grasses to water stress with those of other plants. With increasing water stress stomatal resistance of most C_3 and C_4 plants either shows no response to mild water deficits followed by a sudden increase when a threshold value of leaf water potential is reached (Fig. 3), or it increases continuously as leaf potential decreases from a high value (Doley and Trivett, 1974). Both types of response are found in C_4 grasses. Threshold values vary as much among cultivated C_4 grasses as between C_3 and C_4 plants (Table 3). Both average values for individual species and for plant groups in Table 3, and experimental comparisons of performance by controlled environment-grown and field-grown plants (see Jordan and Ritchie, 1971; McCree, 1974; Turner, 1974) demonstrate the lower threshold values of field-grown material. Even though average values in Table 3 for maize and sorghum grown in controlled environments are similar, when they have been compared in the same experiment, sorghum stomata closed at lower leaf water potentials in both controlled environments and the field (Sanchez-Diaz and Kramer, 1971; Beadle et al., 1973; Turner, 1974). There are insufficient data to make comparisons with C_3 semi-arid shrubs.

f) **Drought Tolerance.** Levitt (1972) proposed that a drought-tolerant species with high dehydration avoidance is one which has a high relative water content at -15 bars leaf water potential, or conversely one which has a low leaf water potential at a relative water content of 50% (Table 4). C_4 grasses grown under controlled conditions or in the field are, on average, more drought-tolerant than herbaceous C_3 species, although the range of values overlap. Herbaceous C_3 plants have a slightly lower wilting point under controlled conditions but the difference disappears under field conditions. In the latter situation C_4 grasses have similar drought tolerance to trees but less tolerance than C_3 semi-arid and arid species. There is a range of tolerance among C_4 grasses varying from that of *Z. mays* grown under controlled conditions to that of the desert grass *Triodia* (Spinifex) grown in the field. The greater drought tolerance of field-grown plants is also clearly evident. Thus, both among C_4 grasses and among all plants drought tolerance seems to be associated more with the environment to which plants have adapted than with any taxonomic, morphological, or other grouping. The same conclusion is reached when the water stress which causes 50% of cells in a leaf to die or 50% of leaves on a plant to die are used as an index of drought tolerance (Oppenheimer, 1960; Levitt, 1972). Furthermore, the "resurrection" plants of South-West Africa, which are among some of the most drought-tolerant higher plants, have both C_3 and C_4 members and they differ both morphologically and taxonomically (Hickel, 1967; Walter and Kreeb, 1970; Gaff, 1971; Levitt, 1972; Gaff and Hallam, 1974).

g) **Transpiration Ratio.** Transpiration ratio (sometimes called water requirement) is the amount of water lost per unit of dry weight produced, and is the reciprocal of water-use efficiency. It is now well-established that all C_4 plants, including grasses, are about twice as efficient as other plants in producing dry matter when soil water supply is adequate (De Wit and Alberda, 1961; Bull, 1969; Downes, 1969; Ludlow and Wilson, 1972; Teare et al., 1973). This is illustrated by data of Schantz and Piemeisel (1927) in which the transpiration ratio of C_4 grasses (310) is half that of a wide range of herbaceous C_3 plants (Table 5a). This difference exists at the leaf, individual plant, and sward levels when comparing C_4 pasture grasses with C_3 legumes (Table 5b). The superiority of C_4 grass leaves occurs over a wide range of environmental and physiological conditions, and is due almost entirely to a higher photosynthetic rate rather

Table 3. Threshold leaf water potential at which stomatal resistance increases markedly. Mean and range (parentheses) of values are given. NT = no threshold, RWC = equivalent relative water contents

	Controlled environments	Field	References[a]
C₄ grasses			
Zea mays	−10	−17	1, 2, 3, 4, 5, 6
Sorghum bicolor	−11	−20	2, 3, 5
Panicum maximum	−6	—	7
Cenchrus ciliaris	−11	—	8
Setaria anceps	NT	—	8
Pennisetum typhoides	—	−20	28
Astrebla lappacea	NT	—	9
C₃ Plants			
Atriplex hastata, cotton, Desmodium uncinatum, Lolium perenne, Macroptilium atropurpureum, soybean, tomato, wheat	−10 (−3 to −24) [76% (65–85) RWC]	—	1, 8, 10, 11, 12, 13, 14, 15, 16, 17, 18, 19
Barley, cotton, Macroptilium purpureum, Phaseolus vulgaris, Stylosanthes humilis, tobacco, Vigna sinensis	—	−14 (−9 to −24) [75% RWC]	5, 8, 12, 19, 20, 21, 22, 23, 29
Trees			
Betula, Picea, Pinus, Populus, Robinia, Vitis	−10 (−7 to −13)	−18 (−15 to −27)	16, 26, 27, 30, 31
Semi-arid and arid shrubs			
Acacia harpophylla	NT	−57	24, 25

[a] *1* Boyer (1970b); *2* Beadle et al. (1973); *3* Sanchez-Diaz and Kramer (1971); *4* Turner and Begg (1973); *5* Turner (1974); *6* Giles et al. (1974); *7* Ludlow and Ng (1976); *8* Ludlow (unpublished); *9* Doley and Trivett (1974); *10* Frank et al. (1973); *11* Slatyer (1970); *12* Jordan and Ritchie (1971); *13* Troughton (1969); *14* Jones (1973); *15* Duniway (1971); *16* Hinckley (1972); *17* Boyer (1970b); *18* Lawlor (1972a); *19* Kanemasu and Tanner (1969); *20* McCown (1973); *21* Millar et al. (1968); *22* Hiler et al. (1972); *23* Millar and Gardner (1972); *24* Tunstall (1973); *25* van den Driessche et al. (1971); *26* Kriedemann and Smart (1971); *27* Kaufmann (1968); *28* Begg et al. (1964); *29* Clark and Hiler (1973); *30* Jarvis and Jarvis (1963); *31* W. R. Watts (personal communication).

Table 4. Drought tolerance of leaves as indicated by relative water content (RWC_{-15}) at -15 bars leaf water potential, and leaf water potential (Ψ_{50}) at 50% relative water content. The leaf water potential at wilting point is also given. Average and range (parentheses) of values are shown

	Controlled environments			Field			References[a]
	RWC_{-15} (%)	Ψ_{50} (bars)	WP (bars)	RWC_{-15} (%)	Ψ_{50} (bars)	WP (bars)	
C_4 grasses							
Zea mays	58	−16	−10	91	—	−16	1, 2, 3, 4, 5, 6, 23
Sorghum bicolor	81	−40	−14	—	—	−19	2, 3, 4, 7, 8, 9
Panicum maximum	59	−20	−9	82	−66	—	10, 11
Cenchrus ciliaris	88	−35	—	87	−73	—	11, 12
Setaria anceps	88	−35	—	—	—	—	12
Pennisetum typhoides	—	—	—	86	−36	—	13
Triodia basedowii	—	—	—	92	−60	—	14
Mean	75	−29	−11	88	−59	−17	
C_3 plants							
Herbaceous							
Barley, beans, *Capsicum*, cotton, *Lotus*, lupins, *Macroptilium atropurpureum*, rape, ryegrass, sunflower, tobacco, tomato	64 (44 to 84)	−23 (−12 to −32)	−15 (−8 to −22)	75 (70 to 92)	−34 (−22 to −35)	−17 (−13 to −22)	2, 5, 7, 8, 11, 15, 16, 17, 18, 19
Trees							
Betula, Cornus, Eucalyptus, Ligustrum, Picea, Pinus, Populus	—	—	—	85 (82 to 90)	−42 (−26 to −65)	—	15, 20
Semi-arid and arid shrubs							
Acacia	95	—	—	94 (92 to 96)	−104 (−60 to −180)	—	21
Artemisia							22

[a] *1* Giles et al. (1974); *2* Neuman et al. (1974); *3* Sanchez-Diaz and Kramer (1971); *4* Beadle et al. (1973); *5* Lawlor (1969); *6* Glinka and Katchansky (1970); *7* Turner (1974); *8* Pasternak (1971); *9* Whiteman and Wilson (1963); *10* Ng et al. (1975); *11* D. C. I. Peake (personal communication); *12* Ludlow (unpublished); *13* Begg et al. (1964); *14* Slatyer (1962); *15* Jarvis and Jarvis (1963); *16* Ehlig and Gardner (1964); *17* Millar et al. (1968); *18* Warren Wilson (1967); *19* Jordan and Ritchie (1971); *20* Knipling (1967); *21* Connor and Tunstall (1968); *22* Kappen et al. (1972); *23* Heichel and Musgrave (1970).

than a lower transpiration rate (Ludlow and Wilson, 1972). Similarly, the superiority at the plant and sward levels is mainly due to higher dry matter increase rather than lower water loss. However, in some C_4 native grasses, it is partly due to a lower water loss associated with higher stomatal resistances; for example, the stomatal resistance of some Australian native C_4 grasses are three times that of introduced C_3 grasses and they use less water (J. R. McWilliam, personal communication).

Table 5. Transpiration ratio (g water per g dr. w.) of (a) a number of C_4 grasses and C_3 herbaceous plants (data from Schantz and Piemeisel, 1927) and (b) of leaves, individual plants and swards of tropical pasture grasses and legumes (from Ludlow and Wilson, 1972)

(a)	C_4 grasses			C_3 herbaceous plants	
	Zea mays		349		
	Sorghum sudanense		305		
	Sorghum spp.		304		
	Panicum miliaceum		267		
	Setaria italica		285	29 species	
	Bouteloua gracilis		338	Range	415–912
		Mean	308	Mean	628
(b)			C_4 pasture grasses		C_3 legumes
	Leaf		50		115
	Plant		203		374
	Sward		304–340		700

The fact that the morphologically—and ecologically—different C_4 dicots (Black, 1971) have a similar transpiration ratio to C_4 grasses indicates that the superiority is associated mainly with the C_4 syndrome. Furthermore, C_3 desert shrubs growing with adequate water supply have values (Hellmuth, 1971) similar to other C_3 plants, demonstrating that the environment to which plants are adapted has less influence on transpiration ratio than it has on other physiological responses discussed in this chapter.

IV. Ecological Implications

The occurrence and distribution of species is determined by many factors in addition to physiological characteristics; for example, reproductive strategies, presence of competitors, possession of extensive root systems, and ability to survive unfavorable periods as seeds and dormant buds. Therefore no attempt will be made to explain the distribution of C_4 grasses. Instead, the possible ecological significance of physiological responses of C_4 grasses to solar radiation, temperature and water stress for growth and survival will be discussed, which may help to explain their distribution.

1. Solar Radiation

Solar radiation is unlikely to influence survival of C_4 grasses in open habitats but C_4 grasses are unable to survive in heavy shade (Hofstra et al., 1972). In Java, grasses which are able to grow in heavy shade of tree canopies are mainly C_3 representatives of the tribe Paniceae which contains predominantly C_4 species. This suggests that in heavy shade there is either no selection pressure for the C_4 syndrome or that it was prevented from developing. Alternatively, these grasses may have only been able to invade and survive in such conditions by losing the C_4 syndrome and reverting to the more primitive C_3 pathway.

2. Temperature

Tolerance of high temperatures and sensitivity to low temperatures probably explain why most C_4 grasses fail to survive in temperate regions and are mainly restricted to the tropics and the sub-tropics. Most C_3 grasses have the opposite characteristics, which probably explains their predominance in temperate regions. However, dicotyledonous C_3 plants which occur from the tropics to the arctic contain representatives which are as tolerant of high temperatures and as intolerant of low temperature as C_4 grasses.

3. Water Stress

Plants with the C_4 syndrome have potentially the highest leaf net photosynthetic rates of all plants, if the supply of water and mineral nutrients is adequate and temperature is optimal. Whether this potential is realized and accompanied by high rates of growth and dry matter production depends upon a number of factors. Some native and cultivated C_4 grasses have lower net photosynthetic rates than other C_4 grasses because their minimum stomatal resistance is greater (Ludlow and Wilson, 1972; Christie, 1974; Gifford, 1974; Doley and Trivett, 1974; J. R. McWilliam, personal communication). High photosynthetic rates, or net assimilation rates of C_4 plants, are not necessarily accompanied by high growth rates if the investment in photosynthetic surface is inferior to C_3 plants (Bull, 1971); for example, *Atriplex spongiosa* (C_4) had a lower growth rate than *Atriplex hastata* (C_3) because the leaf area was smaller even though it had a higher net assimilation rate (Slatyer, 1971), and the C_4 native grass *Astrebla elymoides* had only a 28% higher relative growth rate than the C_3 native grass *Thyridolepis mitchelliana* because it had a lower leaf area ratio, despite the fact that it had a 72% higher net assimilation rate (Christie, 1974). In fact, the peak crop growth rates of C_3 and C_4 species determined under their own preferred environments are similar (Gifford, 1974). Furthermore, the higher dry matter yields of C_4 crop and pasture plants, which have been incorrectly attributed solely to higher growth rates associated with the C_4 syndrome (Black, 1971; Zelitch, 1971; Marx, 1973), are due just as much to the higher temperatures and longer growing season of tropical and sub-tropical areas (Bull, 1971; Gifford, 1974; see also Caldwell, 1974).

In hot, summer-rainfall areas of the tropics and sub-tropics, C_4 grasses have higher rates of photosynthesis and growth, and higher dry matter yields than C_3 plants (Ludlow and Wilson, 1968, 1970, 1972; Stewart, 1970; Heslehurst and Wilson, 1971). These higher growth rates and water-use efficiencies (lower transpiration ratio) of C_4 grasses would aid survival in semi-arid regions with low, variable, and intermittent rainfall, particularly if a certain minimum amount of carbon fixation is necessary for flowering and seed production in annuals or for the formation of dormant buds and "reserve" materials in perennials. Moreover, tolerance of high temperatures allows C_4 plants to receive, without deleterious effects, large heat sums which hasten development more than dry weight increase and result in early flowering and seed production. (This is probably the main reason for the low grain yields of commercial grain sorghum varieties which are adapted to sub-tropical areas, when they are grown in the tropics.) The following examples support these generalizations: along a rainfall and temperature gradient from cool-moist to hot-dry habitats in southern Australia, the growth rate of ecotypes of a C_3 native grass *Danthonia caespitosa* increases and the time to flower and produce seed decreases (K. C. Hodgkinson and J. A. Quin, personal communication); similarly *Atriplex rosea* (C_4) was able to grow, flower, and produce seed on a given amount of soil water in a dry Californian summer where *Atriplex patula* (C_3) died before maturity (Björkman and Berry, 1973).

While the potential advantage of the C_4 syndrome for growth and survival in semi-arid areas during short periods of adequate or near adequate water supply is fairly clear, its advantage during periods of prolonged and sustained stress is less obvious because of the limited amount of information available. Intuitively the tolerance of high leaf temperatures which accompany water stress, the greater drought tolerance, and higher water use efficiencies would appear to give C_4 grasses an ecological advantage over other plants except for C_3 desert shrubs.

Furthermore, if the ability of young leaves to withstand leaf water potentials down to -100 bars and recover unaffected after removal of stress is widespread among C_4 grasses, this would provide a further ecological advantage.

C_4 grasses do not seem to be superior to other plants in the other characteristics discussed in this chapter. However, it is difficult to predict from our current knowledge which of these characteristics might be advantageous for growth and survival. Whether stomatal closure and reduction of water loss at high leaf water potentials and the concomitant reduction of carbon fixation is advantageous or not, depends upon whether survival is limited by water or energy and upon the probability of rain. For example, if a plant has sufficient energy reserves and it stops growing, conserves water, and prevents deleterious water deficits from developing either by death and loss of older leaves to reduce leaf area, or by stomatal closure and rolling of remaining leaves (Stocker, 1972), it will have a better chance of survival if rain does not fall (other things being equal), than one which continues to grow, lose water, and develop deleterious water deficits, unless the additional carbon fixed is put into roots which can exploit larger volumes of soil (McCree, 1974). On the other hand, if rain falls before deleterious water deficits are reached in the latter plant, it will survive and probably produce more dry matter than the former. Moreover, it is difficult to deduce from correlative evidence whether it is more desirable for survival to have a high or low threshold value of stomatal closure, because survival is determined by many factors (such as rooting characteristics, dormant buds, reserve materials etc.). However, correlative evidence does suggest that a high threshold value, per se, is not necessary for survival because the stomata of the C_4 native grass *Astrebla*, and other drought resistant plants from semi-arid and arid regions, do not close completely until leaf water potentials of -50 to -100 bars are reached. Also the threshold value of the more drought-resistant *Sorghum bicolor* is lower than that of *Zea mays*.

Returning now to our original question: do grasses with the C_4 syndrome, which was supposed to have evolved as an adaptation to hot, dry conditions, have superior physiological characteristics which gives them an ecological advantage under those conditions? There is insufficient information to give a conclusive answer, but characteristics such as high drought tolerance of leaves, tolerance of high temperatures, low transpiration ratios, and potentially high growth rates are clearly not disadvantageous.

On the other hand, it must be concluded that the C_4 syndrome is not a prerequisite for growth and survival in semi-arid and arid areas because both "resurrection" plants and grasses of arid regions contain C_3 and C_4 representives (Hartley, 1958a, b; Hartley and Slater, 1960; Oppenheimer, 1960; Stebbins and Crampton, 1961; Christie, 1974; Gaff and Hallam, 1974). In addition, there are situations where particular C_4 grasses only occupy moist niches whereas other C_4 grasses and C_3 grasses occupy the drier niches (Whitman, 1941; Redman, 1975).

V. Conclusions: Future Research

Clearly there is a lack of information on the degree of water stress experienced by C_4 grasses in the field, their responses to water stress, and factors associated with their survival in the field, especially of the ecologically more interesting native and "resurrection" plants. Because of the marked difference in the response of plants grown under controlled conditions and in the field, techniques and equipment must be developed to study response functions and physiological behavior of plants growing in the field, if the goal is to explain or predict field performance. However, controlled environments still offer the best control of environmental conditions to study growth and long term trends of processes, and to elucidate mechanisms. Therefore the extent of, and explanations for, the different behavior of plants grown under controlled conditions and in the field should be determined, so that results obtained under controlled conditions can be extrapolated to a wider range of conditions.

Deficiencies in our knowledge can be summarized by saying that quite a lot is known about the biochemistry and, to a lesser extent, about the physiology of C_4 plants, but the greatest single outstanding problem is the ecological significance of the C_4 syndrome.

References

Acevedo, E., Hsiao, T. C., Henderson, D. W.: Immediate and subsequent growth responses of maize leaves to changes in water status. Plant Physiol. **48**, 631–636 (1971).

Beadle, C. L., Stevenson, K. R., Neuman, H. H., Thurtell, G. W., King, K. W.: Diffusive resistance, transpiration and photosynthesis in single leaves of corn and sorghum in relation to leaf water potential. Can. J. Plant Sci. **53**, 537–544 (1973).

Begg, J. E., Bierhuizen, J. F., Lemon, E. R., Misra, D. K., Slatyer, R. O., Stern, W. R.: Diurnal energy and water exchanges in bulrush millet in an area of high solar radiation. Agr. Meteorol. **1**, 294–312 (1964).

Björkman, O.: Comparative studies on photosynthesis in higher plants. In: Photophysiology (ed. A. C. Giese), Vol. 8, pp. 1–63. New York-London: Academic Press 1973.

Björkman, O., Berry, J.: High-efficiency photosynthesis. Sci. Am. **229**, 80–93 (1973).

Black, C. C.: Ecological implications of dividing plants into groups with distinct photosynthetic production capacities. In: Advances in ecological research (ed. J. B. Cragg), Vol. 7, pp. 87–114. New York-London: Academic Press 1971.

Black, C. C.: Photosynthetic carbon fixation in relation to net CO_2 uptake. Ann. Rev. Plant Physiol. **24**, 253–286 (1973).

Boyer, J. S.: Leaf enlargement and metabolic rates in corn, soybean, and sunflower at various leaf water potentials. Plant Physiol. **46**, 233–235 (1970a).

Boyer, J. S.: Differing sensitivity of photosynthesis to low leaf water potentials in corn and soybean. Plant Physiol. **46**, 236–239 (1970b).

Brix, H.: The effect of water stress on rates of photosynthesis and respiration in tomato plants and loblolly pine seedlings. Physiol. Plantarum **15**, 10–20 (1962).

Bull, T. A.: Photosynthetic efficiencies and photorespiration in Calvin cycle and C_4 dicarboxylic acid plants. Crop. Sci. **9**, 726–729 (1969).

Bull, T. A.: The C_4 pathway related to growth rates in sugar cane. In: Photosynthesis and photorespiration (eds. M. D. Hatch, C. B. Osmond, R. O. Slatyer), pp. 68–75. New York: Wiley Interscience 1971.

Caldwell, M. M.: Adaptability and productivity of species possessing C_3 and C_4 photosynthesis in a cool desert environment. In: Ecophysiological foundation of ecosystems productivity in arid zone. U.S.S.R. Acad. Sci., pp. 27–29. Leningrad: Nauka 1972.

Caldwell, M. M.: Physiology of desert halophytes. In: Ecology of halophytes (eds. R. J. Reimold, W. H. Queen), pp. 355–378. New York-London: Academic Press 1974.

Cary, J. W., Mayland, H. F.: Factors influencing freezing of supercooled water in tender plants. Agron. J. **62**, 715–719 (1970).

Cary, J. W., Wright, J. L.: Response of plant water potential to the irrigated environment of southern Idaho. Agron. J. **63**, 691–695 (1971).

Christie, E. C.: Physiological responses of the semi-arid grasses *Thyridolepis* and *Astrebla* compared with *Cenchrus*. Ph. D. Thesis, Macquarie University, Australia 1974.

Clark, R. N., Hiler, E. A.: Plant measurements as indicators of crop water deficit. Crop Sci. **13**, 466–469 (1973).

Coaldrake, J. E.: Climate. In: Some concepts and methods in sub-tropical pasture research. Bull. 47, pp. 17–26. Commonw. Agr. Bureau, Farnham Royal, England 1964.

Connor, D. J., Tunstall, B. R.: Tissue water relations for brigalow and mulga. Australian J. Botany **16**, 487–490 (1968).

Cooper, J. P.: Potential production and energy conversion in temperate and tropical grasses. Herb. Abstr. **40**, 1–15 (1970).

Cooper, J. P., Tainton, N. M.: Light and temperature requirements for the growth of tropical and temperate grasses. Herb. Abstr. **38**, 167–176 (1968).

De Wit, C. T., Alberda, T.: Transpiration coefficient and transpiration rate of three grain species in growth chambers. JAARB I.B.S. Meded **156**, 73–81 (1961).

Doley, D., Trivett, N. B. A.: Effects of low water potential on transpiration and photosynthesis in Mitchell grass *(Astrebla lappacea)*. Australian J. Pl. Physiol. **1**, 539–550 (1974).

Downes, R. W.: Differences in transpiration rates between tropical and temperate grasses under controlled conditions. Planta (Berl.) **88**, 261–273 (1969).

Duniway, J. M.: Water relations of Fusarium wilt in tomato. Physiol. Pl. Path. **1**, 537–546 (1971).

Ehlig, C. F., Gardner, W. R.: Relationship between transpiration and the internal water relations of plants. Agron. J. **56**, 127–130 (1964).

El-Sharkawy, M. A., Hesketh, J. D.: Effects of temperature and water deficit on leaf photosynthetic rates of different species. Crop Sci. **4**, 514–518 (1964).

Frank, A. B., Power, J. F., Willis, W. O.: Effect of temperature and plant water stress on photosynthesis, diffusion resistance, and leaf water potential in spring wheat. Agron. J. **65**, 777–780 (1973).

Gaff, D. F.: Desiccation-tolerant flowering plants in Southern Africa. Science **174**, 1033–1034 (1971).

Gaff, D. F., Hallam, N. D.: Resurrecting desiccated plants. In: Mechanisms of regulation of plant growth (eds. R. L. Bieleski, A. R. Ferguson, M. M. Cresswell), Bull. **12**, pp. 389–393. Wellington: Royal Soc. New Zealand 1974.

Gandar, P. W., Tanner, C. B.: Leaf growth, tuber growth, and water potentials in potatoes. Agron. Abstr. p. 72 (1974).

Garwood, E. A., Gowman, M. A.: Soil and plant water status and growth of forage species. Grassland Res. Inst. Ann. Rep. 1972, 44–46 (1973).

Gifford, R. M.: The light response of CO_2 exchange: on the source of differences between C_3 and C_4 species. In: Photosynthesis and photorespiration (eds. M. D. Hatch, C. B. Osmond, R. O. Slatyer), pp. 51–56. New York: Wiley Interscience 1971.

Gifford, R. M.: A comparison of potential photosynthesis, productivity and yield of plant species with differing photosynthetic metabolism. Australian J. Pl. Physiol. **1**, 107–117 (1974).

Giles, K. L., Beardsell, M. F., Cohen, D.: Cellular and ultrastructural changes in mesophyll and bundle sheath cells of maize in response to water stress. Plant Physiol. **54**, 208–212 (1974).

Glinka, Z., Katchansky, M. Y.: The effect of water potential on the CO_2 compensation point of maize and sunflower leaf tissue. Israel J. Botany **19**, 533–541 (1970).

Hacker, J. B., Forde, B. J., Gow, J. M.: Simulated frosting of tropical grasses. Australian J. Agric. Res. **25**, 45–57 (1974).

Harlan, J. R.: Theory and dynamics of grassland agriculture. New Jersey: van Nostrand 1956.

Hartley, W.: Studies on the origin, evolution, and distribution of the Gramineae. I. The tribe Andropogoneae. Australian J. Botany **6**, 116–128 (1958a).

Hartley, W.: Studies on the origin, evolution, and distribution of the Gramineae. II. The tribe Paniceae. Australian J. Botany **6**, 343–357 (1958b).

Hartley, W., Slater, C.: Studies on the origin, evolution, and distribution of the Gramineae. III. The tribes of the subfamily Eragrostoideae. Australian J. Botany **8**, 256–276 (1960).

Hatch, M. D., Osmond, C. B., Slatyer, R. O.: Photosynthesis and photorespiration. New York: Wiley Interscience 1971.

Heichel, G. H., Musgrave, R. B.: Photosynthetic response to drought in maize. Philippine Agriculturalist **54**, 102–114 (1970).

Hellmuth, E. O.: Eco-physiological studies on plants in arid and semi-arid regions in Western Australia. III. Comparative studies on photosynthesis, respiration and water relations of ten arid zone and two semi-arid zone plants under winter and late summer climatic conditions. J. Ecol. **59**, 225–259 (1971).

Heslehurst, M. R., Wilson, G. L.: Studies on the productivity of tropical pasture plants. III. Stand structure, light penetration, and photosynthesis in field swards of *Setaria* and green leaf *Desmodium*. Australian J. Agric. Res. **22**, 865–878 (1971).

Hickel, B.: Zur Kenntnis einer xerophilen Wasserpflanze: *Chamaegigas intrepidus* Dtr. aus Südwestafrika. Int. Rev. gesamten Hydrobiol. **52**, 361–400 (1967).

Hiler, E. A., van Bavel, C. H. M., Hossain, M. M., Jordan, W. R.: Sensitivity of southern peas to plant water deficits at three growth stages. Agron. J. **64**, 60–64 (1972).

Hinckley, T. M.: Responses of black locust and tomato plants after water stress. Hort. Sci. **8**, 405–407 (1973).

Hinckley, T. M., Ritchie, G. A.: Reaction of mature *Abies* seedlings to environmental stresses. Trans. Missouri Acad. Sci. **6**, 24–37 (1972).

Hirose, M.: Comparison of physiological and ecological characteristics between tropical and temperate grass species. ASPAC Food and Fertilizer Technology Center, Extension Bull. No. **26** (1973).

Hofstra, J. J., Aksornkoae, S., Atmowidjojo, S., Banaag, Santosa, J. F., Sastrohoetomo, R. A., Thu, L. T. N.: A study on the occurrence of plants with a low CO_2 compensation point in different habitats in the Tropics. Ann. Bogorienses **5**, 143–157 (1972).

Hsiao, T. C.: Plant responses to water stress. Ann. Rev. Plant Physiol. **24**, 519–570 (1973).

Hutton, E. M.: Tropical pastures. Adv. Agron. **22**, 2–73 (1970).

Ivory, D. A.: Effect of temperature on growth of tropical pasture grasses. Ph. D. Thesis, Univ. Queensland, Australia, 1975.

Jarvis, P. G., Jarvis, M. S.: The water relations of tree seedlings. IV. Some aspects of the tissue water relations and drought resistance. Physiol. Plantarum **16**, 501–516 (1963).

Jones, H. G.: Photosynthesis by thin leaf slices in solution. II. Osmotic stress and its effect on photosynthesis. Australian J. Biol. Sci. **26**, 25–33 (1973).

Jordan, W. R.: Growth of cotton seedlings in relation to maximum daily plant-water potential. Agron. J. **62**, 599–701 (1970).

Jordan, W. R., Ritchie, J. T.: Influence of soil water stress on evaporation, root absorption, and internal water status of cotton. Plant Physiol. **48**, 783–788 (1971).

Kanemasu, E. T., Tanner, C. B.: Stomatal diffusion resistance of snap beans. I. Influence of leaf-water potential. Plant Physiol. **44**, 1547–1552 (1969).

Kappen, L., Lange, O. L., Schulze, E.-D., Evenari, M., Buschbom, U.: Extreme water stress and photosynthetic activity of the desert plant *Artemisia herba-alba* Asso. Oecologia Berl.) **10**, 177–182 (1972).

Kaufmann, M. R.: Water relations of pine seedlings in relation to root and shoot growth. Plant Physiol. **43**, 281–288 (1968).

Knipling, E. B.: Effect of leaf aging on water deficit-water potential relationships of dogwood leaves growing in two environments. Physiol. Plantarum **20**, 65–72 (1967).

Kriedemann, P. E., Canterford, R. L.: The photosynthetic activity of pear leaves (*Pyrus communis* L.). Australian J. Biol. Sci. **24**, 197–205 (1971).

Kriedemann, P. E., Smart, R. E.: Effects of irradiance temperature, and leaf water potential on photosynthesis of vine leaves. Photosynthetica **5**, 6–15 (1971).

Laetsch, W. M.: Relationship between chloroplast structure and photosynthetic carbon-fixation pathways. Sci. Prog., Oxford **57**, 323–351 (1969).

Lange, O. L., Schulze, E.-D., Evenari, M., Kappen, L., Buschbom, U.: The temperature-related photosynthetic capacity of plants under desert conditions. I. Seasonal changes of the photosynthetic response to temperature. Oecologia (Berl.) **17**, 97–110 (1974).

Larcher, W.: The effect of environmental and physiological variables on the carbon dioxide gas exchange of trees. Photosynthetica **3**, 167–198 (1969).

Lawlor, D. W.: Plant growth in polyethylene glycol solutions in relations to the osmotic potential of the root medium and the leaf water balance. J. Exp. Botany **20**, 895–911 (1969).

Lawlor, D.: Growth and water use of *Lolium perenne*. I. Water transport. J. Appl. Ecol. **9**, 79–98 (1972a).

Lawlor, D.: Growth and water use of *Lolium perenne*. II. Plant growth J. Appl. Ecol. **9**, 99–105 (1972b).

Lawlor, D.: Water stress and carbon dioxide assimilation. Rothamsted Report for 1972, 99–100 (1973).

Lawlor, D. W., Milford, G. F. J.: The effect of sodium on growth of water stressed sugar beet. Ann. Botany **37**, 597–604 (1973).

Levitt, J.: Responses of plants to environmental stresses. New York-London: Academic Press 1972.

Lipschitz, N., Ilan, A., Eshiel, A., Waisel, Y.: Salt glands of *Chloris gayana* Kunth. Ann. Botany **38**, 459–462 (1974).

Ludlow, M. M.: Effect of water stress on the decline of leaf net photosynthesis with age. In: Environmental and biological control of photosynthesis (eds. R. Marcelle), pp. 123–134. The Hague: W. Junk 1975.

Ludlow, M. M., Jarvis, P. G.: Methods for measuring photorespiration in leaves. In: Plant photosynthetic production, manual of methods (eds. Z. Sestak, J. Catsky, P. G. Jarvis), pp. 294–315. The Hague: W. Junk 1971.

Ludlow, M. M., Ng, T. T.: Water stress suspends leaf ageing. Plant Sci. Letters **3**, 235–240 (1974).

Ludlow, M. M., Ng, T. T.: Effect of water deficit on carbon dioxide exchange and leaf elongation rate of *Panicum maximum* var. *trichoglume*. Australian J. Plant Physiol. **3**, 401–413 (1976).

Ludlow, M. M., Wilson, G. L.: Studies on the productivity of tropical pasture plants. I. Growth analysis, photosynthesis, and respiration of Hamil grass and siratro in a controlled environment. Australian J. Agric. Res. **19**, 35–45 (1968).

Ludlow, M. M., Wilson, G. L.: Growth of some tropical grasses and legumes at two temperatures. J. Australian Inst. Agric. Sci. **36**, 43–44 (1970).

Ludlow, M. M., Wilson, G. L.: Photosynthesis of tropical pasture plants. I. Illuminance, carbon dioxide concentration, leaf temperature, and leaf-air vapour pressure difference. Australian J. Biol. Sci. **24**, 449–470 (1971a).

Ludlow, M. M., Wilson, G. L.: Photosynthesis of tropical pasture plants. III. Leaf age. Australian J. Biol. Sci. **24**, 1077–1087 (1971b).

Ludlow, M. M., Wilson, G. L.: Photosynthesis of tropical pasture plants. IV. Basis and consequences of differences between grasses and legumes. Australian J. Biol. Sci. **25**, 1133–1145 (1972).

Ludlow, M. M., Wilson, G. L., Heslehurst, M. R.: Studies on the productivity of tropical pasture plants. V. Effect of shading on growth, photosynthesis and respiration in two grasses and two legumes. Australian J. Agric. Res. **25**, 425–433 (1974).

Marx, J. L.: Photorespiration: key to increasing plant productivity. Science **179**, 365–367 (1973).

McCown, R. L.: An evaluation of the influence of available soil water storage capacity on growing season length and yield of tropical pastures using simple water balance models. Agr. Meteorol. **11**, 53–63 (1973).

McCree, K. J.: Changes in the stomatal response characteristics of grain sorghum produced by water stress during growth. Crop Sci. **14**, 273–278 (1974).

McPherson, H. G., Slatyer, R. O.: Mechanisms regulating photosynthesis in *Pennisetum typhoides*. Australian J. Biol. Sci. **26**, 329–339 (1973).

McWilliam, J. R., Ferrar, P. J.: Photosynthetic adaptation of higher plants to thermal stress. In: Mechanisms of regulation of plant growth (ed. R. L. Bieleski, A. R. Ferguson, M. M. Cresswell), Bull. 12, pp. 467–476. Wellington: Royal Soc. New Zealand 1974.

Meidner, H.: Further observations on the minimum intercellular space carbon dioxide concentration (Γ) of maize leaves and the postulated roles of "photo-respiration" and glycollate metabolism. J. Exp. Botany **18**, 177–185 (1967).

Millar, A. A., Duysen, M. E., Wilkinson, G. E.: Internal water balance of barley under soil moisture stress. Plant Physiol. **43**, 968–972 (1968).

Millar, A. A., Gardner, W. R.: Effect of the soil and plant water potentials on the dry matter production of snap beans. Agron. J. **64**, 559–562 (1972).

Millar, A. A., Gardner, W. R., Goltz, S. M.: Internal water status and water transport in seed onion plants. Agron. J. **63**, 779–784 (1971).

Monteith, J. L.: Solar radiation and productivity in tropical ecosystems. J. Appl. Ecol. **9**, 747–766 (1972).

Moore, R. T., White, R. S., Caldwell, M. M.: Transpiration of *Atriplex confertifolia* and *Eurotia lanata* in relation to soil, plant, and atmospheric moisture stresses. Can. J. Botany **50**, 2411–2418 (1972).

Neumann, H. H., Thurtell, G. W., Stevenson, K. R., Beadle, C. L.: Leaf water content and potential in corn, soybean, and sunflower. Can. J. Pl. Sci. **54**, 185–195 (1974).

Ng, T. T., Wilson, J. R., Ludlow, M. M.: Influence of water stress on water relations and growth of a tropical (C_4) grass *Panicum maximum* var. *trichoglume*. Australian J. Pl. Physiol. **2**, 581–595 (1975).

Odening, W. R., Strain, B. R., Oechel, W. C.: The effect of decreasing water potential on net CO_2 exchange of intact desert shrubs. Ecol. **55**, 1086–1095 (1974).

Oechel, W. C., Strain, B. R., Odening, W. R.: Photosynthetic rates of a desert shrub, *Larrea divaricata* Cav., under field conditions. Photosynthetica **6**, 183–188 (1972a).

Oechel, W. C., Strain, B. R., Odening, W. R.: Tissue water potential, photosynthesis, ^{14}C-labelled photosynthate utilization, and growth in the desert shrub *Larrea divaricata* Cav. Ecol. Monographs **42**, 127–141 (1972b).
Oppenheimer, H. R.: Adaptation to drought: xerophytism. In: Plant-water relationships in arid and semi-arid conditions. Arid Zone Research Vol. **15**, pp. 105–138. Paris: UNESCO 1960.
Pasternak, D.: Plant and environmental effects on the stomatal regulation of photosynthesis and transpiration in sorghum. Ph. D. Thesis, Univ. Queensland, Australia (1971).
Peake, D. C. I., Stirk, G. B., Henzell, E. F.: Leaf water potentials in pasture plants at Narayen, in southern Queensland. Australian J. Exp. Agric. an. Hus. **15**, 645–654 (1975).
Redman, R. E.: Production ecology of grassland plant communities in western North Dakota. Ecol. Mon. **45**, 83–106 (1975).
Sanchez-Diaz, M. F., Kramer, P. J.: Behaviour of corn and sorghum under water stress and during recovery. Plant Physiol. **48**, 613–616 (1971).
Schantz, H. L., Piemeisel, L. N.: The water requirement of plants at Akron, Colorado. J. Agr. Res. **34**, 1093–1189 (1927).
Shearman, L. L., Eastin, J. D., Sullivan, C. Y., Kinbacher, E. J.: Carbon dioxide exchange in water-stressed sorghum. Crop Sci. **12**, 406–409 (1972).
Slatyer, R. O.: Internal water balance of *Acacia aneura* F. Muell. in relation to environmental conditions. Arid Zone Res. **16**, 137–146 (1962).
Slatyer, R. O.: Carbon dioxide and water vapour exchange in *Atriplex* leaves. In: The biology of *Atriplex* (ed. R. Jones), pp. 23–29. Canberra: C.S.I.R.O. (1970).
Slatyer, R. O.: Relationship between plant growth and leaf photosynthesis in C_3 and C_4 species of *Atriplex*. In: Photosynthesis and photorespiration (eds. M. D. Hatch, C. B. Osmond, R. O. Slatyer), pp. 76–81. New York: Wiley Interscience 1971.
Smith, B. N., Brown, W. V.: The Kranz syndrome in the Gramineae as indicated by carbon isotopic ratios. Am. J. Botany **60**, 505–513 (1973).
Smith, R. L., Boyd, F. T.: A laboratory method for cold resistance evaluation of *Chloris* and other tropical grasses. Proc. Soil and Crop Sci. Soc., Florida. **29**, 175–180 (1969).
Stebbins, G. L., Crampton, B. A.: A suggested revision of the grass genera of temperate North America. In: Recent advances in botany, Vol. 1, pp. 138–145. Toronto: Univ. Toronto Press 1961.
Stewart, G. A.: High potential productivity of the tropics for cereal crops, grass forage crops, and beef. J. Australian Inst. Agr. Sci. **36**, 85–101 (1970).
Stocker, O.: Der Wasser- und Photosynthese-Haushalt von Wüstenpflanzen der Mauretanischen Sahara. III. Kleinsträucher, Stauden und Gräser. Flora **161**, 46–110 (1972).
Taylor, A. O., Craig, A. S.: Plants under climatic stress. II. Low temperature, high light effects on chloroplast ultrastructure. Plant Physiol. **47**, 719–725 (1971).
Taylor, A. O., Rowley, J. A.: Plants under climatic stress. I. Low temperature, high light effects on photosynthesis. Plant Physiol. **47**, 713–718 (1971).
Teare, I. D., Kanemasu, E. T., Powers, W. L., Jacobs, H. S.: Water use efficiency and its relation to crop canopy area, stomatal regulation, and root distribution. Agron. J. **65**, 207–211 (1973).
Troughton, J. H.: Plant water status and carbon dioxide exchange of cotton leaves. Australian J. Biol. Sci. **22**, 289–302 (1969).
Tunstall, B. R.: Water relations of a brigalow community. Ph. D. Thesis, Univ. Queensland, Australia (1973).
Turner, N. C.: Stomatal behaviour and water status of maize, sorghum, and tobacco under field conditions. II. At low soil water potential. Plant Physiol. **53**, 360–365 (1974).
Turner, N. C., Begg, J. E.: Stomatal behaviour and water status of maize, sorghum, and tobacco under field conditions. I. At high soil water potential. Plant Physiol. **51**, 31–36 (1973).
Van den Driessche, R., Tunstall, B. R., Connor, D. J.: Photosynthetic response of brigalow to irradiance, temperature and water potential. Photosynthetica **5**, 210–217 (1971).
Walter, H., Kreeb, K.: Die Hydration und Hydratur des Protoplasmas der Pflanzen und ihre öko-physiologische Bedeutung. Protoplasmatologia, Vol. II C 6. Wien: Springer 1970.

Wardlaw, I. F.: The effect of water stress on translocation in relation to photosynthesis and growth. II. Effect during leaf development in *Lolium temulentum* L. Australian J. Biol. Sci. **22**, 1–16 (1969).

Warren Wilson, J.: The components of leaf water potential. I. Osmotic and matric potentials. Australian J. Biol. Sci. **20**, 329–347 (1967).

Watts, W. R.: Leaf extension in *Zea mays*. III. Field measurements of leaf extension in response to temperature and leaf water potential. J. Exp. Botany **25**, 1085–1096 (1974).

Wesselius, J. C., Brouwer, R.: Influence of water stress on photosynthesis, respiration and leaf growth of *Zea mays* L. Med. Landbouwhogesch. Wageningen **72-33**, 1–15 (1972).

Whiteman, P. C., Koller, D.: Saturation deficit of the mesophyll evaporating surfaces in a desert halophyte. Science **146**, 1320–1321 (1964).

Whiteman, P. C., Wilson, G. L.: Estimation of diffusion pressure deficit by correlation with relative turgidity and betaradiation absorption. Australian J. Biol. Sci. **16**, 140–146 (1963).

Whitman, W. C.: Seasonal changes in bound water content of some prairie grasses. Botan. Gaz. **103**, 38–63 (1941).

Wilson, J. R., Ng, T. T.: Influence of water stress on parameters associated with herbage quality of *Panicum maximum* var. *trichoglume*. Australian J. Agric. Res. **26**, 127–136 (1975).

Zelitch, I.: Photosynthesis, photorespiration and plant productivity. New York-London: Academic Press 1971.

Part 6
Water Relations and Productivity

Preface

There has been considerable recent interest in the assessment of plant productivity and biomass production of the various vegetation types in the world. Since mankind has recognized that the natural resources of the earth are limited, a great effort has been initiated to obtain a world-wide inventory of existing physical and biological bases of life. One large program in this respect was the "International Biological Program", which attempted to determine the biological productivity of the earth at the various trophic levels. In addition it was the aim of such research to obtain insight into the production processes as an important base for predictions and modeling work (Cooper, 1975).

It has been recognized not only in the course of these recent international programs but also during earlier plantgeographical work (Walter, 1968, 1973) that biomass in the different vegetation zones is to a great extent limited by the amount of water available (Kozlowski, 1968, 1972). The physiological basis for this has been explained in the preceding parts of this volume. Especially the chapter by Hsiao et al. (Part 5:F) showed that one of the first processes affected by limited access of water is extension growth. This knowledge is important for various aspects of land use, as vast areas are known where plant distribution is limited by lack of water. These zones are potential reserves for the enlargement of agricultural territories, which will be made necessary in the future by increasing world population. For this reason it is becoming increasingly important to learn more about the complex interaction of water and plant productivity. A qualitative understanding of how plants respond in general to water stress already exists. However, the problems of how different plant types gain their resources photosynthetically and of how photosynthetic productivity can be related to growth under conditions of impaired water relations are far from being solved (Mooney, 1972a).

A general understanding of world plant productivity is given in Part 6:A of this book. Lieth's studies aim at a global survey of primary productivity. Although these mapping efforts are based on correlation models only, it is interesting that the productivity of landscapes and larger geographical entities can be achieved by a correlated technique based on a single factor, namely evapotranspiration. Certainly, the changes in evapotranspiration are in turn correlated to a great number of equally changing climatic factors and there are also many additional unknowns, a complex which needs to be solved if management of a regional scale is to be attempted. Simulations and prediction of plant productivity under optimal growth conditions have proved to be difficult (De Wit, 1965; Penning de Vries, 1972; Shugart et al., 1974). Dealing with arid regions, the limitations in our knowledge of the production process become even more obvious. Part 6:B shows an attempt to solve the problems of modeling crop production in arid zones. Van Keulen had pointed out earlier (1975) that agricultural studies in developing countries are often aimed at immediate, limited solutions of specific problems unter given circumstances. What has been lacking is a systematic and theoretically sound examination of the process. Work with irrigated and fertilized plants as compared to natural growing conditions (Part 6:B) gives the somewhat surprising result that besides water, nitrogen plays a predominant role in plant growth in arid habitats of the Negev. Certainly it might well be that a nitrogen deficit promotes xeromorphic growth similar to the situation known for bog plants (Small, 1973). This, however, would indirectly improve plant water relationships in a desert habitat and limit the effect of drought.

Theoretical considerations provide an important basis for land management. What additional problems arise for agriculture is discussed in Part 6:C by Bierhuizen (see also Yaron et al., 1973). At present it has been realized that one should not only try to achieve the highest yield per unit area. Improvement in economic and efficient water use can be even more important in many areas, and the efficiency of water supply from an engineering point of view and in respect to the plant's regulation in optimizing the transpiration/photosynthesis relationships must also be investigated. Closely related to this problem is the question of how to recognize water stress on a regional basis. Drake gives some information on this very modern field of remote sensing in Part 6:D.

All modeling efforts and process analyses are based on detailed studies of plant production and allocation of products in the various vegetation and plant types (see also Cooper, 1975). An introduction to these problems is given in Part 6:E by Evenari et al. for arid and semi-arid areas. The drier the climate, the greater the demands for the functional adaptations of the growing organisms in order to maintain a hydration level of the protoplast adequate for any active metabolism. In an arid climate these demands are only fulfilled by a relatively small number of morphologically and physiologically highly adapted plant species (see also Vernberg, 1975). It becomes obvious that phytomass and primary production are not constant values. They change considerably in relation to the environmental parameters, since it is in fact the availability of water which fluctuates to a large degree as a limiting factor.

With higher plants the problem exists that a major amount of phytomass is below ground in a heterogenous soil layer. The difficulties arising from this fact have been discussed in detail in Part 2 of this volume. The great unknown in most biomass and productivity studies is the lack of knowledge of production processes and turnover in the root system. This is not only true for temperate zone plants, but even more decisive for plants growing under water stress when the largest part of phytomass may be in the soil. Furthermore, most of the higher plants being "arido-active" (Evenari et al., 1975) have special characteristics that reduce the effects of excessive water loss, and keep the plant metabolically active even during the hot, dry season. The complexity of the interrelationship between water relations and productivity becomes even more clear when dealing with a simpler model plant. "Arido-passive" plants limit their metabolic activity to moist and humid periods only, and there are poikilohydric lower plants that are better adapted than higher plants to repeated changes between activation during favorable and inactivation during unfavorable conditions. Lichens play an exemplary role in this respect (Lange et al., 1975b) and it is therefore aim of Part 6:E to explain the physiological bases for the relations between water content and productivity of lichens. It is also of significance that, for this superficially rather simplified system, the investigations result in a rather complex model which is far from being solved for the essential question of plant existence and interactions, selection, and effects of future changes in the environment.

Plants have adapted in many different ways to conditions of water stress. It would take much more than a short chapter to describe the many different strategies developed during evolution (Vernberg, 1975). It should be obvious, however, that all of these adaptations are not static systems. There is a great amount of acclimation possible also in the developed structure. The acclimation of plants has provided detailed study material for the effect of temperature (Lange et al., 1975a; Pearcy and Harrison, 1974). Some information is also available on water stress for various plant types (Orshan, 1973; Caldwell et al., 1974). This acclimation aspect of water relations and productivity certainly needs to be studied in more detail in the future.

References

Caldwell, M. M., DePuit, E. J., Fernandez, O. A., Wiebe, H. H., Camp, L. B.: Gas exchange, translocation, root growth, and soil respiration of Great Basin plants. US/IBP Desert Biome. Reports of 1973 Progress, vol. 3; Process Studies. Plant Section, 29–60, 1974.

Cooper, J. P. (ed.): Photosynthesis and productivity in different environments. IBP Vol. 3. Cambridge, London, New York, Melbourne: Cambridge Univ. Press 1975.

Evenari, M., Schulze, E.-D., Kappen, L., Buschbom, U., Lange, O. L.: Adaptive mechanisms in desert plants. In: Physiological adaptation to the environment (ed. F. J. Vernberg), pp. 111–129. New York: Intext Educational Publ. 1975.

Keulen, H. van: Simulation of water use and herbage growth in arid regions. Simulation Monographs. Wageningen: Pudoc 1975.

Kozlowski, T. T. (ed.): Water deficits and plant growth, vol. I and II. New York, San Francisco, London: Academic Press 1968.

Kozlowski, T. T. (ed.): Water deficits and plant growth, vol. III. New York, San Francisco, London: Academic Press 1972.

Lange, O. L., Schulze, E.-D., Evenari, M., Kappen, L., Buschbom, U.: The temperature-related photosynthetic capacity of plants under desert conditions. II. Possible controlling mechanisms for the seasonal changes of the photosynthetic response to temperature. Oecologia (Berl.) **18**, 45–53 (1975a).

Lange, O. L., Schulze, E.-D., Kappen, L., Buschbom, U., Evenari, M.: Adaptations of desert lichens to drought and extreme temperatures. In: Environmental physiology of desert organisms (ed. N. F. Hadley), pp. 20–37. Stroudsbourg, Penn.: Dowden, Hutchinson and Ross 1975b.

Mooney, H. A.: Carbon dioxide exchange of plants in natural environments. Botan. Rev. **38**, 455–469 (1972a).

Orshan, G.: Morphological and physiological plasticity in relation to drought. In: Wildland shrubs—their biology and utilization (eds. C. M. McKell, J. P. Blaischell, J. R. Goodin), pp. 245–286. Ogden, Utah: Intermountain Forest and Range, Experiment Station 1973.

Pearcy, R. W., Harrison, A. T.: Comparative photosynthetic and respiratory gas exchange characteristics of *Atriplex lentiformis* (Torr.) Wats. in coastal and desert habitats. Ecology **55**, 1104–1111 (1974).
Penning de Vries, F. W. T.: Respiration and growth. In: Crop processes in controlled environments (eds. A. R. Rees, K. E. Cockshull, D. W. Hand, R. G. Hurd), pp. 327–347. London–New York: Academic Press 1972.
Shugart, H. H., Goldstein, R. A., O'Neill, R. V., Mankin, J. B.: A terrestrial ecosystem energy model for forests. Oecol. Plantarum **9**, 231–264 (1974).
Small, E.: Xeromorphy in plants as a possible basis for migration between arid and nutritionally-deficient environments. Botan. Not. **126**, 534–539 (1973).
Vernberg, F. J. (ed.): Physiological adaptation to the environment. New York: Intext Educational Publ. 1975.
Walter, H.: Die Vegetation der Erde in ökophysiologischer Betrachtung, vol. 2. Stuttgart: Gustav Fischer 1968.
Walter, H.: Die Vegetation der Erde in ökophysiologischer Betrachtung, vol. 1, 3. ed. Stuttgart: Gustav Fischer 1973.
Wit, C. T. de: Photosynthesis of leaf canopies. Agric. Res. Rep. **663**, Pudoc, Wageningen 1965.
Yaron, B., Danfors, E., Vaadia, Y. (eds.): Arid zone irrigation. Ecological Studies, vol. 5. Berlin-Heidelberg-New York: Springer 1973.

A. The Use of Correlation Models to Predict Primary Productivity from Precipitation or Evapotranspiration

H. Lieth

I. Introduction

There is a great demand for assessing the primary productivity of landscapes and larger geographical entities as quickly and conveniently as possible. Statistical assessments of the actual production figures by direct measurements remain point efforts at best and require some means of areal integration afterwards. Such regionalization can be done by dividing the area under investigation into meaningful subdivisions, for which several point measurements (samples) of productivity can be averaged. The averages in combination with the known distribution pattern of the vegetation units may be used to construct either productivity pattern maps or tabulations to assess total average productivity values. This has been done by many authors in the past (see Lieth and Whittaker, 1975), and the most recent tabulation of the net primary productivity of the major vegetation unit is shown in Table 1 (Lieth, 1975a). Such tabulations, together with a pattern map, serve as baseline information against which new approaches of less known reliability are tested. This is illustrated by Fig. 1, a computer map, based originally on the spatial interpretation of tabular data for commonly recognized biological entities on earth (Lieth, 1964).

Initial trials had shown that correlation models to environmental parameters sufficiently correlated to primary productivity may be much more reliable predictors for productivity patterns than vegetation maps. Furthermore, there are certain environmental parameters available in such dense networks across the world that maps resulting from such attempts offer a much better resolution than all classical approaches.

Different environmental factors have an inherent applicability for geographical entities of different sizes; e.g. soil-related factors are good for predicting local variations, water-related factors are suitable for predicting local to regional patterns, and temperature, as well as radiation, is usually acceptable for regional to global patterns. The modeling work presented started with global and regional pattern analysis in which temperature, precipitation, and evapotranspiration were used first.

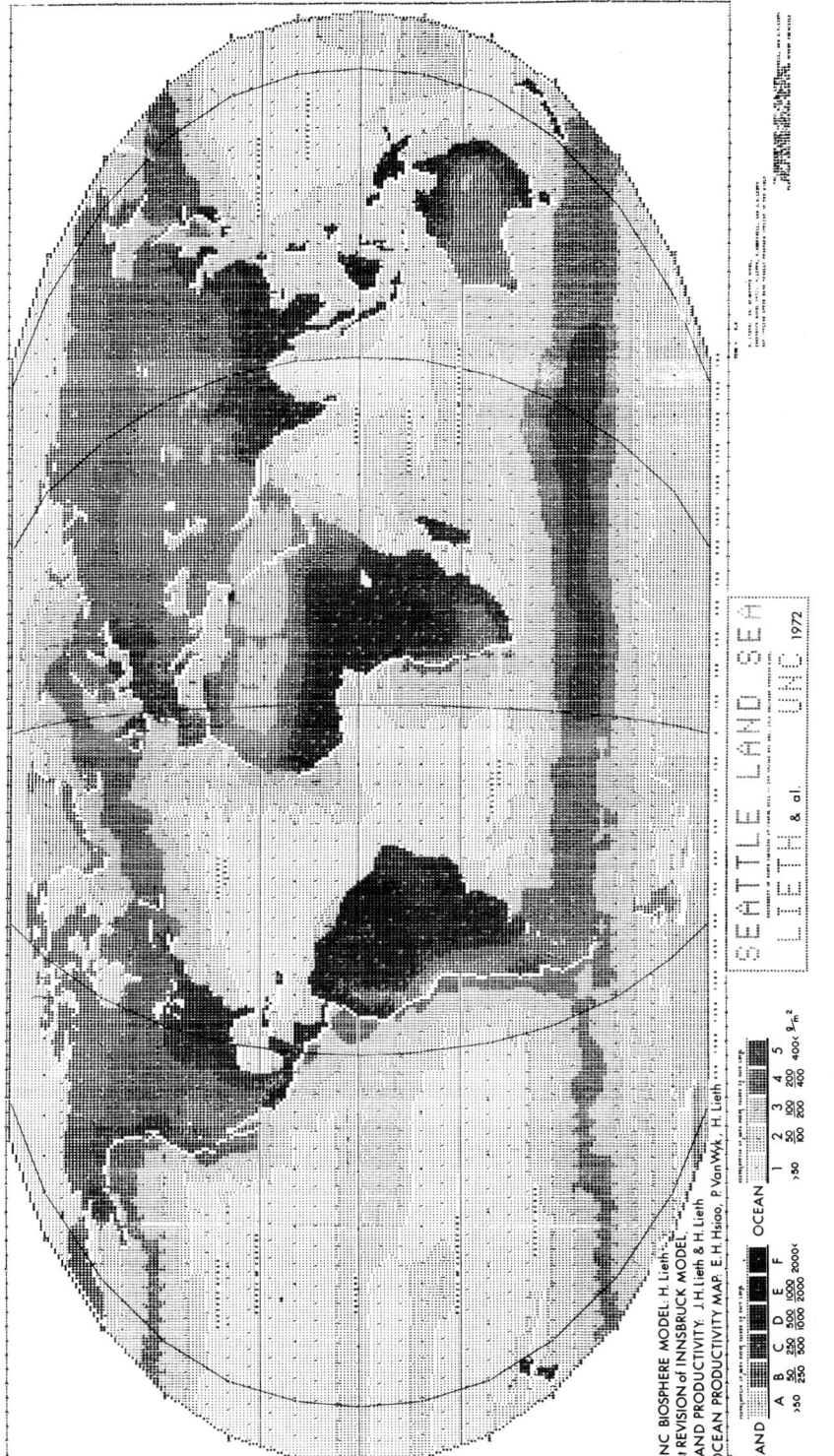

Fig. 1. The Seattle productivity map, a computer simulation of Lieth's (1964) productivity map

Table 1. Net primary productivity and energy fixation of major vegetation units of the world (after Lieth, 1975b). The global values coincide well with the figures calculated from the maps presented in Figs. 4 and 6

Vegetation unit	Area	NPP			Annual energy fixation		
		Range	Approx. mean	Total for area 10^9	Approx. average combustion value	Mean for m^2	Total for area
	10^6 km^2	kg m^{-2} yr^{-1}	kg m^{-2} yr^{-1}	metric tons	kcal g^{-1}	10^6 cal m^{-2}	10^{18} cal
1	2	3	4	5	6	7	8
Forests	**50**			**81.6**			**368.6**
Tropical rainforest	17.0	1–3.5	2.8	47.4	4.1	11.5	195.5
Raingreen forest	7.5	1.6–2.5	1.75	13.2	4.2	7.4	55.5
Summergreen forest	7.0	0.4–2.5	1.0	7.0	4.6	4.6	32.2
Mediterranean Sclerophyll forest (Chaparral)	1.5	0.25–1.5	0.8	1.2	4.9	3.9	5.9
Warm temperature Mixed forest	5.0	0.6–2.5	1.0	5.0	4.8	4.7	23.5
Boreal forest	12.0	0.3–1.2	0.65	7.8	4.6	3.0	36.0
Woodland	**7**	0.2–1.0	0.6	**4.2**	4.6	2.8	**19.6**
Dwarf and open scrub	**26**			**2.6**			**11.0**
Tundra	8.0	0.06–1.3	0.16	1.3	4.5	0.7	5.6
Desert scrub	18.0	0.01–0.25	0.07	1.3	4.5	0.3	5.4
Grassland	**24**			**19.2**			**76.8**
Tropical grassland (including grass dominated savannah)	15.0	0.2–2.9	0.8	12.0	4.0	3.2	48.0
Temperate grassland	9.0	0.07–1.3	0.8	7.2	4.0	3.2	28.8

Models to Predict Primary Productivity from Precipitation or Evapotranspiration

Desert (extreme)	**24**					**0.1**
Dry desert	8.5	0–0.01	0.003	—	4.5	—
Ice desert	15.5	0–0.001	—	—	—	0.1
Cultivated land	**14**	0.1–4.0	0.65	**9.1**	4.1	2.7	**37.8**
Fresh water	**4**			**5.0**			**21.4**
Swamps and marsh	2.0	0.8–4.0	2.0	4.0	4.2	8.4	16.8
Lake and stream	2.0	0.1–1.5	0.5	1.0	4.5	2.3	4.6
Total for continents	**149**			**121.7**			**535.3**
Open ocean	332	2–400	125	41.5	4.9	0.6	199.2
Upwelling zones	0.4	400–600	500	0.2	4.9	2.5	1.0
Continental shelf	26.6	200–600	360	9.2	4.5	1.6	43.1
Algae beds and reefs	0.6	500–4000	2000	1.2	4.5	9.0	3.6
Estuaries	1.4	500–4000	1800	2.5	4.5	8.1	11.3
Total marine	**361**	2–4000	144	**55.0**			**258.2**
Full total	**510**			**176.7**			**793.5**

II. Construction of Correlation Models and Geographical Patterns (Surfaces)

Correlation models are commonly used in biology for a variety of tasks. They are usually applied as an analytical tool to determine possible causal relationships. In this approach the correlation models were used synthetically, as much as the causal relations are accepted a priori. Therefore, the model is concerned primarily with the quantitative aspect of two or more dependent parameters in constructing geographical patterns (surfaces). The procedure can be explained on a specific case for the model which predicts productivity from evapotranspiration (Lieth and Box, 1972): (1) A set of productivity values was collected (net primary productivity NPP in terms of dry matter produced per m^2 and year; data set published in Lieth, 1972, 1973). (2) This data set was paired with actual evapotranspiration values. Geiger's wall map was used to pick the evaporation level for the region where the NPP-data-point was collected. (3) The data pairs were used to calculate a least square nonlinear correlation model over a predetermined curve shape (Fig. 2). (4) The model in the numerical form of Eq. (5) was then used to convert Geiger's actual evaporation map (Fig. 3) into a productivity map (Fig. 4).

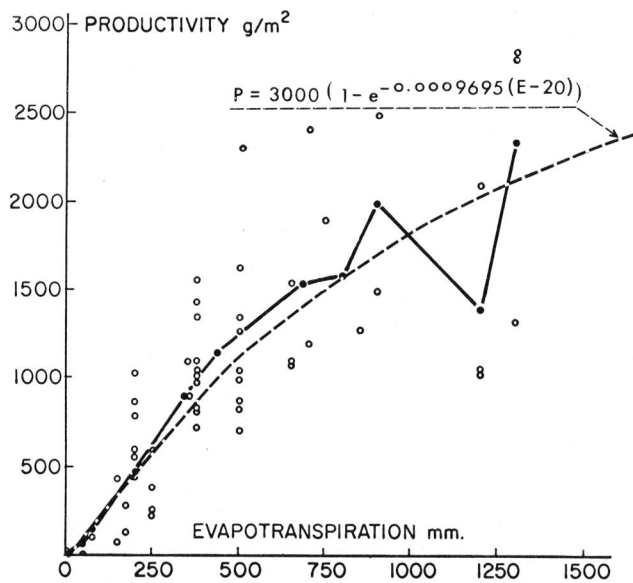

Fig. 2. Annual primary productivity vs. annual actual evapotranspiration. The figure demonstrates the development of Eq. (5). *Open circles:* data-points pairing productivity values from locations for which actual evapotranspiration values were picked from Fig. 3. *Solid circles and solid line:* class averages for 250 and 125 mm evapotranspiration classes. *Interrupted line:* graphical display of Eq. (5) derived from the least-squares regression of the class average points. The numerical form of the equation is shown on the graph. This equation was used to convert the actual evapotranspiration map shown in Fig. 3 into a primary productivity map, the C. W. Thornthwaite Memorial Model (Fig. 4). (After Lieth and Box, 1972)

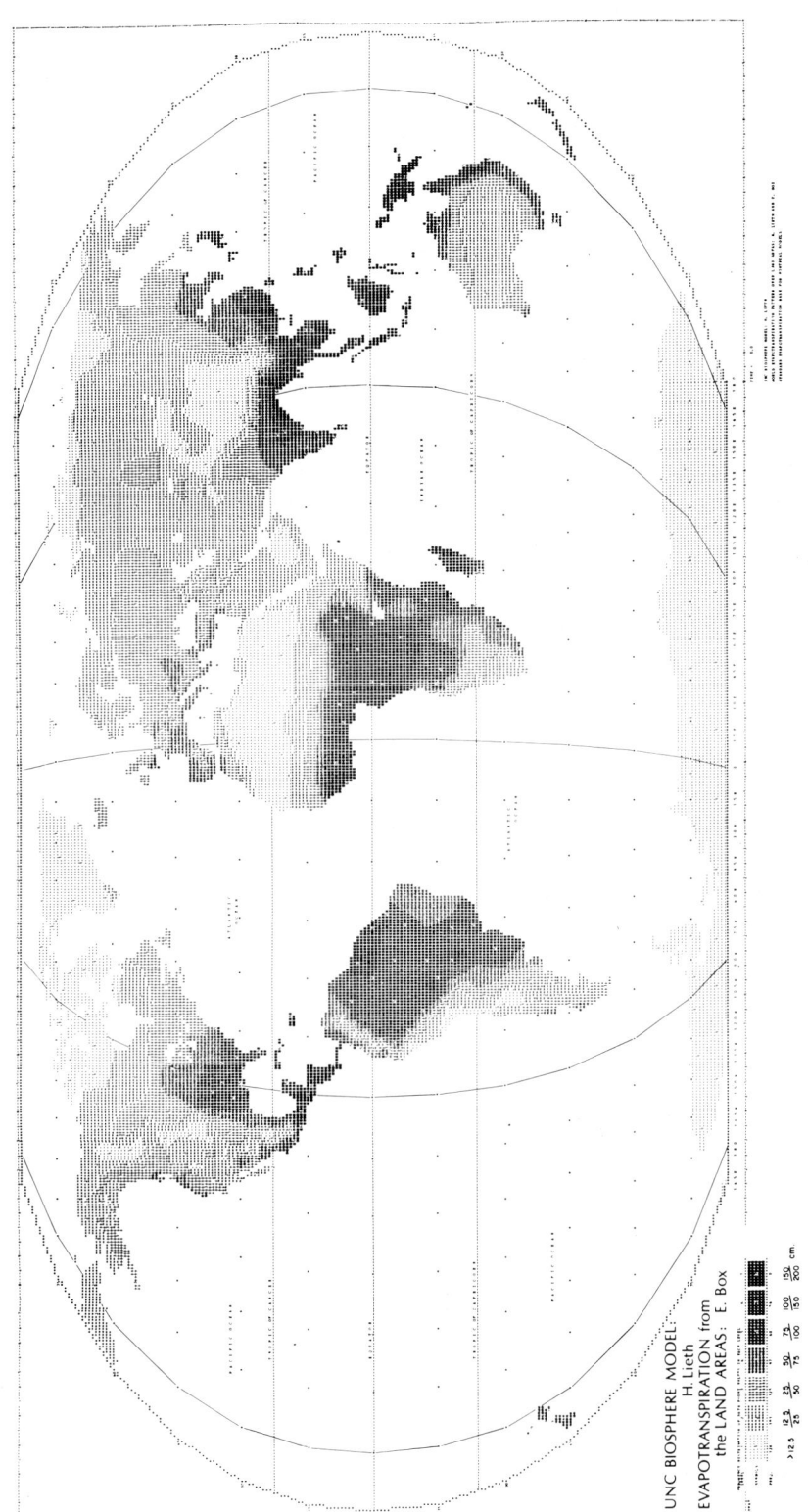

Fig. 3. Actual evapotranspiration from the surface of the globe. Computer simulation of Geiger's wall map by E. Box. (After Lieth and Box, 1972)

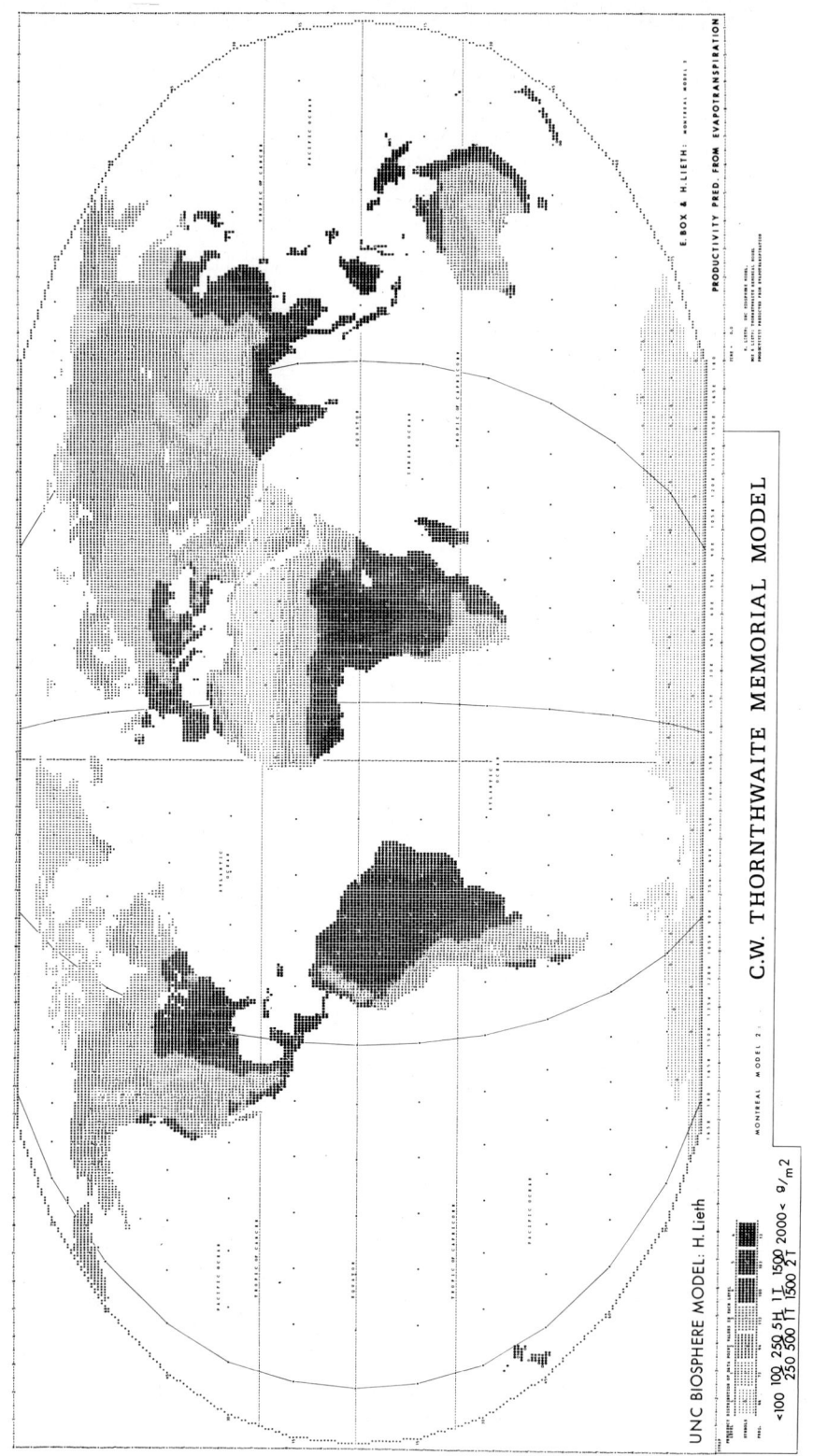

Fig. 4. The productivity map constructed for the land areas using the information presented in Figs. 2 and 3. The map was called Thornthwaite memorial model during a symposium held in Montreal 1971. (After Lieth and Box, 1972)

This procedure is usable to evaluate the productive potential of continents or countries. The approach is specifically convenient because the correlation models are used as input for a computer-mapping routine. The resulting maps can be compared easily with Fig. 1, since the same computer base-map is used. While the prinicple applied is simple, the construction of the entire model is elaborate. In particular, the basic computer map was a major task. Such maps, however, are now available from several sources and the remaining work of constructing the correlation model is easy to do.

III. Some Examples of Correlation Models of Net Primary Productivity versus Water Factor

In constructing significant correlation models one needs to apply ecological experience to judge whether the chosen environmental parameter will enable us to predict NPP adequately or not. Net primary productivity is a process that responds readily to several environmental parameters, of which water is only one. One can circumvent this fact in three ways: (1) by using compound factors in which water is a significant but not single parameter e.g. evapotranspiration; (2) by building multivariable models in which several important parameters are collectively used for calculations, e.g. Ryabchikov's (1968) or Czarnowski's (1973) model, or the one by Terjung et al. (1976); (3) by constructing a set of environmental correlation models and using Liebig's law to restrict its utilization only when the respective factor is the limiting one, e.g. the precipitation submodel in Lieth's Miami model (Lieth, 1973).

Evapotranspiration and net primary production follow similar principles according to ecological and physiological experience. Both depend on radiation, temperature, wind, vapor pressure, and available water. It is therefore a reasonable assumption that both processes appear to a certain degree correlated. If this is the case, the mathematical correlation model can be used to translate information about evapotranspiration into primary productivity and vice versa.

The first guess is that both processes have a linear correlation. The first test of data sets (Lieth, 1961) seemed to indicate that potential evapotranspiration and NPP in humid areas had just such a linear relationship. The data set at that time was so scarce that only a range which is described by the following two equations was assessable:

$$P = 1.5 \ pE - 100 \qquad (1)$$
$$P = 1.25 \, pE - 100 \qquad (2)$$

(P = primary productivity, pE = potential evapotranspiration).

Such a linear regression would make it very easy to convert one printed surface (evapotranspiration) into the other (NPP) by simply changing the scales of the map's legend according to the constants of the Eqs. (1) or (2). When regional data sets became available, Rosenzweig (1968), Whittaker and Niering (1975) developed different correlation graphs. Both curves are shown in Fig. 5. Rosenzweig gave an equation for his curve which reads:

$$\lg P = 1.66 \lg E - 1.66 \tag{3}$$

(P = primary productivity, E = actual evapotranspiration).

Whittaker and Niering (1975) have not yet given an equation for their curve but it may be described in general terms as:

$$P = \frac{K}{1 - e^{-cE}}. \tag{4}$$

Equation (3) describes the exponential acceleration and Eq. (4) describes a sigmoid curve. Whether the curve shape and actual numeric value is more important than the convenience of a fitting curve remains to be discussed. With more production data spread over the whole world and many biome types available, it was sensible to try the approach as was described in Figs. 2 and 4. The equation drawn from the data set was

$$P = 3000(1 - e^{-0.0009696(E-20)}). \tag{5}$$

This curve has the shape of a saturation process.

The difference in the achievements of Eqs. (3) and (5) may be seen in the productivity maps of the US presented by Sharpe (1975). While in general

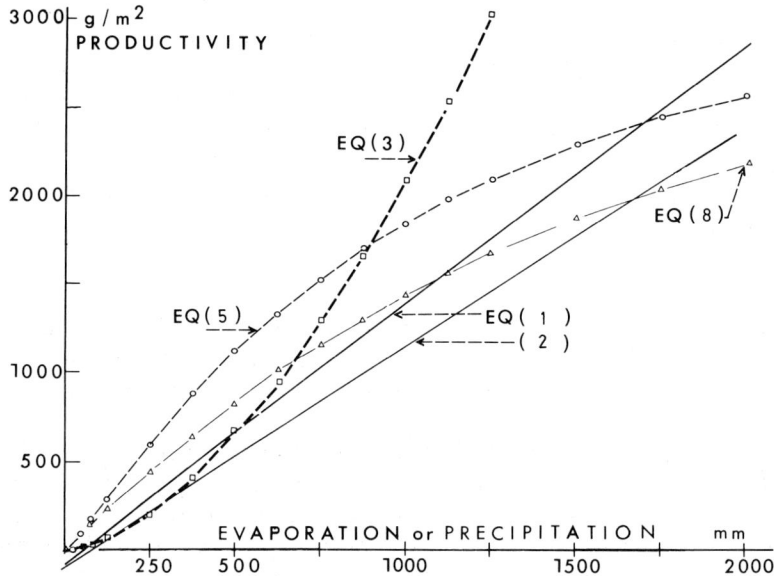

Fig. 5. Comparison of various equations used to express the correlation between primary productivity and evapotranspiration. (1) and (2) Range of potential evapotranspiration relative to productivity taken from Lieth's first model, shown in Fig. 1. (3) Rosenzweig's equation based on actual evapotranspiration. (5) The Box-Lieth model of actual evapotranspiration vs. productivity. (8) The precipitation portion of the Miami model. Comparison of the various equations reveals the weakness of using either a linear equation or an exponential equation. It also shows that the precipitation equation (8) and the Box-Lieth actual evapotranspiration equation (5) run in a logical distance from each other, allowing for about 30–40% runoff of local precipitation

the patterns appear similar, it is evident that Florida shows unrealistically high values. The desert regions appear to be better predicted by Rosenzweig's equation, the moist areas on the other hand by the Thornthwaite memorial model. Altogether for practical purposes it may be just as well to stay with a linear regression as was used in the Eqs. (1) or (2). The comparison with the world map of production shows excessive values only for humid areas in Rosenzweig's model. Since these contribute much more to the total NPP of the world than the dry areas, the errors in his computation become more drastic. More of these problems are discussed in Lieth and Box (1972) and Box (1975).

Building multivariable models is normally the aim of the professional modeler (see van Keulen et al., this volume Part 6:B). A system of differential equations to explain the result of a simple process in an environment with several factors is sometimes the result of such attempts. A few models of this type are to be discussed here. The first model under consideration was built by Ryabchikov and reads

$$\text{Pp} = \frac{W \cdot \text{Tv}}{36 R} \tag{6}$$

(Pp = productive potential, W = effective precipitation, Tv = vegetation period in units of ten days, R = radiation).

The curve describing the NPP predicted from his Pp index is a straight line. The test with empirical data used shows an almost complete fit on a linear scale. The only problem is that some of the parameters used (W, Tv) are hardly measured anywhere and they are subject to biases as well. Ryabchikov's model is based on the data sets compiled by Rodin and Bazilevich (1966) and Rodin et al. (1975).

The next equation using plant-water relations in combinations with other factors is Czarnowski's equation

$$P = c p' L (1 - e^{-g}) \tag{7}$$

(P = primary productivity, p' = is the water vapor pressure for the growing season, g = a humidity index calculated from precipitation over evapotranspiration, L = the "length of the growing season" defined as days above 3°C, e = natural log base and c = a constant value).

Czarnowski (1973) wanted to improve the Miami model (Lieth, 1973). He calculated for four stations the deviation of his model as well as the Miami model against the actual measured values. It is clear that more factors introduced into an equation may improve it but it remains to be seen what Czarnowski's model will achieve if applied generally. Furthermore, there is one shortcoming for quick applications: The number of measured data on L and part of p in Eq. (7) is still limited.

The influence of the individual factor in multivariable models often remains obscure, especially when one or more factors are interactive, either by nature or by their mathematical linkage in the model. One way to by-pass such problems is to develop single-parameter models and connect these additively, by percentage weight or by maximal influence principle. The latter principle was applied in the Miami model (Fig. 6), where it was necessary to couple the influence of

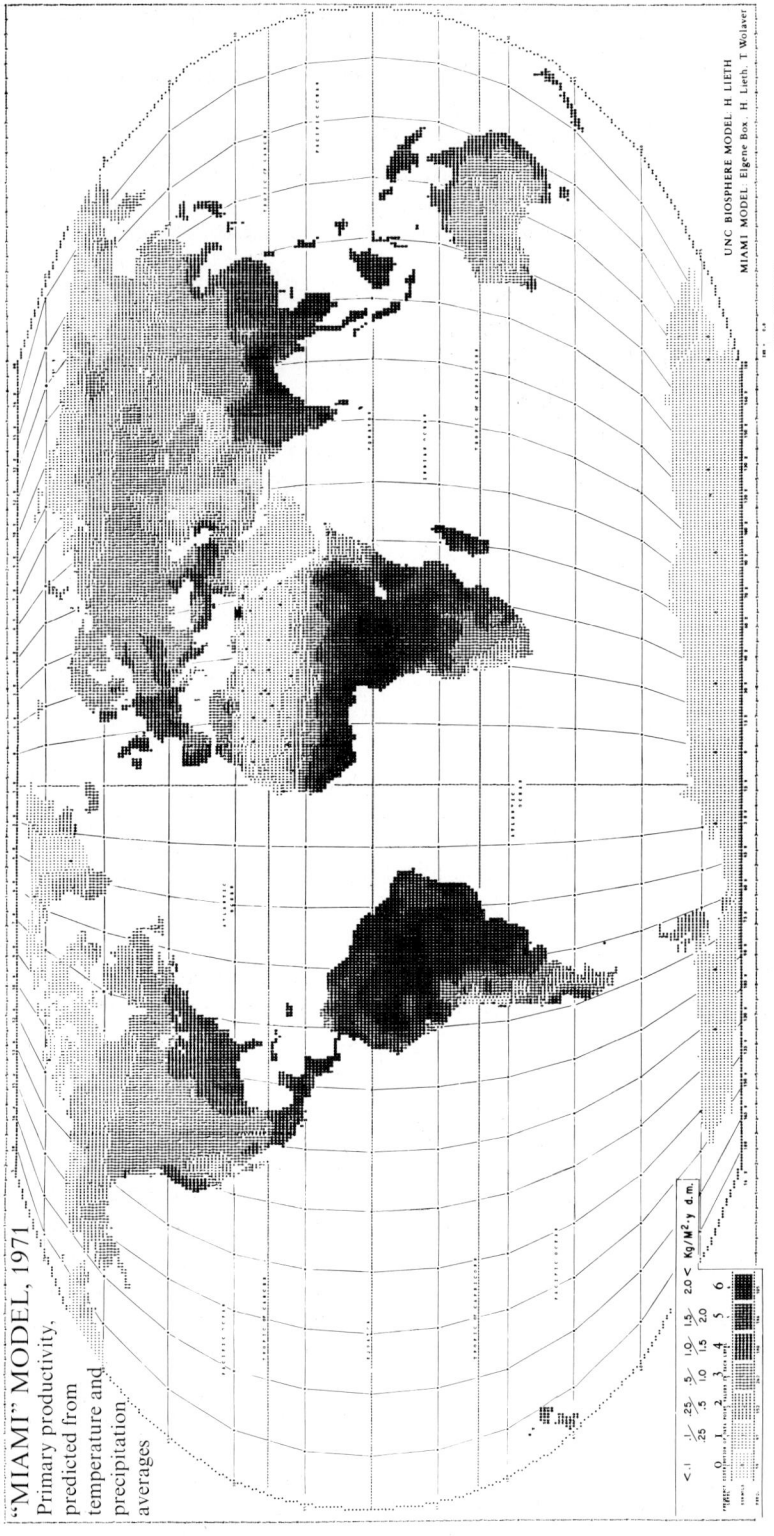

Fig. 6. The 'Miami model': Net primary productivity predicted from average annual temperature and precipitation

temperature and precipitation on NPP for over one thousand stations all over the world.

The two equations had the following form:

$$p = 3000(1 - e^{-0.000644 W}) \tag{8}$$

(p = Net primary productivity, W = annual average precipitation, e = natural log base for precipitation) and

$$p = \frac{3000}{1 + e^{1.315 - 0.119 T}} \tag{9}$$

(p = Net primary productivity, T = annual average temperature, e = natural log base).

The two models constructed individually for each environmental parameter yielded, for many stations, values quite far apart. It was obvious that in the case of a dry desert the precipitation was the crucial limiting factor, whereas in the cold region temperature was decisive. In this case Liebig's law was applied and it was assumed that the lowest value was the correct one. Using this principle the world productivity pattern of the Miami model was constructed. This model certainly requires some improvement, but surprisingly realistic global values for NPP were reached.

IV. Accuracy of Correlation Models

The use of correlation models is not unusual if the regression line and equation are considered only. The most important way in which equations are presently used, is to predict spatial patterns, as was shown in the various maps. If only one predictor is used in a single-variable equation, it could be argued that the map presented shows the patterns of the predictor with an odd drawing of level borders. If, however, the regression line used is curvilinear, the new pattern may have considerable quantitative importance. The depicted maps are quantitatively exploitable and, hence, the conversion of one pattern into another via a curvilinear function is a significant new attribute. The situation becomes even more significant if a new pattern is constructed from more than one predictor. This is the case in the Miami model. The new pattern has, in such maps, a novel property detectable only via the entire set of predictors and the mode of their interconnection. The enormous importance of model-maps for planning purposes becomes evident when one considers the possibility of predicting responses to environmental changes quantitatively. Both possibilities, the quantitative evaluation of maps as well as possible changes, have been exploited by using the map count routine provided by Box (1975). Table 2 shows some examples of global productivity figures assessed in the manner described in this paper in comparison with other attempts. The accuracy of the models depends largely on the quality of the primary data set used for the regression model, the best equation type used, and the density and quality of the predictor's data set.

The main problems in the assurance of the use of the model lie in the applicability of the primary data set to show accurately average and variability

Table 2
(A) Productivity evaluation of the Miami model productivity map[a]. (After Box, 1975)

Level	Productivity range	Number of occurrences	Area estimate (10^6 km^2)	Average (g m^{-2} y^{-1})	Productivity (10^9 g m^{-2} y^{-1})
1	0–100	4 531	15.39	34.9	0.536
2	100–250	2 958	14.15	170.0	2.406
3	250–500	4 505	23.49	380.7	8.943
4	500–1 000	6 122	34.87	727.2	25.357
5	1 000–1 500	3 126	20.43	1 222.2	24.969
6	1 500–2 000	2 884	20.36	1 743.2	35.489
7	2 000–3 000	1 569	11.58	2 313.4	26.778
Total:		25,695	140.22		124.478

[a] All values represent amounts per year. Productivity values are of dry matter, grams per square meter per year or metric tons per square kilometer per year. The average productivity figures (column 6) are the arithmetic averages of the values of all the datum points in the respective intervals. The final productivity figure (column 7) is obtained by multiplying the area estimate (column 5) by the average productivity value column 6). The symbols actually appearing on the Miami map are given in column 3, along with their counted totals in column 4.

(B) Calculation of possible changes of NPP with climatic changes using the Miami model and the Mapcount routine (Box, 1975). The partitioning into latitudinal belts (column 1) shows significant decrease of NPP for 1°C decrease N of 40° latitude only. From 50°N on southward temperature changes of 1°C have little effect. Precipitation changes of 6% (rate of water vapor saturation change in air for 1°C) alter NPP in variable amounts per latitudinal belt. Column 3: Standard Miami model; column 4: 1°C lowered; column 5: difference between columns 4 and 3; column 6: −1°C and −6% precipitation; column 7: difference between columns 6 and 4; column 8: effect of raising temperature by 1°C and precipitation by 6%

1	2	3	4	5	6	7	8
Latit Belt	10^6 km^2 Area land	t/km^2 SMM NPP	SMM −1°C	Δ column 4–3	SMM −1°C −6% P	Δ column 6–4	SMM +1°C +6% P
70–60	13.3	387	356	−31	355	−1	426
60–50	14.7	645	608	−34	600	−8	690
50–40	16.5	804	731	−23	755	−24	849
40–30	15.6	860	854	−6	816	−38	873
30–20	15.2	553	547	−6	519	−28	580
20–10	11.4	894	891	−3	849	−42	956
10–0	10.1	1 823	1 822	−1	1 728	−94	1 905
0–10	10.5	2 024	2 020	−4	1 974	−46	2 069
10–20	9.5	1 275	1 269	−6	1 208	−61	1 351
20–30	9.3	832	823	−9	810	−13	864
30–40	4.1	856	851	−4	815	−36	904
Total land	140.2	124.7 × 10^9 t	122.8 × 10^9 t		118.45 × 10^9 t		130.6 × 10^9 t
% Δ total		—	−1.6		−5		+4.8

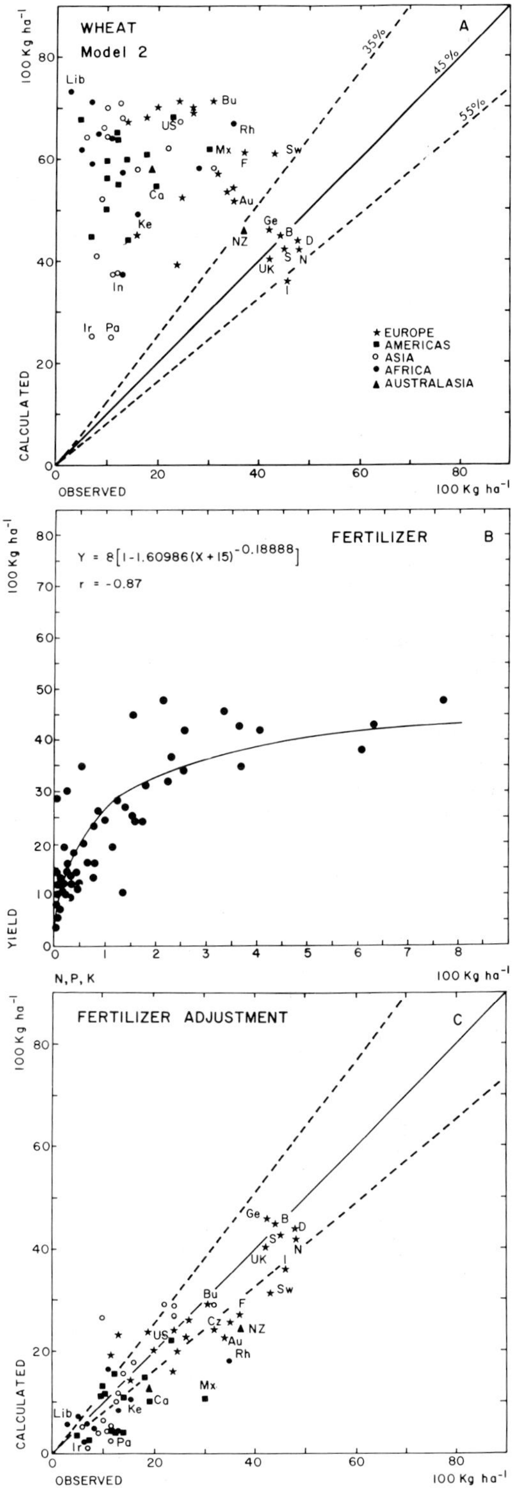

Fig. 7 A–C. The prediction of wheat yield from various countries from energy budget elements (A) in combination with fertilizer application (B). The resulting prediction (C) shows a reasonable fit with the observed levels. Abscissa (B)—Average amount of fertilizer applied. Further explanation in text. (Courtesy of Terjung, Louie, and O'Rourke)

of the measured values against the predictor scale. Summarizing the experiences with the global productivity predictions it can be stated that r^2 values between 0.65 and 0.8 are achievable and result in significant patterns and quantitative evaluations of large area productivity. How accurate the modeling can be is best demonstrated by applying the multivariable model on single crops where statistical data are abundant. An example was recently represented by Terjung and Louie (1973) in which global wheat yields were predicted from environmental data, assuming that about 33 % of the net primary productivity is the average percentage of this grain yield. The result if the step-wise modeling, presented in Fig. 7A, shows the correlation of wheat yield predicted from climatological environmental parameters versus the observed figures from various countries as published in the FAO statistics. Fig. 7B shows the correlation of yield versus fertilizer applications (Mitscherlich-Baule's yield law). Fig. 7C shows the result of the combination between Fig. 7A and B. The improvement of the prediction from Fig. 7A to C is obvious.

V. Conclusions

In summary, it can be stated that all present experience with pattern analysis by correlation models shows them to be highly useful tools. They can provide the basic levels for predicting productivity on a global as well as a regional scale.

References

Box, E.: Quantitative evaluation of the global productivity models generated by computers. In: Primary productivity of the biosphere. Ecological Studies, vol. 14 (eds. H. Lieth, R. H. Whittaker), pp. 256–284. Berlin-Heidelberg-New York: Springer 1975.
Czarnowski, M. S.: W sprawie mapy i modeln siedliskowej zdoldosci prodikcyinej Ziemi. Przeglad geograficzny **45**, 295–308 (1973).
Lieth, H.: La produccion de sustancia organica por la capa vegetal terrestre y sus problemas. Acta Cient. Venezolana **12**, 107–114 (1961).
Lieth, H.: Versuch einer kartographischen Erfassung der Stoffproduktion der Erde. Geographisches Taschenbuch 1964/65, pp. 72–80. Wiesbaden: F. Steiner 1964.
Lieth, H.: Über die Primärproduktion der Pflanzendecke der Erde. Z. f. Angew. Botan. **46**, 1–37 (1972).
Lieth, H.: Primary production: Terrestrial ecosystems. J. Human Ecology **1**, 303–332 (1973).
Lieth, H.: Comparative analysis of some biomass properties on the ecosystem level. In: Primary productivity of the biosphere. Ecological Studies, vol. 14 (eds. H. Lieth, R. H. Whittaker), 352 pp. Berlin-Heidelberg-New York: Springer 1975a.
Lieth, H.: The primary productivity in ecosystems: Comparative analysis of global patterns. In: Unifying concepts in ecology (eds. W. H. Van Dobben, R. H. Lowe-McConnell), pp. 67–88. The Hague: W. Junk 1975b.
Lieth, H., Box, E.: Evapotranspiration and primary productivity; C. W. Thornthwaite Memorial Model. In: Papers on selected topics in climatology (ed. J. R. Mather), vol. 2, pp. 36–44, Elmer C. W. Thornthwaite's Assoc. (1972).
Lieth, H., Whittaker, R. H. (eds.): Primary productivity of the biosphere. Ecological Studies **14**, 352 pp. New York: Springer 1975.
Rodin, L. E., Bazilevich, N. I.: Production and mineral cycling in terrestrial vegetation. Edinburgh: Oliver and Boyd 1966.

Rodin, L. E., Bazilevich, N. L., Rozov, N. N.: Productivity of the world's main ecosystems. In: Productivity of world ecosystems, pp. 13–26. Washington, D. C.: Nat. Acad. Sci. 1975.

Rosenzweig, M. L.: Net primary productivity of terrestrial communities, prediction from climatological data. Am. Midland Natur. **102,** 67–74 (1968).

Ryabchikov, A. M.: Hydrothermal conditions and the productivity of plant mass in the principal landscape zones. Vestnik MGV, Geogr. Moscow **5,** 41–48 (1968).

Sharpe, D. M.: Methods for studying the primary productivity of regions. In: Primary productivity of the biosphere. Ecological Studies, vol. 14 (eds. H. Lieth, R. H. Whittaker), pp. 147–166. Berlin-Heidelberg-New York: Springer 1975.

Terjung, W. H., Louie, S-F.: Energy budget and photosynthesis of canopy layers. Ann. Assoc. Amer. Geogr. **63,** 109–130 (1973).

Terjung, W. H., Louie, S-F., O'Rourke, P. A.: Global photosynthesis model. Intern. J. Biometeorol. **20,** in press (1976).

Whittaker, R. H., Niering, W. A.: Vegetation of the Santa Catalina Mountains, Arizona. 5. Biomass, Production, and diversity along the elevation gradient. Ecology **56,** 771–791 (1975).

B. The Use of Simulation Models for Productivity Studies in Arid Regions

H. van Keulen, C. T. de Wit, and H. Lof

I. Introduction

A large area of the world's surface consists of arid and semi-arid regions. Although the exact magnitude of the area covered by these climates depends on their definition, it is around 30% of the total land surface. The occurrence of disasters, especially drought and starvation, have caused increasing interest in the agricultural problems and potentialities of these regions. Many developed countries take an active part in agricultural research in the arid zone, through the investment of either money or knowledge for the establishment of special projects or the development of specific regions. This research is often aimed at the solution of a special problem, or at least at spectacular results through an immediate increase in production. Despite the efforts of international organizations and agencies, little progress seems to have been made towards a coordinated research program, in which a systematic and theoretically sound examination of the processes critical for the level of herbage production in arid zones is carried out. In particular, the interaction between the various processes and their relative importance in the arid zone have not been studied thoroughly. In the framework of a joint Dutch-Israeli research project on "Actual and Potential Herbage Production under Semi-Arid Conditions" special attention was paid to the development of tools to integrate the existing knowledge into meaningful systems and to indicate ways of using these for extrapolation to areas or circumstances that have not been studied in detail (see Evenari et al., this volume Part 6:E). This is done by developing computer models based on physical, physiological, and chemical processes, that can be applied in regional development programs in the arid zone. Such models can also be used in deciding on the allocation of limited funds available for development of arid regions or in the evaluation of economical advantages that can be gained from irrigation or fertilizer programs.

In this paper a simulation model 'ARID CROP' is treated. This model calculates the course of dry-matter production of a crop canopy and the distribution of moisture in the soil below that crop, from basic or derived physical and physiological properties of plant and soil. Meteorological data from standard weather stations are given as input. The basic assumption is that the crop is optimally supplied with nutrients and that moisture is the main factor liming growth.

A presentation of the complete model and a detailed treatment of the processes is given by van Keulen (1975).

II. The Structure of the Model

ARID CROP is a state-variable model. Such models are based on the assumption that the state of each system at any moment may be quantitatively characterized and that changes in state may be described by mathematical equations. This leads to models in which state-, rate- and driving variables are distinguished. State variables are quantities such as the amount of biomass, the leaf area index, the depth of the root system, the amount of water in various soil layers, and so on. Driving variables characterize the interactions at the boundary of the system and are continuously measured, like rainfall, temperature, and radiation. Each state variable is associated with rate variables that characterize their rate of change at a certain moment. Their value depends on the state and driving variables according to rules that are preferably formulated on the basis of the knowledge of the physical, chemical, and biological processes that take place.

After calculating all rates, these are realized over a short time interval according to the scheme: state variable on time $t+\Delta t$ is equal to state variable on time t plus the rate of time t times the time interval Δt. Obviously the time interval must be so short that even the relatively fast rates of change are materially constant during the time interval Δt. In its most elementary form this is a process of numerical integration over time and the simulation model may be replaced by an analytical solution in cases where the equations are simple enough.

However, most models are too complicated and contain too many discontinuities and even random processes to apply straightforward numerical integration methods, and therefore various simulation techniques for digital computers have been developed to handle the models. The model ARID CROP is written in CSMP, a simulation language that is widely used at present in plant physiology, ecology, and soil science.

The number of state variables that may be distinguished in plant growth models is depressingly large. For instance, not only the weight and surface of the leaves, but also their nitrogen and mineral content, their enzymes and other biochemical characteristics may be considered state variables. Models are, therefore, always simplified presentations of the real system and the simplification manifests itself by the limited number of state variables that are treated. It is assumed that considerable reduction of this number may be obtained by limiting the purpose of the model, it being tacitly assumed that processes may then be ordered with respect to their importance and that only those within focus of the purposes need to be handled in detail. For this reason ARID CROP is restricted to situations with optimal mineral and nitrogen supply.

It is further assumed that for each purpose there is an optimum in the number of state variables that should be considered. At first the applicability of the model increases with increasing number of state variables, but then it decreases again because the addition of new state variables diverts the attention from those that were introduced at first and considered the most important. The heuristic process of obtaining a set of state variables ordered in accordance with their importance takes much time in each modeling effort, and the more so, because a reduction in the number of state variables necessitates a hierarchical approach to model building. In such an approach, models of important sub-systems are developed on basis of sound physical, chemical, and biological principles, which are then used for the construction of smaller and more descriptive models for further use.

There are also aspects of the system that are not sufficiently understood to apply this approach. Sub-models are then based on an experimental analysis of the system in situations where the model will be applied, so that generality may be lost. Such a procedure is justified when it is shown that the behavior of the model is relatively insensitive with respect to these aspects. In the case of ARID CROP, this concerns especially the processes that govern the morphogenesis of the leaf- and root apparatus.

III. Description of the Model ARID CROP

1. Driving and State Variables

The main driving variables of ARID CROP are the macro-meteorological data on a daily basis: daylength (latitude and date), total global radiation (relative duration of sunshine), rain, maximum and minimum temperature, dewpoint in morning and afternoon, and average windspeed.

The soil is characterized by its physical properties of which the water content at field capacity, at wilting point, and in air-dry condition are the most important.

The physiological characteristics that govern the growth of the plant are net CO_2-assimilation and transpiration of the leaves, distribution of dry matter, leaf area growth, and the rate of advance of the rooting front.

The main state variables that are distinguished are the water content of the soil up to 200 cm, the temperature of the soil, the live biomass of the plant above and below ground, the leaf area, and the rooting depth. The plant water status is eliminated as a state variable to avoid time steps smaller than a day.

2. Soil Physical Processes

For the description of the physical processes in the soil (see this volume Part 2), the total soil depth is divided into an arbitrary number of compartments, not necessarily of the same size, each one of which is considered to be homogeneous (de Wit and van Keulen, 1972).

Moisture transport between compartments under the influence of developing potential gradients is not simulated, but descriptive formulations for redistribution of the water are used, based on the result of detailed models for infiltration and evaporation. It has been shown that the use of these simplifications has little influence of the availability of water for plant roots, which is the main interest here, rather than detailed soil moisture dynamics.

The total amount of water infiltrating into the soil, which consists of the precipitation corrected for the influence of run-off or run-on, is distributed over the various layers, in such a way that they are subsequently filled up till "field capacity" from the top one down, until all the water is dissipated. Possible surplus is lost by drainage below the potential rooting zone.

Potential evaporation from the soil surface is obtained from the evaporative demand of the atmosphere, calculated with the Penman equation (1956) and the division of energy between canopy and bare soil. The actual rate of evaporation is calculated taking into account the reduction due to drying of the upper soil compartment. The reduction factor, as a function of soil moisture content in the top compartment, is obtained from a detailed model of soil evaporation. The total moisture loss by evaporation is withdrawn from the various compartments, taking into account the physical properties of the soil and the actual moisture distribution.

Soil temperature is calculated as the running average of the air temperature. A detailed treatment of soil heat flow seems unnecessary, as the influence of the temperature on root growth and water uptake is not known accurately. Moreover, during the greater part of the season these processes take place in the deeper soil layers, where the fluctuations are less pronounced. Uptake of water by the root system depends on both the distribution of roots and the distribution of moisture in the soil. The root system is considered to be homogeneous in horizontal direction. Allowance is made for a partial compensatory effect when part of the roots are in dry soil layers, by assuming a greater uptake by those roots which are in wet soil. Actual moisture withdrawal is determined by the potential crop transpiration and the moisture status in the soil, taking into account the effect of soil temperature on water viscosity and root permeability.

3. Growth of the Crop

The germination process of species from natural vegetation is very complicated, because both the properties of the available seed stock and the micro-environment in the upper soil layers show great fluctuations. The importance of these factors for germination of winter annuals has been shown by Janssen (1974). As in this model, however, the main interest is in total dry matter production rather than in the botanical composition of the vegetation, such a detailed treatment is not included. Germination proceeds when the moisture content in the upper 10 cm of the soil is above wilting point and establishment is assumed after a temperature sum of 150 days °C above zero is obtained (Tadmor et al., 1968). When the soil is dried by evaporation before the required temperature sum is reached, the seedlings die and a new wave of germination starts only after rewetting. At the establishment the biomass is set equal to the initial biomass, for which a constant value of 100 kg ha^{-1} is assumed, irrespective of the germination conditions. Total seasonal dry matter production is not very sensitive to this value, but the growth pattern may change, which is an important factor when the vegetation is used for grazing. Initialization therefore remains a major difficulty in this model.

The rate of growth of the vegetation is determined by the transpiration rate and the water use efficiency. The latter is obtained as the ratio between potential growth and potential transpiration (see Bierhuizen, this volume Part 6:C), assuming that its value is independent of soil moisture conditions or canopy properties. The independence of soil moisture status may not be true for some of the species, which may adopt a more efficient use of the available water, when under stress (Lof, 1975; see Hall et al., this volume Part 3:D). However, the absolute amounts of water transpired during such periods are so small, that the effect of a different efficiency on dry matter production is negligible.

The potential rate of transpiration is calculated from the evaporative demand of the atmosphere, determined by radiation intensity and the combined effect of wind speed and humidity, and the leaf area of the canopy.

The equations to calculate the potential transpiration from daily meteorological data are derived from comparison with the detailed model mentioned before, whose results showed good agreement with measured data (van Keulen and Louwerse, 1975).

The potential rate of growth, being defined as the increase in dry weight, is obtained from the rate of gross photosynthesis, taking into account losses for maintenance respiration, depending on amount of dry matter present and temperature, and efficiency of conversion of primary photosynthates into structural material (growth respiration) (Penning de Vries, 1974). Gross photosynthesis as a function of radiation intensity and leaf area index is read from tabulated functions, which are obtained as outputs of the higher resolution model (hierarchical approach). The actual rate of growth is obtained by multiplying the water use efficiency with the actual rate of transpiration.

It is assumed that plant material is continuously dying, under favorable conditions at a relative rate of 0.005 day^{-1}, going up to 0.1 day^{-1}, when either severe water stress develops or when the crop reaches maturity. When, as a result of dying material, the amount of living biomass drops below a limiting value, the canopy is assumed to be completely dead and growth will continue only after a new wave of germination.

The development pattern of the canopy, characterized by the rate and order of appearance of vegetative and generative organs, is governed by genetic and environmental factors. For a given crop at a certain location i.e. at a given daylength, temperature is the most important external factor (van Dobben, 1962). The relation between air temperature and development rate applied in the model is constructed from field experience, assuming that it is linear in the normal range of temperatures (van Schaik and Probst, 1958). The development stage of the crop influences the growth rate through an effect on the potential transpiration rate, is being reduced from 1 to 0 between development stage 0.75 and 1 (ripeness). Although the numerical values are chosen somewhat arbitrary, the qualitative effect, causing cessation of growth at a certain time, even when water is still available in the soil, is in agreement with observed phenomena.

The morphological characteristics of the canopy, governing the relation between weight and leaf area, are difficult to simulate, as the underlying processes are not yet fully understood. However, the leaf area is a factor in the calculation of water use efficiency as well as in the ratio between evaporation and transpiration. Therefore, leaf area growth is simulated from the growth rate, assuming a leaf area ratio dependent on air temperature. Data are used derived from classical growth analysis experiments with wheat. Although this may seem an over-simplification in a situation where the canopy consists of a mixture of species quite different in morphological properties, it is justified by the fact that deviations do not have a decisive influence on the final result.

The rooting system of the canopy is defined by both its vertical extension and its dry weight. It is assumed that a "root front" moves down at a constant rate when conditions are favorable. Allowance is made for the influence of temperature on the rate of vertical extension, and growth stops when the roots reach a dry soil layer, because of the high mechanical resistance. The root system is considered to be homogeneous in the horizontal direction, so that moisture uptake is governed by the average moisture content in each compartment and not by gradients developing around individual roots. The latter is based on evidence that new roots can develop within a very short time (1–2 days). Root weight is obtained from the growth rate by imposing a ratio of division of the material between above ground and underground plant parts, depending on the development stage of the canopy. When development proceeds, progressively less of the newly-formed material is transferred to the roots, leading to an increasing shoot to root ratio towards maturity. Under normal growing conditions a ratio of 3 to 4 is reached at the end of the growing season. Because of lack of data there is no feedback of root weight to vertical extension or root activity.

The model is executed with time steps of one day, this being small enough as compared to the time constant of crop growth, which is equal to the reciprocal of its relative growth rate (de Wit and Goudriaan, 1974). Moreover it is very unlikely that more detailed meteorological data are available.

Integration is performed according to the simple rectilinear method.

IV. Validation of the Model

Experiments to collect validation data for the model were carried out at the "Tadmor Experimental Farm" near Beersheva in the Northern Negev Desert of Israel. It is an area with a mediterranean type climate, having warm dry summers and cool moist winters. The long-term average rainfall is about 250 mm year^{-1}, the fluctuations between seasons being considerable: 42 mm in 1962/63 and 414 mm in 1964/65.

The soil consists of a homogeneous thick layer of 'Löss' (young eolean material) with practically no profile development. The physical properties are favorable for plant growth, with a water-holding capacity of $\pm 0.15\,\text{cm}^3\,\text{cm}^{-3}$ between field capacity and wilting point, and no mechanical barriers for root growth.

The vegetation is an abandoned crop land vegetation, predominantly made up of herbaceous annual species. The predominant species are the grasses *Phalaris minor*, *Hordeum murinum* and *Stipa capensis*, the Cruciferae *Erucaria boveana* and *Reboudia pinnata*, the Compositae *Arthemis melaleuca* and *Centaurea iberica*, and the Fabaceae *Trigonella arabica* and *Medicago polymorpha*. The botanical composition may change considerably both from year to year and from place to place, as a result both of germination conditions and composition of the available seed stock. That seems, however, to have very little effect on the production potential, as no great differences in water use efficiency between species could be established.

The experiments were carried out in the seasons 1971/72, 1972/73 and 1973/74, which were wet years with 315, 245 and 350 mm of rain respectively, and with a favorable distribution. Standing crop was measured via a double-sampling technique based on visual estimates (Tadmor et al., 1974, 1975). The measurements were performed about every fortnight during the growing season. Soil moisture was determined also every two weeks and after each sufficiently large fall of rain with the neutron moderation technique in 30 cm increments to a depth of 180 cm. This was combined with gravimetric sampling of the top 30 cm, where the neutron method is not reliable because of scatter to the atmosphere. In Figs. 1–6 the measured and simulated values of above ground dry matter production and total soil moisture are given for the three growing seasons. When comparing the results of dry matter production, the difference between the season 1971/72—where measured and simulated data never show differences exceeding ±5%—and the next two seasons is striking. The reason for this discrepancy points to an inherent difficulty in model building. During development of the model, the 1971/72 data have always been used to test the behavior of the model. In the situation where both modeling and experimentation are carried out by the same scientist—or team—it is almost impossible to avoid continuous interaction between the two, leading to the—subconscious—introduction of subjective functions in the model, which may be difficult to pinpoint. It is, therefore, of great importance that in the evaluation phase of the model completely independent data are used to test the validity of the opinion expressed in the model (van Keulen, 1974). Another difference between 1971/72 and the other seasons is the initial biomass that must be assumed to simulate the measured growth pattern. In 1971/72, 25 kg of dry matter ha^{-1} was introduced, which, when applied in 1972/73, lead to a much lower growth rate than measured (Fig. 3). It is most likely that the disking, which was carried out before the season 1971/72 to ensure proper mixing of the fertilizer that was applied, had a strong negative effect on germination, probably through burying of the seeds to such a depth that the seedlings did not reach the surface (the disking also caused a rather abnormal botanical composition of the vegetation). The different initial biomass values have, however, only a small effect on the final yield, indicating that eventually the available amount of water determines the total production, the differences being caused by somewhat higher evaporative losses when full cover is postponed, and by differences in water use efficiency in different periods. Initialization remains, however, a major problem (see Evenari et al., this volume Part 6:E), especially when the model is to be applied in connection

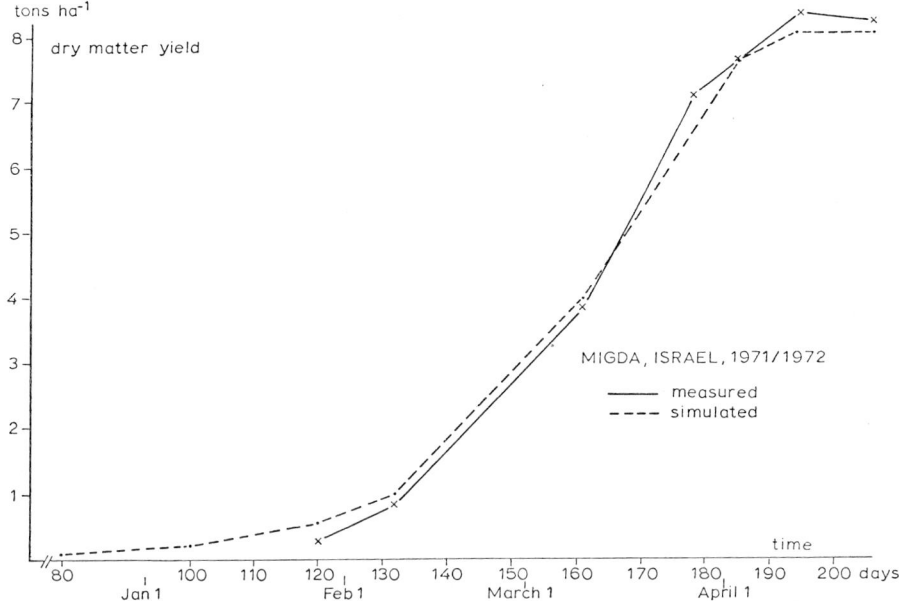

Fig. 1. Comparison between measured and simulated dry matter production of natural vegetation in Migda 1971/72. (After van Keulen, 1975)

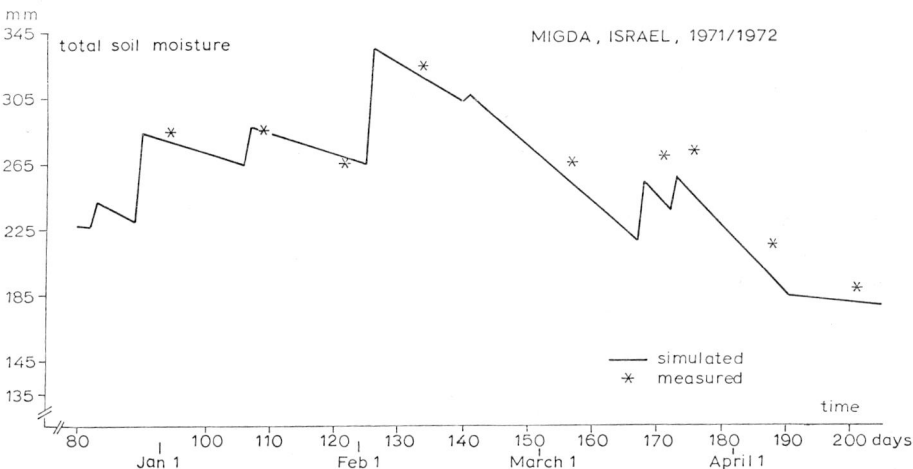

Fig. 2. Comparison between measured and simulated course of total soil moisture under natural vegetation in Migda 1971/72. (After van Keulen, 1975)

with grazing models. For sheep, which are pregnant during that period, the amount of food available in the beginning of the growing season is of primary importance. In such situations the best solution seems to be to measure standing crop at some early stage and apply that as initial value in the model.

The differences in measured and simulated growth rate in 1972/73 (day 130–day 145) cannot satisfactorily be explained. There may be some speculation

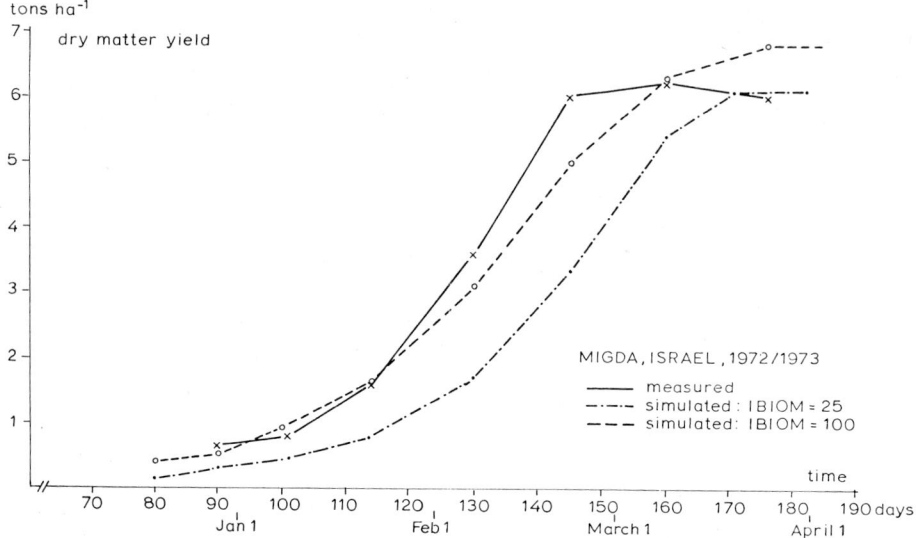

Fig. 3. Comparison between measured and simulated dry matter production of natural vegetation in Migda 1972/73. (After van Keulen, 1975)

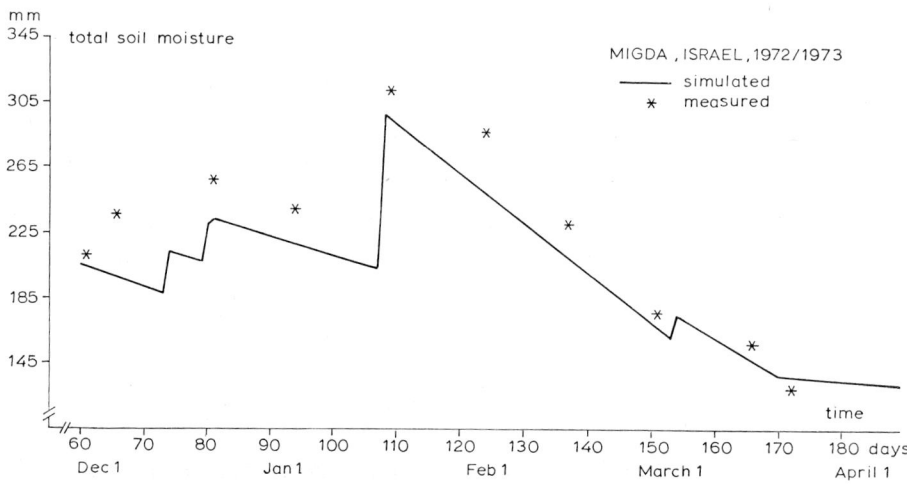

Fig. 4. Comparison between measured and simulated course of total soil moisture under natural vegetation in Migda 1972/73. (After van Keulen, 1975)

about a different division of dry matter between shoot and root, due to the very favorable conditions in the soil, but without more information about the process of root growth no conclusion can be drawn. In 1973/74 the deviations between simulation and experiment must be attributed to nitrogen status in the field. In the first two seasons 440 kg N ha^{-1} was applied, of which 400 kg

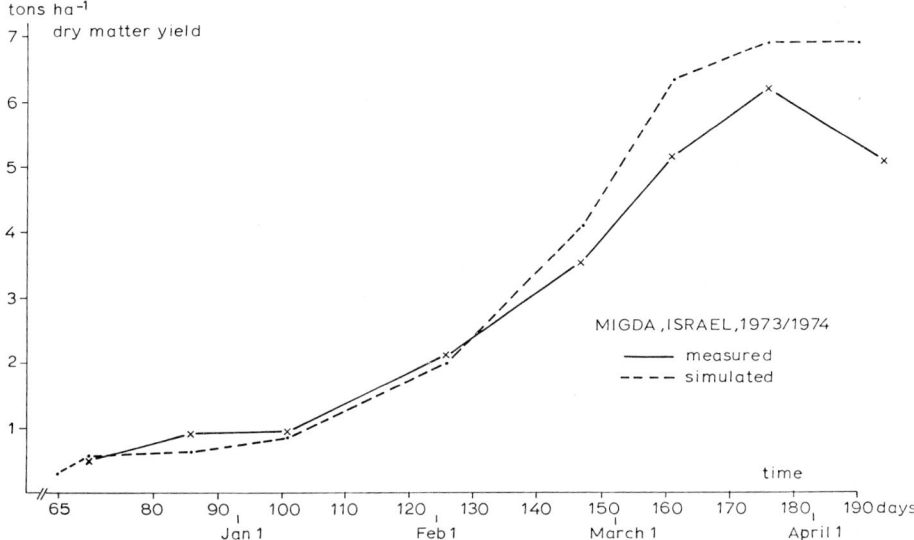

Fig. 5. Comparison between measured and simulated dry matter production of natural vegetation in Migda 1973/74. (After van Keulen, 1975)

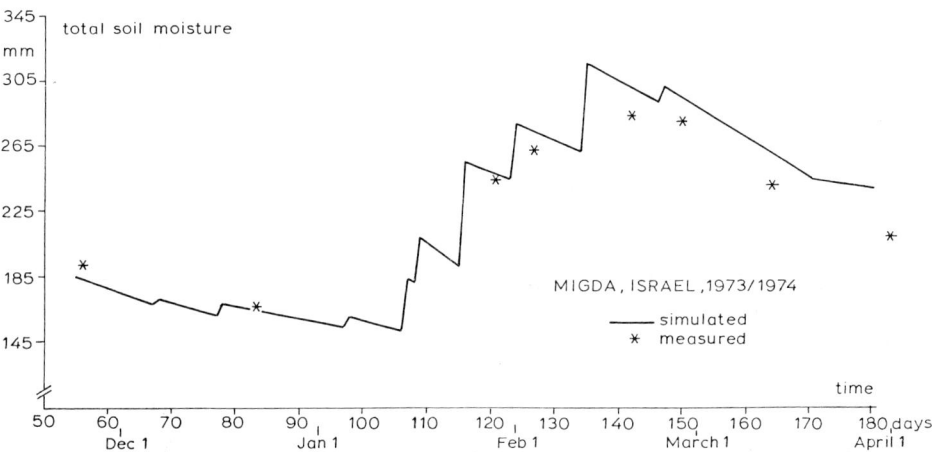

Fig. 6. Comparison between measured and simulated course of total soil moisture under natural vegetation in Migda 1973/74. (After van Keulen, 1975)

was in the form of ammonia, which is subject to volatilization. In these two seasons 350 kg N ha^{-1} was removed by the canopy, so that the store of N in the soil could have been depleted by the end of 1972/73. The amount of 100 kg N ha^{-1} applied at the beginning of 1973/74 would not be sufficient to maintain potential growth throughout the season. This is supported by the N-analysis of the vegetation, which showed lower N-percentages at the end of 1973/74 than in the previous seasons.

The soil moisture data show in all three seasons good agreement between simulation and experiment, taking into account the great variability in the field, restricting the accuracy of measurement to about 10 % of the field mean.

From the overall comparison between the simulated and measured data, it may be concluded that the model simulates rather accurately the dry matter production of the vegetation and the water balance in the soil below it, under conditions of optimal nutrient supply and moisture limitation. It is, however, also clear from the description of the model that, despite the reasonable results, there are still some areas where our knowledge is only fragmentary and where improvements are possible. However, such a model should not reach a static state, as the growing insight in the relevant processes and the research which could be initiated by its results must lead to continuous improvement of the model.

V. Application of the Model

In order to get an insight in the production capacity of the northern Negev, the model ARID CROP was executed with thirteen years of historical weather data, collected at the Gilat Experimental Station, ± 8 km from the experimental site.

During this period the experimental area was used for grazing experiments on the main vegetation types of the region. As part of this research program standing crop on ungrazed plots was determined at about the moment of peak production (Tadmor et al., 1974). The results of the simulation runs are summarized in Table 1, together with some data on weather and actual production.

The first striking phenomenon is, that in 8 out of 13 years, the calculated production is higher than the measured production. This is caused by the fact that the calculation assumes optimum supply with nutrients, while under field conditions without fertilization nitrogen is very often the limiting factor. This

Table 1. Summary of results of simulation model ARID CROP (for explanation see text)

Season	Rainfall mm	Aboveground Simulated production kg ha^{-1}	Aboveground Measured production kg ha^{-1}	TRCT g H$_2$O (g DM)$^{-1}$	TRCR g H$_2$O (g DM)$^{-1}$
1961/62	120	280	—	422	4285
1962/63	72	185	—	343	3890
1963/64	357	6305	3400	310	566
1964/65	414	6171	3100	299	671
1965/66	220	1675	1600	354	1314
1966/67	284	5571	3600	312	510
1967/68	235	4420	2800	272	532
1968/69	212	2473	2500	254	857
1969/70	172	1193	1200	306	1442
1970/71	260	2529	1100	312	1028
1971/72	315	6242	3600	239	505
1972/73	245	5922	3500	291	414
1973/74	351	5833	3800	228	602

has been shown in experimental fields, where in the last three seasons calculated and measured yield are the same with fertilization. Moreover, recent work by Harpaz (1975) shows that when nitrogen is also introduced in a model as a possible limiting factor, the simulated yields are in fair agreement with the measured ones. This shows that, under the weather conditions prevailing in the Negev, in most cases not moisture but nitrogen is the main factor limiting production. This could also be the case in other semi-arid regions with comparable climatological conditions. In these areas, therefore, increased production may be obtained by introducing nitrogen into the system, either through fertilization or by the introduction of leguminous species in the sward.

The next point of interest is the calculation of the transpiration coefficient (TRC), which is given both with respect to total calculated transpiration (TRCT) and with respect to total rainfall (TRCR). The variation in TRCT is almost a factor 2, which reflects the different external conditions during the actual growth period. In periods with a low humidity of the air and a rather high radiation level, more water is lost at the same level of photosynthesis. Such periods prevail in the northern Negev, especially when the wind is blowing from the east, in extreme conditions leading to the so-called sharav (chamsin).

The variation in TRCR is even greater, as this is also influenced by the distribution of the rainfall. A larger number of showers causes a greater loss of water through direct evaporation from the soil surface, thus decreasing the overall efficiency of water use.

These results show that, even when nutrition is eliminated as a decisive factor, it is very difficult to predict productivity on the basis of statistical analysis. It seems, thus, more promising to use simulation models for the calculation of the production potential of new areas. It should, however, be kept in mind, that prediction in the actual situation is still completely dependent on the quality of the meteorological forecasts.

VI. Conclusions

It has been firmly established in theory and practice that the potential growth rate of natural grassland species is as high as that of cultivated species during periods that water and nutrients happen to be available, so that yield potential, as such, is now reason for the introduction of so-called improved species. This is the more so, because some of the natural annual species already start to form ripe seeds within a few months after germination, which may safeguard the crop of next year.

Eco-physiological studies of some winter annuals show that, when conditions happen to be optimal, little difference exists between species in terms of productivity and water use efficiency. However, under stress conditions a general distinction may be made between savers and spenders (Lof, 1975), but this distinction is of small practical importance as long as all available water is used during the growth period.

Under good nutritional conditions in the Negev, yields of 6000 kg ha^{-1} with a reasonable nitrogen content may be obtained with a rainfall of only

250 mm during the winter season. However, in 8 out of 13 years, it appeared that the yields were considerable lower than could be expected, because of nitrogen shortage. Simulated and measured yields in these years are also in fair agreement when nitrogen is introduced as a limiting factor into the model (Harpaz, 1975).

Especially in years with little rain, several germination flushes occur and a considerable fraction of the water may be lost by evaporation. It appears then that the yield may be considerably improved by soil heterogeneity that promotes the occurrence of local run-off/run-on.

Systems analyses, followed by model synthesis that covers the whole field or primary production, grazing, animal production, herd management, and marketing seem a promising way to develop new grazing methods and options for new ways of life in arid regions. The experience with only a small part of the system (primary production), is that only thorough attempts, in which scientific analyses and field experimentation are closely linked, may lead to trustworthy results and that quick results, however attractive on first sight, may be very costly in the long run.

References

Dobben, W. H. van: Influence of temperature and light conditions on dry matter distribution, development rate and yield in arable crops. Neth. J. Agr. Sci. **10**, 377–389 (1962).

Harpaz, Y.: Simulation of the nitrogen balance in semi-arid regions. Ph. D. Thesis, Wageningen 1975.

Janssen, J. G. M.: Simulation of germination of winter annuals in relation to microclimate and micro distribution. Oecologia (Berl.) **14**, 197–228 (1974).

Keulen, H. van: Evaluation of models. Proc. 1st Intern. Congr. Ecol., The Hague. Wageningen: Pudoc 1974.

Keulen, H. van (ed.): Simulation of water use and herbage growth in arid regions. Simulation Monographs. Wageningen: Pudoc 1975.

Keulen, H. van, Louwerse, W.: Simulation models for plant production. Proc. W. M. O. Symp. On agrometeorology of the wheat crop, Braunschweig, 1973 Offenbach: Deutscher Wetterdienst, pp. 196–205, 1975.

Lof, H.: Water use efficiency and competition of arid zone annuals, with special reference to the grasses *Phalaris minor* Retz. and *Hordeum murinum* L. Agric. Research Report **853**, pp. 1–109. Wageningen: Pudoc 1976.

Penman, H. L.: Evaporation: An introductory survey. Neth. J. Agr. Sci. **4**, 9–29 (1956).

Penning de Vries, F. W. T.: Substrate utilization and respiration in relation to growth and maintenance in higher plants. Neth. J. Agr. Sci. **22**, 40–44 (1974).

Schaik, P. H. van, Probst, A. H.: Effects of some environmental factors on flower production and reproductive efficiency in soy beans. Agron. J. **50**, 192–197 (1958).

Tadmor, N. H., Brieghet, A., Noy-Meir, I., Benjamin, R. W., Eyal, E.: An evaluation of the calibrated weight-estimate method for measuring production in annual vegetation. J. Range Manage **28**, 65–69 (1975).

Tadmor, N. H., Eyal, E., Benjamin, R. W.: Plant and sheep production on semi-arid annual grassland in Israel. J. Range Manage. **27**, 427–432 (1974).

Tadmor, N. H., Hillel, D., Cohen, Y.: Establishment and maintenance of seeded dryland range under semi-arid conditions. Final. Tech. Rep. U.S.D.A. Proj. A10-CR-45 (1968).

Wit, C. T. de, Goudriaan, J.: Simulation of ecological processes. Simulation Monographs. Wageningen: Pudoc 1974.

Wit, C. T. de, Keulen, H. van: Simulation of transport processes in soils. Simulation Monographs. Wageningen: Pudoc 1972.

C. Irrigation and Water Use Efficiency

J. F. BIERHUIZEN

I. Introduction

In the preceding decades knowledge regarding water supply, water uptake, transpiration, and plant growth has increased enormously. Various handbooks have been written on this subject from either a physical, ecophysiological, agronomic, or engineering point of view (Slatyer, 1967; Kozlowski, 1972; Kovda et al., 1973; Monteith, 1973). Moreover, field and laboratory methods for measuring plant water relations have recently been greatly improved (Slavík, 1974).

Problems regarding irrigation and water use efficiency have become increasingly important because in large areas water is scarce and/or of a poor quality. At present it is realized that one should not only try to achieve the highest yield per unit area. Improvement in economic and efficient water use can nowadays be even more important. To achieve this, models have been developed (Stewart and Hagan, 1973). It is not the purpose of this chapter to deal with all aspects of irrigation and water use efficiency, for which the reader is referred to the above-mentioned surveys and other chapters of this book (see Hall et al., this volume Part 3:D; van Keulen et al., this volume Part 6:B). Only three topics have been selected viz. the efficiency of water supply from an engineering point of view, transpiration/photosynthesis relationships, and some agronomic aspects. In the next paragraphs some current results and ideas will be presented.

II. Efficiency of Water Supply

At present the estimated irrigated area in the world in which either basin, furrow, border, sprinkler, or subsurface irrigation is applied, is approximately 200 million ha. The overall water supply efficiency depends on losses in (1) the conveyance of water from a storage reservoir to the farm, (2) the distribution of water within the farm, and (3) the application on the field.

1. Conveyance Efficiency

The conveyance efficiency of an irrigated area (ε_c) is the ratio between the quantity of water received by all farms (V_f) and the total quantity supplied (V_t) according to:

$$\varepsilon_c = \frac{V_f}{V_t}. \qquad (1)$$

It is obvious that the conveyance efficiency depends on evaporation and leakage.

In a closed supply system the evaporation is negligeable. Often a pressure exists between 3 and 4 atm, which, even in the case of a small leakage, leads to considerable spilling of water. In practice the efficiency may vary between 0.8 and 0.9. Closed systems are mainly applied in connection with sprinkler irrigation for high value crops.

In an open system the water loss due to evaporation depends on the climatic conditions, for which data are usually available, and on the area of the water surface, which is often 2% or less of a total irrigation area. The water loss via leakage depends on the pressure difference, the surface of the canal bed, and the permeability. In a sandy soil lining of the canals is often applied. The permeability may decline after some time due to flocculation of silt and organic material. The latter process is enhanced by the availability of nitrogen in the irrigation water.

Using questionnaires Bos and Nugteren (1974) analyzed the water utilization efficiency of 91 areas in the world (approximately 3 million ha) varying in climate and management. They observed that the conveyance efficiency was high (0.90) and independent of the irrigation area when delivery to the farm was continuous. With an intermittent supply to the farm, however, an efficiency of only 0.85 was observed for an area between 3000 and 7000 ha, and this was even lower for smaller or larger areas. It was assumed that below 3000 ha lack of qualified personnel, and above 7000 ha communication problems, are the main causes for the reduction in conveyance efficiency. With an intermittent system ε_c could be improved by using rotational unit areas between 70 and 300 ha. In their inquiry Bos and Nugteren (1974) could not demonstrate a favorable effect of lining of the canals on the efficiency because the majority of the canals with a high soil permeability (sand) in general were lined, and those with a low permeability (clay) were not lined.

2. Farm Ditch Efficiency

The farm ditch efficiency (ε_{fd}) is the ratio between the quantity applied to the field (V_a) and the total quantity applied to the farm (F_f) according to:

$$\varepsilon_{fd} = \frac{V_a}{V_f}. \qquad (2)$$

Bos and Nugteren (1974) observed that a continuously supplied paddy farm had a high ε_{fd} value of 0.92, which was independent of farm size. In the case of intermittent irrigation, however, ε_{fd} was approximately 0.6 on a farm size of 1 ha, increasing to 0.8 with an increase in farm size up to 20 ha. The increase in efficiency with intermittent irrigation in larger farms could be ascribed to the longer delivery period. A delivery period of 10 h resulted in an ε_{fd} of 0.58, increasing to a maximum of 0.88 for a 200 h delivery. In general, an increment in efficiency of 0.1 to 0.15 was obtained with pipelines or lined ditches.

3. Application Efficiency

The application efficiency (ε_a) is the ratio between the quantity retained in the root zone (V_r) and the total quantity applied to the field (V_a) according to:

$$\varepsilon_a = \frac{V_r}{V_a}. \tag{3}$$

The low efficiency of 0.32 of a paddy crop with a continuously flooded basin is due to heavy percolation and uncontrolled run-off according to Bos and Nugteren. With intermittent flooding this value may rise to 0.4 or 0.5. Continuously flooded paddy cultivation is usually practised on clay or silty clay, as on a sandy soil percolation losses would be too high. It was observed that the average efficiency of intermittent basin, furrow border, and sprinkler irrigation was 0.58, 0.57, 0.53, and 0.67 respectively. The efficiency of sprinkler irrigation is lower than would be expected. This may be caused by the high rate of application (more than 10 mm h^{-1}), which, especially in clay soils, may exceed the infiltration rate and thus enhance surface run-off. It is known that trickle irrigation is highly efficient but often too expensive. For surface irrigation the optimal efficiency depends on the quantity applied and the stream used by the farmer. Approximate figures are an application of 600 to 800 m^3 per ha and a stream of $2-3 \text{ l s}^{-1}$ per m width in the case of borders and $1-2 \text{ l s}^{-1}$ per m width in the case of furrows.

The average values of ε_c, ε_{fd} and ε_a in the survey of Bos and Nugteren were 0.74, 0.77, and 0.53 respectively, from which other parameters can be calculated such as:

farm efficiency $\varepsilon_{fd} \cdot \varepsilon_a$ which is 0.44
distribution efficiency $\varepsilon_c \cdot \varepsilon_{fd}$ which is 0.58
overall efficiency $\varepsilon_c \cdot \varepsilon_{fd} \cdot \varepsilon_a$ which is 0.30

It is obvious from these low figures that, at present, the improvement of farm and distribution management is a more important means of saving water than is a direct reduction of plant transpiration.

III. Transpiration/Photosynthesis Relationships

The water use efficiency of a crop, a plant, or a leaf is generally represented as the ratio: dry matter production/evapotranspiration or net photosynthesis/transpiration both expressed in the same units. Other ratios used are the reciprocal value of the water use efficiency i.e. the transpiration coefficient or transpiration ratio, which for most crops varies between 100 and 1000 (Miller, 1938). This wide variation can often be explained by a critical analysis of the environmental effects on transpiration and net photosynthesis.

Transpiration (T) and net photosynthesis (P) can be described in the general form of diffusion equations depending on a gradient and a resistance according to:

$$T = \frac{\Delta e}{r_a + r_s} \qquad (4)$$

$$P = \frac{\Delta CO_2}{r'_a + r'_s + r'_m} \qquad (5)$$

$$T/P = \frac{\Delta e}{\Delta CO_2} \cdot \frac{r'_a + r'_s + r'_m}{r_a + r_s} \qquad (6)$$

in which Δe and ΔCO_2, show the difference between water vapor concentration and the carbondioxide concentration; r_a, r_s and r_m is the laminary boundary layer resistance, the stomatal resistance, and the mesophyll resistance; the accents r_a and r'_a, r_s and r'_s denote the difference in the diffusion coefficient for H_2O and CO_2 respectively.

Constants have to be included for the exact dimensions and further details are described elsewhere in this book (see Gates, this volume Part 3:A; Waring and Running, this volume Part 3:F; Hall et al., this volume Part 3:D).

Some aspects of Δe, ΔCO_2 and the resistance pathway will be discussed below.

1. Effect of Water Vapor Concentration Difference

Instead of the water vapor concentration, often the vapor pressure expressed in mm Hg is applied in Eq. (4), for which a conversion factor has to be applied (Slatyer, 1967). It is generally assumed that at the wall surface of the substomatal cavity, the vapor pressure is maximal (e_{max}), its value depending on the leaf temperature. The actual vapor pressure (e_a) in the ambient air depends on the prevailing temperature and relative humidity. With a constant resistance pathway ($r_a + r_s$), T plotted against Δe ($e_{max} - e_a$) gives a linear relationship, its slope being the reciprocal value of the resistance. It is mentioned in more detail elsewhere in this volume (see Hall et al., this volume Part 3:D) that r_a depends on windspeed and leaf geometry and r_s on light intensity, CO_2 concentration, air humidity, and the internal water relations of the plant.

It is obvious from Eq. (6) that climatic zones representing various vapor pressure deficits may demonstrate different transpiration ratios T/P. It has been shown by Bierhuizen and Slatyer (1965) that for a number of crops grown in different parts of the world a linear relation exists between T/P and Δe (Fig. 1). The data were originally collected by Briggs and Shantz (1913) and a crude estimation was made of saturation deficits for these localities from meteorological tables. Bierhuizen and Slatyer (1965) demonstrated also from simultaneous measurements of net photosynthesis and transpiration of cotton leaves this linear relation between T/P and Δe. It should be realized that the

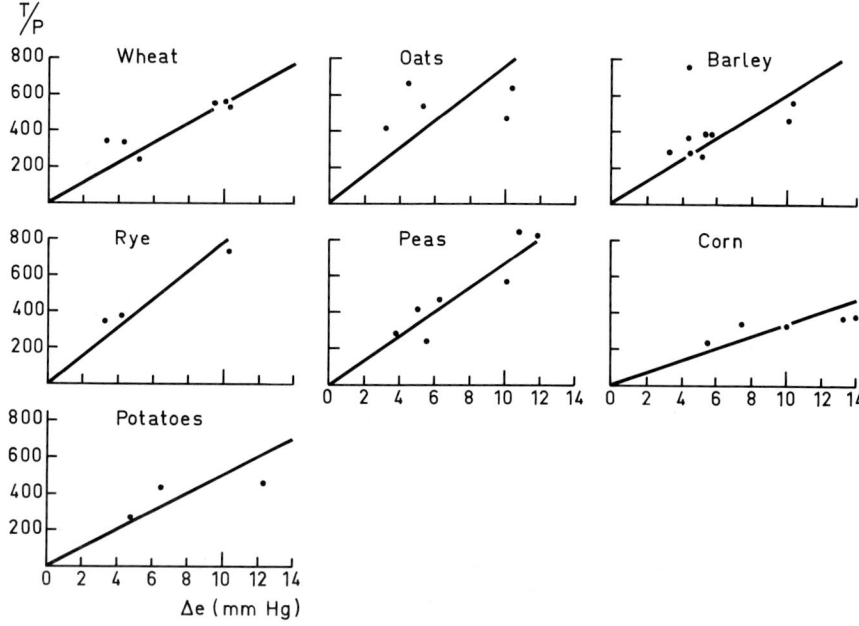

Fig. 1. The effect of saturation deficit of different locations in the world (Δe) on transpiration ratio T/P for various crops. (Redrawn from Bierhuizen and Slatyer, 1965)

leaf temperature is crucial. With a high saturation deficit of the air (desert condition) leaf temperature is often lower than air temperature, the actual Δe thus being lower than the saturation deficit. In temperate zones with a high humidity, on the other hand, leaf temperature and actual Δe are often higher than that of the ambient environment. The actual difference between leaf and air temperature depends on the difference in energy between net radiation and transpiration (latent heat) and on the windspeed for which an energy balance has to be considered. This aspect has been discussed elsewhere (see Gates, this volume Part 3:A). It should be mentioned that sometimes other relationships between yield and transpiration have been developed, mainly on the basis of solar radiation or latitude (De Wit, 1958; Stanhill, 1960), which generally proved to be inadequate, e.g. when applied to humid and arid locations at similar latitudes.

2. Effect of Carbon Dioxide Concentration

The external CO_2 concentration of the ambient air is usually in the order of 0.03% or 300 ppm. It shows a diurnal variation, the amplitude depending on the height above the canopy, with a maximum occurring during the night and a minimum during the day.

Depending on the species, an increment in $(CO_2)_{ext}$ to a value between 1000 and 2000 ppm increases net photosynthesis. In greenhouses an artificial supply of CO_2 may increase yield up to 30% without an appreciable increase in water use. This fact can be explained on the basis of Eqs. (4) and (5). Artificial supply of CO_2 in greenhouses is rather important. For example, with a leaf area index of 3 and a light intensity above 0.2 cal cm^{-2} min^{-1}, net photosynthesis might be in the order of between 700 and 3000 cm^3 CO_2 m^{-2} h^{-1}. The storage of CO_2 per m^3 of volume in air is approximately 300 cm^3, which means that in a greenhouse without any CO_2 source, a height of 2.5 m is exhausted easily within an hour. The CO_2 production of a gravel substrate, a soil without manure, and a strawbale substrate, is in the order of 0, 50, and 1000 cm^3 m^{-2} h^{-1} respectively (Bierhuizen, 1973). Before artificial CO_2 supply was introduced in practice, it had been observed that cucumber plants grown in hydroponic gravel cultures yielded less than those on strawbales, a fact which now can be ascribed to differences in CO_2 supply from the substrate.

The supply of CO_2 from a human being is fairly substantial and can be in the order of 60,000 cm^3 CO_2 h^{-1}. The ventilation of a greenhouse might vary between 1000 and 30,000 cm^3 m^{-2} h^{-1} CO_2 depending on number, spacing, and opening of the windows, whereas wind speed and temperature gradients are important as well. Although CO_2 fertilization increases the water use efficiency (reducing the transpiration coefficient) to a great extent, application in the open is not yet feasible.

The internal CO_2 concentration depends on the balance between photosynthesis and respiration and thus on $(CO_2)_{ext}$, light intensity and temperature. It has been demonstrated that large differences in $(CO_2)_{int}$ exist between species.

During the night photosynthesis is at zero point, and at a certain respiration rate an outflux of CO_2 occurs from the leaf because $(CO_2)_{int}$ exceeds $(CO_2)_{ext}$. With an increase in light intensity $(CO_2)_{int}$ declines until a minimum is attained. With an increase in temperature this minimum will rise, its increment depending on the species (Fig. 2A). It has been observed that in C_4 plants (see Huber and Sankhla, this volume Part 5:C; Ludlow, this volume Part 5:D) CO_2 fixation is rather effective, which leads in general to an extreme low value of $(CO_2)_{int}$

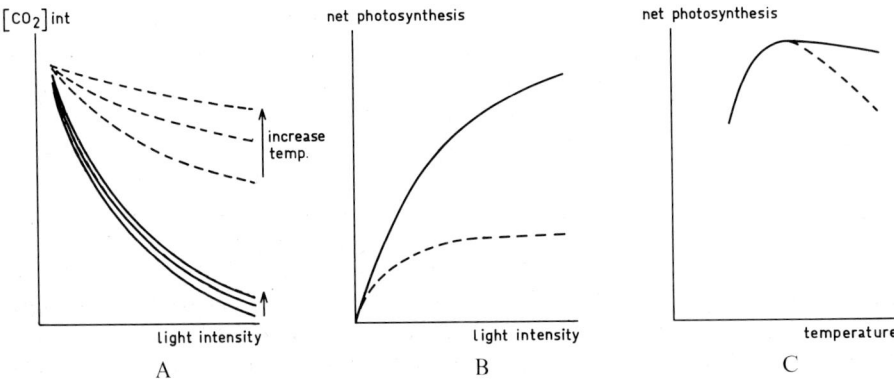

Fig. 2. The behavior of efficient (——) and nonefficient species (---) on $(CO_2)_{int}$ as affected by light intensity and temperature, on net photosynthesis as affected by light intensity, on net photosynthesis as affected by temperature

(Black, 1973). In contrast to this, in C_3 plants an increase in temperatures enhances $(CO_2)_{int}$ to a great extent. As a consequence, the effect of temperature and light intensity on net photosynthesis is different, as is shown in Fig. 2 B and C. It is obvious that, due to the low $(CO_2)_{int}$ and thus a high ΔCO_2 [see Eq. (6)], corn, a C_4 species, has a low transpiration coefficient (high water use efficiency), as is shown in Fig. 1.

It has been demonstrated by Nunez et al. (1968) that in a shadow plant such as coffee, an increase in day/night temperature from 20°/12°, to 25°/20°, and then up to 32°/25°C increases the transpiration, a result which can be ascribed to an increase in Δe [see Eq. (4)]. At full light intensity its level is higher than at 1/20 of full light intensity because more energy is available which changes Δe and possibly r_s. Differences in leaf growth between full and 1/20 of full light intensity were small, whereas optimal leaf growth was observed at a day/night temperature of 25°/20°C, a result which could be expected from Fig. 2b and c. It is obvious that at a high light intensity and a temperature above approximately 28°C the transpiration coefficient increases (lower water use efficiency).

An interesting observation by Nunez et al. (1968) is the fact that net photosynthesis after a dark period is reasonably high, but declines rapidly in light. It may be assumed that the formation of assimilates via photosynthesis is higher than the transport of assimilates from the chlorophyllic centers towards the growing points. This may lead to accumulation of assimilates after a prolonged illumination, which in turn increases photorespiration and thus $(CO_2)_{int}$.

3. Importance of Mesophyll Resistance in Relation to Laminary, Boundary, and Stomatal Resistance

It is obvious from Eqs. (5) and (6) that species with a low mesophyll resistance (r'_m) will have a relatively high net photosynthesis and thus a low transpiration ratio. Slatyer (1973) demonstrated that r'_m of maize and millet is relatively low and in the order of 1 s cm^{-1} as compared with wheat and cotton (approx. 3 s cm^{-1}). Calculated values of r'_m in *Amaranthus cruentus* (C_4) and *Celosia argenta* (C_3) were 1 and 5 s cm^{-1} respectively at light saturation (Grubben, 1975). A fairly high level of 23 s cm^{-1} was observed for citrus by Stanhill (1972). It was shown by Slatyer (1973) that r'_m, although depending on the species, remains relatively unchanged towards drought induction.

In an irrigation interval, the gradual decline in soil moisture in general induces a stomatal closure, thus increasing r_s. With a high value of r'_m, stomatal closure reduces transpiration more than net photosynthesis, thus increasing the water use efficiency. Stanhill (1972) observed a decline in dry matter production of 10% and in water use of 30% with citrus, when the soil moisture potential changed from -0.1 bar till -3.0 bar. By contrast, in species with a low value of r'_m it can be expected that the percentage decline in water use and net photosynthesis is about the same. Above, general assumptions are made, because stomatal closure leads to a rise in leaf temperature, thus increasing Δe, and also $(CO_2)_{int}$.

Data concerning sorghum and alfalfa, obtained in Tunesia (van Hoorn et al., 1969) were used to calculate gross and net water use efficiency. Daily evapotranspiration during summer was in the order of 7 mm per day. The irrigation cycle varied between 12 and 16 days, the amount of water applied being either in excess or lower than evapotranspiration. The exact values are given in Table 1. It is obvious from the table that the net water consumption for alfalfa (column 3) is lower than the total amount of water supplied on a basis of 8 and 10 mm per day. Especially in the case of sorghum, the loss due to percolation is extremely high at 10 mm/day. With a low water supply, the net water consumption is higher than the amount applied, thus indicating an extraction of water from the soil. As a consequence, the gross water use efficiency declines considerably with an abundant water supply. The net water use efficiency is independent of the irrigation treatment in sorghum, but in alfalfa a lower water supply increases water use efficiency. This result could be expected on the basis of the magnitude of r'_m, which is low in sorghum and high in alfalfa. This result suggests that in the case of water shortage, only crops with a high r'_m value should be limited in their water supply. Such a decision, however, depends on other agronomic aspects.

Table 1. The effect of water supply (in mm per day) with irrigation intervals varying between 14 and 16 days on yield and water use efficiency of alfalfa (12th June–4th September) and sorghum (21st April–19th September)

Water supply in mm per day	Total amount of irrigation in mm	Net water consumption in mm	Yield in tons green matter/ha	Production in kg green matter m^{-3} water	
				gross	net
		alfalfa			
4	332	387	21.5	6.5	5.8
6	499	542	24.0	4.8	4.6
8	664	599	26.1	3.9	4.4
10	831	608	23.9	2.9	3.9
		sorghum			
4.3	649	552	58.0	8.9	10.5
5.8	875	630	65.2	7.5	10.4
8.6	1 297	681	70.7	5.4	10.4
10.7	1 619	668	69.5	4.3	10.4

IV. Some Agronomic Aspects

In the previous sections the water use efficiency has been discussed from an engineering and an ecophysiological point of view. In addition some agronomic aspects are important, such as the effect of moisture depletion on yield and water use and the occurrence of sensitive periods to drought.

1. Effect of Moisture Depletion on Yield and Water Use

In the past there have been many arguments as to whether water retained in the soil between field capacity and wilting point is equally available or, in other words, whether irrigation should be applied at a soil moisture potential of −16 bar or at a much higher value. The threshold value at which the yield of marketable products declines, depends on weather, soil, and plant conditions. It is realized at present that in certain circumstances a pF of 4.2 could be allowed as a threshold value and in other conditions a pF of 3.0 (−1 bar) or even higher is critical.

Crop yield is not increased by irrigation at low soil moisture stress during weather conditions with relatively low temperatures and low light intensities causing low evaporation rates. In practice, planting and major growth well ahead of a hot dry weather period will result in a higher pF threshold value (Hagan, 1955; Bierhuizen and de Vos, 1959).

Soil conditions are important as well, especially those providing root development and water transport towards the roots (see Caldwell, this volume Part 2:A; Greacen et al., this volume Part 2:B). Favorable conditions are those where a soil has a good structure to a great depth, thus providing good aeration and infiltration, nonsaline conditions, and a relatively low fertility level with an equal distribution of the nutrients to a great depth. It has been shown by van der Schaaf (1955) that, on soil profiles comprising a clay layer of varying thickness (20 to 120 cm) on sandy subsoil, the depth of rooting almost completely depended on the thickness of the top layer. With a root depth of less than 60 cm, a reduction in yield could occur because of insufficient moisture availability in the root zone. In such a case root depth could be increased by means of deep ploughing. A soil moisture retention curve may give useful information regarding available water and aeration. In loess soils with a small volume of air at field capacity optimal yield cannot be achieved with frequent irrigation because of the lack of aeration (Bierhuizen and de Vos, 1959).

Plants with fast-growing roots, distributed densely to a great depth, require less frequent irrigation (Hagan, 1955). From an agronomic point of view, however, a low threshold value in pF is necessary in cases where fresh weight and quality are important characteristics of the marketable product. When the primary characteristics are the quantity of dry matter or the content of certain substances (sugar, oil, etc.) irrigation can be less frequent.

From the general aspects discussed above, it appears that threshold values of pF at which water should be applied depend on many conditions. Threshold values for various crops are given by Salter and Goode (1967).

2. Effect of Sensitive Periods

Although some crops have a great tolerance towards drought, there may nevertheless, during their life cycle, be periods during which they are more sensitive. In particular the change from the vegetative to the generative phase may cause root growth to lag behind, which increases the sensitivity to drought (see Caldwell, this volume Part 2:A). It has been shown by van der Post (1968) that in cucumbers and tomatoes root growth stopped during flowering, leading to a large increase in the top/root ratio. De Stigter (1969) showed that growth of the root tip of cucumber was inhibited by even a single flower on the plant. After removal of the fruit the roots rapidly resumed growth.

Since senescence of roots continues, cessation of root growth from the vegetative to the generative phase decreases water uptake and thus increases the sensitivity to drought. It has sometimes also been observed that under equal climatic conditions, evapotranspiration is higher during the generative phase, thus increasing the water requirement (Anonymus, 1970). It is obvious that under these circumstances irrigation should be more frequent.

V. Conclusions

1. The overall water use efficiency of an irrigation area is rather low and an improvement of farm and distribution management in order to save water is important.

2. The water use efficiency depends on the gradient of vapor pressure and carbon dioxide, and on the magnitude of the mesophyll resistance. In general a C_4 species has a higher water use efficiency (dry matter per unit water use) than a C_3 species. Closing of stomates due to suboptimal water supply increases the water use efficiency in a C_3 plant but less in a C_4 plant.

3. The threshold value in soil moisture at which water should be applied depends on climate, soil, and plant conditions. During the growth cycle, periods of sensitivity to drought occur during the shift from the vegetative to the generative phase.

Acknowledgements. The author is greatly indebted to Ir. Bos, International Institute for Land Reclamation and Improvement, Wageningen and to Dr. J. W. van Hoorn, Department of Land and Water Use. Agricultural University, Wageningen, for valuable advice and criticism.

References

Anonymus: Research and training on irrigation with saline water. Techn. Rpt./UNESCO/UNDP/(SF) (1970).
Bierhuizen, J. F.: Carbon dioxide supply and net photosynthesis. Acta Hort. **32**, 119–127 (1973).
Bierhuizen, J. F., Slatyer, R. O.: Effect of atmospheric concentration of water vapour and CO_2 in determining transpiration photosynthesis relationships of cotton leaves. Agr. Meteorol. **2**, 259–270 (1965).
Bierhuizen, J. F., de Vos, N. M.: The effect of soil moisture on the growth and yield of vegetables. In: Supplemental Irrigation Comm. VI Copenhagen Conference Report (ed. E. W. Schierbeck), pp. 83–92. Wageningen: Institute for Land and Water Management Research 1959.
Black, C. C.: Photosynthetic carbon fixation in relation to net CO_2-uptake. Ann. Rev. Plant Physiol. **24**, 253–286 (1973).
Bos, M. G., Nugteren, J.: On irrigation efficiencies. Intern. Inst. Land Reclamation and Improvement. I.L.R.I. **19**, 1–90 (1974).
Briggs, L. J., Shantz, H. L.: The water requirements of plants. 1. Investigations in the Great Plains. U.S. Dept. Agr. Bur. Plant. Ind. Bull. **284**, 1–96 (1913).
Grubben, G. J. H.: La culture de l'amaranthe, légume-feuilles tropicales avec référence spéciale au Sud-Dahomey. Med. Landbouwhogesch. Wageningen **75–6**, 1–223 (1975).
Hagan, R. M.: Factors affecting soil moisture—plant growth relations. In: 14th Int. Hort. Congress (ed. J. P. Nieuwshalen), Vol. 1, pp. 82–103. Wageningen: H. Veenmann en Linen 1955.

Hoorn, J. W. van, Bierhuizen, J. F., Combeau, A., Combremont, R., Ollat, C. H.: Besoin en eau des cultures irriguées à l'eau saumatre. Int. Comm. Irr. Drainage, 7th Congr. **23,** 775–795 (1969).

Kovda, V. A., Berg, C. van den, Hagan, R. M.: Irrigation, drainage and salinity. Paris: Hutchinson, F. A. O., UNESCO 1973.

Kozlowski, T. T.: Water deficits and plant growth. New York-London: Academic Press 1972.

Miller, E. C.: Plant physiology 2. New York: McGraw-Hill 1938.

Monteith, J. L.: Principles of environmental physics. London: Edward Arnold 1973.

Nunez, M. A., Bierhuizen, J. F., Ploegman, C.: Studies on productivity of coffee. 1. Effect of light, temperature and CO_2 concentration on photosynthesis of *Coffea arabica*. Acta Botany. Neerl. **17,** 93–102 (1968).

Post, C. J. van der: Simultaneous observations on root and top growth. Acta Hort. **7,** 138–144 (1968).

Salter, P. J., Goode, J. E. (eds.): Crop responses to water at different stages of growth. Comm. Agric. Wageningen: Bur. Farnham Royal Buchs 1967.

Schaaf, D. van der: Beworteling van de grasmat in het randgebied van de Noord Oost Polder. Versl. Landb. Onderzoek **617,** 57–70 (1955).

Slatyer, R. O.: Plant-water relationships. New York-London: Academic Press 1967.

Slatyer, R. O.: The effect of internal water status on plant growth, development and yield. In: Plant Response to Climatic Factors (ed. R. O. Slatyer) Proc. Uppsala Symposium, pp. 177–188. Paris: UNESCO 1973.

Slavik, B.: Methods of studying plant water relations. Ecological Studies, Vol. 9. Berlin-Heidelberg-New York: Springer 1974.

Stanhill, G.: The relationship between climate and the transpiration and growth of pastures. Proc. 8th Intern. Grassland Congr., Reading, pp. 293–296 (1960).

Stanhill, G.: Recent developments in water relation studies. Some examples from Israel citriculture. Proc. 18th Int. Hort. Congr. IV (eds. N. Goren, K. Mendel), pp. 367–383. Rehobot, Israel: Ministry of Agriculture 1972.

Stewart, J. I., Hagan, R. M.: Functions to predict effects of crop water deficits. J. Irrig. Drain. Div. I. R. **4,** 421–439 (1973).

Stigter, H. C. M. de: Growth relations between individual fruits and between fruits and roots in cucumber. Neth. J. Agric. Sci. **17,** 234–240 (1969).

Wit, C. T. de: Transpiration and crop yields. Verlag Landbouwk. Onderzoek **646,** 1–88 (1958).

D. Estimating Water Status and Biomass of Plant Communities by Remote Sensing

B. G. DRAKE

I. Introduction

Before remotely sensed spectral data can be applied to studies of plant water stress, it is necessary to understand the relationship between reflectance, the ratio of the radiant energy reflected from an object to that incident upon the object, and the relevant plant parameters. This paper will briefly summarize selected data for the effect of water stress on reflectance of single leaves and on the relationship between reflectance of communities and biomass. Methodology will not be discussed but reviews may be found in Holmes, 1970; Johnson, 1969; Fuchs and Tanner, 1966; Myers and Allen, 1968; Myers, 1970.

II. Water Stress, Reflectance, and Temperature of Single Leaves

1. Reflectance and Scattering

Reflectance of incident radiant energy from a green leaf can be described by comparing its behavior in three broad bands of the solar spectrum. In the visible region (350–700 nm) reflectance is less than 10%, it is about 50% in the far red (700–1000 nm), and it declines as the wavelength increases in the infrared beyond 1000 nm (Willstatter and Stoll, 1913; Gates et al., 1965). There is a small peak of reflectance in the visible at 550 nm and two minima at 1430 and 1950 nm in the infrared (Fig. 1).

Simmler (1862 cited after Mestre, 1935) was one of the first to recognize the dependence of the spectral properties of leaves on their structure. Differences between the refractive index for air (1.0) in numerous air spaces within the palisade and spongy parenchyema and the refractive indexes for hydrated cell walls (1.4) (Woolley, 1975) and cellular organelles result in scattering of virtually all radiant energy penetrating the epidermis. Chlorophyll and accessory pigments absorb strongly in the visible region of the spectrum, resulting in low reflectance. Increasing water absorption accounts for decreasing reflectance with increasing wavelength in the infrared (Knipling, 1970; Gausman et al., 1969, 1970). The

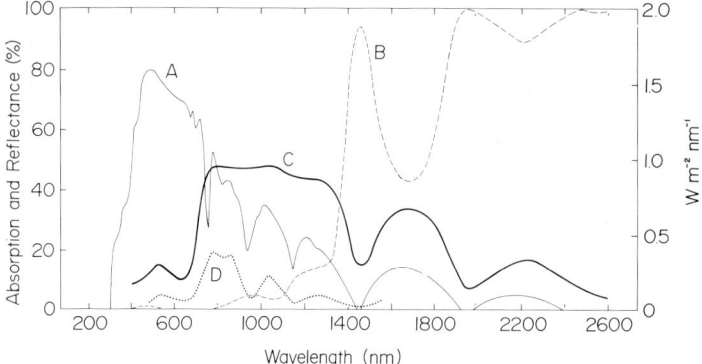

Fig. 1. *A* Spectral radiant flux density (W m^{-2} nm^{-1}) of sunlight normal to earth's surface at sea level. Molecular absorption by water vapor occurs in the region of 710, 815, 930, 1120, 1430, and 1950 nm. (Redrawn from Thekaekara, 1965.) *B* Percent absorption by 1 mm of liquid water. *C* Reflectance of healthy, turgid, green leaf (*Liriodendron tulipifera* L.). (*B* and *C* redrawn from Hoffer and Johannsen, 1969). *D* Spectral radiant flux (W m^{-2} nm^{-1}) reflected from a cotton canopy. (Redrawn from Myers and Allen, 1968)

absence of substantial amounts of any absorber and the high degree of scattering explain the high reflectance in the far red.

2. Reflectance and Short-term Water Stress

The reflectance of a healthy leaf increases as it loses water. Thomas et al. (1966) using cotton, and Sinclair (1968) using corn and soybeans report that during the initial stages of drying (100 to 80% relative water content, RW[1] there was very little change in reflectance. Below 80% RW, reflectance increased at all wavelengths, and especially in the infrared water absorption bands at 1430 and 1950 nm. Pearman (1966) obtained similar results with *Acacia cyanophylla*. Hoffer and Johannsen (1969) reported increases in reflectance of corn leaves at all wavelengths (500 to 2300 nm) when RW was below 66%. Since wilting occurs at RW >80%, it seems probable that there would be no detectable change in remotely sensed reflectance of canopies experiencing early stages of water stress. However, pre-wilting water stress will influence leaf temperature, which may be detected radiometrically (Tanner, 1963), and orientation of leaves, which influences image density on infrared film (Thomas et al., 1966).

3. Reflectance and Long-term Water Stress

The effects of long-term water stress on the reflectance from single leaves may be more readily detected by current remote sensing methods than are

---

[1] $RW = [(w_f - w_d) \times 100]/(w_t - w_d)$ where w_f = fresh weight, w_d = dry weight, w_t = turgid weight.

the consequences of brief water deficits. Leaves from plants subjected to prolonged or even intermittent water stress take on xeric characteristics; the palisade is more compact, cell walls are thicker, and more inner surface is exposed to air per unit outer surface than in mesic leaves of the same species (Turrell, 1936). Dadykin and Bedenko (1960) reported increased reflectance in English oak (*Quercus robur* L.) and basswood (*Tilia cordata* Mill.) subjected to moisture stress during growth; the more severe the stress, the greater the reflectance in the visible and far red spectral regions. Gausman et al. (1970) found a large increase in reflectance of far red in cotton leaves as they developed during the first ten days after emergence. Changes in reflectance were related to changes in compactness of the tissue; reflectance of far red was low in young, emergent leaves which were also very compact, had small cells and few intercellular spaces. When the same leaves were eight days older cells were larger and more loosely arranged and reflectance was higher. Billings and Morris (1951) found that the drier the habitat, the greater the reflectance of visible radiation from single leaves. Higher reflectance is also associated with thickened palisade layers. Gates et al. (1965) report higher visible and far red reflectance in sun leaves than shade leaves. Sun leaves typically have a thicker palisade layer than do shade leaves (Esau, 1965).

Visible reflectance of nitrogen-deficient leaves (which had very little chlorophyll) was several-fold greater than visible reflectances in leaves not deficient in nitrogen, but there were only slight differences for far red reflectance (Myers, 1970). Chlorophyll content is sensitive to water stress (Bourque and Naylor, 1970; Virgin, 1965).

4. Leaf Temperature

The temperature of a leaf is determined by the exchange of heat with its environment through radiant, convective, and evaporative processes (Raschke, 1958). Stomatal control of transpiration, especially at elevated temperatures, can sharply influence leaf temperature. The temperature of *Xanthium strumarium* leaves, in which closed stomata caused reduced transpiration, was 6–10° C above the temperature of similar leaves that had open stomata and were transpiring normally (Drake and Salisbury, 1972). Tanner (1963) reported that the temperature difference between well-watered and water-stressed potatoes was 1.5 to 3.0° C and that the temperature of potatoes rose 11.5° C above air temperature as a result of effects of water stress on stomatal control of transpiration. He observed no visual signs of these increased canopy temperatures. The temperature of individual leaves or of leaf canopies can be measured with radiation thermometers (Fuchs and Tanner, 1966).

III. Reflectance and Biomass of Communities

Although reflectance spectra obtained from canopies are more complex because of interaction with other leaves and soil, they are qualitatively similar to those obtained with single leaves (Myers and Allen, 1968). However, reflectance from canopies is approximately 50% of reflectance of single leaves in the visible and 75% of the reflectance of single leaves in the far red, the infrared water vapor troughs are broader, and there are additional minima in the far red at 830 and 980 nm (Myers and Allen, 1968; Holmes, 1970; Knipling, 1970). Reflectance patterns for well-watered and water-stressed plants vary due to differences in the structure of individual leaves as well as the canopy morphology.

In addition to effects on leaf anatomy, chlorophyll content, and orientation, water stress also effects total water content and biomass (Walter, 1973). There are few studies of canopy reflectance and water stress. However, since water stress effects biomass, it is relevant to discuss studies related to reflectance and the biomass of green standing crop.

Kanemasu (1974) found an increase in far red reflectance coincident with an increase in leaf area index in wheat and soybeans. No attempt was made to relate his results to biomass, but leaf area index increased coincidentally with increased far red reflectance. Pearson (1973) studied spectral reflectance (350–950 nm) in prairie grass *(Bouteloua gracilis)*. Measurements were made before greening in the spring and at the peak of summer growth. There was a reduction in the reflectance of red (650–700 nm) and an increase in reflectance of far red as the biomass increased. Pearson (1973) reported a linear relationship between green standing biomass, removed step-wise and at random from individual sample plots, and the ratio of far red to red measured for each step in removing plant material on these plots. The slope of this relationship increased as chlorophyll content declined throughout the season. When data on biomass and the ratio of reflectance of far red to red from many different plots sampled throughout the growing season were pooled, the relationship between the reflectance ratio and green biomass was non-linear above 300 g m^{-2} dr. wt.

Figure 2 shows the seasonal trend in biomass and community reflectance of red and far red in two salt marsh communities; one dominated by the shrub *Iva frutescens* and the other by the sedge *Scirpus olneyi*. Red reflectance decreases abruptly with the onset of growth when greening occurs in the spring. At the same time, far red reflectance sharply increases in the shrub but not the sedge community, which indicates differences between community types. The similarity between the annual course of reflectance and standing crop of green biomass holds only for the early and late part of the growing season, and the decline in red reflectance stops well before peak biomass is reached in either community. Attempts to utilize reflectance as a direct measure of biomass in these communities have met with limited success.

IV. Conclusions

Remote sensing as a tool in the study of water stress in plant communities is yet in its infancy; the effects of water deficits on reflectance in single leaves have been investigated but, except for studies of canopy temperature, there have been too few reports of these effects on reflectance from large areas of plants to make even tentative conclusions possible. The effects of water stress on community biomass are well known but the relationship between biomass

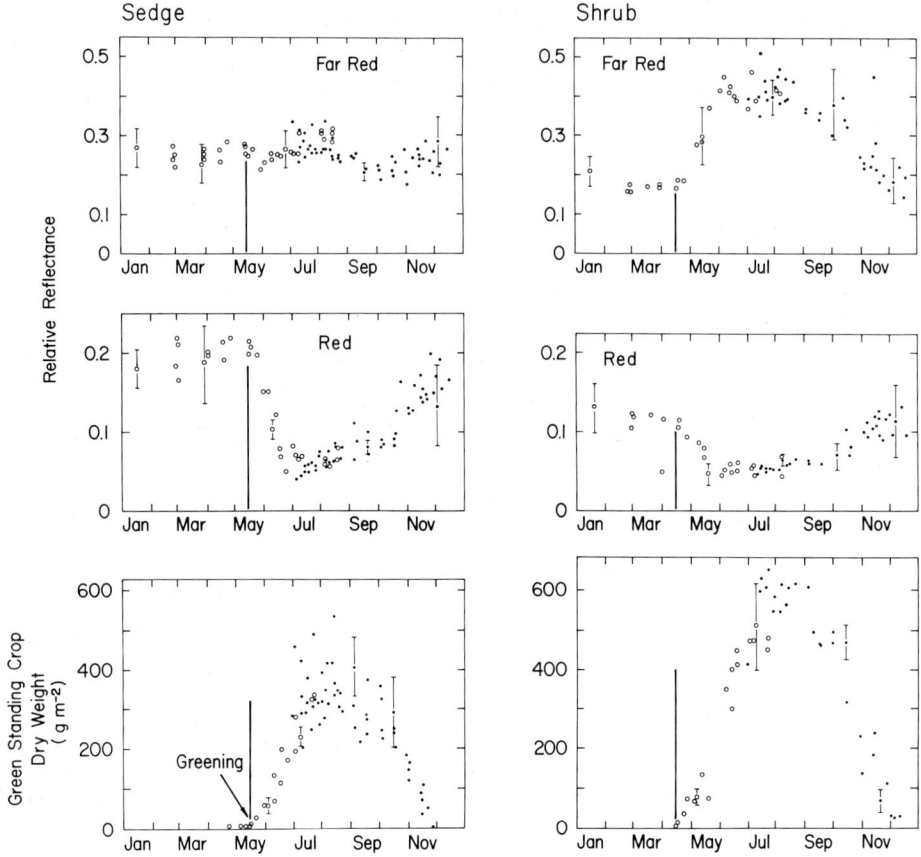

Fig. 2. Seasonal trend of green standing crop biomass and relative reflectance of red (655–705 nm) and far red (775–825 nm) in two salt marsh communities. Sedge consists of approximately 50% *Scirpus olneyi* and mixture of *Distichlis spicata* and *Spartina patens*. Shrub is approximately 80% *Iva frutescens* and about 20% mixed grasses and sedges. Green standing crop for the two communities is dry weight of total green matter. *Solid points* (●) data collected in 1973, *open circles* (○) data collected 1974. Relative reflectance is ratio of radiant energy reflected from canopy to radiant energy reflected from a white plate suspended over canopy. The data are means of 10 values for reflectance data and means of 5 values for biomass. Representative standard deviations are plotted for each set of values

and reflectance of visible, far red, and infrared has not yet been adequately exploited. The few results that are available reveal that the hope of obtaining a measure of standing crop by measuring reflectance from a canopy has been realized only in the case of a grass community of low biomass.

Acknowledgements. Research was supported in part by a grant from the Program for Research Applied to National Needs of the National Science Foundation to the Chesapeake Research Consortium and by a grant from the Smithsonian Fluid Research Fund.

Published with the approval of the Secretary of the Smithsonian Institution.
The assistance of Vete Clements, Roger Ford, and Tom Roman in data collection and Douglass Hayes in the preparation of figures is gratefully acknowledged.

References

Billings, W. D., Morris, R. J.: Reflection of visible and infrared radiation from leaves of different ecological groups. Am. J. Botany **38**, 327–331 (1951).

Bourque, D. P., Naylor, A. W.: Large effects of small water deficits on chlorophyll accumulation and ribonucleic acid synthesis in etiolated jack bean (*Canavalia ensiformis* L. D. C.). Plant Physiol. **47**, 591–594 (1971).

Dadykin, V. P., Bedenko, V. P.: The connection of the optical properties of plant leaves with soil moisture. Dokl. Acad. Nauk SSSR **134**, 965–968 (1960).

Drake, B. G., Salisbury, F. B.: Aftereffects of low and high temperature pretreatment on leaf resistance, transpiration, and leaf temperature in *Xanthium*. Plant Physiol. **50**, 572–575 (1972).

Esau, K.: Plant Anatomy. p. 465. New York: John Wiley and Sons 1965.

Fuchs, M., Tanner, C. B.: Infrared thermometry of vegetation. Agron. J. **58**, 597–601 (1966).

Gates, D. M., Keegan, H. J., Schleter, J. C., Weidner, V. R.: Spectral properties of plants. App. Opt. **4**, 11–20 (1965).

Gausman, H. W., Allen, W. A., Cardenas, R.: Reflectance of cotton leaves and their structure. Rem. Sens. Env. **1**, 19–22 (1969).

Gausman, H. W., Allen, W. A., Cardenas, R., Richardson, A. J.: Relation of light reflectance to histological and physical evaluations of cotton leaf maturity. App. Opt. **9**, 545–552 (1970).

Hoffer, R. M., Johannsen, C. J.: Ecological potentials in spectral signature analysis. In: Remote sensing in ecology (ed. P. Johnson), p. 1–16. Athens: Univ. Georgia Press 1969.

Holmes, R. A.: Field spectroscopy. In: Remote sensing of environment with special reference to agriculture and forestry, pp. 298–323. Washington, D. C.: Nat. Acad. Sci. and Nat. Res. Council 1970.

Johnson, P.: Remote sensing in ecology. Athens: Univ. Georgia Press 1969.

Kanemasu, E. T.: Seasonal canopy reflectance patterns of wheat, sorghum, and soybean. Rem. Sens. Env. **3**, 43–47 (1974).

Knipling, E. B.: Physical and physiological basis for the reflectance of visible and near-infrared radiation from vegetation. Rem. Sens. Env. **1**, 155–159 (1970).

Mestre, H.: The absorption of radiation by leaves and algae. Cold Spring Harbor Symp. Quant. Biol., vol. **III**, 191–209 (1935).

Myers, V. I.: Soil, water, and plant relations. In: Remote sensing of environment with special reference to agriculture and forestry, pp. 253–297. Washington, D. C.: Nat. Acad. Sci. Nat. Res. Council 1970.

Myers, V. I., Allen, W. A.: Electrooptical sensing methods as non destructive testing and measuring techniques in agriculture. App. Opt. **7**, 1 819–1 838 (1968).

Pearman, G. I.: The reflection of visible radiation from leaves of some western Australian species. Australian J. Biol. Sci. **19**, 97–103 (1966).

Pearson, R. L.: Remote multispectral sensing of biomass. Ph. D. Thesis, Colorado State University 1973.

Raschke, K.: Über den Einfluß der Diffusionswiderstände auf die Transpiration und die Temperatur eines Blattes. Flora **146**, 546–578 (1958).

Sinclair, T. R.: Pathway of solar radiation through leaves. M. S. Thesis, Purdue 1968.

Tanner, C. B.: Plant temperatures. Agron. J. **55**, 210–211 (1963).

Thekaekara, M. P.: The solar constant and spectral distribution of solar radiant flux. Solar Energy **9**, 7–20 (1965).

Thomas, J. R., Myers, V. I., Heilman, M. D., Wiegand, C. L.: Factors affecting light reflectance of cotton. Proc. 4th Symp. Rem. Sens. Env., Univ. Mich. IST Rep. No. 4864-11-x, pp. 305–312 (1966).

Turrell, F. M.: The area of internal exposed surface of dicotyledon leaves. Am. J. Botany **23**, 255–264 (1936).
Virgin, H. I.: Chlorophyll formation and water deficit. Physiol. Plantarum **18**, 994–1000 (1965).
Walter, H.: Vegetation of the earth in relation to climate and the ecophysiological condition. London: English Univ. Press Ltd. and Berlin-Heidelberg-New York: Springer 1973.
Willstätter, R., Stoll, A.: Untersuchungen über die Assimilation der Kohlensäure. Berlin: Springer 1918.
Woolley, J. T.: Refractive index of soybean leaf cell walls. Plant Physiol. **55**, 172–174 (1975).

E. Plant Production in Arid and Semi-Arid Areas

M. EVENARI, E.-D. SCHULZE, O. L. LANGE, L. KAPPEN,
and U. BUSCHBOM

I. Introduction

The determination of phytomass and the assessment of actual and potential primary production of a natural ecosystem has become an important part of modern ecological research. In order to understand the functioning of an ecosystem, to analyse it, and to build valid models (see van Keulen et al., this volume Part 6:B), exact phytomass and production data must be available. This general aim, however, is difficult to approach, since phytomass and primary production are not constant values (see Lieth, this volume Part 6:A). Even for the same habitat they depend upon the total environmental conditions which are changing from year to year. Therefore, the fluctuations of phytomass and production and their relation to the environmental parameters need to be considered. This problem may be of minor importance for ecosystems living in humid environments. It is of great significance, however, in arid and semi-arid areas, where the main factor limiting phytomass and production is the availability of water. It is precisely this factor which in these regions fluctuates to a very large degree from year to year.

The special aim of this chapter is to demonstrate these fluctuations of phytomass and production of ecosystems in arid and semi-arid areas, and to show to what degree various plants and life forms of such ecosystems contribute to these changes. The results of such considerations could lead towards optimal management practices of arid and semi-arid ecosystems, which are more and more affected by man's activities (see van Keulen et al., this volume Part 6:B; Bierhuizen, this volume Part 6:C). In the following chapter, which should serve as an example, data are presented mainly from experiments conducted in the Negev desert of Israel (Avdat Desert Research Station located in the Central Negev Highlands; for detailed description and climatic conditions see Evenari et al., 1971). The boreal cold deserts will not be considered. The lichen and algae production will also be omitted, since such data are known only for very few desert ecosystems (see Harris, this volume Part 6:F; Kappen et al., 1975).

II. Survey of Phytomass, Net Annual and Relative Annual Production of Some Main Vegetation Units of the Globe

Deserts are characterized by the lowest phytomass and net annual phytomass production of all vegetation units shown in Table 1. However, their relative primary production, i.e. the net annual primary production divided by their maximum standing phytomass, is very high. The opposite is true for all vegetation units in which the dominant life form is represented by trees. Such vegetation

Table 1. Phytomass, net annual primary and relative primary production (net annual primary production/phytomass) of some of the main vegetation units of the world. All figures represent dry matter. The number in brackets indicates the source from which the figures are taken: (1) Whittaker (1970), Table 4-2 and the authors cited there; (2) Lieth (1972, 1973), Tables 2 and 3 and the authors cited there; (3) Bazilevich et al. (1971)

Vegetation type	Phytomass of mature stands (t ha^{-1})	Net annual primary production (t ha^{-1} yr^{-1})	Relative primary production
Tropical forest	600–700 (3)	27 –35 (3)	0.04–0.05
	60–800 (1)	10 –50 (1)	0.06–0.17
Broadleaf forest (subtropical, subboreal)	370–450 (3)	12 –20 (3)	0.03–0.06
Boreal forest	200–500 (2)	2 –15 (2)	0.01–0.03
	60–400 (1)	4 –20 (1)	0.05–0.07
	100–350 (3)	4 –10 (3)	0.03–0.04
Tundra woodland	125 (3)	5 (3)	0.04
Savanna	20–150 (2)	2 –20 (2)	0.1 –0.14
Tropical grass and shrub savanna	20– 40 (3)	7 –12 (3)	0.3 –0.35
Temperate grassland	20– 50 (2)	1.5–15 (2)	0.08–0.3
Herbaceous prairie (subboreal)	35 (3)	15 (3)	0.43
Semi-arid steppe and shrub steppe (subboreal, and subtropical)	20– 35 (3)	6 –13 (3)	0.3 –0.37
Tundra	7– 28 (3)	0.7– 2.5 (3)	0.09–0.1
	1– 30 (2)	1 – 4 (2)	0.13–1.0
Steppified desert (subtropical)	12 (3)	10 (3)	0.83
Desert scrub	1– 40 (1)	0.1– 2.5 (3)	0.06–0.1
Desert (subtropical, subboreal, tropical)	1– 4.5 (3)	0.5– 1.5 (3)	0.33–0.5

units have the highest phytomass of all ecosystems and the lowest relative primary production irrespective of the climatic conditions under which they grow. The figures of relative primary production for the tundra woodlands and for the tropical forests are of the same order of magnitude. The reason for this is that, in comparison with all other life forms, a very large percentage of the phytomass of trees consists of photosynthetically unproductive tissue. Desert ecosystems, however, cannot sustain a large phytomass. Their relative primary production is high, since the dominant life forms are either arido-passive[1]

---

[1] 'arido-active' plants are metabolically active during the dry, rainless season; 'arido-passive' plants are either dormant during the dry season or dying at its beginning and surviving it in the form of propagules (Evenari et al., 1975b).

therophytes and perennial ephemeroids (geophytes and hemicryptophytes) or arido-active[1] shrubs (nanophanerophytes), dwarf shrubs (chamaephytes), and transition forms between hemicryptophytes and chamaephytes. In comparison with trees, the phytomass of these life forms contains a smaller percentage of photosynthetically inactive tissue.

III. Phytomass and Production of Some Arid and Semi-Arid Vegetation Units and Their Annual Fluctuations

Table 2 shows data of phytomass and production of some specific arid and semi-arid vegetation units of the Near East and of Algeria at average conditions of rainfall. The figures for total phytomass of the *Zygophyllum dumosum* community of the Negev being less than 2.5 t ha^{-1} fit well into the lowest category (represented by the subtropical deserts) of the phytomass map of Bazilevich and Rodin (1971). With less than 1 t ha^{-1} the *Artemisia herba-alba* community of the Negev has even a phytomass which is only little above that of extreme deserts (Bazilevich et al., 1971), whereas the phytomass of the *Hammada scoparia* community and the Algerian *Artemisia* community fall into the second lowest category of the Russian authors (2.5–5.0 t ha^{-1}, highland deserts). The Syrian *Artemisia sieberi* community belongs to their third class (5.1–12.5 t ha^{-1}, subtropical semi deserts).

The relative primary production of all plant communities of Table 2 is much higher than that of any formation with trees as dominant elements (Table 1). The low values of the relative primary production of the *Zygophyllum*, the mixed *Artemisia-Zygophyllum*, and the *Hammada* communities contrast with the high values of all *Artemisia* communities because their dominant plants are dwarf shrubs which contain a much larger percentage of permanent woody tissue than *Artemisia herba-alba*. This plant is a transition form between chamaephyte and hemicryptophyte. At the beginning of the growing season, new green shoots emerge from temporary renewal buds of a permanent woody base. During the dry season these shoots are shed and thus add very little to the living permanent woody phytomass (Orshan, 1953, 1963). The *Artemisia* communities therefore have a value of relative primary production which is in the same order of magnitude as in plant communities where herbs or grasses are predominant.

The figures given in Table 2 are based on measurements of only one year, and it may be supposed that the data given in Table 1 for deserts are also based mostly on data collected during one or two years. Yet, in arid areas rain and runoff vary very much from one year to the next. In Avdat, for example, the quotient of rainfall variation, i.e. maximum to minimum annual rainfall, amounts to 8 (Shanan et al., 1967). Consequently, the annual fluctuations of phytomass and production play an important part in the overall productivity of ecosystems living under such conditions.

Table 2. Phytomass, net annual primary and relative primary production of some arid and semi-arid vegetation units. All figures represent dry matter. The numbers in brackets in column "vegetation unit" indicate the source from which the figures are taken: (1) Evenari et al. (1975a); (2) Orshan and Diskin (1968); (3) Rodin et al. (1972)

Locality	Vegetation unit	Rainfall (mm)	Total peak phytomass (t ha^{-1})	Total net annual primary production (t ha^{-1} yr^{-1})	Relative primary production
Negev Highland	*Zygophyllum dumosum* community (1)	93.6 (1970/71)	1.45 (1971)	0.25	0.17
	community (2)	92.2 (1963/64)	1.56 (1964)	0.35	0.22
Negev Highland	*Artemisia herba-alba* community (1)	93.6 (1970/71)	0.89 (1971)	0.44	0.49
Northern Negev	community (2)	164.5 (1963/64)	0.96 (1964)	0.36	0.38
Algeria	community (3)	300.0	4.19	1.77	0.42
Negev Highland	Mixed *Artemisia herba-alba–Zygophyllum dumosum* community (1)	93.6 (1970/71)	1.61 (1971)	0.47	0.29
Syria	*Artemisia sieberi* community (3)	120–190	6.11	2.38	0.39
Negev Highland	*Hammada scoparia* community (1)	93.6 (1970/71)	3.32 (1971)	1.43	0.44

Table 3 demonstrates this fact for a *Hammada* community at Avdat. Its peak phytomass fluctuates from one year to the next between 8.24 t ha^{-1} and 1.86 t ha^{-1}. The main component in the vegetation responsible for this remarkable change are the annuals. In a year with a rainfall much below average (which is about 80 mm) very few annuals germinate and their phytomass is practically nil.

Table 3. Peak phytomasses (dry matter) of *Hammada scoparia* community at Avdat in years with different precipitation

Year	Rain mm	Plant	Peak above ground phytomass (t ha^{-1})	Peak below ground phytomass (t ha^{-1})	Total peak phytomass (t ha^{-1})
1970/71	94	Hammada scoparia	0.35	0.37	0.72
		Peganum harmala	0.15	0.02	0.17
		Carex pachystylis	0.44	1.90	2.34
		Annuals	0.08	0.01	0.09
		Total	**1.02**	**2.30**	**3.32**
1971/72	163	Hammada scoparia	0.73	0.41	1.14
		Peganum harmala	0.24	0.08	0.32
		Carex pachystylis	0.53	2.30	2.83
		Annuals	2.70	1.25	3.95
		Total	**4.20**	**4.04**	**8.24**
1972/73	54	Hammada scoparia	0.29	0.34	0.63
		Peganum harmala	0.10	0.02	0.12
		Carex pachystylis	0.00	1.11	1.11
		Annuals	0.00	0.00	0.00
		Total	**0.39**	**1.47**	**1.86**
1973/74	134	Hammada scoparia	0.40	0.38	0.78
		Peganum harmala	0.14	0.02	0.16
		Carex pachystylis	0.56	2.32	2.88
		Annuals	1.15	0.46	1.61
		Total	**2.25**	**3.18**	**5.43**

In years with a rainfall much above average, the phytomass of annuals constitutes almost half of the total phytomass. The perennial, ephemeroid, arido-passive geophyte *Carex pachystylis* behaved in a dry year (1972/73) like the annuals as far as its above ground parts are concerned. The rhizomes, which grow together with the roots in the upper 2–3 cm of the soil, were dormant. The arido-active plants of *Hammada s.* and *Peganum harmala* restrict their production as well, but the change from one year to another is less drastic. This example shows the importance of annuals and geophytes for the phytomass and production of certain desert plant communities. In years with high precipitation throughout the rainy season, the cover of the annuals may amount to 80–100% (cf. also p. 446). Table 3 demonstrates also that the fluctuations of the below-ground phytomass of the whole community are considerably smaller during the four observation years than those of the above-ground parts. The ratio of the above-ground phytomass of a dry year (1972/73) and that of a humid year (1971/72) is 1 to 10.8, while the respective figure for the below-ground phytomass is 1 to 2.7. Thus, the phytomass fluctuations of the whole system are mainly due to the fluctuations in the above-ground parts, whereas the below-ground phytomass stays more or less constant.

The large phytomass fluctuations of annuals might be typical for many desert communities, but there are also other desert vegetation types for which this rule does not apply. In contrast to the *Hammada* community, the annuals play no, or only a minor, role in the *Artemisia* and *Zygophyllum* communities growing in the Avdat region. This is true even in years with high precipitation. In the Avdat area these communities are restricted to hilltops and slopes. The seed reserves of annuals in these soils are very meagre. In the *Zygophyllum* community 120 seeds of annuals were found on the surface per m^2 and 730 were found in the upper 2 cm of soil per m^2. The corresponding figures for the *Artemisia* community were 780 and 1,580. In contrast to this, the respective figures rise in the *Hammada* community to 22,660 and 27,180. The reasons for the poor seed reserves of the *Artemisia* and *Zygophyllum* communities are rather complex. Since the soil is shallow and stony and since much of the rainwater is lost by runoff, the water conditions are unfavorable for germination of annuals even in a rainy year. Only few seeds succeed in germinating, which is true for annuals as well as for perennials (Evenari et al., 1971). In addition, in deserts there are great differences in germinability of seeds of the same population (Koller, 1969). Only a small percentage of the seed reserve germinates at the same time, leaving always a large percentage of viable seeds dormant in the soil. Since seedling mortality also is high (Evenari et al., 1971), only few plants reach maturity for new seed formation. Seed transport between the different habitats is apparently quite restricted, since many of the propagules of desert annuals are antitelechoric (Zohary, 1937). In contrast to the situation at Avdat, annuals and their phytomass fluctuations play an important role for the productivity in the *Zygophyllum* and *Artemisia* communities of the Northern Negev, in areas with higher rainfall, deeper soils, less runoff, and higher infiltration rates (cf. Table 4).

In Table 3 the total phytomass of the *Hammada* community fluctuates by a factor of about 4.5 (1.86 to 8.24 t ha^{-1}). The corresponding fluctuations in rainfall amount only to a factor of about 3 (55 to 163 mm of rain). In a hydrological analysis Shanan (1975) calculated the rainfall-runoff relationships. It was shown that in a dry year (1972/73) the soils of the *Hammada* community received only 11.34 mm of water, originating mainly from runoff floods. This corresponds to 21% of the total seasonal precipitation of 54 mm. In contrast, in a wet year (1971/72) a total of 48.9 mm of water, originating from runoff floods, was available, which is 31% of the total precipitation. The proportion of these infiltration rates from runoff floods, which is 4.3, is obviously much more important for the water balance of the plants in the *Hammada* community than the scattered direct rainfall (see van Keulen et al., this volume Part 6:B). Therefore, this factor explains much better the biomass fluctuations of these two years than the rainfall proportion alone. For desert ecosystems the figure of direct rainfall, as expressed by the annual amount of rain, is of only questionable importance as a measure of the water source for plants. A relatively large percentage of the precipitation falls with low hourly intensities of short duration and long time gaps between these rains. These small amounts of water moisten only the soil surface. This moisture evaporates rapidly. For desert ecosystems growing on flood plains, rains of high intensities, which produce runoff from the surrounding slopes, are decisive. The runoff collects alone of 1 to 3 yearly floods are the main source of water and wet the soil sufficiently.

The phytomass fluctuations shown in Table 3 are certainly not the biggest occurring in the desert. During the driest year known in recent times in the Negev (1962/63), the rainy season had a total of only 29.5 mm of precipitation. During this, and the following year, the production of green phytomass of an *Artemisia* and a *Zygophyllum* community were measured by Orshan and Diskin (1968) in areas having a higher average annual rainfall than Avdat (Table 4). In the *Artemisia* community the production in the wet year (1964) was seven times higher than in the dry year (1963); the corresponding figure for the *Zygophyllum* community is six. Similar to the case of the *Hammada* community, the large fluctuations of the total phytomass are mainly due to the fact that in one year no annuals grew at all, whereas in the following year they represented 54.1% of the phytomass of the *Artemisia*, and even 70.6% of that of the *Zygophyllum* community.

Where the habitat conditions become more extreme, i.e. wherever soil is too shallow, too saline, or availability of water is restricted to either very short times or only to the upper few centimeters, the only life form capable of existing are certain arido-passive annuals, which have a very low phytomass. Because of complete biomass turn-over they have a relative primary production of 1. In the northern Dead Sea valley, on highly saline soils, annuals were the only plants found growing in 1974 (118 mm of rain 1973/74) with an overall phytomass and annual production of 0.4 t ha^{-1} (Danin, personal communication).

IV. Permanent Phytomass

The knowledge of phytomass, annual production, and their fluctuations from year to year is not sufficient for understanding the primary production of arid ecosystems, since only a certain amount of the annual production is added to the permanent phytomass of the perennial components of the ecosystem. This will be demonstrated for the case of *Zygophyllum d.*, which according to the number of plants and its phytomass, is the dominant species of the *Zygophyllum* community at Avdat (Table 5). As described above, this community is characterized by a lack of annuals. The growth of the above-ground phytomass of *Zygophyllum d.* from March to October (year 1971 with 94 mm rain) is shown in Table 6. The green phytomass produced in 1971 reaches its peak with 0.049 t ha^{-1} in May. In late April the lignification of the formed biomass begins. By the end of June the production of lignified phytomass of this year reaches its maximum. From May onwards the green phytomass becomes smaller, due to lignification and to the continuous reduction of the metabolically active surface by leaf shedding (Evenari et al., 1971). This process continues until the end of the dry season. In September the green phytomass consists only of petioles, and at the beginning of the next growing season no green phytomass of the preceeding growing season will be left.

The question is: which amount of the annual plant production persists permanently? It is difficult to calculate how much of the new phytomass of a

Table 4. The peak green phytomass (dry matter) of an *Artemisia herba-alba* and a *Zygophyllum dumosum* community in different years. (After Orshan and Diskin, 1968)

Year	Rain mm	Plant	Green phytomass (t ha^{-1})
		Artemisia herba-alba community	
1961/62	38.0	*Artemisia herba-alba*	0.1433
		Noaea mucronata	0.1479
		Gymnocarpos fruticosum	0.0773
		Helianthemum ventosum	0.0149
		Asphodelus microcarpus	0.0354
		Annuals	0
		Total	**0.4188**
1962/63	164.5	*Artemisia herba-alba*	0.1695
		Noaea mucronata	0.3810
		Gymnocarpos fruticosum	0.3700
		Helianthemum ventosum	0.0122
		Asphodelus microcarpus	0.1178
		Annuals	1.2370
		Total	**2.2875**
		Zygophyllum dumosum community	
1961/62	51.9	*Zygophyllum dumosum*	0.3498
		Gymnocarpos fruticosum	0.0157
		Helianthemum kahiricum	0.0110
		Annuals	0
		Total	**0.3765**
1962/63	92.9	*Zygophyllum dumosum*	0.5842
		Gymnocarpos fruticosum	0.0330
		Helianthemum kahiricum	0.0050
		Annuals	1.4910
		Total	**2.1132**

growing season is invested in the formation of a new annual ring in the main stem and all the branches, thus adding to the permanent phytomass. There are estimations which show that this amount is only very small. According to the measurements of Fahn et al. (1963), the average width of the annual rings of the main stems of 4 *Zygophyllum d.* plants, varying in age from 100 to 128 years, was 349 μ. The narrowest ring of 135 μ apparently indicated the 'worst', the widest ring of 748 μ the 'best' year. Measurements on 11 *Zygophyllum d.* plants, ranging in age from 4 to 145 years, have shown that the addition of one annual ring to all the permanent lignified parts of all the *Zygophyllum d.* plants in a community amounts to only about 1.8 to 8.4 10^{-4} t ha^{-1} of dry weight (Evenari et al., 1971). This figure indicates only the order of magnitude, yet shows that the amount of phytomass invested each year in the annual rings of older plant parts is very small. The biomass of newly lignified branches, however, is high. In the example described in Table 5, it amounts to 8.9% of the initial permanent above-ground phytomass. However, usually only a small part of this increase will remain permanently.

The process which reduces the so-called 'permanent' above-ground phytomass in arid habitats has been described as 'survival by partial death' (Evenari et al., 1971). The stems of mature plants of some of the arido-active dwarf shrubs (*Artemisia h.-a.* or *Zygophyllum d.*) can split into a number of branches which are physiologically independent of each other. In drought years when not enough

Table 5. Number of plants and dry weight of above-ground peak phytomass in 1971 of the main components of a *Zygophyllum dumosum* community at Avdat; ap = annual growth of the shoots grown in 1971; pp = perennial parts of the shoots

Species	Number of plants per ha	Peak above-ground phytomass (t ha^{-1})		
		ap	pp	Total
Zygophyllum dumosum	104.5	0.09	0.540	0.63
Hammada scoparia	19.0	0.021	0.039	0.060
Artemisia herba-alba	58.8	0.007	0.007	0.014
Reaumuria negevensis	91.3			0.050
Gymnocarpos fruticosum	6.0			0.009

Table 6. Dry weight at various dates of 1971 of the above-ground phytomass of *Zygophyllum dumosum* produced during that year; (a) green, photosynthesizing parts; (b) lignified parts

Dates	(a) (t ha^{-1})	(b) (t ha^{-1})
12.III	0.0154	0
25.III	0.0255	0.0020
7.IV	0.0356	0.0067
18.V	0.0490	0.0139
22.VI	0.0295	0.0480
20.VII	0.0141	0.0480
13.IX	0.0128	0.0480

water is available for all the branches from one individual plant, a number of branches may die and only a few survive. In wet years the survivors may restore a part or all of the phytomass lost by this process. Observations at Avdat permit to estimate, at least for one year, the amount of phytomass lost through this 'survival by partial death'. In a dry year (1972/73 with 54 mm of rain), 3–5% of the 'permanent' lignified above-ground phytomass of the *Zygophyllum d.* plants of a *Zygophyllum* community died and did not revive in the following wet year (1973/74, with 134 mm of rain). If the peak permanent above-ground phytomass of *Zygophyllum d.*, shown in Table 5, was taken as a basis for an estimation, then, under a similar rainfall situation as described for the dry year 1972/73, the plants would have lost 0.016–0.027 t ha^{-1}. This amounts to about half of the phytomass added in a 'normal' year. The phytomass losses of the community would be even higher in an extremely dry year (e.g. 1962/63 with 29.5 mm rain). It was estimated at Avdat that at least 15–20% of the permanent above-ground phytomass of *Zygophyllum d.* died in such a year. It would take a number of consecutive years with high precipitation to repair this damage, because only then germination and development of new, young plants could take place. The indicated losses of above-ground phytomass are even higher for *Artemisia h.-a.* For a dry year, e.g. 1972/73, the estimated losses are 26–28% and these increase under extreme situation (1963) to about 40–50%.

V. Potential Production

In addition to measurements of phytomass and production of desert plant communities under natural conditions, it is of interest to determine these quantities when water as the main limiting factor is eliminated. For this purpose, plants of *Zygophyllum d.* and *Hammada s.* were artificially irrigated. The irrigation during 16 months increased the above-ground phytomass of *Zygophyllum d.* by a factor of 2.3 and that of *Hammada s.* by the factor of 11.8. These figures show the order of magnitude by which the elimination of water stress can increase production of these desert plants.

For *Zygophyllum d.*, other experiments show that this increase is not a continuous process. During the first five years of irrigation the phytomass of *Zygophyllum d.* increased by a factor of about 6.7. During the next four years the peak phytomass fluctuated around this value with deviations of not more than about ±7 to 8%. The plants apparently had reached a new equilibrium with their environment.

VI. Recovery

It is evident that in an arid habitat with extreme differences in yearly precipitation, losses in permanent phytomass are possible which can cause almost complete destruction of a plant community. Therefore, an important question arises concerning the time of reconstruction necessary to build up a desert community again. This process is certainly mainly dependent on the dominant life form in that vegetation. Experiments at Avdat show for a pure stand of *Artemisia h.-a.* that, after removal of all plants, this site went through a series of successions during the first four years (Evenari et al., 1971). Only after a rainy season with precipitation high above average were many seeds of *Artemisia h.-a.* able to germinate. It then took another 10 years for reestablishment of the community with its components. At this time, the phytomass was not different from that of control plots. The reason for this surprisingly fast recovery is the high relative primary production of this special community (Table 2), and the fact that the dominant species *Artemisia h.-a.* has, in comparison with *Zygophyllum d.* or *Hammada s.*, a low permanent lignified phytomass. Therefore, a similar regeneration experiment with a *Hammada* or *Zygophyllum* community certainly would take a much longer period. This is especially true for *Zygophyllum d.* with its low relative annual production and large permanent lignified phytomass. Long time experiments at Avdat confirm this supposition.

VII. Conclusions

In regions with less than 120–150 mm rainfall, the relationship between annual precipitation and primary production is frequently overruled by local runoff. On hill sites most of the rainfall can be lost and water infiltration is rather low, whereas on flood plains water is gained. This became obvious from the production in a *Hammada* community which was more related to the water gain by runoff floodings than to the actual amount of precipitation in this habitat. In future work on desert ecosystems much more attention has to be paid to measurements of runoff and infiltration rates.

Hot deserts permit the coexistence of various plant communities that differ considerably in their phytomass, primary production, and relative primary production. In spite of all these differences, there is one feature typical for all arid zone plant communities. These are the large relative fluctuations of phytomass and production above and below very low mean values. The relative changes of phytomass and production of desert ecosystems are certainly the largest or one of the largest in any ecosystem of the globe.

The large fluctuations of phytomass and production can be considered from two different aspects. On the one hand, they mean a permanent danger for the existence of desert ecosystems. This becomes true in times when there are too many fluctuations on the minus side, causing accumulative debits of phytomass, especially if this occurs during 3–4 consecutive dry years. On the other hand, the large fluctuations cause special adaptations of desert plants and desert ecosystems to their harsh environment. This means that all the special morphological, anatomical, and physiological adaptations of desert plants to their general environment, and to the special environmental conditions of each year, cumulate in one final result: A dry matter production geared quantitatively to the amount of water available to the single plant and the whole ecosystem (see also Noy-Meir, 1973). This can certainly be true also for non-desert ecosystems. But desert ecosystems, at least those studied in the Negev, are in this respect special in two ways. They are characterized by a low species diversity and a pronounced diversity of adaptive control mechanisms (Evenari et al., 1975a). These systems have a low degree of stability and a high degree of resilience and persistence. Holling (1973) has stated the case precisely: 'Resilience determines the persistence of relationships within a system and is a measure of the ability of these systems to absorb changes of state variables, driving variables, and parameters, and still persist'. It is exactly this type of instability and resilience which gives the desert ecosystems persistence, i.e. the power to survive under most adverse environmental conditions. For the resilience of these ecosystems as a whole, the diversity of survival strategies and not the diversity of species, is an important factor. In the cases dealt with in this chapter species diversity decreases still further with increasing environmental stresses.

It seems that future research aimed at understanding the functioning of desert ecosystems should be directed, inter alia, towards a combination of long-range measurements of the fluctuations of phytomass and primary production of these systems and, simultaneous measurements of the main eco-physiological parameters governing and regulating production, such as e.g. photosynthesis and water balance, and their dependence of the environmental conditions, which in deserts are so variable.

One gap in our knowledge on desert ecosystems, which future research should fill, concerns the below-ground phytomass. There are few direct measurements of annual net production of the below-ground phytomass and the amount of below-ground litter.

Another aspect of future research should be directed towards the problems of recovery of arid land ecosystems. For greater understanding of the vegetation types in deserts it is important to know the time span and the successional stages until a vegetation reaches some steady state (climax). Such research could be connected with measurements of phytomass and production in order to determine the initial high degree of 'productivity' (term used in the sense of Stocker, 1969; Duvigneaud, 1974), which in the mature ecosystems approaches zero.

References

Bazilevich, N. J., Rodin, L. E.: Geographical regularities in productivity and the circulation of chemical elements in the earth's main vegetation types. Am. Geogr. Soc. (Review and translation) **12**, 24–52 (1971).

Bazilevich, N. J., Rodin, L. E., Rozov, N. N.: Geographical aspects of biological productivity. Soviet Geography (Review and translation) **12**, 293–317 (1971).

Duvigneaud, P.: La synthèse écologique. Paris: Ed. Doin 1974.

Evenari, M., Bamberg, S., Schulze, E.-D., Kappen, L., Lange, O. L., Buschbom, U.: The biomass production of some higher plants in Near-Eastern and American deserts. In: Photosynthesis and productivity in different environments (ed. J. P. Cooper), IBP, vol. 3, pp. 121–127, Cambridge-London-New York-Melbourne: Cambridge Univ. Press 1975a.

Evenari, M., Schulze, E.-D., Kappen, L., Buschbom, U., Lange, O. L.: Adaptive mechanisms in desert plants. In: Physiological adaptations to the environment (ed. F. J. Vernberg), pp. 111–129. New York: Intext Education Publ. 1975b.

Evenari, M., Shanan, L., Tadmor, N.: The Negev. The challenge of a desert. Cambridge, Mass.: Harvard Univ. Press 1971.

Fahn, A., Wachs, N., Ginzburg, C.: Dendrochronological studies in the Negev, Israel. Explor. J. **13**, 291–299 (1963).
Holling, C. S.: Resilience and stability of ecological systems. Ann. Rev. of Ecol. Systematics **4**, 1–23 (1973).
Kappen, L., Lange, O. L., Schulze, E.-D., Evenari, M., Buschbom, U.: Primary production of lower plants (lichens) in the desert and its physiological basis. In: Photosynthesis and productivity in different environments (ed. J. P. Cooper), IBP, vol. 3, pp. 133–143. Cambridge-London-New York-Melbourne: Cambridge Univ. Press 1975.
Koller, D.: The physiology of dormancy and survival of plants in desert environments. In: Dormancy and survival (ed. H. W. Woolhouse), Symp. Soc. Exper. Biol. vol. 23, pp. 449–469. Cambridge: Univ. Press 1969.
Lieth, H.: Über die Primärproduktion der Pflanzendecke der Erde. Angew. Botan. **46**, 1–37 (1972).
Lieth, H.: Primary production: Terrestrial ecosystems. Human Ecology **1**, 303–332 (1973).
Noy-Meir, I.: Desert ecosystems: Environment and producers. Ann. Rev. Ecol. Systematics **4**, 25–51 (1973).
Orshan, G.: Note on the application of Raunkiaer's system of life forms in arid regions. Palestine J. Bot. **6**, 1–3 (1953).
Orshan, G.: Seasonal dimorphism of desert and Mediterranean chamaephytes and its significance as a factor in their water economy. In: The water relations of plants (eds. A. J. Rutter, F. H. Whitehead), pp. 206–222. Oxford: Blackwell Sci. Publ. 1963.
Orshan, G., Diskin, S.: Seasonal changes in productivity under desert conditions. In: Functioning of terrestrial ecosystems at the primary production level (ed. F. E. Eckardt), pp. 191–201. Paris: UNESCO 1968.
Rodin, L. E., Bazilevich, N. J., Miroshnichenko, Y. M.: Productivity and biogeochemistry of Artemisieta in the Mediterranean area. In: Ecophysiological foundation of ecosystem productivity in arid zone (ed. L. E. Rodin), pp. 193–198. Leningrad: Nauka 1972.
Shanan, L.: Rainfall and runoff relationship in small watersheds in the Avdat region of the Negev desert. Ph. D. Thesis, Jerusalem 1975.
Shanan, L., Evenari, M., Tadmor, N. H.: Rainfall pattern in the central Negev desert. Israel Explor. J. **17**, 163–184 (1967).
Stocker, O.: Die "Stoffproduktion" in Urwäldern und anderen Pflanzengesellschaften im Gleichgewicht. Mitt. Florist.-Soziol. Arb. Gem., NF. **14**, 422–434 (1969).
Whittaker, R. H.: Communities and ecosystems. New York: MacMillan 1970.
Zohary, M.: Die verbreitungsökologischen Verhältnisse der Pflanzen Palästinas I. Die antitelechorischen Erscheinungen. Beih. Botan. Centralbl. **66**, 1–155 (1937).

F. Water Content and Productivity of Lichens

G. P. Harris

I. Introduction

In many respects lichens are unique organisms. The dual nature of the lichen thallus with its algal and fungal components ensures that many of the features of lichen morphology, physiology and reproduction have no parallels in any other group of plants. The general biology and physiology of lichens has been adequately reviewed by Smith (1962), Kershaw (1963), Ahmadjian (1967), Hale (1968), Ahmadjian and Hale (1973), and Farrar (1973). The thallus morphology is invariably simple with the algal cells either scattered throughout the thallus or organized into a distinct layer. There are no morphological adaptations such as cuticular layers and stomata for controlling water loss, and water uptake is a passive process.

The internal state of the lichen thallus fluctuates markedly in passive response to external conditions, and lichens have evolved in such a way that physiological processes are adapted to and tolerant of such fluctuations. Much recent research has been concentrated in the areas of photosynthesis, productivity and water relations. This has largely resulted from the rather recent realization that lichens are not intrinsically intractable and that, although photosynthesis rates are lower than in higher plants, they make good experimental material and can be kept alive in the laboratory for considerable periods (Kershaw and Millbank, 1969; Harris and Kershaw, 1971). Developments in microclimatological methods have also enabled research workers to attempt to measure the fluctuation of thallus water contents in the field (Kershaw and Rouse, 1971a; Harris, 1972) and to construct simple models of such changes from climatological data (Rouse and Stewart, 1972; Hoffman and Gates, 1970; Harris, 1972).

II. Productivity of Lichens

There have been a number of recent measurements of lichen photosynthesis (Table 1). It can be seen that the range of values lies between about 0.1 and 5.2 mg CO_2 g dr. wt.$^{-1}$ h^{-1}. Schulze and Lange (1968) compared the productivity rates of higher plants with those of lichens and showed that lichens have markedly lower net photosynthesis rates. The low net photosynthesis rates per gram are explained by the fact that the algal component forms only a small fraction of the total thallus biomass (Bednar, 1963). This also explains the high respiration rates (Table 1) as the thallus is predominantly fungal hyphae. We do not know, however, what proportion of the fungal hyphae are actively respiring. Work by Wilhelmsen (1959) is of interest as it demonstrated that the chlorophyll content of lichen thalli is only one quarter to one tenth that of leaves. Also Ertl (1951) demonstrated that the upper fungal cortex of the lichen thallus has a lower transparency (57–74%) than the leaf epidermis of higher plants (87–96%). This will ensure that the effective light intensity which impinges on the algal chloroplasts is reduced.

Hill and Woolhouse (1966) recorded sun/shade adaptations in lichens taken from different habitats; both chlorophyll contents and morphological characteristics were observed to change. Adaptive changes in the overall photosynthetic efficiency of lichens have also been recorded (Harris, 1971 b). What we do not know is the rate at which individual thalli can adapt to changed environmental conditions and whether or not such adaptive changes as occur in the lichenized algae can only take place after cell division.

Smith (1961) observed seasonal changes in the respiration rate of *Peltigera polydactyla*, and Harris (1971 b) recorded seasonal fluctuations in photosynthetic

Table 1. Some recent photosynthesis rate measurements. Figures are quoted as mg CO_2/g dr. thallus wt. h (P_{max}, maximum observed photosynthesis rate; R respiration). The numbers in brackets following the respiration rates are the temperatures (°C) at which the determinations were made

Reference	Date	Species	mg CO_2/g h^{-1}		
			P_{max}	R	°C
Lechowicz and Adams	1974	*Cladonia rangiferina*	3.0	0.7	(25)
Lechowicz and Adams	1973	*Cladonia mitis*	1.13	0.47	(22.5)
Wirth and Türk	1973	*Hypogymnia physodes*	2.6	1.2	(12)
		Cetraria sepincola	2.2	0.6	(8)
Nowak	1973	*Hypogymnia physodes*	5.0	1.5	(25)
		Parmelia sulcata	5.2	2.3	(25)
		Parmelia fuliginosa	5.2	4.4	(25)
		Parmelia caperata	2.7	1.5	(25)
		Evernia prunastri	5.2	2.0	(27)
		Ramalina farinacea	4.0	1.7	(25)
		Parmelia borreri	1.5		
Eickmeyer and Adams	1973	*Cladonia ecmocyna*	0.2	0.2	(15)
Lange and Kappen	1972	*Lecanora melanopthalma*	0.12	0.07	(2)
		Neuropogon acromelanus	0.24	0.08	(5)
Rundel	1972	*Cladonia subtenuis*	1.5	1.0	(23)
Harris	1971 a, b	*Parmelia sulcata*	1.4	0.3	(20)
		Parmelia caperata	1.4	0.25	(20)
		Hypogamia physodes	1.3	0.2	(20)
Peet and Adams	1972	*Cladonia subtenuis*	0.5		
		Cladonia rangiferina	0.5		
Kallio and Heinonen	1971	*Nephroma arcticum*	0.55		
		Cetraria nivalis	0.25		
		Parmelia olivacea	0.30		
Lange et al.	1970a	*Ramalina maciformis*	1.6	0.3	(10)

rates and algal numbers. In *Parmelia caperata* the rate of photosynthesis was shown to be proportional to the number of algal cells per cm^2. This cell number was influenced by environmental conditions. So it is certain that both the overall thallus physiology and the algal content of the thallus change with time in a dynamic fashion. There is a direct response to the local microclimatological conditions (Nowak, 1973; Harris, 1971 b).

Considerable difficulty has been experienced in calculating growth rates from photosynthesis rates. Harris (1972) found that simulating "growth" in *Parmelia caperata* did not give realistic comparisons with observed rates in the field. Evidently as thalli increase in size, growth continues from the tip and the thallus may die or at least lose its photosynthetic potential in older areas (Hale, 1973; Armstrong, 1974). Kershaw and Harris (1971 b, c) found that in three species of *Cladonia* the algae were only present in about the top third of the podetia. The growth of each podetial tip or marginal lobe of the thallus is independent of other lobes and there is no evidence for translocation of photosynthates from central to marginal areas (Phillips, 1969; Hale, 1970, 1973) or of intercalary growth (Hale, 1970). Armstrong (1974) concludes from his studies, and from a review of the literature, that there is an initial exponential growth phase (up to about 3 mm in diameter) followed by a phase of linear growth by the margin of the thallus.

Overall lichen growth rates are low, partly because of their low net photosynthesis rates and partly because of their dependence of favourable environmental conditions. Hale (1973) has reviewed the literature on growth rates and has shown that most lichens grow from 1 to 5 mm per year. There is considerable variation from year to year as would be expected in organisms so dependent on the vagaries of the environment.

Given that there is variability from lobe to lobe in the thallus, samples of lichen thalli used for photosynthetic rate measurements should therefore be carefully chosen; the age and/or position from which the thallus or piece of thallus was collected should be known and stated.

III. Water Relations of Lichens

Water uptake and loss in lichens appears to be a purely physical process. Despite the assertions of Blum (1973) we have no data on the active influence of metabolism on water uptake or loss. At present there is no data which is not incompatible with physical mechanisms. The water content of the thallus is dependent on water input from rain, runoff, dew formation and air humidity; and losses by evaporation. There is no evidence that the rhizinae of lichens act as roots to take up water from the substrate. Although Kershaw and Rouse (1971 b) found an apparent correlation between soil moisture and the growth of *Cladonia alpestris*, the true mechanism is the control of lichen growth and soil moisture by the net radiation at the site. Higher net radiation levels in open sites lead to a reduction of lichen growth as the lichens rapidly dry out. Thus the depth of the lichen mat is reduced in open areas, the mulching effect is reduced, and soil moisture decreases.

1. Water Uptake

Hydration in lichens, therefore, appears to be a purely physical process. Uptake of liquid water is rapid: most thalli become fully hydrated in a matter of 1 to 2 min if water is freely available (Stocker, 1927; Smith, 1962; Blum, 1973). The process of water uptake appears to be by capillary action between the fungal hyphae and by absorption by hyphal walls. The initial water content is rapidly achieved. Blum (1973), however, records "oversaturated" water contents after a period of hours in water. This presumably is due to a further, slower hydration of all thallus air spaces and cell walls. The fully hydrated water content of most lichens varies from 1 to 4 g of water per g dr. wt. (cf. Blum, 1973, Table 1). As the absolute amount of water held by the thallus determines the degree of hydration after rain and the amount by which the lichens dry out under evaporative conditions, and because different thallus growth forms of the same species (Hill and Woolhouse, 1966) have different saturated water contents, the water contents of lichen thalli should be expressed in such absolute terms.

Dew formation and humid air have been demonstrated to be important sources of water for some species. Perhaps the best account of the importance of dew fall and air humidity is that of Lange et al. (1970a) working on *Ramalina maciformis* in the Negev desert. In this instance the lichens become sufficiently hydrated by dew before dawn to become actively photosynthetic during the first hours of the morning. As the sun rose the thalli were dried out and photosynthesis ceased (cf. Butin, 1954). These lichens can therefore grow well in the complete absence of rain. Smith (1962) records the existence of lichens which are so heavily encrusted with non-wettable lichen substances as to be completely unable to take up liquid water; such lichens are assumed to depend entirely on atmospheric humidity as a water supply.

Whether or not uptake of water from humid air is ecologically important for all species is debatable. Certainly all thalli are in equilibrium with the surrounding air, but it would appear that the air must be close to saturation for significant uptake to occur (Stocker, 1927, 1954; Bertsch, 1966). The process of uptake is slow, taking hours rather than minutes for liquid water. Certainly many species can reach 50% thallus saturation in air of 100% relative humidity (Stocker, 1954). Whether or not this water content is ecologically significant will depend on the frequency of exposure to saturated air and the physiological response to low water contents (see below). It is a common observation that rich lichen floras occur in areas where high relative humidities are prevalent.

The water which is absorbed by the thallus appears to be largely stored in extracellular regions. The heavily thickened hyphal walls and the gelatinous sheaths of some of the lichenized algae are the major storage structures. There are no data on the cytoplasmic water contents of lichens under different conditions. The cytoplasmic water content of the thallus must, however, be small compared to the total water content. Because of this essentially extracellular water uptake, the process is very similar to that observed in agar gels (Blum, 1973).

Kershaw and Rouse (1971 a) and Harris (1972) have used small cellulose paper conductivity grids to simulate lichens in the field and have monitored both wetting and drying cycles. The paper increases in conductivity when wet in the same way as do the lichens (Harris, 1969), so that conductivity measurements can be taken and water contents inferred. These grids have demonstrated rapid saturation by rain (Harris, 1972) and the importance of dew fall to sub-arctic lichens (Kershaw and Rouse, 1971a).

2. Water Loss

Early experiments showed that lichens rapidly lost water when placed in a drying environment (Stocker, 1954; Butin, 1954). Lichens have no morphological mechanisms for resisting desiccation, and these observations were confirmed when Harris (1969) measured the internal and air resistances of lichens exposed in a wind tunnel. The air resistance proved to be of the order of 0.05–0.1 sec cm^{-1} depending on thallus size and morphology. The internal resistance to dessication remained low (<1 sec cm^{-1}) until about 20% saturation (0.2 g H$_2$O/g dr. wt.) when it rose rapidly to >20 sec cm^{-1} (Fig. 1). It was suggested that the process of evaporation initially removes water from interstitial spaces and finally from hyphal walls.

Because of the apparent repeatability of these resistance measurements it is possible to use microclimatological methods to attempt to predict water contents in the field. Such methods for prediction of evaporation are well-developed (Penman, 1956; Monteith, 1965, 1973) as are energy balance approaches (Hoffman and Gates, 1970). Rouse and Stewart (1972) have clearly demonstrated the importance of lichen thalli, with their high mulching effects when dry, to the overall water balance of high latitude regions where terricolous lichens are often dominant. Other field studies have been carried out by Kershaw and Rouse (1971a), Harris (1972), and Rouse and Stewart (1972).

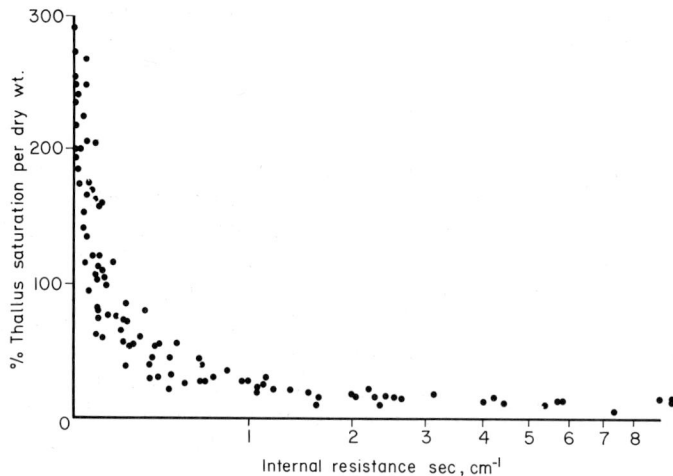

Fig. 1. % thallus saturation per g dr. wt. plotted against total internal resistance of *Parmelia* sp.

The microclimatological approach points out a clear problem for those interested in water relations and productivity of lichens. Work by Harris (1969), Lange et al. (1970a, b) and Kershaw and Larson (in press, 1975) demonstrates that lichen surface temperatures change markedly as the thallus water content changes. The ratio of H, the sensible heat flux, to LE, the evaporative heat flux, will change markedly as the water content of the lichen falls. At high water contents LE is high enough to cool the lichen thallus at least 5°C below air temperature. As the water content falls and the evaporation rate is slowed by the rise in internal resistance, H becomes greater than LE and finally LE becomes small (Rouse and Stewart, 1972), allowing the thallus temperature to rise. The papers of Lange et al. (1970a, b) show this effect well. Great care must therefore be taken when the results of physiology experiments are interpreted. Thallus temperatures fluctuate by as much as 10°C as the thallus water content varies, so experiments at constant temperature cannot be compared to experiments or field observations in which temperatures vary. Also evaporation from experimental material must be controlled or at least accurately known. A further point of considerable interest concerns thallus colour. Lichen thallus colours vary from light grey to nearly black, and colour has a large effect on the radiant energy balance. Hence temperature and the evaporation rate of the thallus are affected. Dark lichens are warmer than light ones. Kershaw and Larson (1975) have shown that dark terricolous lichens can be warm enough to photosynthesize under freezing temperatures, as by radiant energy absorption the thallus temperature is raised above 0°C and water is absorbed from melted snow and ice.

IV. Thallus Water Content and Physiological Response

This section will be treated as two inter-related parts. Firstly the initial response to wetting of the thallus and secondly the response to fluctuating water contents assuming that the thallus has been prewetted.

1. Wetting a Dry Thallus

There appears to be no physiological activity exhibited by the lichen in a dry state ($<10\%$ thallus saturation). Thus when a dry lichen is first wetted two distinct phenomena occur. Firstly an immediate release of CO_2 from a non-metabolic source, and secondly a metabolic release of CO_2 (Farrar, 1973). As yet there is no evidence to suggest what the source of the non-metabolic CO_2 might be. Smith and Molesworth (1973) surmise that it may come from carbonates and organic acids left in the tissue upon dessication, which react when wetted. More research is clearly required. The metabolic release of CO_2 is not necessarily just an increase in the basal respiration rate. In *Peltigera p.* Smith and Molesworth (1973) found that the metabolic burst upon rewetting was, unlike respiration, azide- and cyanide-sensitive. This might indicate that the reason for the CO_2 release and the burst of metabolism was the re-establishment of cell structure and function after drying. Certainly the metabolic rate increase depends on the minimum level to which the thallus is dried; drying *Xanthoria aureola* to less than 40% of maximum saturation was required to produce the effect. Smith and Molesworth (1973) reported that the "wetting burst" of metabolic CO_2 release in *Peltigera p.* continued for 8–10 h after wetting and accounted for about twice the CO_2 release of previously wetted thallus discs respiring at their basal rate.

There is still some controversy in this area, as Farrar (1973) in a postscript to his valuable review insists that the "wetting burst" of CO_2 is purely physical in origin. Harris (1971 b) in his work with *Parmelia c.* and *Hypogymnia physodes* recorded no significant wetting burst and Smith and Molesworth (1973) showed that the wetting burst in *Xanthoria a.* from xeric habitats was reduced, as compared with *Peltigera p.* Ried (1960 b) observed the phenomenon in lichens from different habitats and concluded that the magnitude of the effect was a function of the habitat, the effect being reduced in lichens from xeric habitats as compared with those from mesic sites. More work is evidently required as, if the CO_2 release is proved to be of at least partly metabolic origin, frequent wetting and drying will cause considerable loss of carbon from sensitive species. Such losses will affect their overall productivity in different habitats (Farrar, 1973). It is interesting to note that Lange (1969) observed the release of CO_2 upon rewetting *Ramalina maciformis* with water but did not observe the phenomenon if the thallus was rewetted with water vapour.

2. The Physiological Response to Changes in Thallus Water Content

There have been a number of observations of the physiological response of lichens to changes in their thallus water contents (Ried, 1960 a, b, c; Bliss and Hadley, 1964; Kershaw and Rouse, 1971 a; Harris, 1971 b; Kershaw, 1972; Lechowicz and Adams, 1974).

a) Photosynthesis and Respiration. The photosynthetic response to the full range of thallus water contents is remarkable. Photosynthesis can be detected at water contents as low as $0.1 g H_2O/g$ dr. wt. The rate of photosynthesis continues to increase as the water content increases, until a maximum rate is reached at some water content less than the saturated value. Above the maximum the net photosynthesis rate decreases to full saturation (Fig. 2). We do not know how lichens manage to maintain photosynthetic ability at such low water contents. There is, however, undoubtedly, a correlation between dessication resistance and other mechanisms such as frost resistance. Farrar (1973) has suggested that the polyols produced in lichens play an important role in conferring resistance to environmental stress.

Harris (1969, 1971 b) developed a hydrometer technique for simultaneous measurements of CO_2 flux and weight which made measurements of the water content and productivity relationship relatively simple. Kershaw (1972) used the same technique on lichens collected in South Ontario. Eickmeier and Adams (1973) used large pieces of the lichen *Cladonia ecmocyna* which were continuously weighed with a top-loading balance. Other workers (cf. Lange et al., 1970 a) weighed the lichen to determine its water content before and after a brief period in the measurement chamber of a gas analyser system. There may be problems with such techniques as the rate of water loss from the entire thallus may not be even if a large piece of thallus is used. The ecological significance of such experiments should therefore be interpreted with care.

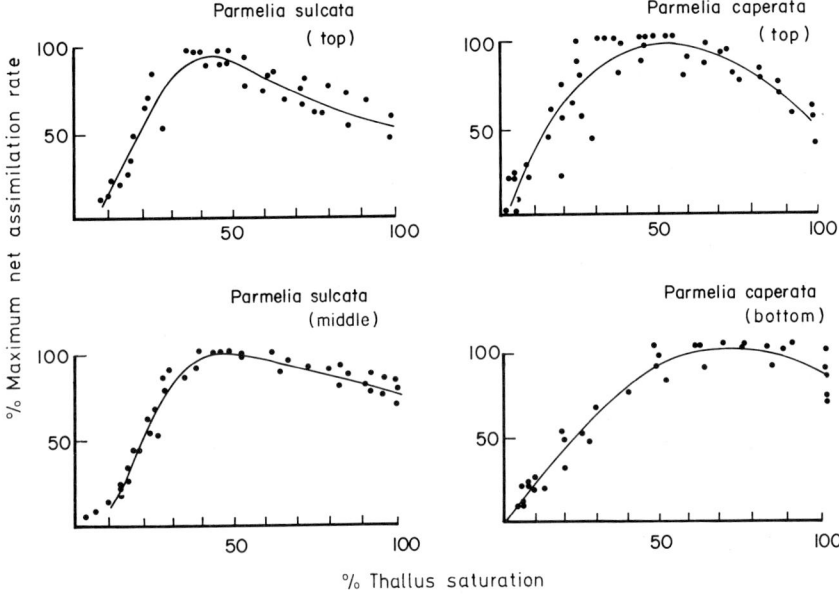

Fig. 2. % maximum net assimilation rates for *Parmelia sulcata* and *P. caperata* plotted against relative thallus saturation

Such experiments as have been done have indicated that the response of both net photosynthesis and respiration to changes in water content were non-linear (Figs. 2 and 3). They also showed that the differences in the net photosynthesis/water response, which appeared to be ecologically adaptive in nature, could be accounted for by alterations in the physiology of both the algal and fungal components. Adding back the respiration rate to the net photosynthesis rates produced "gross" photosynthesis curves for each species (Figs. 4 and 5). This procedure of course assumes that fungal respiration is the same in light and dark (see below).

The algal "gross" photosynthesis curve shows a steadily rising photosynthesis rate as water contents increase and then a decrease to full saturation. The rate of photosynthesis by the algae is high at remarkably low water contents. This represents a singular adaptation to growth in xeric environments. Unfortunately we have no idea how this feat is accomplished. It is possible that at low water contents the little water present is preferentially absorbed by the algal cells, so that their water content is above that of the thallus as a whole. Even allowing for such a speculative mechanism, the algal cells show remarkable adaptation to low water contents.

Physiological control by the alga and the fungus accounts for two observed phenomena. Firstly: the decline of net photosynthesis at water contents approaching saturation is accounted for by the increase in fungal respiration at high water contents and a decrease in the algal photosynthesis rate. Net photosynthesis is the sum of two non-linear processes. Secondly: the optimum water content (at which maximum net photosynthesis occurs) is a function of both the respective shapes of the photosynthesis and respiration curves as well as of the relative rates of the two processes. As the overall respiration rate increases in relation to the "gross" photosynthesis rate, the net rate of photosynthesis will decrease and the optimum water content will appear to decrease. Thus the existence of an optimum and the water content at which it occurs appear to be at least partly under fungal control.

As the fungus exerts some control over the net photosynthesis rate, temperature, by affecting fungal respiration, is important also. At high light and low temperature the net photosynthesis/water response approximates to the "gross" photosynthesis curve (Figs. 4 and 5). At low light and high temperature the net response is similar to the respiration response with perhaps a low rate of net photosynthesis at low water contents (Peet and Adams, 1972). Thus care must be taken to carry out net photosynthesis measurements at a range of water contents and temperatures otherwise only a part of the whole thallus response will be detected. Kershaw and Harris (1971a) inferred from simulation and experimental results that the lichen response to light, temperature and water was buffered to give a near-constant response. The conclusion is erroneous and was caused by a fault in the simulation model used. This conclusion was

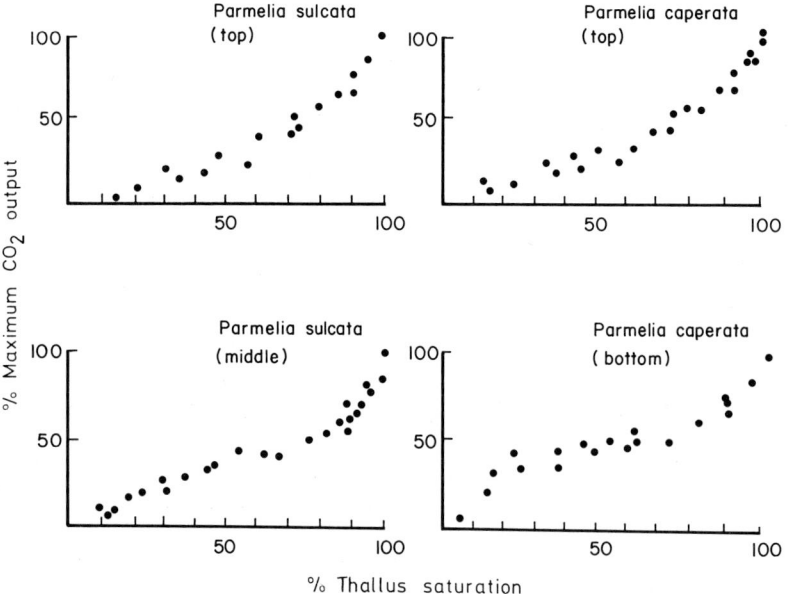

Fig. 3. % maximum dark respiration rates for *Parmelia sulcata* and *P. caperata* plotted against relative thallus saturation

reached because the simulation model employed (Harris, 1972) used relative, not absolute water contents for its physiological calculations and hence assumed a much too simplistic relationship between productivity and water content.

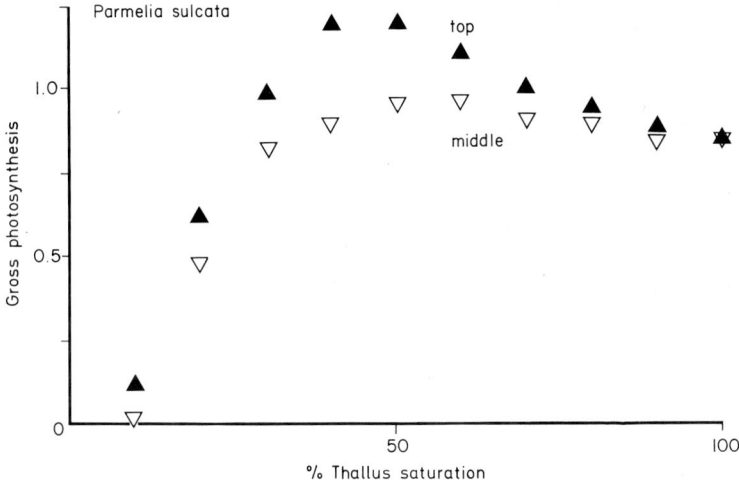

Fig. 4. Calculated 'gross' photosynthesis (mg CO_2 g^{-1} dr. wt. h^{-1}) as related to relative thallus saturation for *Parmelia sulcata* originating from a varying tree hight (top, middle)

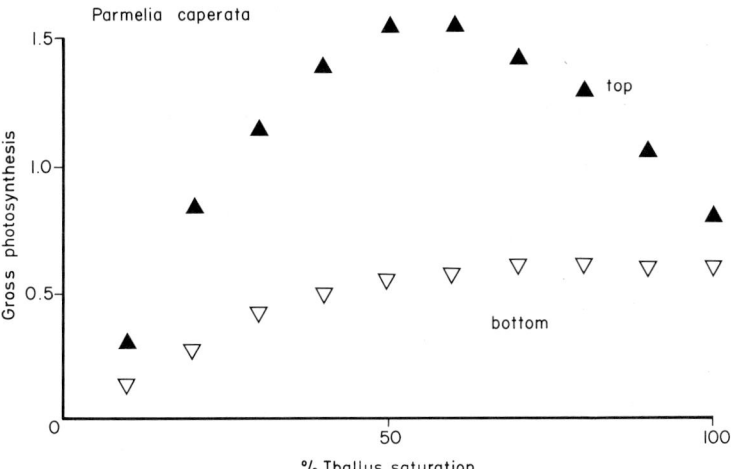

Fig. 5. Calculated 'gross' photosynthesis (mg CO_2 g^{-1} dr. wt. h^{-1}) as related to relative thallus saturation for *Parmelia caperata* originating from a varying tree hight (top, bottom)

b) Ecological Considerations. The optimum water content for net photosynthesis clearly varies in an ecologically adaptive fashion (Figs. 2, 6 and 7). Kershaw and Rouse (1971a), Kershaw and Harris (1971a, b) and Harris (1971b) have all shown that mesic environments lead to high optima (80 % thallus saturation

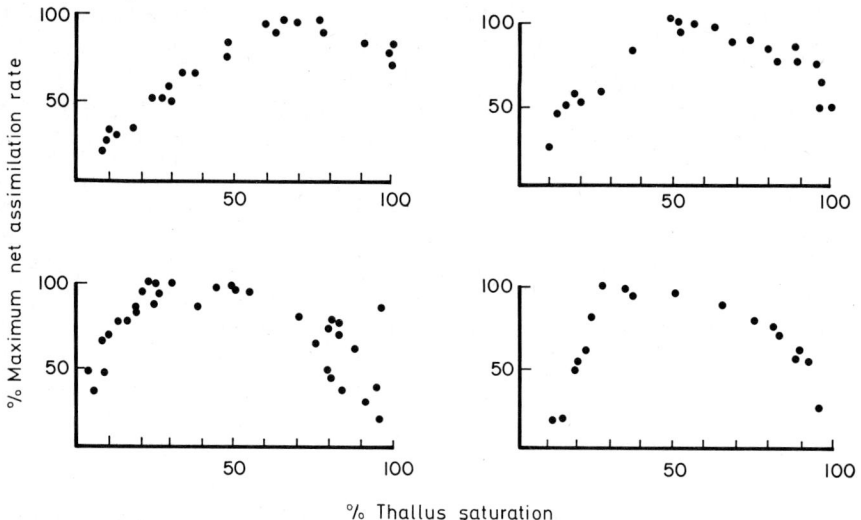

Fig. 6. The ecotypic response of altered physiology in *Parmelia caperata*. From top left to bottom right thalli were collected from increasingly xeric sites in Great Britain. (After Kershaw and Harris, 1971b)

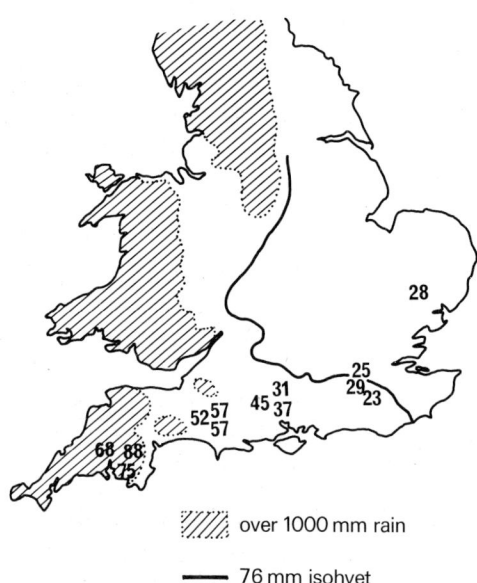

Fig. 7. Map of southern England showing correlation between optimum water contents for net assimilation and annual rainfall. (After Kershaw and Harris, 1971b)

in relative units), xeric habitats to low optima (as low as 20 % thallus saturation in relative units). We do not know if these adaptive responses are the result of physiological adaptation or environmental selection of viable propagules. Given the dynamic response of lichens to environmental light intensity changes (Harris, 1969), there is every reason to expect that such responses are truly adaptations. Examination of the physiological response of lichen thalli to light, temperature and water at different times of the year would make this distinction clear. It is worth restating here that changes in the optimum water content for net photosynthesis can be brought about by changes in the fungal respiration rate. Smith (1961) has demonstrated such seasonal changes so we might expect water content optima to alter seasonally.

Lichen thalli require moisture for growth but continuous saturation leads to rapid death of the thalli in most species. Harris and Kershaw (1971) showed clearly that lichens (*Parmelia* sp.) grow well for limited periods if they are continually wetted and allowed to dry. The suggested explanation for this phenomenon was as follows. During the drying period there is a time when the algal photosynthesis rate is high and the fungal respiration rate is reduced. Under these conditions the algae can photosynthesize, store, and metabolise photosynthates in the dark for their own use, the transfer of carbohydrates to the fungus being restricted. These conditions favour algal growth. At saturation fungal metabolism and growth are favoured, along with a rapid transfer of photosynthates from alga to fungus (Smith and Drew, 1965). These conditions favour the fungus. Harris and Kershaw (1971) found that fluctuating water contents, wet to dry, gave the best growth rates in their lichens under experimental conditions and the greatest longevity. Storage and metabolism of algal photosynthetic products was demonstrated by electron microscopy. The experimental thalli died within three months in continuous light whereas growth was active and sustained under 12-h L and D cycles. Harris and Kershaw (1971) argued that alternating periods of photosynthesis and respiration under conditions which favour first one component and then the other are necessary for the continuation of the symbiotic balance between the alga and the fungus.

V. Conclusions: Water Relations and Productivity A Synthesis

It is now clear that water is of vital importance to lichens, as it is to all plants. However, the strategy which the lichens have evolved is not to conserve water and to use the soil as a reservoir, but to be entirely dependent on the more ephemeral water supplies such as rain and dew fall. To do this they have achieved some remarkable physiological adaptations, not the least of which is an ability to remain dormant whilst dry and to become metabolically active again when water is available. This strategy ensures survival in extremely hostile environments such as the Negev desert (Lange et al., 1970a, b). The strategy does not, however, ensure constant growth rates from year to year. Kärenlampi (1971) demonstrated this clearly in Finland when he showed that the growth of various *Cladonia* species was linearly related to the rainfall and not so well

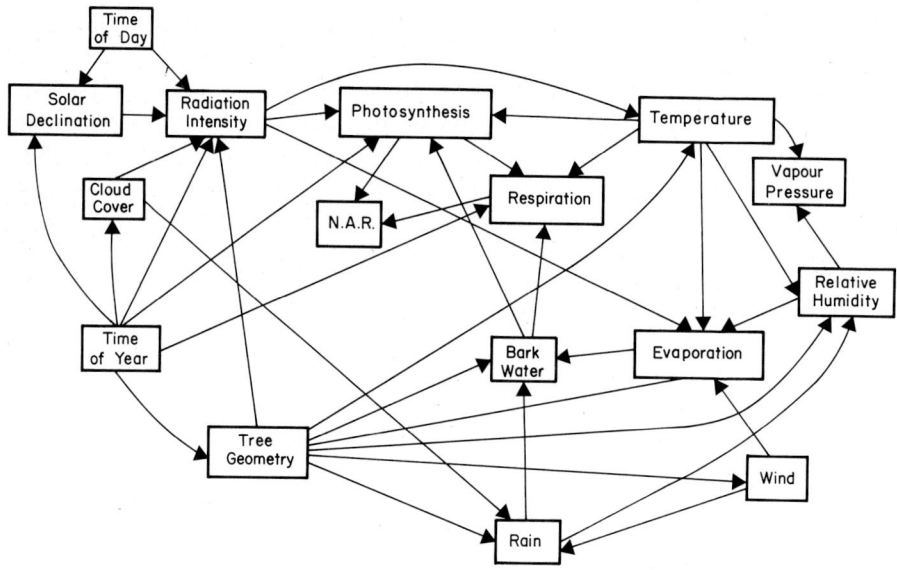

Fig. 8. Diagram of simulated component model. Not all possible interactions shown were included in the model. (After Harris, 1972)

correlated with light and temperature. Armstrong (1974) concluded that there was considerable variability in growth rates from year to year in many lichens. The period of growth and the productivity will depend directly on water availability.

In a simulation study, Harris (1972) attempted to test directly the hypothesis that light intensity and water availability were the controlling factors affecting the vertical distribution of corticolous lichens. The component model (Fig. 8) calculated water availability at six heights in a model tree for each 15-min period over two years. By reference to measured relationships between productivity, thallus water content and light intensity a productivity value was computed and summed. After two years of simulated growth (Fig. 9) the final vertical distribution was compared to the observed field distribution (Fig. 10) (Harris, 1971a).

Although there is undoubtedly some correlation between the simulated and observed results, the study serves a most useful function as a means of defining what further knowledge is required. Firstly there is insufficient data on the light regime in and around tree canopies. Secondly there is insufficient data to allow accurate calculations of evaporation to be made for sites in different parts of tree canopies. Thirdly, as yet we only have a rough idea of how quickly thalli become wet in different parts of a tree when rain commences. These first three problems relate to the microclimatology of the site chosen. It would perhaps have been wiser to start with a simpler situation in this respect. With current microclimatological techniques and theory, accurate statements can be made about physical environmental parameters in simple systems only.

The rest of the problems relate to the physiological response of the lichens. Harris (1972) did not carry out the full matrix of light, water and temperature responses. Furthermore his data on seasonal variation in response were incomplete to say the least. Finally errors were introduced into simulated physiological parameters because relative water contents and photosynthesis rates were used. Absolute quantities are important. The assumed feed-back from alga to fungus was introduced to make the model conform to experimental data (Kershaw and Harris, 1971a). Although such a mechanism may well exist, its introduction becomes unnecessary if the model is calculated with absolute rates. The full three-dimensional response matrix is necessary.

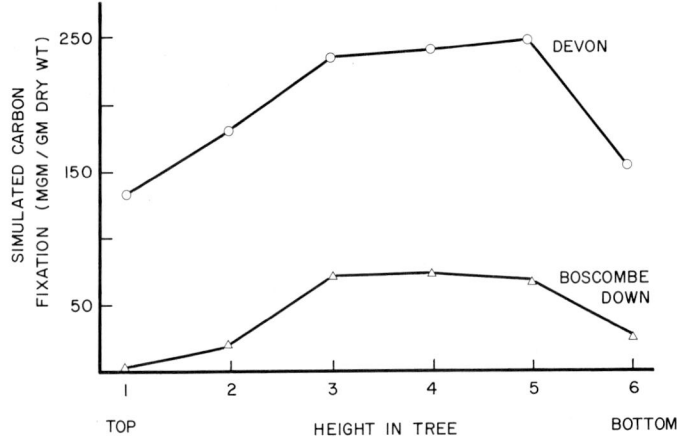

Fig. 9. Simulated growth of *Parmelia caperata* at different heights in oak tree at two sites. (After Harris, 1972). Boscombe Down has a lower annual rainfall than Devon

Fig. 10. Observed distribution of *Parmelia caperata* in Oaks in southern England. (After Harris, 1971a)

Despite these problems, however, the model did demonstrate certain phenomena clearly. It demonstrated the importance of water availability to lichens as controlled by a vertical evaporation gradient and it demonstrated that lichen productivity is irregular and directly correlated with periods of rainfall. Furthermore it showed that rare combinations of high water availability at night followed by drying days can lead to periods of net carbon loss of as long as three weeks duration.

Although some workers (Lechowicz and Adams, 1974) consider that such models, when refined and elaborated, represent the best possible explanatory method for production ecology it should not be forgotten that a simple correlation between a model prediction and a field distribution remains simple. There are many complexities of ecological situations which make simulation modelling impossible. Productivity models merely tell us if and when survival is possible; they give no information on interactions between individuals, on selection or on adaptation to future changes in the environment. The relationship between lichen productivity and thallus water content is one facet of an overall survival strategy.

References

Ahmadjian, V. (ed.): The lichen symbiosis. Blaisdell. Waltham, Mass.: 1967.
Ahmadjian, V., Hale, M. E. (eds.): The lichens. New York-London: Academic Press 1973.
Armstrong, R. A.: Growth phases in the life of a lichen thallus. New Phytologist **72**, 913–918 (1974).
Bednar, T. W.: Physiological studies on the isolated components of the lichen *Peltigera aphthosa*. Ph. D. Thesis, Madison, Wisconsin, 1963.
Bertsch, A.: Über den CO_2-Gaswechsel einiger Flechten nach Wasserdampfaufnahme. Planta **68**, 157–166 (1966).
Bliss, L. C., Hadley, E. B.: Photosynthesis and respiration of alpine lichens. Am. J. Botany **51**, 870–874 (1964).
Blum, O. B.: Water Relations. In: The lichens (eds. V. Ahmadjian, M. E. Hale), pp. 381–400. New York-London: Academic Press 1973.
Butin, H.: Physiologisch-ökologische Untersuchungen über den Wasserhaushalt und die Photosynthese bei Flechten. Biol. Zentr. **73**, 459–502 (1954).
Eickmeier, W. G., Adams, M. S.: Net photosynthesis and respiration of *Cladonia ecmocyna* (Ach.) Nyl from the Rocky Mountains and comparison with three eastern alpine lichens. Am. Midland Naturalist **89**, 58–69 (1973).
Ertl, L.: Über die Lichtverhältnisse in Laubflechten. Planta **39**, 245–270 (1951).
Farrar, J. F.: Lichen physiology: progress and pitfalls. In: Lichens and air pollution (eds. B. W. Ferry, M. S. Braddeley, D. L. Hawksworth), pp. 238–282. Univ. Toronto, 1973.
Hale, M. E. (ed.): Lichen handbook. Washington: Smithsonian Inst. Press 1968.
Hale, M. E.: Single-lobe growth-rate patterns in the lichen *Parmelia caperata*. Bryologist **73**, 72–81 (1970).
Hale, M. E.: Growth. In: The lichens (eds. V. Ahmadjian, M. E. Hale), pp. 473–492. New York-London: Academic Press 1973.
Harris, G. P.: A study of the ecology of corticolous lichens. Ph. D. Thesis, Univ. London, 1969.
Harris, G. P.: The ecology of corticolous lichens. I. The vertical zonation on Oak and Birch in S. Devon. J. Ecol. **59**, 431–439 (1971a).
Harris, G. P.: The ecology of corticolous lichens. II. The relationship between physiology and environment. J. Ecol. **59**, 441–452 (1971b).

Harris, G. P.: The ecology of corticolous lichens. III. A simulation model of productivity as a function of light intensity and water availability. J. Ecol. **60**, 19–40 (1972).

Harris, G. P., Kershaw, K. A.: Thallus growth and the distribution of stored metabolites in the phycobionts of the lichens *Parmelia sulcata* and *P. physodes*. Can. J. Botany **49**, 1367–1372 (1971).

Hill, D. J., Woolhouse, H. W.: Aspects of the autecology of *Xanthoria parietina* agg. Lichenologist **3**, 207–214 (1966).

Hoffman, G. R., Gates, D. M.: An energy budget approach to the study of water loss in cryptogams. Bull. Torrey Botan. Club **97**, 361–366 (1970).

Kallio, P., Heinonen, S.: Influence of short-term low temperature on net photosynthesis in some subarctic lichens. Rep. Kevo. Subarctic Res. Stat. **8**, 63–72 (1971).

Kärenlampi, L.: Studies on the relative growth rate of some fruticolous lichens. Rep. Kevo. Subarctic Res. Stat. **7**, 33–39 (1971).

Kershaw, K. A.: The lichens. Endeavour **22**, 65–69 (1963).

Kershaw, K. A.: The relationship between moisture content and net assimilation rate of lichen thalli and its ecological significance. Can. J. Botany **50**, 543–555 (1972).

Kershaw, K. A., Harris, G. P.: A technique for measuring the light profile in a lichen canopy. Can. J. Botany **49**, 609–611 (1971a).

Kershaw, K. A., Harris, G. P.: Simulation studies and ecology. I. A simple defined system and model. In: Statistical ecology (eds. G. P. Patil, E. C. Pielou, W. E. Waters), Vol. 3, pp. 1–21. Penn. State Univ. Press 1971b.

Kershaw, K. A., Harris, G. P.: Simulation studies and ecology. II. Use of the model. In: Statistical ecology (eds. G. P. Patil, E. C. Pielou, W. E. Waters), vol. 3, pp. 23–42. Penn. State Univ. Press 1971c.

Kershaw, K. A., Larson, D. W.: Studies on lichen-dominated systems. IX. Topographic influences on microclimate and species distribution. Can. J. Botany **52**, 1935–1945 (1974).

Kershaw, K. A., Millbank, J. W.: A controlled environment lichen growth chamber. Lichenologist **4**, 83–87 (1969).

Kershaw, K. A., Rouse, W. R.: Studies on lichen-dominated systems. I. The water relations of *Cladonia alpestris* in spruce-lichen woodland in Northern Ontario. Can. J. Botany **49**, 1389–1399 (1971a).

Kershaw, K. A., Rouse, W. R.: Studies on lichen-dominated systems. II. The growth pattern of *Cladonia alpestris* and *Cladonia rangiferina*. Can. J. Botany **49**, 1401–1410 (1971b).

Lange, O. L.: Experimentell-ökologische Untersuchungen an Flechten der Negev-Wüste. I. CO_2-Gaswechsel von *Ramalina maciformis* (Del.) Bory. unter kontrollierten Bedingungen im Laboratorium. Flora (Jena) **158**, 324–359 (1969).

Lange, O. L., Kappen, L.: Photosynthesis of lichens from Antarctica. In: Antarctic Terrestrial Biology (ed. G. A. Llano) Antarctic Res. Series **20**, pp. 83–95. Washington, D. C.: Am. Geophys. Union 1972.

Lange, O. L., Schulze, E.-D., Koch, W.: Experimentell-ökologische Untersuchungen an Flechten der Negev-Wüste. II. CO_2-Gaswechsel und Wasserhaushalt von *Ramalina maciformis* (Del.) Bory. am natürlichen Standort während der sommerlichen Trockenperiode. Flora (Jena) **159**, 38–62 (1970a).

Lange, O. L., Schulze, E.-D., Koch, W.: Experimentell-ökologische Untersuchungen an Flechten der Negev-Wüste. III. CO_2-Gaswechsel und Wasserhaushalt von Krusten- und Blattflechten am natürlichen Standort während der sommerlichen Trockenperiode. Flora (Jena) **159**, 525–538 (1970b).

Lechowicz, M. J., Adams, M. S.: Net photosynthesis of *Cladonia mitis* from sun and shade sites on the Wisconsin pine barrens. Ecology **54**, 413–419 (1973).

Lechowicz, M. J., Adams, M. S.: Ecology of *Cladonia* lichens. II. Comparative physiological ecology of *C. mitis*, *C. rangiferina*, and *C. uncialis*. Can. J. Botany **52**, 411–422 (1974).

Monteith, J. L.: Evaporation and environment. In: The State and Movement of Water in Living Organisms (ed. G. E. Fogg), Symp. Soc. Exp. Biol., vol. 19, pp. 205–234. London: Cambridge Univ. Press 1965.

Monteith, J. L. (ed.): Principles of Environmental Physics. London: Arnold 1973.

Nowak, R.: Vegetationsanalytische und experimentell-ökologische Untersuchungen über den Einfluß der Luftverunreinigung auf rindenbewohnende Flechten. Dissertation Tübingen 1973.
Peet, M. M., Adams, M. S.: Net photosynthesis and respiration of *Cladonia subtenuis* (Abb.) Evans, and comparison with a northern Lichen species. Am. Midland Naturalist **88,** 446–454 (1972).
Penman, H. L.: Evaporation, an introductory survey. Neth. J. Agric. Sci. **4,** 9–29 (1956).
Phillips, H. C.: Annual growth rates of three species of foliose lichens determined photographically. Bull. Torrey Botan. Club **96,** 202–206 (1969).
Prince, C. R.: Growth rates and productivity of *Cladonia arbuscula* and *Cladonia impexa* on the Sands of Forvie Scotland. Can. J. Botany **52,** 431–433 (1974).
Ried, A.: Thallusbau und Assimilationshaushalt von Laub- und Krustenflechten. Biol. Zentr. **79,** 129–151 (1960a).
Ried, A.: Stoffwechsel und Verbreitungsgrenzen von Flechten. II. Wasser- und Assimilationshaushalt, Entquellungs- und Submersionsresistenz von Krustenflechten benachbarter Standorte. Flora (Jena) **149,** 345–385 (1960b).
Ried, A.: Nachwirkungen der Entquellung auf den Gaswechsel von Krustenflechten. Biol. Zentr. **79,** 657–678 (1960c).
Rouse, W. R., Stewart, R. B.: A simple model for determining evaporation for high-latitude upland sites. J. Appl. Meteorol. **11,** 1063–1070 (1972).
Rundel, P. W.: CO_2 exchange in ecological races of *Cladonia subtenuis*. Photosynthetica **6,** 13–17 (1972).
Schulze, E. D., Lange, O. L.: CO_2-Gaswechsel der Flechte *Hypogymnia physodes* bei tiefen Temperaturen im Freiland. Flora (Jena) **158,** 180–184 (1968).
Smith, D. C.: The physiology of *Peltigera polydactyla*. Lichenologist **1,** 209–226 (1961).
Smith, D. C.: The biology of lichen thalli. Biol. Rev. **37,** 537–570 (1962).
Smith, D. C., Drew, E. A.: Studies in the physiology of lichens. V. Translocation from the algal layer to the medulla in *Peltigera polydactyla*. New Phytologist **64,** 195–200 (1965).
Smith, D. C., Molesworth, S.: Lichen Physiology. XIII. Effects of rewetting dry lichens. New Phytologist **72,** 525–533 (1973).
Stocker, O.: Physiologische und ökologische Untersuchungen an Laub- und Strauchflechten. Flora (Jena) **121,** 334–415 (1927).
Stocker, O.: Wasseraufnahme und Wasserspeicherung bei Thallophyten. In: Handbuch der Pflanzenphysiologie (ed. W. Ruhland), pp. 160–172. Berlin-Göttingen-Heidelberg: Springer 1954.
Wilhelmsen, J. B.: Chlorophylls in the lichens *Peltigera*, *Parmelia* and *Xanthoria*. Botan. Tidsskr. **55,** 30–39 (1959).
Wirth, V., Türk, R.: Über Standort, Verbreitung und Soziologie der borealen Flechten *Cetraria sepincola* (Ehrh) Ach. und *Parmelia olivacea* s. ampl. in Mitteleuropa. Veröff. Landesstelle Naturschutz u. Landespflege Baden-Württ. **41,** 88–117 (1973).

Part 7
Water and Vegetation Patterns

Preface

Nearly one third of the terrestrial surface of the earth consists of arid regions—the steppe formations, the savannas, the semi-deserts and the deserts—and there are additionally numerous habitats where, by local climate and soil, the water factor influences plant distribution and composition (Schmitthüsen, 1968). There is no other factor which determines the vegetation aspect of the earth as greatly as water. It is also true that temperature is of great importance for the distribution of various plant species (Meusel et al., 1965) and the polar termination of forests is well known. But the earth's zonation of vegetation types geographically on a broad scale is primarily dependent on the water factor. Troll (1948) and Stocker (1964) showed the distribution of the different vegetation types on an idealized continent bordering with its north-south coast on an ocean and extending from an oceanic climate at the coast into an increasingly continental climate in its interior. It is quite obvious from this model continent that actually only a very small area of the earth is favored by an oceanic climate with sufficient supply of water and low evaporation rate (see also Walter, 1968, 1973).

In the first Parts of this volume (1–5) the physiological bases of plant adaptation to water conditions have been discussed extensively, and in Part 6 the effect of water on the biomass production was explained on a global basis as well as for certain vegetation types and life forms. It is the purpose of this last Part to try to relate these findings to the geographical distribution patterns of vegetation types. Ecological conclusions about functional adaptations and distributions of vegetation are difficult to ascertain when the plant or vegetation type in question occurs in a region of optimal environmental conditions. In this case the occurrence and growth of species are eventually determined by various competition effects. The functional bases of plant existence are often easier to recognize under extreme environmental conditions, which makes the ecology of plants in extreme habitats of special interest. In the following Part the effect of water is studied in some typical examples of environmental situations.

The alpine timberline is an example of a very distinct and abrupt zone where the extension of forests is ultimately determined by a natural climatic barrier. The causes for such altitudinal zonation have been sought for a long time. It was suggested earlier that timberline is correlated to some temperature factor of the habitat. However, it soon became obvious that no such simple causal relationship exists. It took many years of experimental work to solve this complex interaction of various factors. In Part 7:A Tranquillini shows that probably a primary factor causing a restriction of altitudinal tree growth is water. Water relations during winter and early spring seem to be of decisive significance for survival. These water relations are dependent in a complicated manner on primary productivity and on the length of the growing period.

During plant evolution different forms have developed which are adapted to the different habitats, and the phenomenon of convergence of plant life-forms in areas of similar climate is one of the classical observations in plant geography. Grisebach tried as early as 1872 to explain the vegetation of the earth by the correlation of climate to morphological and anatomical features, especially the convergence of leaf size and form. Later it became obvious that this relationship is not so simple, and that not only the morphological and anatomical structure but also physiological performance are important for an understanding of the adaptations of the vegetation of the earth. The interactions between plants and their surroundings are based on the functional adaptations of various morpho-genotypes. The convergence of life-forms and structures is perhaps most dramatic in the mediterranean-type climate (di Castri and Mooney, 1973). This phenomenon is studied by Dunn et al. in Part 6:B of this volume. They try to explain the evolutionary strategies and the adaptive significance of the presence of evergreen sclerophyllous leaves in the mediterranean type environment.

Water as limiting factor for plant growth is certainly most decisive in the desert, and numerous experiments were performed to study plant life and adaptations in this most extreme habitat. However, work is still necessary to explain on a functional basis the geographical patterns of desert vegetation in the world. Stocker has devoted almost all his life not only to explaining details of functional adaptations in single species but also to giving a general survey of the probable causes of a distinct zonal asymmetric distribution of plant life-forms and morpho-genotypes in the large deserts of the world (Part 6:C). In the Sahara, the "diffuse" dwarf-shrub desert of the north stands in sharp contrast to the "contracted" tree vegetation of the south. This phenomenon is typical not only for North Africa, but also for the desert belt in North America and in a homologous form also in South Africa. Following detailed studies on the physiological adaptations of the different morpho-genotypes, Stocker comes to the conclusion that this phenomenon might be caused by the evolutionary origin of the different floral elements. In the desert zones near the Tropic of Cancer, the holarctic floras of the cold arid regions in the north, where the tree life-form is not dominant, meet the palaeotropic vegetation in the south, where the tree life-form is typical.

It is quite clear that these Parts on water and vegetation patterns can only be an example of an attempt to connect physiological adaptations and global productivity and biomass to the vegetation patterns of the earth. It should be an aim of future studies (see also Müller-Dombois and Ellenberg, 1974) to explain the vegetation patterns of the earth as they have most recently been classified for example by Whittaker (1973) on such functional bases.

References

Castri, F. di, Mooney, H. A. (ed.): Mediterranean type ecosystems. Origin and structure. Ecological Studies, Vol. 7. Berlin-Heidelberg-New York: Springer 1973.

Grisebach, A.: Die Vegetation der Erde nach ihrer klimatischen Anordnung. Leipzig: Wilhelm Engelmann 1872.
Meusel, H., Jäger, E., Weinert, E.: Vergleichende Chorologie der Zentraleuropäischen Flora. Jena: VEB Gustav Fischer 1965.
Müller-Dombois, D., Ellenberg, H.: Aims and methods of vegetation ecology. New York-London-Sydney-Toronto: John Wiley and Sons 1974.
Schmitthüsen, J.: Allgemeine Vegetationsgeographie. Berlin: Walter de Gruyter 1968.
Stocker, O.: A plant-geographical climatic diagram. Israel J. Bot. **13**, 154–165 (1964).
Troll, C.: Der asymmetrische Aufbau der Vegetationszonen und Vegetationsstufen auf der Nord- und Südhalbkugel. Jahresber. Geobot. Forschungsinst. Rübel f. 1947, 46–63 (1948).
Walter, H.: Die Vegetation der Erde in ökophysiologischer Betrachtung. Vol. II. Stuttgart: Gustav Fischer 1968.
Walter, H.: Die Vegetation der Erde in ökophysiologischer Betrachtung. Vol. I, 3. Aufl. Stuttgart: Gustav Fischer 1973.
Whittaker, R. H.: Ordination and classification of communities. Handbook of vegetation science **5**. The Hague: W. Junk 1973.

A. Water Relations and Alpine Timberline

W. Tranquillini

I. Introduction

Where mountains attain sufficient height, the upward extension of forest is ultimately determined by a natural climatic barrier. The upper forest limit, i.e. timberline (Waldgrenze) is defined as the line joining the highest occurrence of closed forest.

The timberline lies at greatly varying altitudes in different regions of the earth (Hermes, 1955). As a generalization it increases in altitude from the poles to the equator (Daubenmire, 1954). This increase is well illustrated in the mountain systems of the North American Cordilleras which stretch in a north-south direction. In coastal Alaska at latitude 65° N the timberline lies at 200 m above sea level. It then rises continuously southwards along the main mountain axes reaching 4000 m on the highest Mexican volcanoes at a latitude of approximately 20° N. The timberline lies higher on the central mountain chains with continental climate than on the outer mountains with a more oceanic climate.

These rules reflect the influence of temperature and the general climatic character of various global and mountain zonations (Brockmann-Jerosch, 1919). This is also recognizable on a small scale in local differences of timberline height. The forest generally climbs higher on warm sunny faces than on adjoining cooler shady faces (Köster and Mayer, 1970). Depression of timberline toward glaciated valley heads in the European Central Alps (the valley-head phenomenon) also probably results from the convergence of cold air currents (Scharfetter, 1938; Friedel, 1967).

The altitude of the timberline and its physiognomy are, however, also determined by the tree species. This results in such enormous diversity in the various mountains of the earth (Hermes, 1955; Baig, 1972) that it cannot be described comprehensively here. Common to all mountains, however, is zonation of the vegetation (Troll, 1961; Stocker, 1963). In temperate mountain regions we find tree species that require good-quality sites at low altitude. Their distribution is determined by forest management or competition between the species. At higher altitudes these tree species are replaced by others with more modest site demands, i.e. such as can adapt to a harsher climate. In this zone the upper limit varies from species to species, but these individual species limits are usually not conspicuous since the forest as a whole continues to higher levels. Only when the limit for existence of the species reaching the highest altitude is attained, and the closed forest ends more or less abruptly, does the timberline become clearly visible (Fig. 1).

Above the timberline there can be an adjoining altitudinal belt of low-growing woody plants, e.g. stands of mountain pine *(Pinus mugo)* or dwarf heath shrubs. They reach their upper limit a little higher up the slope and then give way to alpine grass-heath.

Where near-natural conditions still prevail, the upper timberline is a sharp discontinuity (Ellenberg, 1966; Schiechtl, 1966). Above this, there is a narrow transition zone where the forest opens out, which is termed 'combat zone' (Kampfzone). This zone usually only stretches over 50–100 m altitude and near its lower edge large upright trees occur singly or in small groups (Fig. 2). Its upper boundary is the treeline or tree limit (Baumgrenze), defined as a line joining the uppermost erect trees. Above this there is the 'Krummholz'-zone with mainly severely deformed trees, and at its upper limit there are only low-growing stunted forms which either hug the ground surface or attain at most the height of the snow cover.

The causes for altitudinal zonation of the vegetation in mountains and for the limits of forest (i.e. timberline), Krummholz and dwarf shrubs have been sought for a long time (Schröter, 1926; Böhm, 1969). In previous investigations

Fig. 1. Sharp timberline at 2100 m above sea level in the Iseltal (Hohe Tauern, Austria). Timberline comprises cembran pine *(Pinus cembra)*, above which a closed scrub belt contains *Calluna vulgaris*. Below timberline a band of cembran pine/larch forest *(Larix decidua)* reaches down into the subalpine spruce forest *(Picea abies)* at 1700 m. On glacial terraces various species are found which demand better sites; some of these occur up to 1500 m in montane spruce forest. (Photo: R. Stern)

the attempt was made to compare timberline regionally with isolines of various climatic factors or other climatic indices (Marek, 1910; Gams, 1931). This led to the recognition that summer temperatures play an important role in the height of timberline and that the overall climatic character of a region can modify this basic effect (Brockmann-Jerosch, 1919).

Fig. 2. Combat zone at timberline field station near Obergurgl (Tirol, Austria). The hut is at 2190 m above sea level. Up to 2140 m (treeline) the dominant cembran pine *(P. cembra)* still develops more or less erect trees which are, however, damaged so that stand closure no longer occurs

A clearer insight into the causal relationships between climate, forest-, tree- and Krummholz-limits was not forthcoming until the start of ecophysiological investigations (Däniker, 1923; Michaelis, 1934a, b). These are superior to the previously cited methods of correlation, since the life processes of trees are measured in situ and response to climatic factors can be recognized directly (Tranquillini, 1967).

Surprisingly these ecophysiological studies showed that, contrary to the conclusions from older works, the water relations during winter are of decisive significance in limiting the altitudinal range of trees. Therefore, in the following the water balance of timberline trees during the winter will be described and any connection existing between winter water relations and summer temperatures sought.

II. Water Relations of Trees at the Timberline

To characterize the water relations of plants one can, for example, determine the time sequence of osmotic potential, water potential and water content of leaves. Marked diurnal or seasonal changes in these parameters lead one to suspect perturbations in water balance (see also Waring and Running, this volume Part 3:E). The crossing of certain limits gives warning of drought damage.

From July 1955 to May 1956 the osmotic potential and water content of young and old cembran pine *(Pinus cembra)* at the forest and Krummholz limits were determined (Fig. 3). The small fluctuations in needle water content

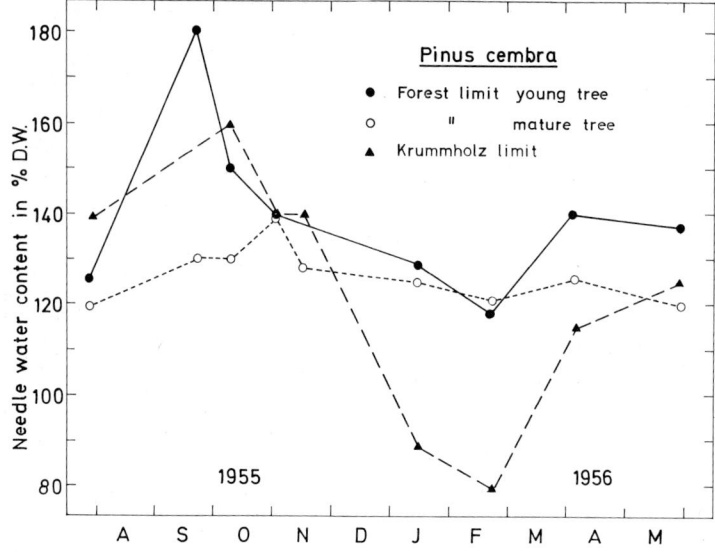

Fig. 3. Seasonal changes in water content (in % dr. wt.) of 1–3 year needles of cembran pine *(Pinus cembra)* at timberline and at Krummholz limit near Obergurgl (Tirol, Austria). Note heavy water loss at Krummholz limit in January and February

of adult trees at timberline on a site with ample snow (2070 m) shows that their water relations were at no time seriously disturbed. At the same site there were slightly greater fluctuations in young trees with their uppermost twigs stretching up above the snow surface during winter. These young trees were less deeply rooted and contained smaller reserves of water in stem and branches and were therefore also occasionally subject to greater desiccation during the summer. In winter these trees lost up to 36% of their autumn water content. Exposed small trees in the Krummholz zone (2230 m) with little snow cover suffered even greater water losses of more than half the autumn water content, and during the most severe drought in February exhibited a water content of only 80% of foliar dry weight. The osmotic potential of these needles was below -40 atm (Tranquillini, 1957).

Similar results were obtained in the winter 1972/73 with cembran pine at and above timberline (Baig et al., 1974). Once again maximum needle drying was attained at all altitudes in February. At the upper krummholz limit at 2140 m above sea level water content fell to 85 % of dry weight. In the following winter (1973/74) needles of the same sample trees dried to a considerably smaller extent (Fig. 4).

As a comparison with cembran pine, water content was determined for shoot tips of spruce *(Picea abies)* taken from various altitudes on Patscherkofel (Fig. 5). The data of the first winter showed an extremely sharp decrease in water content with increase in altitude. Shoots at the Krummholz limit lost water particularly rapidly, so that by February they were completely desiccated. In the following winter the change in water balance was less dramatic; shoots up to the treeline maintained moderate water deficits and severe desiccation

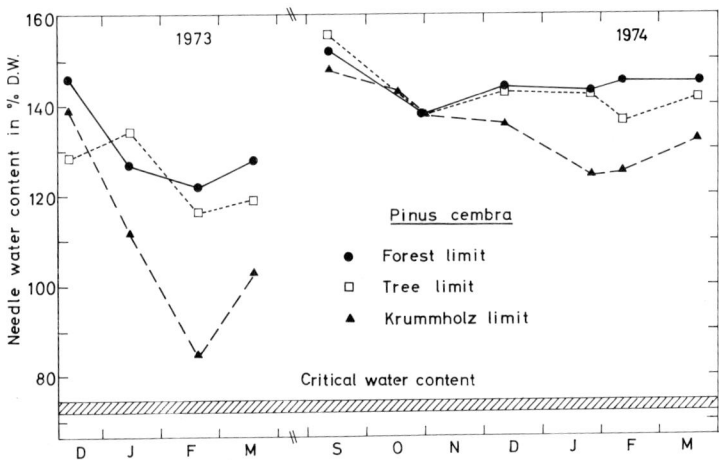

Fig. 4. Water content (% dr. wt.) of first-year needles of cembran pine *(Pinus cembra)* during the course of the 1972/73 and 1973/74 winters at various altitudes on Patscherkofel near Innsbruck, Austria. Critical water content below which needles begin to be damaged taken from work of Pisek and Larcher (1954)

occurred only in the Krummholz zone. The differences between the two winters will be discussed later.

This series of experiments confirmed that needles of spruce and cembran pine, and thus most probably all evergreen conifers at timberline, are subject to marked winter desiccation. The degree of this desiccation increases with increase in altitude and at the last uppermost outposts of Krummholz the lethal limit is crossed in some year in the late winter. Similar results were observed at timberline on various mountains (Goldsmith and Smith, 1926; Michaelis, 1934a, b; Steiner, 1935; Schmidt, 1936; Müller-Stoll, 1954; Wardle, 1968; Lindsay, 1971).

If one compares these minimum winter water contents from above timberline with limiting values determined experimentally when needles of spruce and cembran pine showed drought damage (Pisek and Larcher, 1954), then it can

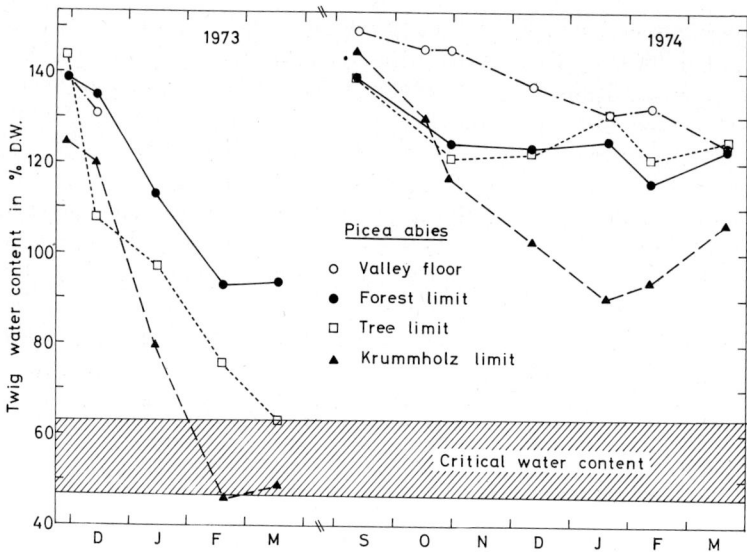

Fig. 5. Water content (% dr. wt.) of first-year shoots of spruce *(Picea abies)* during 1972/73 and 1973/74 winters at various altitudes on Patscherkofel near Innsbruck, Austria. Range of water content at which needles begin to be droped is hatched

be concluded that these winter minima come close to the drought limit or even exceed it. There can be no doubt that damage to twigs and needles exposed above the snow in the combat and Krummholz zone is caused primarily by winter desiccation (frost-drought). It is more rarely caused by freezing. The frost resistance of twigs from all natural stands at timberline reaches well below winter temperatures (Ulmer, 1937; Schwarz, 1970; Sakai and Weiser, 1973), and the late breaking of dormancy ensures that the frost-sensitive growth phases occur in the period with least frost.

III. Causes of Winter Desiccation of Trees at Timberline

1. Microclimate

The factors determining the water relations of trees become more critical with the approach of winter. The fall in soil temperature during autumn progressively impedes water uptake by trees. Uptake ceases when frost penetrates the soil and reaches the roots. Whether the soil freezes or not depends, apart from the severity of the winter, largely on snow cover. A snow depth of only a few tenth of a meter effectively insulates the soil against the cold. This was concluded from soil temperature recordings on sites with heavy and light snow-cover at timberline (Aulitzky, 1961). The temperature under a snow-cover fell below 0°C only in the upper 20 cm of soil, and even there it was still above -1°C. In lower soil layers temperatures were always above 0°C. In contrast, soil temperature on a wind-exposed snow-free site immediately above timberline dropped below 0°C down to a depth of 1 m from December to May ($< -1°C$ from February to March). Closer to the soil surface the temperature even fell below $-4°C$ (Fig. 6).

Wardle (1968) reported similar results for the Front Range, Colorado. Soil temperature under the canopy of a high altitude forest stand scarcely fell below 0°C, whereas in the Krummholz with a paucity of snow it remained under 0°C from mid-November to mid-April with an extreme low of $-10°C$.

Since the water available to plants, occurring mainly in the medium soil pores, freezes at about $-1°C$ (Larcher, 1957) one can assume that trees on sites with ample snow always have access to liquid water. Their water uptake could only be restricted by the low temperatures of the soil (Havranek, 1972). In contrast, on sites which have little or no snow in the wind-exposed combat

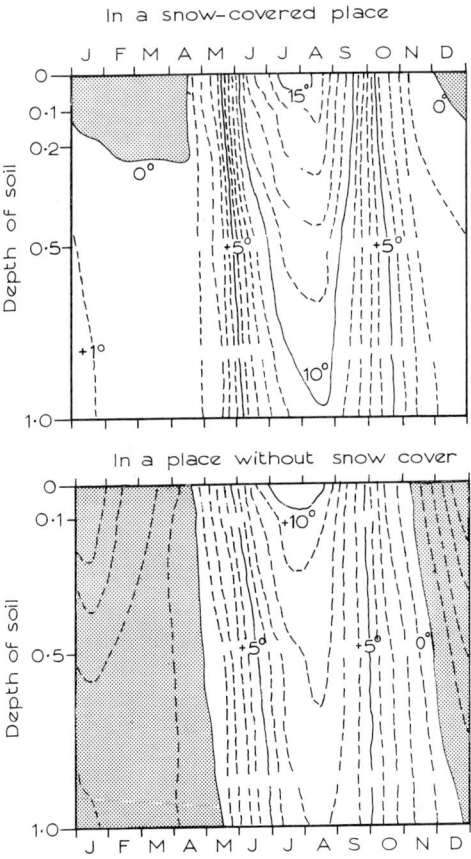

Fig. 6. Isotherms of monthly mean soil temperatures at 0–1 m depth on a snow-covered site, as well as on a mainly snow-free site at timberline (2070 m) near Obergurgl, Austria. (From Aulitzky, 1961)

zone, water uptake by trees from the soil may be interrupted for five months from December to April, even when roots extend down to 1 m below the surface.

Unlike soil frost, ice formation in conducting xylem of stem and branches plays no greater a role in the winter water balance in the combat zone than at lower altitudes. Certainly in severe winters ice frequently forms in stems and branches (Michaelis, 1934a, b; Larcher, 1957, 1963; Sakai, 1966, 1968; Zimmermann, 1964), obstructing water supply to the needles at least temporarily. This applies to trees in tall stands as much, if not more than to trees in the combat zone, since their shaded stems and branches (Michaelis, 1934a), and especially the bases of stems in deep snow (Wardle, 1968) remain frozen longer.

In the combat zone conditions for evaporation are particularly extreme in late winter. As a result of reflection from snow-covered mountain slopes, radiation is up to twice as great as in the lowland; spring (April) intensities equal those in mid-summer (Turner, 1961). This leads to needle temperature rising considerably above ambient temperature (see Gates, this volume Part 3:A). Needle temperature rose above 0°C on 83% of days during a mild winter and 74% of days during the following harsher winter, and absolute needle maximum attained 18.4°C in March and 29.7°C in April (Tranquillini and Turner, 1961). This April figure is almost as high as the mid-summer maximum. At the same time the air is still relatively cool and consequently holds little water vapor. The resulting steep water vapor gradient encourages evaporation, which close to the snow surface on cold winter days with high insolation can reach the same rate as on a dry site in summer (Michaelis, 1934a).

The conditions detailed above explain why the water relations are balanced up to timberline and then significantly worsen in the combat zone. However, the rapid increase of damage from frost-drought within the narrow combat zone suggests that there may be factors other than climatic ones that are of considerable importance.

Michaelis (1934a, b) was the first to suggest that the drought resistance of trees declines rapidly in the combat zone because above a certain altitude new shoots mature insufficiently as a result of the reduced length of the growing season. The resistance of needles to transpiration is then inadequate to protect them from drying.

2. Resistance of Needles to Cuticular Transpiration

Trees at timberline react to the increasing difficulty of water uptake in the autumn by successive reduction of water loss and eventually by complete stomatal closure (Cartellieri, 1935; Tranquillini, 1957). Work to date shows that stomata of trees at timberline remain closed from November to April (Michaelis, 1934a; Tranquillini and Machl-Ebner, 1971). The rate at which water reserves are depleted when water uptake is restricted, i.e. the rate at which shoots desiccate, consequently depends largely on the rate of cuticular transpiration (see also Schönherr, this volume Part 3:B). Under comparable climatic conditions this is dependent on the cuticular resistance to transpiration which is determined by the thickness and structure of the cuticular layers.

In view of these considerations the course of transpiration of excised shoot tips taken from spruce and cembran pine during winter from various altitudes was measured while they were drying under constant conditions in a climate chamber. By this procedure, it was intended to simulate late winter conditions in the combat zone and establish any differences in cuticular resistance to transpiration under extreme drought (Baig et al., 1974; Tranquillini, 1974). Transpiration declined sharply in the first 24 h to a low level, after which there was only a slight further decrease. As is known from similar work by Pisek and Berger (1938) and Pisek and Winkler (1953) the decline in transpiration is caused by hydroactive closure of the stomata. Once the falling transpiration curve approaches the horizontal, stomata are closed and water loss is solely cuticular. The values for the latter part of the curve, i.e., for cuticular transpiration, are presented in Figs. 7 and 8.

Spruce twigs sampled in December 1972 varied remarkably at different altitudes (Fig. 7). Shoots from the valley (1000 m above sea level) transpired least, while those from timberline (1940 m) clearly showed much higher transpiration rates. The greatest water loss was from near treeline (2090 m) and from the wind-exposed side of trees in the Krummholz (2140 m). Transpiration rate of the latter after 25 h in the climate chamber was 14 mg g^{-1} shoot dr. wt. h^{-1} and was thus seven times greater than for twigs from the valley. Their high cuticular transpiration apparently led to a rapid dehydration of the needle epidermis resulting in a marked reduction in water loss; even so, just prior to desiccation damage it was still three times the rate of valley twigs.

Differences in cuticular transpiration between the same sites were much less in 1971 and 1973 than in 1972 (Fig. 7), thus indicating a higher transpiration

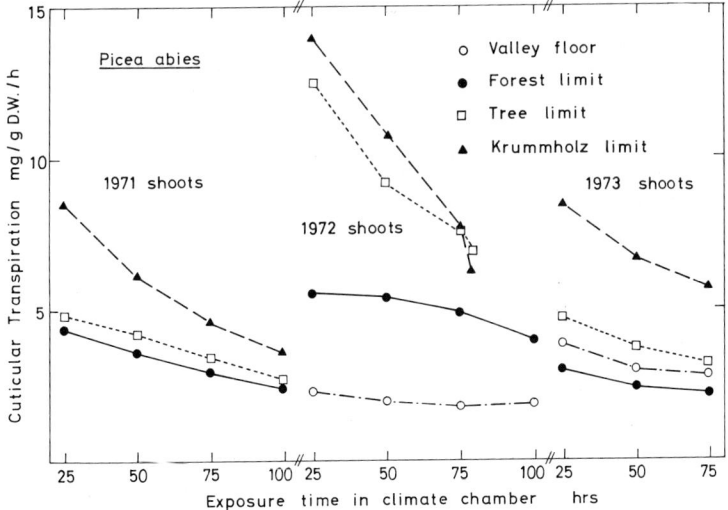

Fig. 7. Water loss from excised but rewatered shoots of spruce trees (*Picea abies*) subsequent to stomatal closure in a controlled-climate wind tunnel (mg H$_2$O g^{-1} shoot dr. wt. h^{-1}, at 15°C, 43% rH, 10 klx, 4 m sec^{-1} wind). Shoots were taken from various altitudes on Patscherkofel, Austria. 1st and 2nd year needles of 1972 tested December 3, 1972, and 1st year needles of 1973 tested February 12, 1974

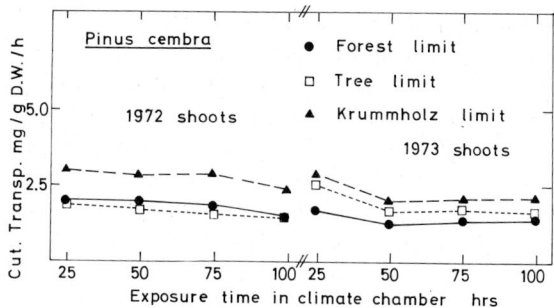

Fig. 8. Water loss from excised cembran pine *(Pinus cembra)* shoots subsequent to stomatal closure in a controlled-climate wind tunnel (mg H_2O g^{-1} needle dr. wt. h^{-1} at 15°C, 43% rh, 10 klx, 4 m sec^{-1} wind). Shoots were taken from various altitudes on Patscherkofel, Austria. 1st year needles tested December 9, 1972 and February 12, 1974

resistance. Only needles from the Krummholz transpired distinctly more than needles from other sites. The differences between transpiration rates from the tree limit, forest limit and valley were not significant.

If one compares cuticular transpiration of the samples, measured under uniform laboratory conditions (Fig. 7), with the course of their shoot water content in the field at the various altitudes in the winters 1972/73 and 1973/74 (Fig. 5), then a distinct correlation between these two parameters emerges. The greater the rate of transpiration (i.e. the smaller the cuticular transpiration resistance) in the laboratory, the faster and further did the needles desiccate during winter at the field site. This means that the altitudinal increase in needle water loss depends not only on altitudinal deterioration of climatic factors, as has been presumed until now, but that the reduction in transpiration resistance is also decisive.

A similar correlation was found for needles of cembran pine. Their cuticular transpiration was considerably lower than for spruce shoots (Fig. 8) and they were subject to less desiccation in the field (Fig. 4). But even in this extremely resistant tree species transpiration resistance is clearly reduced at the top site (2 140 m), so that needle desiccation is correspondingly greatest there in winter. Higher cuticular resistance to transpiration in cembran pine makes this species more resistant to drought than spruce. Cembran pine thus grows to higher altitudes in the Central Alps than spruce and, together with the deciduous larch, it dominates at timberline.

The question now arises whether differences in transpiration resistance of spruce and cembran pine needles from various altitudes are explicable in terms of varying cuticle thickness.

3. Thickness of Cuticular Layers

Insufficient transpiration resistance of needles at high altitude could be the result of inadequate maturation of new shoots, as was suggested by Michaelis

(1934a, b). It is possible that at high altitude sites late commencement of growth and early cessation of tissue differentiation as a result of cold temperatures prevent the full cuticular protection of needles from developing. Lange and Schulze (1966) investigated the cutinization of cell walls in spruce needles and found that even in the lowlands it takes 3 months from the time of bud burst for the cuticular layers to reach their final thickness.

There are a few observations from arctic and alpine timberlines showing that many shoot tips do not fully mature, especially in climatically unfavorable years, and that in winter these shoots are particularly susceptible to "frost-drought" damage (Wardle, 1968; Holtmeier, 1971). Measurements of cuticle thickness at higher altitudes were, however, lacking recently.

Table 1. Thickness of cuticular layers (µm) of spruce and cembran pine needles from a range of altitudes measured in 1972. (From Baig and Tranquillini, 1976)

	Valley floor	Forest limit	Treeline	Krummholz limit
Picea abies				
Thickness of total cuticular layers				
1st-year needles	6.0	5.37	4.12	3.45
2nd-year needles	6.3	6.26	4.5	4.7
Cuticle				
1st-year needles	1.0	0.6	0.3	0.25
2nd-year needles	1.5	1.3	0.8	0.8
Cutinized layers				
1st-year needles	5.0	4.8	3.7	3.2
2nd-year needles	4.8	5.0	3.7	3.9
Pinus cembra 1st-year needles				
Total cuticular layers		5.04		3.3
Cuticle		1.6		0.6
Cutinized layers		3.5		2.6

The data in Table 1 show that total thickness of cuticular layers decreases progressively from valley floor to the Krummholz limit (Baig and Tranquillini, 1976). Differences were particularly distinct in the 1972 needles where total thickness from the valley was nearly twice that from the Krummholz zone. The difference in cuticle thickness of 75% was even greater than for total thickness.

1971 needles had thicker cuticular layers than the younger 1972 needles at the same site. Although the total thickness of the cuticular layer is not greater in cembran pine than in spruce needles, the cuticle proper is considerably thicker in pine. Thus cuticle thickness appears to be of special significance to water permeability. Therefore, in Fig. 9 the thickness of the cuticle is correlated with cuticular transpiration in a climate chamber 50 h after the start of the experiment. Cuticular water loss, i.e. the permeability to water vapor, increases more than linearly with decrease in cuticular thickness. The exponential increase in transpiration explains why water loss and hence desiccation damage increase so rapidly in the combat zone. Cembran pine needles lose less water through cuticular layers of similar thickness. Besides thickness one must, therefore, take the fine structure into consideration (Sitte and Rennier, 1963; see Schönherr, this volume Part 3:B).

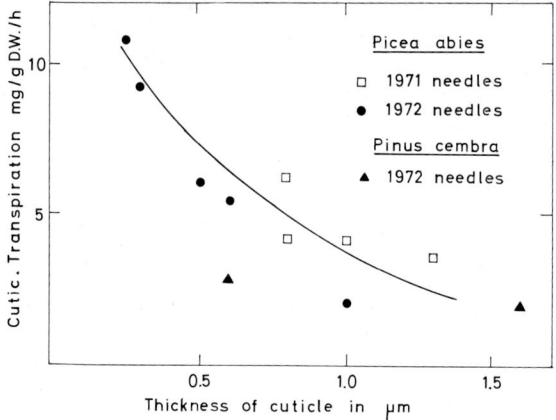

Fig. 9. Dependence of cuticular transpiration of spruce *(Picea abies)* and cembran pine *(Pinus cembra)* needles on cuticle thickness. Needles of various ages from different altitudes were maintained under controlled conditions in a climate chamber for 50 h for measurement

4. Temperature Regime and Length of Growing Season

The question has still to be answered why needles mature differently at diverse altitudes and in different seasons and thus develop varying cuticle thicknesses. If one compares cuticular transpiration of 1971, 1972 and 1973 needles, then the high rate of 1972 needles is immediately obvious. Differences between 1971 and 1973 needles are minor (Fig. 7). A comparison of air temperatures (Table 2) between these three years during the growing season (May–October) shows that 1972 was abnormally cool, whereas temperatures in the summers of 1971 and 1973 were above average. Cool short summers apparently do not allow new shoots to develop fully, especially at high altitude sites.

In order to determine the significance of length of growing season for shoot maturation and cuticular resistance to transpiration, 3-year-old potted seedlings of spruce were placed at different altitudes. The growing season for some of

Table 2. Mean monthly air temperature (°C) during the growing season on the summit of Patscherkofel, Austria (2245 m) in 1971, 1972, and 1973, as well as the mean for 1967–1973

Month	1971	1972	1973	1967–73
V	3.4	0.4	2.5	1.6
VI	3.7	4.4	6.1	4.7
VII	8.5	6.7	6.4	7.4
VIII	9.4	6.7	9.4	7.4
IX	3.2	1.0	6.0	4.5
X	3.4	0.0	0.8	3.0
Mean	5.3	3.2	5.2	4.8

these plants was artificially shortened by exposure to cold temperatures in a climate chamber (Tranquillini, 1974).

The varying rate of shoot development at different altitudes is reflected in the mean shoot length measured on August 13 (Table 3). Shoot length decreased with increasing altitude and was least in plants at 1950 m with an artificially reduced growing season. By the end of September these shoot-length differences were largely eliminated among plants placed at high altitude sites. However, saturation water contents, which can be used as an indicator for the degree of completion of shoot development (Pharis, 1967), show that in mid-September and to a lesser extent even in mid-October considerable differences existed in shoot maturation. Shoots of plants with an artificially reduced growing season always contained most water and were least developed.

Table 3. Mean current year shoot length and saturation water content of young spruce from a range of altitudes

Site	Length of top shoot (cm)		Saturation water content (% dr. wt.)	
	August 13	September 20	September 12	October 16
Valley floor (700 m)	5.73	5.85	171	159
Forest limit (1950 m)	3.36	3.86	205	161
Krummholz limit (2150 m)	2.84	3.17	204	156
Forest limit with growing season shortened artificially by 5 weeks	1.17	3.27	257	212

The cuticular transpiration of plants from the above treatments was as expected. The higher the altitude of the site and the shorter the growing season, the greater the transpiration. After 16 h shoots from 700 m showed a rate of only 4.5 mg g^{-1} h^{-1}, whereas for those from 1950 m with the artificially shortened growing season the rate was three times as high, i.e., 13.9 mg (Fig. 10).

This experiment clearly demonstrates that the development of an effective diffusive resistance of the cuticle necessitates a growing season of adequate length. The results of a reduction of the growing season at timberline by 5 weeks was that winter caught the new shoots in an immature state, in which high transpiration rates led to rapid desiccation. Their cuticular transpiration was comparable to shoots formed in the 1972 season on spruce growing naturally at the Krummholz limit in Austria (Patscherkofel). The summer of 1972 was short and cool (Fig. 7) and in the following winter of 1972/73 shoots were badly damaged by frost-drought (Fig. 5). From this, one can estimate approximately the critical minimum length of growing season required by spruce for survival at timberline. Seedlings were held in winter dormancy in a growth chamber at -2 to $-3°C$ until June 27, 1973. They were subsequently exposed to a continuously warm, frost-free climate until September 22, 1973. At the end of September the first night frosts of down to $-5°C$ and light snow falls probably reduced plant activity. Activity ceased by October 20, 1973 with the onset of more severe frosts down to $-12°C$ and a permanent snow cover (cf. Tranquillini, 1967). The length of the period with full shoot activity was 3 months, with

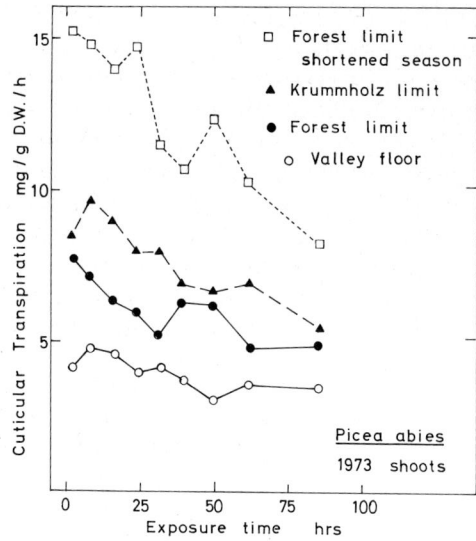

Fig. 10. Water loss from excised spruce *(Picea abies)* shoots subsequent to stomatal closure (mg H_2O g^{-1} shoot dr. wt. h^{-1}) in a controlled-climate wind tunnel. 1973 shoots were taken from 4-year-old seedlings placed at a range of altitudes in October 1972. Some of the timberline seedlings were subjected to artificial reduction in growing season of 5 weeks

an additional month of reduced activity. The experiments showed that this was insufficient for spruce shoots to mature fully and develop an effective protection against excessive transpiration.

This agrees well with much older work indicating that trees require a minimum period of 3.5 months for completion of the annual growth ring (Marek, 1910) and that tall trees require at least a growing season of 3 months for seasonal growth processes to be completed (Kerner, 1863, cited after Marek, 1910). In more recent work by Lange and Schulze (1966) with spruce at low altitude, 3 months were needed for the cuticular cell wall layers to develop fully and this also corroborates the above findings.

Still little is known about exact correlations between climatic factors and the sequence of tree growth and development processes. Certainly temperature plays a very important role. Temperatures about freezing point probably halt shoot development for some time, even after the weather has become warmer again. The repeated spells of cold weather during summer that occur in poor growing seasons at timberline are particularly unfavorable for completion of needle development. In this respect cembran pine is less demanding than spruce; it apparently requires less heat for producing mature needles with adequate transpiration resistance and thus tolerates a shorter and cooler growing season.

The extent of cuticle formation must also depend on photosynthetic production by twigs during the summer. The formation of adequately thick and dense cuticle layers is not just a problem of time but also requires organic material. In the Krummholz zone on Patscherkofel it was found that twigs on the south side of spruce crowns, which receive more radiation than on the shaded north side, were not as far matured and accordingly showed higher cuticular transpiration (Baig et al., 1974). At high altitude high insolation intensities lead to destruction of chlorophyll, especially in spruce, and this results in reduced photosynthetic rates (Ronco, 1970; Benecke, 1972). Furthermore, the southern aspect is the windward side on Patscherkofel and strong winds also restrict photosynthesis (Tranquillini, 1969; Caldwell, 1970). Both effects are cumulative and lead to reduced primary production by exposed twigs. The resultant carbohydrate deficiency prevents the needles from attaining their full protection against excessive transpiration.

Cuticular transpiration of shoots on the windward side of crowns could also be increased by removal of wax and cuticular layers from the needles, due to abrasion caused by wind rubbing twigs against each other or against the snow surface. This increases the permeability of needles to water vapor. Such an effect has been demonstrated in a wind tunnel with grasses (Hall and Jones, 1961; Grace, 1974). This phenomenon was not apparent in the investigations with spruce because an increase in cuticular permeability of the same magnitude as that occurring naturally on wind-exposed twigs at the Krummholz limit was obtained by shortening the growing season (cf. Figs. 7 and 10).

IV. Conclusions: Ecophysiological Analysis of the Alpine Timberline and Its Dynamics

The experiments have shown quite clearly that the increase in frost-drought damage above the timberline, and hence the upper limit of tree growth, are determined not so much by increasing climatic rigor as by the rapidly decreasing drought resistance of the trees themselves. The main cause of this decrease in drought resistance is the decrease in resistance to cuticular transpiration, which in turn results from a decrease in thickness of the cuticle. Thickness and resistance of the cuticle are strongly correlated with the length and warmth of the growing season in which needles are formed. Both decrease with increasing altitude more rapidly in a cool summer than in a warm summer. This explains why the altitude of timberline correlates so well with summer isotherms.

A similar causal sequence can be assumed for the polar timberline, where tree growth is likewise limited by frost-drought (Holtmeier, 1971). In fact the relationship is even more apparent, since the increase in frost-drought is solely dependent on latitudinally determined shortening of the growing season and decrease in summer heat. In the Alps, as a high mountain-range at temperate latitudes, the effect is aggravated by radiation that increases the evaporative demand in winter.

The results also point to a possible explanation for the fluctuations in the altitude of timberline that have occurred in historical times following the clearance of forest and during the post-glacial period through climate changes (Patzelt and Bortenschlager, 1973). Starting with a deforested slope (Fig. 11), danger from frost-drought is assumed to commence at altitude A and to increase at a progressively faster rate with further increases in altitude. This is caused on the one hand by increasing severity of drought and on the other by the decrease in length of growing season.

Up to altitude A, tree development is unrestricted and a dense stand forms. In the zone A to B damage by frost-drought is limited, especially in favorable seasons, so that regeneration still develops into upright trees and eventually to a closed canopy stand. However, this is a slower process than below altitude A. The resulting stand itself diminishes the original frost-drought susceptibility because wind within the stand is reduced, leading to more even deposition of snow which in turn reduces penetration of the soil by frost. A sharply-defined forest limit thus develops at altitude B, above which a sudden increase in frost-drought susceptibility occurs. Immediately above timberline tree regeneration is damaged so severely that canopy closure is no longer achieved. Only individuals in particularly favorable spots develop to full size, while the majority remain stunted. The degree of damage increases rapidly until at altitude C the limit of stunted growth (Krummholz limit) is attained. This lethal limit lies at a different altitude for each subalpine tree species, because the drought resistance of species varies according to the heat demand for development of full cuticular protection. In the Austrian situation the lethal limit for spruce *(Picea abies)* lies lower than for cembran pine *(Pinus cembra)* and mountain pine *(Pinus mugo)*.

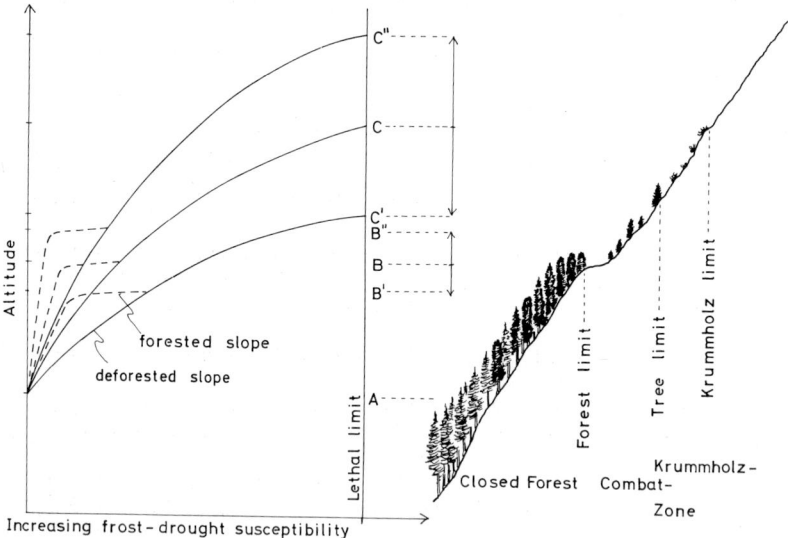

Fig. 11. Schematic representation of increasing susceptibility to frost-drought of trees on a deforested slope in optimal, normal and unfavorable seasons (solid line). Up to point B (forest limit), regeneration develops into a closed stand which itself reduces danger of frost-drought (broken line). Frost-drought susceptibility increases suddenly above point B, i.e. in combat zone. Krummholz limit fluctuates between *C'* and *C"* according to climate of several consecutive seasons

Comparable differences between species exist elsewhere. In the Rocky Mountains of Colorado where at 3150 m altitude the drought-resistant species *Pinus flexilis* and *Pinus aristata* grow as upright trees, surrounding *Picea engelmanni* and *Abies lasiocarpa* are depressed into badly damaged Krummholz (Wardle, 1965). Likewise, in the Rila and Pirin Mountains of south-east Europe the drought-resistant pines *(Pinus peuce, Pinus leucodermis)* attain greater altitudes than spruce *(Picea abies)*.

The diagram in Fig. 11 explains why the degree of damage above timberline can vary from year to year according to the prevailing climate. If several favorable seasons succeed one another then tree and Krummholz limits extend upward (C″). Conversely, after a succession of unfavorable years these limits begin to be depressed (C′) and can even affect the stable forest limit (B′). It is suggested that this mechanism also caused depressions of timberline, estimated to be about 200 m, in the post-glacial times during periods of climatic deterioration (Bortenschlager, 1975).

Acknowledgements. The author wishes to express his appreciation to Dr. Udo Benecke for his translation of the German manuscript. He is grateful to Dr. W. H. Havranek and Dr. P. Wardle for reading the manuscript and for helpful discussions relating to timberline terminology.

References

Aulitzky, H.: Die Bodentemperaturen in der Kampfzone oberhalb der Waldgrenze und im subalpinen Zirben-Lärchenwald. Mitt. Forstl. Bundesversuchsanst. Mariabrunn **59**, 153–208 (1961).
Baig, M. N.: Ecology of timberline vegetation in the Rocky Mountains of Alberta. Ph. D. Thesis, Calgary 1972.
Baig, M. N., Tranquillini, W.: Causes of timberline: morphology and anatomy of needles in relation to cuticular transpiration resistance. Can. J. Botany (in press, 1976).
Baig, M. N., Tranquillini, W., Havranek, W. M.: Cuticuläre Transpiration von *Picea abies*- und *Pinus cembra*-Zweigen aus verschiedener Seehöhe und ihre Bedeutung für die winterliche Austrocknung der Bäume an der alpinen Waldgrenze. Cbl. ges. Forstwesen **91**, 195–211 (1974).
Benecke, U.: Wachstum, CO_2-Gaswechsel und Pigmentgehalt einiger Baumarten nach Ausbringung in verschiedene Höhenlagen. Angew. Botan. **46**, 117–135 (1972).
Böhm, H.: Die Waldgrenze der Glocknergruppe. Wiss. Alpenvereinshefte (München) **21**, 143–167 (1969).
Bortenschlager, S.: Ursachen und Ausmaß postglazialer Waldgrenzschwankungen in den Ostalpen (in press, 1975).
Brockmann-Jerosch, H.: Baumgrenze und Klimacharakter. Zürich: Rascher and Cie. 1919.
Caldwell, M. M.: Plant gas exchange at high wind speeds. Plant Physiol. **46**, 535–537 (1970).
Cartellieri, E.: Jahresgang von osmotischem Wert, Transpiration und Assimilation einiger Ericaceen der alpinen Zwergstrauchheide und von *Pinus cembra*. Jb. wiss. Botanik **82**, 460–506 (1935).
Däniker, A.: Biologische Studien über Baum- und Waldgrenze, insbesondere über die klimatischen Ursachen und deren Zusammenhänge. Vierteljahresschrift Naturforsch. Ges. Zürich **68**, 1–102 (1923).
Daubenmire, R.: Alpine timberlines in the Americas and their interpretation. Butler Univ. Bot. Stud. **11**, 119–136 (1954).
Ellenberg, H.: Leben und Kampf an den Baumgrenzen der Erde. Naturwiss. Rundschau **19**, 133–139 (1966).

Friedel, H.: Verlauf der alpinen Waldgrenze im Rahmen anliegender Gebirgsgelände. Mitt. Forstl. Bundesversuchsanst., Wien **75**, 81–172 (1967).

Gams, H.: Die klimatische Begrenzung von Pflanzenarealen und die Verteilung der hygrischen Kontinentalität in den Alpen. Z. Ges. f. Erdkunde (Berlin) **9/10**, 321–346 (1931).

Goldsmith, G. W., Smith, J. H.: Some physico-chemical properties of spruce sap. Colorado College Publ. Sc. Ser. **13**, 13 (1926).

Grace, J.: The effect of wind on grasses. 1. Cuticular and stomatal transpiration. J. Exp. Botany **25**, 542–551 (1974).

Hall, D. M., Jones, R. L.: Physiological significance of surface wax on leaves. Nature (London) **191**, 95–96 (1961).

Havranek, W.: Über die Bedeutung der Bodentemperatur für die Photosynthese und Transpiration junger Forstpflanzen und für die Stoffproduktion an der Waldgrenze. Angew. Botan. **46**, 101–116 (1972).

Hermes, K.: Die Lage der oberen Waldgrenze in den Gebirgen der Erde und ihr Abstand zur Schneegrenze. Kölner Geogr. Arbeiten Heft **5** (1955).

Holtmeier, F. K.: Waldgrenzstudien im nördlichen Finnisch-Lappland und angrenzenden Nordnorwegen. Rep. Kevo Subarctic Res. Stat. **8**, 53–62 (1971).

Köstler, J. N., Mayer, H.: Waldgrenzen im Berchtesgadener Land. Jb. Ver. Schutze Alpenpfl. Tiere (München) **35**, 1–35 (1970).

Lange, O. L., Schulze, E.-D.: Untersuchungen über die Dickenentwicklung der kutikularen Zellwandschichten bei der Fichtennadel. Forstwiss. Zentr. **85**, 27–38 (1966).

Larcher, W.: Frosttrocknis an der Waldgrenze und in der alpinen Zwergstrauchheide. Veröff. Museum Ferdinandeum Innsbruck **37**, 49–81 (1957).

Larcher, W.: Zur spätwinterlichen Erschwerung der Wasserbilanz von Holzpflanzen an der Waldgrenze. Ber. Naturwiss. Med. Ver. Innsbruck **53**, 125–137 (1963).

Lindsay, J. H.: Annual cycle of leaf water potential in *Picea engelmannii* and *Abies lasiocarpa* at timberline in Wyoming. Arctic and Alpine Res. **3**, 131–138 (1971).

Marek, R.: Waldgrenzenstudien in den österreichischen Alpen. Petermanns Mitt., Erg. Heft **168** (1910).

Michaelis, P.: Ökologische Studien an der alpinen Baumgrenze. IV. Zur Kenntnis des winterlichen Wasserhaushaltes. Jb. wiss. Botan. **80**, 169–247 (1934a).

Michaelis, P.: Ökologische Studien an der alpinen Baumgrenze. V. Osmotischer Wert und Wassergehalt während des Winters in den verschiedenen Höhenlagen. Jb. wiss. Botan. **80**, 337–362 (1934b).

Müller-Stoll, W. R.: Beiträge zur Ökologie der Waldgrenze am Feldberg im Schwarzwald. Angew. Pflanzensoziol., Festschrift Aichinger II, 824–847 (1954).

Patzelt, G., Bortenschlager, S.: Die postglazialen Gletscher- und Klimaschwankungen in der Venedigergruppe (Hohe Tauern, Ostalpen). Z. Geomorph. N. F. **16**, 25–72 (1973).

Pharis, R. P.: Seasonal fluctuation in the foliage-moisture content of well-watered conifers. Botan. Gaz. **128**, 179–185 (1967).

Pisek, A., Berger, E.: Kutikuläre Transpiration und Trockenresistenz isolierter Blätter und Sprosse. Planta (Berl.) **28**, 124–155 (1938).

Pisek, A., Larcher, W.: Zusammenhang zwischen Austrocknungsresistenz und Frosthärte bei Immergrünen. Protoplasma **44**, 30–46 (1954).

Pisek, A., Winkler, E.: Die Schließbewegung der Stomata bei ökologisch verschiedenen Pflanzentypen in Abhängigkeit vom Wassersättigungszustand der Blätter und vom Licht. Planta (Berl.) **42**, 253–278 (1953).

Ronco, F.: Influence of high light intensity on survival of planted Engelmann spruce. Forest Sci. **16**, 331–339 (1970).

Sakai, A.: Temperature fluctuation in wintering trees. Physiol. Plantarum **19**, 105–114 (1966).

Sakai, A.: Mechanism of desiccation damage of forest trees in winter. Contrib. Inst. Low Temp. Sci. Ser. B **15**, 15–35 (1968).

Sakai, A., Weiser, C. J.: Freezing resistance of trees in North America with reference to tree regions. Ecology **54**, 118–126 (1973).

Scharfetter, R.: Das Pflanzenleben der Ostalpen. Wien und Leipzig: Deuticke 1938.

Schiechtl, H. M.: Physiognomie der Waldgrenze im Gebirge. Allgem. Forstz. **77**, 105–111 (1966).

Schmidt, E.: Baumgrenzenstudien am Feldberg im Schwarzwald. Thar. Forstl. Jb. **87**, 1–43 (1936).

Schröter, C.: Das Pflanzenleben der Alpen. Zürich: Albert Raustein 1926.

Schwarz, W.: Der Einfluß der Photoperiode auf das Austreiben, die Frosthärte und die Hitzeresistenz von Zirben und Alpenrosen. Flora **159**, 258–285 (1970).

Sitte, P., Rennier, R.: Untersuchungen an cuticularen Zellwandschichten. Planta (Berl.) **60**, 19–40 (1963).

Steiner, M.: Winterliches Bioklima und Wasserhaushalt der Pflanzen an der alpinen Baumgrenze. Bioklim. Beibl. **2**, 57–65 (1935).

Stocker, O.: Das dreidimensionale Schema der Vegetationsverteilung auf der Erde. Ber. Deut. Botan. Ges. **76**, 168–178 (1963).

Tranquillini, W.: Standortsklima, Wasserbilanz und CO_2-Gaswechsel junger Zirben *(Pinus cembra* L.) an der alpinen Waldgrenze. Planta (Berl.) **49**, 612–661 (1957).

Tranquillini, W.: Über die physiologischen Ursachen der Wald- und Baumgrenze. Mitt. Forstl. Bundesversuchsanst. Wien **75**, 457–487 (1967).

Tranquillini, W.: Photosynthese und Transpiration einiger Holzarten bei verschieden starkem Wind. Cbl. ges. Forstwesen **86**, 35–48 (1969).

Tranquillini, W.: Der Einfluß von Seehöhe und Länge der Vegetationszeit auf das cuticuläre Transpirationsvermögen von Fichtensämlingen im Winter. Ber. Deut. Botan. Ges. **87**, 175–184 (1974).

Tranquillini, W., Machl-Ebner, I.: Über den Einfluß von Wärme auf das Photosynthesevermögen der Zirbe (*Pinus cembra* L.) und der Alpenrose (*Rhododendron ferrugineum* L.) im Winter. Rep. Kevo Subarctic Res. Stat. **8**, 158–166 (1971).

Tranquillini, W., Turner, H.: Untersuchungen über die Pflanzentemperaturen in der subalpinen Stufe mit besonderer Berücksichtigung der Nadeltemperaturen der Zirbe. Mitt. Forstl. Bundesversuchsanst. Mariabrunn **59**, 127–151 (1961).

Troll, C.: Klima und Pflanzenkleid der Erde in dreidimensionaler Sicht. Naturwissenschaften **48**, 332–348 (1961).

Turner, H.: Jahresgang und biologische Wirkungen der Sonnen- und Himmelsstrahlung an der Waldgrenze der Ötztaler Alpen. Wetter und Leben **13**, 93–113 (1961).

Ulmer, W.: Über den Jahresgang der Frosthärte einiger immergrüner Arten der alpinen Stufe, sowie der Zirbe und Fichte. Jb. Wiss. Botanik **84**, 553–592 (1937).

Wardle, P.: A comparison of alpine timber lines in New Zealand and North America. New Zeal. J. Botany **3**, 113–135 (1965).

Wardle, P.: Engelmann spruce (*Picea engelmannii* Parry.) at its upper limits on the Front Range, Colorado. Ecology **49**, 483–495 (1968).

Zimmermann, M. H.: Effect of low temperature on ascent of sap in trees. Plant Physiol. **39**, 568–572 (1964).

B. The Water Factor and Convergent Evolution in Mediterranean-type Vegetation

E. L. Dunn, F. M. Shropshire, L. C. Song, and H. A. Mooney

I. Introduction

One of the classical observations in plant geography is the phenomenon of convergence of plant life forms in areas of similar climate. This convergence is perhaps most dramatic in the mediterranean[1]-climatic areas of the world: the Mediterranean region proper, California, central Chile, southwest Africa and southwest Australia. All of these areas are characterized by a distinct climatic pattern with six months of cool, but not cold, winters with moderate rainfall followed by hot, dry summers. In response to this climate a similar vegetation occurs consisting of dense stands of shrubs and low trees with small, thick, evergreen leaves.

Recently there has been a renewal of interest in convergent evolution in mediterranean climates (Naveh, 1967; Specht, 1969a, b; Mooney and Dunn, 1970a; Mooney et al., 1970), with a previous volume in this series devoted to the subject (di Castri and Mooney, 1973).

The adaptive physiological basis for convergent evolution of the evergreen sclerophyllous shrub form in the mediterranean-type climates of California and Chile has been studied by several authors (Mooney and Dunn, 1970a, b; Dunn, 1970, 1975; Harrison, 1971; Morrow, 1971; Shropshire, 1972; Morrow and Mooney, 1974). This chapter summarizes the results and describes an experimental approach toward understanding the interaction between plants and environmental stress. It focuses on convergence of form and function in response to drought and high temperature stresses characteristic of mediterranean climates. Experimental evidences from field studies in California and Chile are used as examples of plant responses in mediterranean-type climates, although ecophysiological studies have been made in other mediterranean climates (Grieve and Hellmuth, 1970; Hellmuth, 1971; Larcher, 1961a, b).

The California field site was located about 45 km east of Los Angeles at the Tanbark Flat lysimeter installation in the San Dimas Experimental Forest

---

[1] As used in this chapter, Mediterranean with a *capital letter* refers to the geographical Mediterranean Sea region, and when written with a *small letter*, mediterranean refers to similar characteristics of other regions.

at an elevation of 855 m (Mooney and Parsons, 1973). The plant species studied (nomenclature after Munz and Keck, 1963) were representative of the California chaparral and could be compared with particular Chilean species. They included *Adenostoma fasciculatum* H. and A., *Heteromeles arbutifolia* M. Roem., and *Prunus ilicifolia* (Nutt.) Walp. in the Rosaceae and *Rhus ovata* Wats. in the Anacardiaceae.

The Chile field site was located about 30 km southwest of Santiago on a south-facing slope of the Quebrada de la Plata at an elevation of approximately 600 m (Schlegel, 1966). Ecologically equivalent but taxonomically unrelated species (nomenclature after Muños-Pizarro, 1959) studied in Chile were *Kageneckia oblonga* R. and Pav. and *Quillaja saponaria* Mol. in the Rosaceae; *Lithraea caustica* H. and Arn. in the Anacardiaceae; and *Colliguaya odorifera* Mol. in the Euphorbiaceae.

II. Environmental Stresses in Mediterranean-type Climates

Mediterranean climates combine moderate rainfall in winter at cooler temperatures below optimum for plant growth followed by hot, dry summers with no rainfall during the periods in which temperatures are favorable for plant activities. Aschmann (1973) recently discussed the unique features of mediterranean climates, and the worldwide distribution of these climates according to his restricted definition is shown in Fig. 1. Di Castri (1973), using modifications of Walter's climadiagrams, compared the latitudinal climatic gradients in both California and Chile covering the range of mediterranean climates in these areas.

The basic climatic patterns of the two specific study areas used as examples in this chapter are compared by Thornthwaite climate diagrams of the water balance (Thornthwaite and Mather, 1955, 1957) in Fig. 2. The general similarities between the two areas are evident as well as some important differences relevant to the physiological responses observed during the specific study periods discussed in this chapter. The seasonal distribution of rainfall is similar, but the average yearly totals in the two study sites are different (Table 1). The rainfall preceding the three summer-fall drought periods in the two sites was also quite different and reflects the inherent variability in this climatic type.

The calculated values of potential evapotranspiration (PE) in the Thornthwaite system (Thornthwaite and Mather, 1955, 1957) depend primarily on the monthly mean temperatures corrected for latitude and day length, and reflect the similarity in these values between the two areas. With Thornthwaite's calculated value of actual evapotranspiration (AE) as an indicator of vascular plant activity (Major, 1963), the general limitations of this climatic type become apparent. The limitation to plant activity in the cooler months as a result of low energy input, compared to the restriction of activity in the warmer summer months due to low water availability is evident (Fig. 2).

The difference in environmental stress at the two field sites and especially in the three summer–fall drought periods is evident in the calculated actual

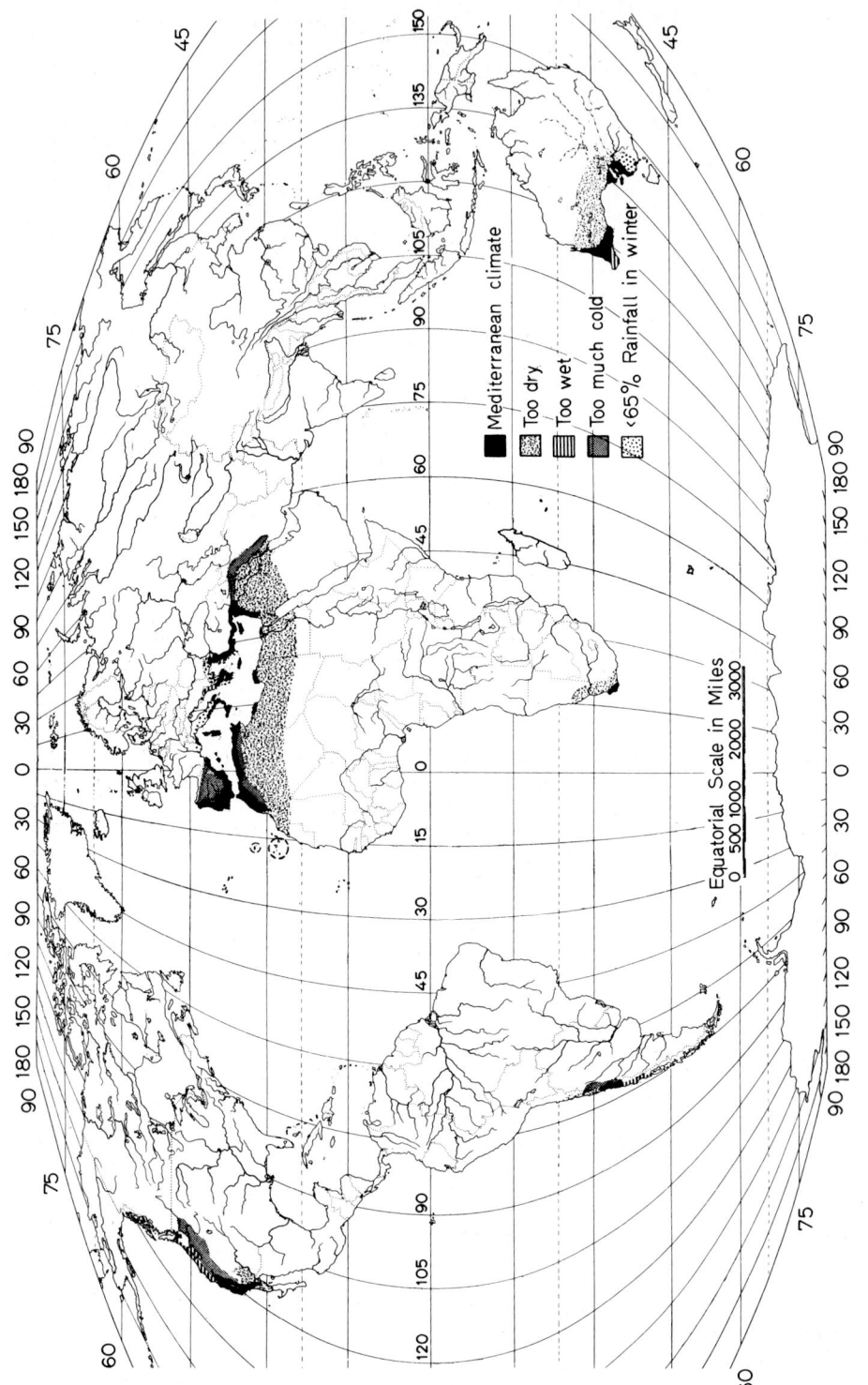

Fig. 1. World distribution of the areas with mediterranean climates. (From Aschmann, 1973)

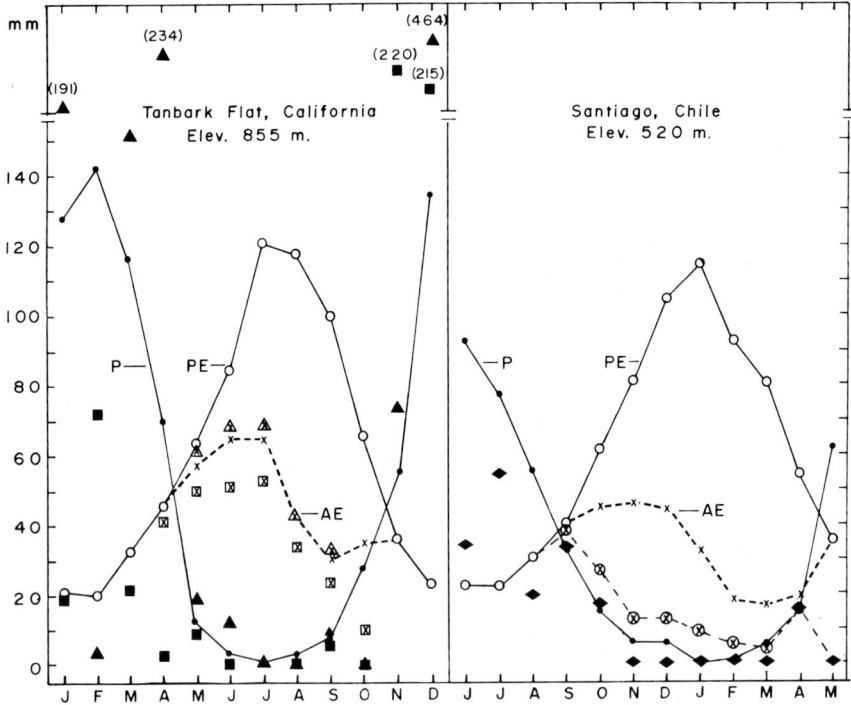

Fig. 2. Thornthwaite climate diagrams of California and Chile study areas. Long-term means for precipitation, P (●—●), potential evapotranspiration, PE (o—o), and actual evapotranspiration, AE (×---×) based on soil moisture storage value of 300 mm. Monthly measured precipitation prior to and during study periods indicated as follows: California, July 1965 to June 1966 (■); July 1966 to June 1967 (▲); Chile, June 1967 to May 1968 (◆). Calculated AE for the three summer drought periods during the study as follows: California, April to October 1966 (⊠); May to September 1967 (△); Chile, September 1967 to May 1968 (⊗)

Table 1. Precipitation characteristics of the California and Chile study areas

	Rainfall, mm	% of mean
Tanbark Flat climatic station, California[a]		
27-year mean[b]	693	100
July 1965–June 1966[c]	578	83.4
July 1966–June 1967[c]	1142	164.8
Quinta Normal weather station, Chile[d]		
91-year mean[e]	372	100
January–December, 1967[f]	172	46.2

[a] Elevation 855 m, latitude 34° 12′ N, longitude 117° 46′ W.
[b] From Sinclair et al. (1958).
[c] From U. S. Weather Bureau records.
[d] Elevation 520 m, latitude 33° 27′ S, longitude 70° 2′ W.
[e] From Mather (1965).
[f] From Chilean Weather Bureau records.

evapotranspiration values for the specific time periods of the study (Fig. 2). Prior to the summer measurements in California in 1966, less than average rainfall had occurred and much of that precipitation had come as early winter storms (November, 220 mm, and December, 215 mm) with much less than normal rainfall in late winter and early spring.

Prior to the measurements in the summer of 1967, however, there had been 65% more than average rainfall with much of this precipitation occurring in late spring (April, 234 mm). In contrast, the situation in Chile before the field measurements from January to May, 1968, was entirely different. Less than half of the normal rainfall had occurred during 1967 near Santiago, resulting in a much greater environmental drought stress and a much reduced calculated actual evapotranspiration value.

The difference in rainfall amount and distribution and resulting environmental water stress is also evident in the pattern of soil moisture depletion at different depths measured at the field site in California during the study period (Fig. 3). It is apparent from these data that the deep-rooted characteristic of these evergreen shrubs is an extremely important adaptive strategy to extend physiological activity into and to withstand the periods of summer drought (Hellmers et al., 1955; Shachori et al., 1967).

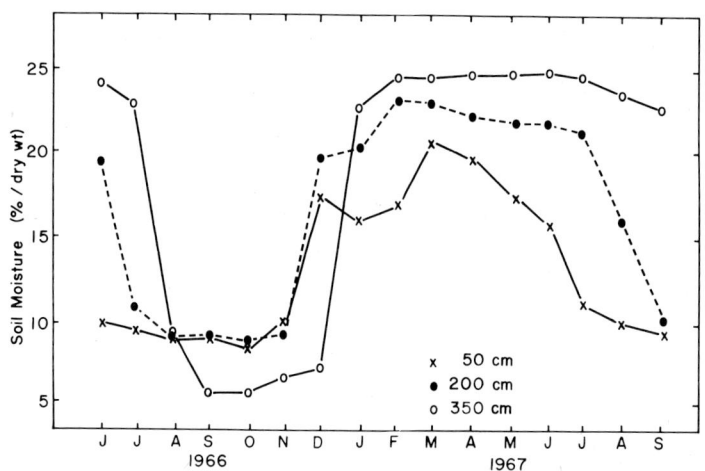

Fig. 3. Seasonal pattern of soil moisture content at different depths at Tanbark Flat, California, during the study period

III. Ecological Significance of Leaf Structure

The small, thick, evergreen leaf is one of the most characteristic features of mediterranean-climate vegetation. A comparison of a quantitative measure of this feature, the "index of sclerophylly" or "Hartlaubcharakter" (Stocker,

1931) reveals some significant ecological features of these sclerophylls (Table 2). It is obvious that these species invest a significant amount of their carbon economy in an evergreen leaf structure with the potential for photosynthesis all year round.

The sclerophylly index value is not constant through the year, but the leaves continued to gain weight even in the severe drought conditions in Chile in 1968. The value also depends on the environmental conditions in which the leaf is produced.

There appear to be both beneficial and detrimental aspects of this thick leaf structure. One obvious benefit is the ability to avoid or delay the onset of drought stress. The significance of the sclerophyll leaf in this regard is reflected in the relationship between tissue water potential and relative water content (Fig. 4). As a drought adaptation, this characteristic response provides low water potentials to maintain gradients for water movement from the soil to the leaf with relatively small changes in relative cell volumes (water content) and therefore small structural changes at the sites of metabolic activity (Cowan and Milthorpe, 1968). In the California and Chile species, tissue water potentials of about -37 bars were observed at relative water contents of about 70%. These values are similar to the values of the most xerophytic species reported by Slatyer (1960) and others (see Ehlig, 1966). The different degrees of water

Table 2. Seasonal sclerophylly index values for various species

	Spring[a]	Summer[b]	Winter[c]	Spring[d]	Summer[e]	Mean
		g dr. wt. dm^{-2} leaf area (one side)				
Heteromeles arbutifolia						
1966–1967	1.59	1.92	1.85	2.28	—	1.91
1967–1968	1.10	1.59	—	—	—	1.35
Rhus ovata						
1966–1967	1.71	2.19	2.47	3.10	—	2.37
1967–1968	0.86	1.73	—	—	—	1.30
Prunus ilicifolia						
1966–1967	—	1.30	1.43	1.46	—	1.40
1967–1968	0.81	1.13	—	—	—	0.97
Lithraea caustica	—	2.20	2.78	—	—	2.49
Kageneckia oblonga	—	1.66	1.78	—	—	1.72
Quillaja saponaria	—	1.72	1.93	—	—	1.83
Colliguaya odorifera	—	2.45	2.60	—	—	2.53
Quercus ilex[f]	—	1.52	1.66	1.68	1.74	1.65
Olea europaea[f]	1.08	1.64	1.84	1.90	2.02	1.70
Beta vulgaris[g]	—	0.56	—	—	—	—
Helianthus annuus[g]	—	0.27	—	—	—	—

[a] Prior to June of current year's growth.
[b] July–October of current year's growth; January in Chilean species.
[c] November–February, same leaf tissue as 1 and 2; May in Chilean species.
[d] March–May, same leaf tissue as 1 and 2.
[e] Same leaf tissue as 1 and 2; greater than one year old.
[f] From Larcher, 1961a.
[g] From Geisler and Stocker, 1966.

stress observed in California compared to Chile are also apparent from these data (Fig. 4).

Some of the detrimental aspects of the sclerophyll leaf structure are related to restrictions on CO_2 uptake rates. The leaf structure with a thick cuticle and effective stomatal responses have been shown to restrict rates of water loss and simultaneously, of CO_2 uptake (Harrison, 1971; Morrow, 1971; Morrow and Mooney, 1974). In addition, there appears to be a considerable intrinsic resistance to CO_2 uptake, a "mesophyll" or "intracellular" resistance, in these evergreen sclerophylls, which is a significant proportion of the total resistance to CO_2 uptake (Dunn, 1970, 1975). It remains to be determined how much of this intrinsic intracellular resistance is a direct result of the sclerophyllous structure necessary to avoid or delay drought stress.

IV. Seasonal Patterns of Photosynthesis, Water Relations, and Productivity

The seasonal patterns of photosynthesis and tissue water relations in several evergreen sclerophylls in both California and Chile, which have been discussed by Dunn (1970, 1975), showed a reduction in daily CO_2 uptake as well as potential photosynthetic activity as the summer drought progressed, with a

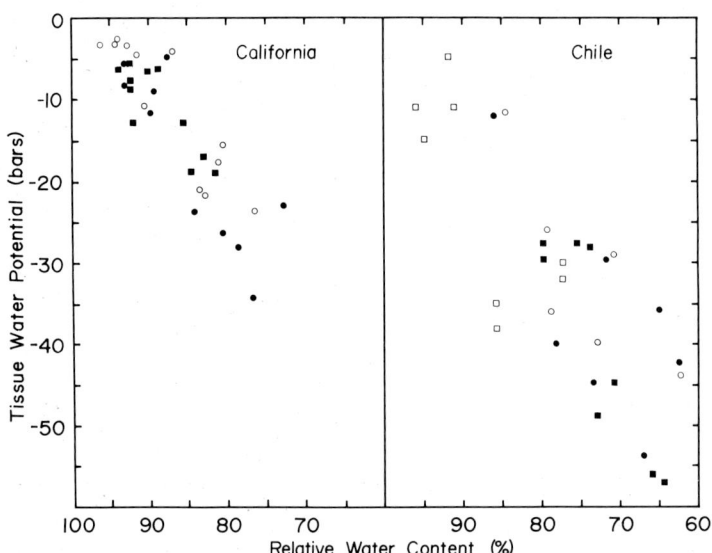

Fig. 4. Relationship between tissue water potential and relative water content for some Californian and Chilean sclerophyll shrubs. Points represent monthly maximum and minimum values observed for .left: *Prunus ilicifolia* (○), *Adenostoma fasciculatum* (●), *Heteromeles arbutifolia* (■); right: *Lithraea caustica* (□), *Kageneckia oblonga* (■), *Quillaja saponaria* (○), and *Colliguaya odorifera* (●)

recovery in photosynthetic activity after winter rains had restored available soil moisture (Fig. 5). The reduction in CO_2 uptake capacity was related to the level of tissue water stress, which in turn was affected by the length and severity of the drought period.

In the specific periods covered by these field studies, plants showed the least stress in the summer of 1967, with a greater reduction in photosynthetic

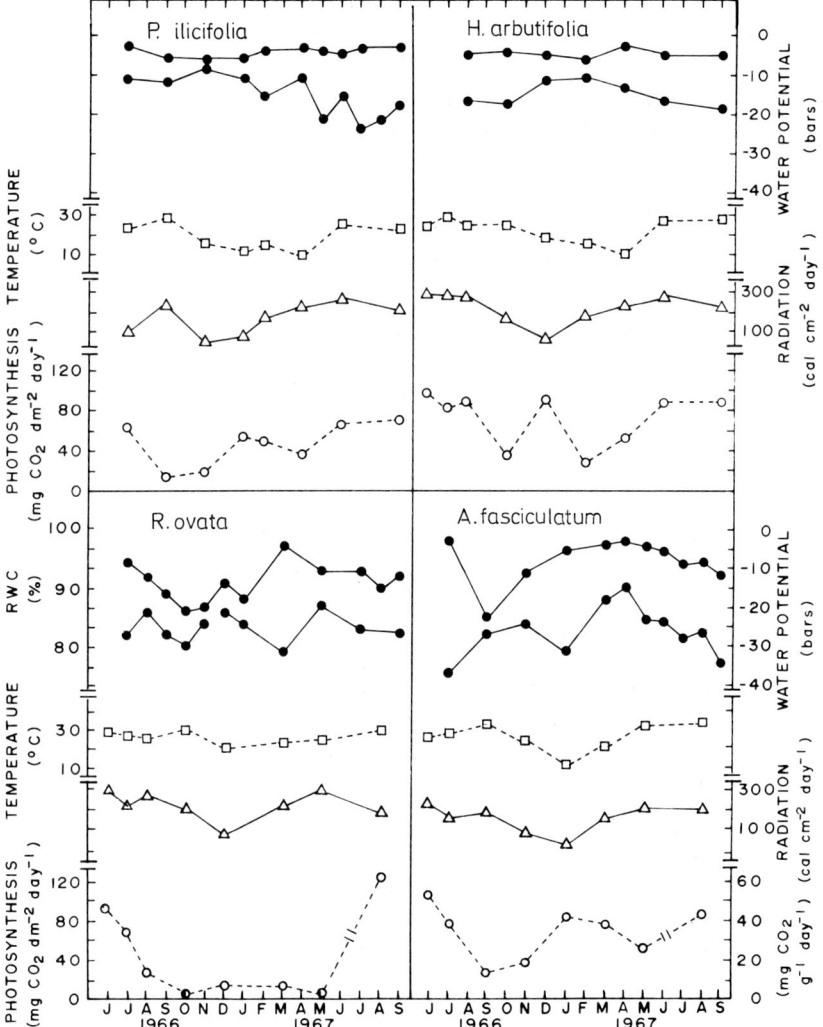

Fig. 5. Summary of seasonal changes in net daily photosynthesis and maximum and minimum daily water status in relation to mean chamber temperature and daily visible radiation for four California chaparral shrubs. Water status expressed as percent relative water content for *Rhus ovata* and as water potential measured with pressure chamber in other species. Values of water potential before April 1967 were extrapolated from water potential–relative water content relationship. Break in photosynthesis line indicates shift from old to new leaf tissue

activity during the summer and fall of 1966. The Chilean species showed the greatest reduction in photosynthetic capacity during the extended drought of the southern hemisphere summer of 1968 (Fig. 6).

All four Californian species were able to maintain positive daily CO_2 balances during all seasons of the year, even though they showed significant reductions in CO_2 uptake by the end of the drought in 1966. All species except *Rhus ovata* showed a recovery in daily photosynthetic rates after the winter rains. There also appeared to be some environmental limitations to photosynthetic capacity at the cooler temperatures of winter and early spring (Fig. 5).

The reduction in photosynthetic capacity was much more marked in the Chilean species under more severe drought conditions. Three of the four species showed negative daily CO_2 balances on some days, and on the days with positive balances, the values were very low (Fig. 6). Only the photosynthetic activity of *Lithraea caustica* was similar to that of the Californian species, but it also showed a reduction as the severe drought progressed.

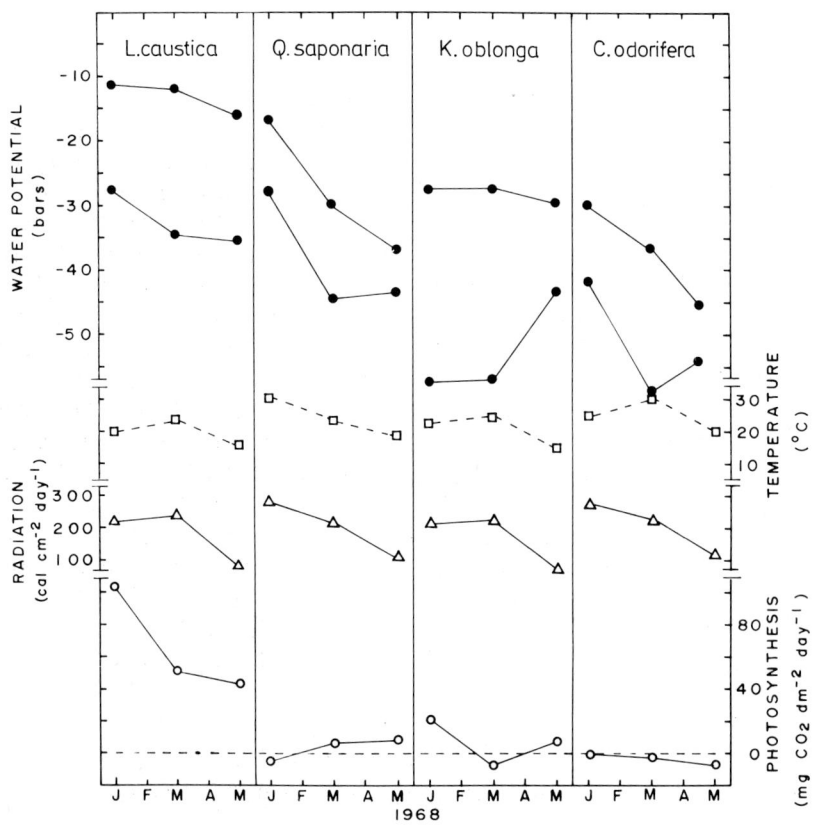

Fig. 6. Summary of seasonal changes in net daily photosynthesis and maximum and minimum daily water potential in relation to mean chamber temperature and daily visible radiation for four Chilean matorral shrubs. Water potential measured with pressure chamber technique

Table 3. Seasonal variations in the carbon balance of some California and Chile sclerophyll shrubs

	Dry summer-fall[a]	winter[b]	Wet summer-fall[c]	Extremely dry summer-fall[d]
	g CO_2 g^{-1} dry weight per six months			
Heteromeles arbutifolia	7.5	5.1	12.0	—
Rhus ovata	3.5	0.7	17.4	—
Prunus ilicifolia	4.5	6.0	11.2	—
Adenostoma fasciculatum	5.3	6.4	7.8	—
Lithraea caustica	—	—	—	4.7
Kageneckia oblonga	—	—	—	0.8
Quillaja saponaria	—	—	—	0.3
Colliguaya odorifera	—	—	—	−0.3

[a] June–Nov. 1966.
[b] Dec. 1966–May 1967.
[c] June–Nov. 1967.
[d] Jan.–June 1968.

The degree of reduction in daily CO_2 uptake observed in all the species was related to the degree of tissue water stress, especially the maximum observed water potential or relative water content, and not necessarily to the minimum observed daily values (Figs. 5 and 6). This pattern was observed both on a daily as well as a seasonal basis (Dunn, 1970, 1975). These results suggested, and more recent field studies (with simultaneous measurements of leaf diffusive resistance and net photosynthesis) have shown, that the observed reductions in photosynthetic activity in these evergreen sclerophylls were primarily due to stomatal closure alone and not to direct effects of drought on photosynthetic metabolism (Harrison, 1971; Morrow, 1971; Morrow and Mooney, 1974).

Seasonal and yearly carbon gains calculated from the daily CO_2 balances (Table 3) showed that in typical years these evergreen sclerophylls remained photosynthetically active all year long. Approximately half of the yearly carbon fixation occurred in the summer–fall drought period and half in the winter and spring. In the very favorable conditions of a wet spring and summer (1967) these species were capable of significant increases in carbon fixation. However, under very severe drought conditions (Chilean summer, 1968), only one of the four species maintained an adequate carbon balance, and one of the species actually showed a net carbon loss during our period of observations.

V. Evolutionary Consequences of Mediterranean-type Environmental Stresses

The interaction of environmental factors, physiological consequences and evolutionary responses by evergreen sclerophyllous vegetation characteristic of mediterranean climates is summarized in Fig. 7. Mooney and Dunn (1970a) have discussed these ideas in more detail. An evolutionary "solution" to any one selective force must be functionally compatible with operational solutions to other environmental factors. As the number and sequence of environmental stresses increase, such as in mediterranean climates, the possible plant responses or "strategies" that are compatible become limited, resulting in the extreme degree of convergence in form as well as function of species in these widely separate but similar environments.

An analysis of the palaeohistory of the vegetation of western North America (Axelrod, 1973) provides another interpretation of the evolution of the sclerophyllous vegetation in mediterranean climates. Axelrod suggests that many of the present evergreen sclerophyll species are similar, if not identical, to species found in the fossil record which existed under more moderate environmental conditions with warmer winters and summer rainfall. According to Axelrod (1973), the present species composition and growth-form convergence is one of differential survival of preadapted forms in response to these selective forces rather than evolution of these growth forms in response to the mediterranean climate, since many of these species existed long before the specific climatic pattern of today.

The evolutionary end result is the same, however: an evergreen, deep-rooted sclerophyllous shrub capable of metabolism at high water stress in the summer as well as at cooler temperatures and lower light intensities in the winter. This combination of solutions to the pattern of environmental stresses characteristic of present-day mediterranean climates appears to be the most effective compromise from a range of possible plant responses, with the result that everywhere this climatic pattern has developed, species, independent of taxonomic affinity, with these characteristics have been the most successful survivors.

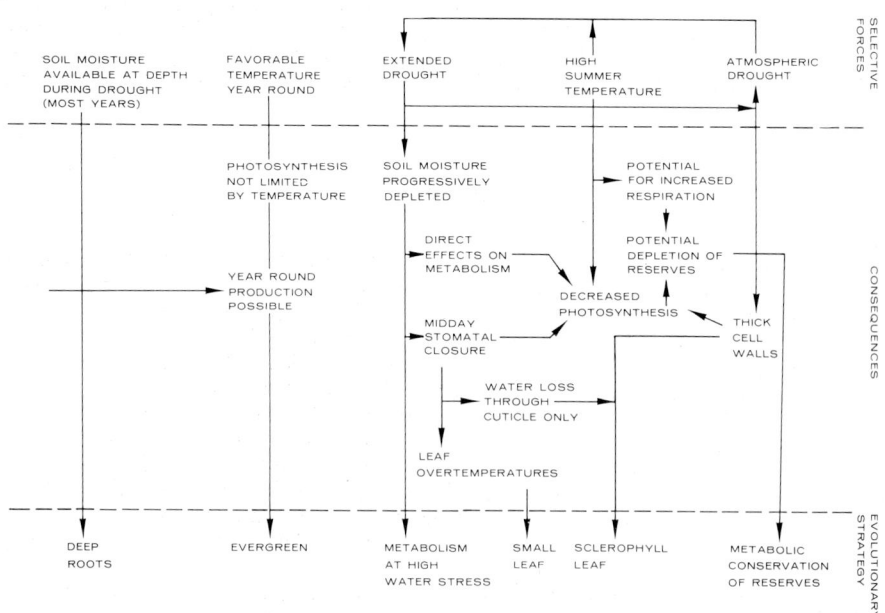

Fig. 7. Summary of plant responses and adaptive strategy of evergreen sclerophylls in response to environmental stresses characteristic of mediterranean-type climates

VI. Conclusions

As a result of the environmental pattern characteristic of mediterranean-type climates, several stresses are imposed on the capacity of an evergreen plant to maintain a positive carbon balance. Typical evergreen sclerophylls from California and Chile showed similar responses to these stresses. High evaporative demands and progressively depleted soil moisture resulted in internal water deficits. A thick leaf allowed these species to avoid or delay drought stress. Severe water deficits were also delayed or prevented by effective stomatal closure and cuticular restriction of water loss. Where adequate soil moisture was unavailable due to low precipitation, limited root systems, poor soil characteristics, or for other reasons, internal water stress increased to a critical point at which the evergreen leaf began to lose more carbon from respiratory activities than could be fixed in photosynthesis.

Significant photosynthesis rates were maintained during winter, contributing approximately half of the yearly CO_2 uptake.

Evergreen, sclerophyllous species, by exploiting soil water at depth and enduring drought periods of unpredictable but not excessive duration, can extend their period of net CO_2 uptake through the entire year. Species with these characteristics effectively exploit all the environmental resources of temperature, light and moisture. This adaptive mode, therefore, is a more efficient competitor than other growth-form strategies which utilize some of these environmental resources for only limited periods of time.

Acknowledgements. Financial support for this study was provided by National Science Foundation grants GB-5223 and GB-8184 to H. A. Mooney, an NSF Graduate Fellowship and a General Biological Supply House (Turtox) Scholarship to E. L. Dunn, the University of California-University of Chile Cooperative Program, and S. R. Bailey. The help of S. R. Bailey, A. T. Harrison, J. Kummerow, J. A. Martinez, P. A. Morrow, and F. A. Stevens was, and still is, greatly appreciated.

References

Aschmann, H.: Distribution and peculiarity of mediterranean ecosystems. In: Mediterranean type ecosystems: origin and structure. Ecological Studies, vol. 7 (eds. F. di Castri, H. A. Mooney), pp. 11–19. Berlin-Heidelberg-New York: Springer 1973.

Axelrod, D. I.: History of the mediterranean ecosystem in California. In: Mediterranean type ecosystems: origin and structure. Ecological Studies, vol. 7 (eds. F. di Castri, H. A. Mooney), pp. 225–277. Berlin-Heidelberg-New York: Springer 1973.

Cowan, I., Milthorpe, F. L.: Plant factors influencing the water status of plant tissue. In: Water deficits and plant growth: development, control and measurement (ed. T. T. Kozlowski) vol. I, pp. 137–193. New York-London: Academic Press 1968.

di Castri, F.: Climatographical comparisons between Chile and the western coast of North America. In: Mediterranean type ecosystems: origin and structure. Ecological Studies, vol. 7 (eds. F. di Castri, H. A. Mooney), pp. 21–36. Berlin-Heidelberg-New York: Springer 1973.

di Castri, F., Mooney, H. A. (eds.): Mediterranean type ecosystems: origin and structure. Ecological Studies, vol. 7. Berlin-Heidelberg-New York: Springer 1973.

Dunn, E. L.: Seasonal patterns of carbon dioxide metabolism in evergreen sclerophylls in California and Chile. Ph. D. Thesis, Los Angeles, Univ. California 1970.

Dunn, E. L.: Environmental stresses and inherent limitations affecting CO_2 exchange in evergreen sclerophylls in mediterranean climates. In: Perspectives of biophysical ecology. Ecological Studies, vol. 12 (eds. D. M. Gates, R. B. Schmerl), pp. 159–181. New York-Heidelberg-Berlin: Springer 1975.

Ehlig, C. F.: Water retention characteristics: plant leaves. In: Environmental biology (ed. P. L. Altman, D. S. Dittmer), p. 463. Bethesda, Maryland: Fed. Am. Soc. Exp. Biol. 1966.

Geisler, G., Stocker, O.: Leaf structure and water availability: Angiosperms. In: Environmental biology (ed. P. L. Altman, D. S. Dittmer), pp. 468–469. Bethesda, Maryland: Fed. Am. Soc. Exp. Biol. 1966.

Grieve, B. J., Hellmuth, E. O.: Eco-physiology of western Australian plants. Oecol. Plant. **5**, 33–68 (1970).

Harrison, A. T.: Temperature related effects on photosynthesis in *Heteromeles arbutifolia* M. Roem. Ph. D. Thesis. Stanford, 1971.

Hellmers, H., Horton, J. S., Juhren, G., O'Keefe, J.: Root systems of some chaparral plants in southern California. Ecology **36**, 667–678 (1955).

Hellmuth, E. O.: Eco-physiological studies on plants in arid and semi-arid regions in western Australia III. Comparative studies on photosynthesis, respiration and water relations of ten arid zone and two semi-arid zone plants under winter and late summer conditions. J. Ecol. **59**, 225–259, 1971.

Larcher, W.: Jahresgang des Assimilations- und Respirationsvermögens von *Olea europea* ssp. *sativa* Hoff. et Link., *Quercus ilex* L. und *Quercus pubescens* Willd. aus dem nördlichen Gardaseegebiet. Planta (Berl.) **56**, 575–606 (1961 a).

Larcher, W.: Zur Assimilationsökologie der immergrünen *Olea europea* und *Quercus ilex* und der sommergrünen *Quercus pubescens* im nördlichen Gardaseegebiet. Planta (Berl.) **56**, 607–617 (1961 b).

Major, J.: A climatic index to vascular plant activity. Ecology **44**, 485–498 (1963).

Mather, J. R. (ed.): Average climatic water balance data of the continents. VIII. South America. Climatology **18**, 297–433. Centerton, N. J., Laboratory of Climatology 1965.

Mooney, H. A., Dunn, E. L.: Convergent evolution of mediterranean-climate evergreen sclerophyll shrubs. Evolution **24**, 292–303 (1970a).

Mooney, H. A., Dunn, E. L.: Photosynthetic systems of mediterranean-climate shrubs and trees of California and Chile. Am. Naturalist **104**, 447–453 (1970b).

Mooney, H. A., Dunn, E. L., Shropshire, F., Song, L.: Vegetation comparisons between the mediterranean climatic areas of California and Chile. Flora **159**, 480–496 (1970).

Mooney, H. A., Parsons, D. J.: Structure and function of the California chaparral—an example from San Dimas. In: Mediterranean type ecosystems: origin and structure. Ecological Studies, vol. 7 (eds. F. di Castri, H. A. Mooney), pp. 83–112. Berlin-Heidelberg-New York: Springer 1973.

Morrow, P. A.: The eco-physiology of drought adaptation of two Mediterranean climate evergreens. Ph. D. Thesis, Stanford 1971.

Morrow, P. A., Mooney, H. A.: Drought adaptations in two Californian evergreen sclerophylls. Oecologia (Berl.) **15**, 205–222 (1974).

Muñoz-Pizarro, C.: Sinopsis de la Flora Chilena (2nd ed.). Santiago, Chile: Univ. Chile Press 1959.

Munz, P. A., Keck, D. D.: A California Flora. Berkeley, California: Univ. California Press 1963.

Naveh, Z.: Mediterranean ecosystems and vegetation types in California and Israel. Ecology **48**, 445–459 (1967).

Schlegel, F.: Pflanzensoziologische und floristische Untersuchungen über Hartlaubgehölze im La Plata-Tal bei Santiago de Chile. Ber. Oberhessische Ges. Natur- und Heilkunde zu Giessen, New Series. Naturwissenschaften **34**, 183–204 (1966).

Shachori, A., Rosenzweig, D., Poljakoff-Mayber, A.: Effect of Mediterranean vegetation on the moisture regime. In: Forest hydrology (eds. W. E. Sopper, H. W. Lull), pp. 291–311. Oxford: Pergamon Press 1967.

Shropshire, F. M.: Convergent evolution of two chaparral shrubs under similar environments on two continents. Ph. D. Thesis, Los Angeles, California 1972.

Sinclair, J. D., Hamilton, E. L., Waite, M. N.: A guide to the San Dimas Experimental Forest. U. S. Forest Service, Calif. Forest and Range Exp. Sta. Misc. Paper **11** (1958).

Slatyer, R. O.: Aspects of the tissue water relationships of an important arid zone species (*Acacia aneura* F. Muell.) in comparison with two mesophytes. Bull. Res. Counc. Israel **8 D,** 159–168 (1960).

Specht, R. L.: A comparison of the sclerophyllous vegetation characteristic of mediterranean type climates in France, California, and southern Australia I. Structure, morphology, and succession. Australian J. Botany **17,** 277–292 (1969a).

Specht, R. L.: A comparison of the sclerophyllous vegetation characteristic of mediterranean type climates in France, California, and southern Australia II. Dry matter, energy, and nutrient accumulation. Australian J. Botany **17,** 293–308 (1969b).

Stocker, O.: Transpiration und Wasserhaushalt in verschiedenen Klimazonen I. Untersuchungen an der arktischen Baumgrenze in Schwedisch-Lappland. Jb. Wiss. Bot. **75,** 494–549 (1931).

Thornthwaite, C. W., Mather, J. R.: The water balance. Climatology **8,** 1–86. Centerton, N. J.: Drexel Inst. Tech. Laboratory of Climatology 1955.

Thornthwaite, C. W., Mather, J. R.: Instructions and tables for computing potential evapotranspiration and the water balance. Climatology **10,** 181–311. Centerton, N. J., Laboratory of Climatology 1957.

C. The Water-Photosynthesis Syndrome and the Geographical Plant Distribution in the Saharan Deserts

O. STOCKER

I. Introduction

Desert plants have to navigate between two cliffs: the danger of desiccation and the danger of starvation; their problem is, how to reduce water consumption and how to maintain photosynthesis at the same time. Between these two strategies, a physiological balance, that can be called the 'water-photosynthesis syndrome', must be achieved.

This problem was the subject of two activities in the Northern Sahara at Beni Ounif in the Southalgerian desert at the foothills of the Sahara-Atlas mountain range, in the Southern Sahara of Atar, and in the Sahel of Mauritania. In this north–south transect of the Sahara, there meet the holarctic and the palaeotropic floristic regions. At this borderline an obvious change takes place, not only in the floristic composition, but also in the vegetational physiognomy: The diffuse dwarf-shrub desert of the North Sahara stands in sharp contrast to the contracted desert with trees of the South Sahara (Stocker 1970, 1971, 1972, 1974).

Such a vegetational zonation in deserts is by no means a unique case, but a common phenomenon in the desert zones of the world. To give only two examples: on the North American continent the holarctic 'northerly' dwarf-shrub desert of the Great Basin adjoins the neotropic 'southerly' thorn-shrub and columnar cactus desert of the Sonora and of Mexico (cf. McGinnies, 1955; Knapp, 1965), and on the South African continent, the palaeotropic thorn-shrub desert borders the South African succulent desert (cf. Acocks, 1953).

II. The Floristic and Physiognomic Aspects of the Sahara

Among the great desert zones of the world the North African Sahara, with its Asiatic extension through Arabia to India, has special importance (Fig. 1). Stretching over two continents, it covers an area of more than ten million km^2 and provides particularly extreme conditions, representing the best and most uniform model of north-south zonation in a warm desert at the latitude of the tropic of Cancer.

The zonation there results from the decrease in precipitation towards the central parts of the Sahara, where rain can be a centenary event (Fig. 2). In

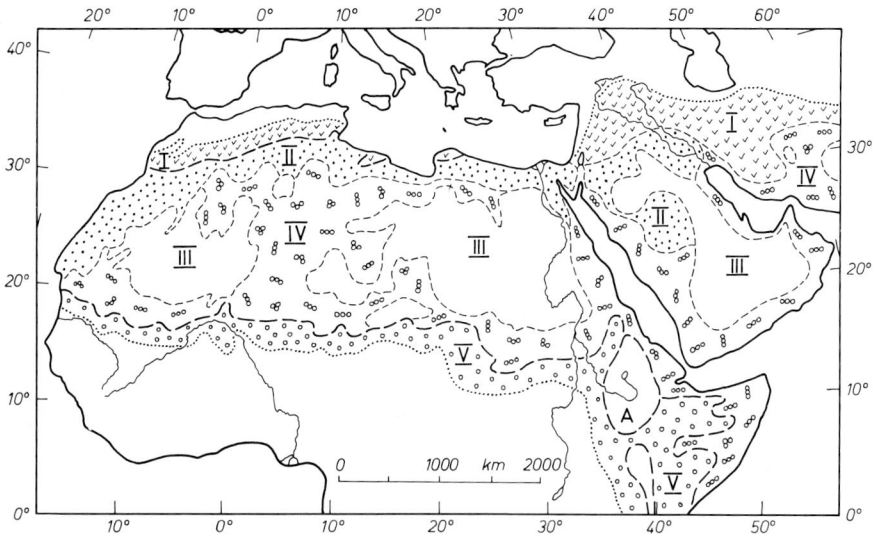

Fig. 1. The Saharan desert complex. *I* Irano-Turanian steppe. *II* Diffuse dwarf-shrub desert of the Northern Sahara. *III* Central absolute desert. *IV* Contracted tree desert of the Southern Sahara. *V* Thorn and succulent Savanna of the Sahel. The enclaves of mountain vegetation are not marked, with the exception of Abyssinia. (After UNESCO-FAO, 1969; Zohary, 1973; Quézel, 1965; Keay, 1959)

the physiognomical aspect, this gradual decrease of precipitation is manifested in a diminution and thinning of the vegetation up to the emptiness of the absolute desert.

When approached from the north via the Irano-Turanian steppe, the North Sahara desert zone (Fig. 1, I) starts at 150 mm of winter rainfall. Floristically this part belongs to the holarctic region; plant geographically it is called the Saharo-Arabian region[1] (Zohary, 1963, 1973).

Trees are lacking, only dwarf shrubs of not more than some ten cm height are scattered 'diffusely' over the area with large distances between each shrub; only in moister places, such as wadis and washes, does the vegetation become more dense (Barry and Celles, 1972/73).

The absolute desert (Fig. 1, III) begins at precipitations below ten to five mm, where perennial plants cannot exist and only annuals may grow for a short lifespan after the rare event of a sufficient rainfall. The floristic composition of the vegetation indicates that the absolute desert belongs to the South Sahara (Fig. 1, III). A transitional range, where the species and physiognomical forms of the holarctic region intermingle with those of the palaeotropics, and from where some species penetrate far into the other region, is situated along a strip with a rainfall of 30–50 mm.

The South Sahara belongs plant geographically to the Sudanian region. The vegetation consists of elements of the savanna flora, with families such

---

[1] The formerly used term 'saharo-indian' is wrong in its second part (Zohary 1963, 1973).

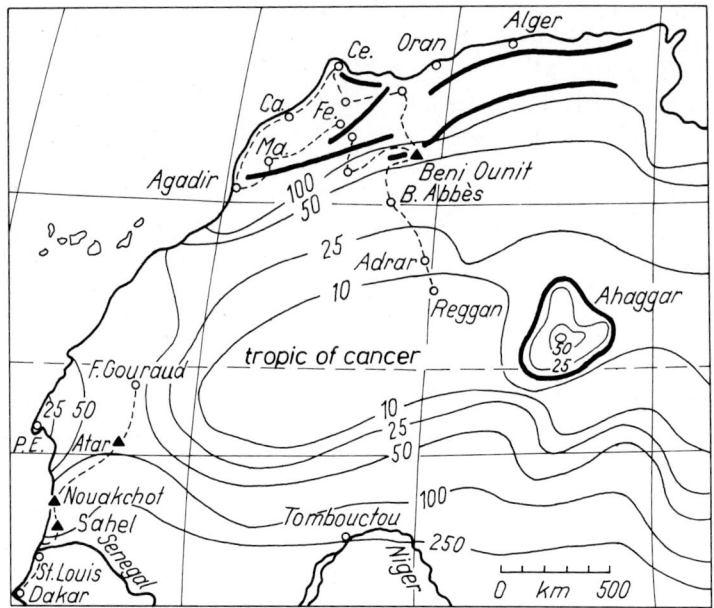

Fig. 2. Isohyets of the Western Sahara, and travelling routes with working fields. (From Stocker, 1974)

as *Mimosaceae, Caesalpiniaceae, Capparidaceae,* and *Asclepiadaceae* being the most important. The physiognomical aspect differs greatly from that in the North Sahara. While the vegetation there is scattered 'diffusely' over the area, it is 'contracted' here at the same amount of precipitation. Plant growth is confined to the wadis and their flood plains with no, or only very sporadical, perennial plant coverage (Monod, 1954). This contracted vegetation is, surprisingly enough, dominated by trees and big shrubs.

With increasing summer rainfall, the contracted vegetation becomes gradually denser towards the south. Where the precipitation amounts to about 150 mm, the southern border of the desert is reached with a zone of diffuse, open thorn and succulent bushland, built up by genera of *Acacia, Balanites, Salvadora, Euphorbia, Adenium* etc. This zone is named the Sahel (Fig. 1, V). It is the southern counterpart of the northern steppe and turns into savanna at a precipitation of around 500 mm.

It is an obvious suggestion to correlate the asymmetry of the zonation with the climatic conditions (Fig. 2). In the western Sahara, where the present investigations were carried out, the annual mean of precipitation is between 100 and 50 mm, in Beni Ounif as well as in Atar. But for an interpretation of the biological aridity, the 'xerothermic index'[2], developed especially for the Mediterranean

---

[2] The number of days without rain, fog, or dew is multiplied by the factor H, representing a criterion of air humidity. Days with rain enter the calculation as full days, days with fog or dew as half-days. H is 1.0 at a relative humidity below 40%, it is 0.9 at 40–60%, 0.8 at 60–80%, 0.7 at 80–90%, and 0.6 at 90–100%.

region and the Sahara by Emberger (UNESCO-FAO, 1963), is more applicable. In this index, the number of biologically dry days per dry season amounts to 345 in Beni Ounif and 350 in Atar (UNESCO-FAO, 1963). Since, according to Emberger, the desert climate begins with 300, and absolute desert climate with 355 days, the indices demonstrate for both places a real desert climate, in Atar coming close to that of absolute deserts. The fact that Beni Ounif has winter rains, whereas Atar has summer rains, possibly aggravates this situation, since it is widely assumed that the latter are biologically less effective. Thus, it is conceivable that the lack of a widespread diffuse vegetation in the Southern Sahara results from a higher biological aridity. In contradiction to such an assumption, however, is the presence of trees in the wadis just in this region. An explanation for this phenomenon could be given in the not-yet-proven presumption that the water run-off in the plantless areas of the southern Sahara is better than in areas with some vegetation, thus leading to a better wetting of the wadis (cf. Evenari et al., 1971; Stocker, 1971). After all, it is questionable whether the problem of asymmetry can be explained by climatology only. From the aspect of the plant constitution the problem will be approached in the following paragraph.

III. The Water-Photosynthesis Syndrome in the Northern and in the Southern Sahara

1. Equivalent Plant Types

Our analysis of the water-photosynthesis syndrome is founded on field measurements of the diurnal variations of transpiration and photosynthesis[3].

Figure 3 gives the diurnal curves of two shrubs, which morphologically exhibit the same type of malacomorphous, deciduous leaves, but belong to different zones of the Sahara. These are *Ziziphus lotus (Rhamnaceae)* from the Northern and *Cassia aschrek (Caesalpiniaceae)* from the Southern Sahara. The difference between the two physiological constitutions is evident. In the North Sahara, *Ziziphus l.* remains far behind the South Saharan *Cassia a.* in the physiological efficiencies. In spite of the more xeromorphous leaves with a double upper epidermis, and in spite of the more severe atmospheric drought (measured by evaporation), the transpiration of *Ziziphus l.* does not follow the potential evaporation in the forenoon hours, and remains on an almost constant level for the rest of the day. Such 'flag curves' are typical in the North Sahara. In such a manner the plant adjusts its water consumption to an existence minimum, which is needed to prevent the danger of a heat death, using the transpiration cold (Lange, 1959, 1962). However, this restriction is not sufficient

---

[3] Measurements were performed with the Darmstadt 'momentary method'. For transpiration this is based on rapid weighing of detatched plant parts (Stocker, 1929, 1956a, b), for photosynthesis and respiration on comparing the conductivity of CO_2 absorbing sodium hydroxid in air currents, running directly or over no detatched plant parts, enclosed in cuvettes during short term (Stocker and Vieweg, 1960).

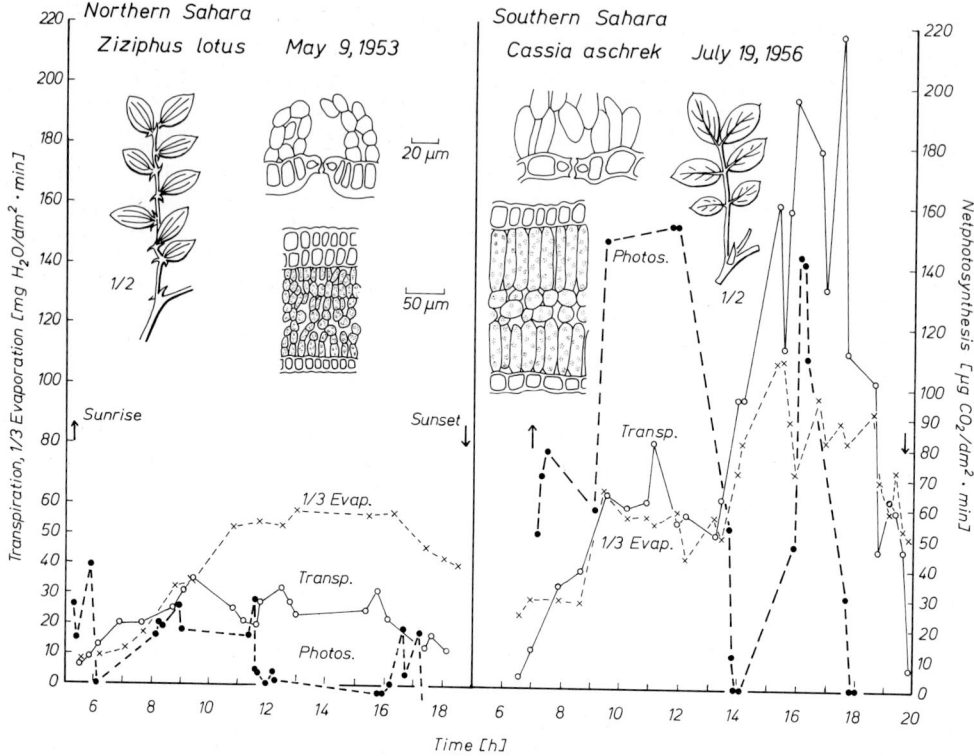

Fig. 3. Diurnal courses of transpiration, evaporation and net photosynthesis by the North and South Saharan shrubs *Ziziphus lotus* (*Rhamnaceae*), and *Cassia aschrek* (*Caesalpiniaceae*), with morphological and anatomical views of the active organs

to keep the water relations balanced. Increasing water deficits lead to a deterioration of the water potential, which, from a certain critical point on, results in a rapid decrease of photosynthesis (Stocker, 1956c); this is obvious between 11:00 and 12:00 a.m. Net photosynthesis reaches the compensation point or drops below it, and recovers only at decreasing evaporation and transpiration in the late afternoon.

In the South Sahara, this process is quite different with *Cassia a.* Although its leaf is less xeromorphous and the evaporation rate is higher, nevertheless, water supply is sufficient to keep the transpiration in step with the evaporation rate almost unrestrictedly until the early afternoon, and to keep it on a high level for the rest of the day. The water potential remains favorable, and the afternoon reduction of photosynthesis is of only a short duration. The daily gains of CO_2 uptake are compiled in Table 1[4]. Related to surface area, for *Cassia a.* the daily rate of transpiration is 3.8 times that of *Ziziphus l.* and the photosynthetic production is even 6.6 times greater.

---

[4] Abbreviations in Table 1 to 4: fr. wt. = fresh weight, dr. wt. = dry weight, d = daylight period.

Table 1. Daily sums of evaporation, transpiration, and net-photosynthesis in malacophyllous shrubs

	Evaporation $\frac{g\ H_2O}{dm^2\ d}$	Transpiration related to		Photosynthesis related to	
		surface $\frac{g\ H_2O}{dm^2\ d}$	dr. wt. $\frac{g\ H_2O}{g\ d}$	surface $\frac{mg\ CO_2}{dm^2\ d}$	dr. wt. $\frac{mg\ CO_2}{g\ d}$
Ziziphus lotus (Northern Sahara)	101.5	17.8	44.4	8.2	20.5
Cassia aschrek (Southern Sahara)	154.5	66.8	166.3	54.3	135.2

The dry weight is another reference system[5]. Ecologically, the relations to surface area and to dry weight have different meanings. For the former the photosynthetic production of carbohydrates is referred to the exchange surface between leaf and environment. In commercial terms this means the 'efficiency of the production' regardless of the 'profitableness of the invested capital'. For a calculation of the ecological profitableness the dry weight is the conformable basis, as expression of the 'carbohydrate capital', which has been 'invested' in building up the leaf. These considerations, referring first of all only to the producing parts of the plant, are also of fundamental importance for the interpretation of the general ecological situation.

In Fig. 4 the Central European deciduous forest region is incorporated into the comparison. The summer-green malacophyllous beech *(Fagus silvatica)* stands as a representative (Schulze, 1970). It is compared to *Ziziphus l.*, a malacophyllous shrub, green throughout the dry period in the North Sahara, and *Acacia raddiana*, a rain-green malacophyllous tree in the South Sahara. Transpiration, net photosynthesis and, as a measure for aridity, the saturation deficit of the air are shown for sunny days in the beginning and the middle of each growing season.

The environmental conditions of each of the three habitats are very different. In the growing season of *Fagus s.*, light and temperature are of significance; aridity does not play a role. *Ziziphus l.* is subject to the same winter-summer rhythm as *Fagus s.*, but is, from the time of sprouting, stressed by a high saturation deficit of the air and the soon-beginning dryness of the soil. *Acacia r.* lives in accordance with the dry and rainy seasons, but it is not influenced by cold as an obstructive factor.

The vegetation cycle of *Fagus s.* begins with low rates of transpiration and of photosynthesis, which rise to very steady summer rates without any difficulties in water relations. The water relations of *Ziziphus l.* are put under stress by severe dryness from the beginning of the vegetation period onward. Transpiration passes in form of a characteristic 'flag curve', and the deterioration of the water potential injures photosynthesis. The stress becomes even more

---

[5] The agreement of the surface and dry weight values by *Ziziphus l.* and *Cassia a.* is accidental, caused by the same dry weight/surface area quotients.

severe in the summer dry period; transpiration is further reduced and net photosynthesis drops to zero in the afternoon. Moreover, *Acacia r.* is subject to rapidly changing conditions in the rainy season; the rates of transpiration and of photosynthesis, which are already rather high in the dry season, become still higher, irrespective of moderate depressions during the noon peaks of the saturation deficit of the air.

The daily sums of transpiration and photosynthesis that result from the curves of Fig. 4 are compiled in Table 2. In order to understand their ecological significance, it is necessary to take into account the total duration of time during which they are realizable in nature. This applies especially to the approximately equally high values of *Fagus s.* in the summer and *Acacia r.* in the rainy period. With *Fagus s.*, the efficiency stays constant for at least four months, with some deductions for days of bad weather. But with *Acacia r.* it is impaired by the irregularity of the moist period, which shortens the time of possible

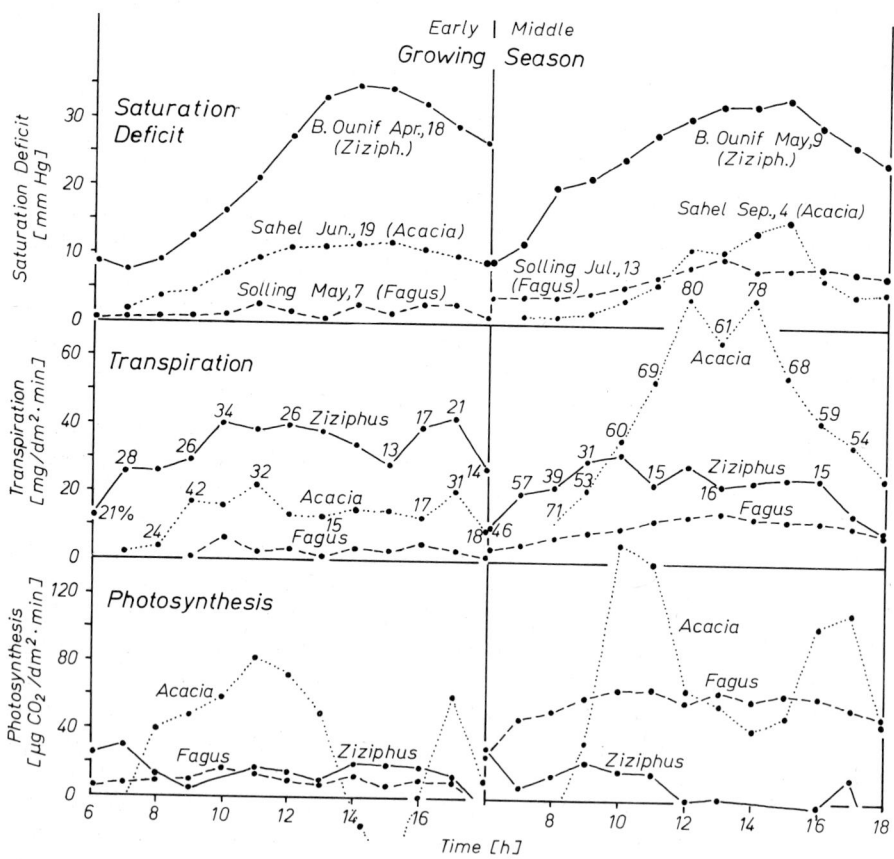

Fig. 4. Diurnal courses of transpiration and net photosynthesis, compared with the saturation deficit of the air, of *Fagus silvatica* in the Solling (German upland), *Acacia raddiana* in the Sahel (Mauritania), and *Ziziphus lotus* in Beni Ounif (Southern Algeria). The numbers with the transpiration curves indicate the relative transpiration

Table 2. Daily sums in the beginning (B) and in the middle (M) of the growing period

	Mean saturation deficit mm Hg		Transpiration g H_2O / dm^2 d		Photosynthesis mg CO_2 / dm^2 d	
	B	M	B	M	B	M
Fagus silvatica (Central Europe)	1.4	6.9	1.8	13.8	8.1	44.7
Ziziphus lotus (Northern Sahara)	22.1	24.4	25.7	17.8	10.5	8.2
Acacia raddiana (Sahel)	8.4	6.2	9.8	29.2	19.0	49.0

photosynthesis. Considering all of this, it becomes obvious why Central Europe has closed tall forests, whereas the vegetation in the Sahel consists of open thorn and succulent shrubs, and in the North Sahara of dwarf shrubs only.

2. Other Constitution Types

From the special case of the malacomorphous deciduous leaf the discussion may be extended now to other constitutional plant types of the North and South Sahara.

The categories of Table 3 are based upon morphological characteristics of the growth form and of the leaf structure; with some species the classification is uncertain. Criteria for the leaf structure are the 'dimension quotients': the surface development (surface area per g fr. wt.), the sclerophyllous character (g dr. wt. per surface area), and the degree of succulence (water content per surface area). The figures in the Table are mean values (Stocker, 1972, 1974).

A survey of the constitution types of the Northern Sahara (Beni Ounif) reveals that the switch and the thorn shrubs are the most prominent components of the perennial vegetation. Of these, *Retama retam, Zilla macroptera, Launaea arborescens*, and *Rhanterium adpressum* were thoroughly examined. The 'massive construction' of the evergreen, long-lived shoots of the switch and thorn shrubs stands in contrast to the 'lightweight construction' of the deciduous malacophyllous leaves. This is manifested in the quotients of leaf dimensions (Table 3): The surface development per fresh weight decreases greatly (0.70 to 0.24 dm^2 g^{-1}), whereas the sclerophyllous character and the degree of succulence increase (0.50 to 1.61 g dm^{-2} and 1.05 to 2.05 g dm^{-2}). Since the macrophyllous type is of only minor importance in the North Sahara, the trend to a massive construction can be regarded as a general tendency among the desert plants (Stocker 1928, 1970–1974).

In respect to transpiration, experience proves that there are desert plants with strong, and others with weak transpiration. Besides the malacophyllous shrubs *(Ziziphus l., Rhus tripartita, Capparis spinosa)*, also the hemicryptophytes and cryptophytes *(Peganum harmala, Citrullus colocynthis)*, as well as a remarkable group of sclerophyllous hemicryptophytes *(Echinops spinosus, Centaurea pungens)*, belong to the strongly transpiring type (19.5, 29.1, and 38.5 g dm^{-2} d^{-1}

Table 3. Constitution types of the Northern and the Southern Sahara

Constitution types (species cf. text)	Number of species	Dimension quotients				Transpiration related to			Photosynthesis related to	
		Surface developm. $\frac{dm^2}{g}$	Scleroph. character $\frac{g}{dm^2}$	Degree of succulence $\frac{g}{dm^2}$	Relative %	Surface $\frac{g}{dm^2 d}$	dr. wt. $\frac{g}{g d}$	Water content $\frac{g}{g d}$	Surface $\frac{mg\ CO_2}{dm^2 d}$	dr. wt. $\frac{mg\ CO_2}{g d}$
Northern Sahara (Beni Ounif)										
Malacophyllous shrubs	(3)	0.70	0.50	1.05	22	19.5	40.5	28.0	10.5	21.8
Switch and thorn shrubs	(4)	0.24	1.61	2.50	20	13.4	12.0	5.4	9.7	8.7
Malacophyllous hemicrypto- and geophytes	(2)	0.38	0.44	2.19	31	29.1	67.6	13.1	5.7	13.2
Sclerophyllous cushion shrub	(1)	0.23	1.42	2.84	16	11.5	8.1	4.1	4.6	3.2
Sclerophyllous hemicryptophytes	(2)	0.34	0.72	2.31	50	38.5	63.0	17.1	—	—
Xerohalic succulent	(1)	0.21	1.60	3.30	17	12.4	5.2	3.5	5.1	2.0
Hygrohalic succulents	(2)	0.36	0.68	2.85	17	8.0	11.9	4.3	4.9	7.3
Mean		**0.34**	**1.00**	**2.45**	**25**	**18.9**	**29.8**	**10.8**	**6.8**	**8.9**
Southern Sahara (Atar)										
Malacophyllous trees and shrubs	(5)	0.45	0.77	1.96	43	37.4	53.6	24.3	29.3	45.0
Evergreen, broad-leaved sclero-phyllous trees	(2)	0.24	1.35	2.28	14	20.9	14.8	13.5	25.1	17.8
Switch trees	(2)	0.15	2.94	3.80	25	31.3	13.0	8.3	19.4	8.6
Malacophyllous geophytes	(2)	0.24	1.11	3.02	31	43.2	40.0	14.5	13.9	14.5
Mean		**0.27**	**1.54**	**2.77**	**28**	**33.2**	**30.4**	**15.2**	**21.9**	**21.5**

respectively). The massive and succulent types exhibit much lower transpiration rates: the thorn and succulent shrubs 13.4, the sclerophyllous hemispherical cushion shrub *Anabasis aretioides* 11.5, the xerohalic succulent *Limoniastrum feei* 12.4, the hygrohalic succulents *Zygophyllum album* and *Frankenia thymoides* 8.0 g dm^{-2} d^{-1}. When transpiration is based on dry weight and on water content, the differences become greater. Noteworthy is that the switch and the thorn shrubs are comparable to the succulents in regard to the degree of succulence, as well as in regard to the better utilization of the water reserved for transpiration.

The rates of net photosynthesis, related to surface area, are all in a range between 10.5 and 4.6 mg CO_2 dm^{-2} d^{-1}, with the switch and the thorn shrubs reaching almost the same values as the malacophyllous shrubs. For the ecological effect the duration of the growing season has to be taken into account, as then, in the annual production, the evergreen thorn and succulent shrubs are greatly favored. The tendency of some malacophyllous shrubs to keep their leaves beyond the season may be understandable in this connection; they hibernate, at least in part; for instance in *Rhus t.* and *Capparis s.*, both inhabitants of rock crevices, which are relatively well-watered habitats.

For the vegetation of the South Sahara the conditions are much different. The change of the sociological structure is already expressed in the list of the examined constitution types (Table 3). Predominant is the malacophyllous type with a tendency to persistent leaves. There are very divergent types in this group: tree- and shrub-like *Leguminosae (Acacia r., Cassia a.)*, dwarf shrubs with big, soft or with small, succulent leaves *(Abutilon muticum, Chrozophora brocchiana)*, the succulent salt shrub *Nitraria retusa* with double leaf change for the rainy and the dry season (Shmueli, 1948), and the peculiar, tree-shaped shrub *Calotropis procera* with its huge, succulent leaves. In contrast to the North Sahara, there are evergreen sclerophyllous broadleaved trees *(Boscia senegalensis, Salvadora persica)*. *Leptadenia pyrotechnica* and *Capparis decidua* are representatives of the switch tree type, and *Citrullus colocynthis* and *Cucumis melo* of malacophyllous geophytes. The predominance of trees and big shrubs in the physiognomical aspect becomes evident from this enumeration.

The dimension quotients indicate the same relations as in the North Sahara: the evergreen broadleaved sclerophyllous trees are spaced between the malacophyllous and the switch trees. However, comparing North and South Saharan constitution types quantitatively, the surface development is smaller in the South, for the special types as well as on average (0.27 versus 0.34). On the contrary, the sclerophyllous character and the degree of succulence are more pronounced (1.54 versus 1.00 and 2.77 versus 2.45). Thus, the structural change follows the increasing environmental aridity and is compatible with the assumption of a higher aridity in the South Sahara.

The continuity of the dimensional quotients stands in sharp contrast to the discontinuity of the physiological water-photosynthesis syndrome. What has been shown for the individual case of the malacophyllous type (Figs. 3, 4 and Tables 1 and 2), now turns out to be of general validity (see Table 3); not only the water turnover, but also the photosynthetic production especially is very much higher in the Southern Sahara. When related to surface area, the average CO_2 uptake there is 3 times higher (21.9 versus 6.8 mg CO_2

$dm^{-2} d^{-1}$), and when related to dry weight, it is 2.3 times higher (21.5 versus 8.9 mg CO_2 g^{-1} d^{-1}). The production, as well as the profitableness increase drastically in direction north to south, in opposition to the environmental conditions. This manifests the constitutional contrariety of the two vegetations, different in their floristic origin.

IV. Holarctic and Palaeotropic Constitution Types

1. Morphological Structure and Water-Photosynthesis Syndrome

The investigation will now be widened by the additional consideration of the holarctic and the palaeotropic floral elements. For this purpose Table 4 gives some available, although fragmentary, material. For the Central European holarctic region a compilation of data is found in Stocker (1974, p. 516 and following pages). As for the African palaeotropic region, there is a lack of reliable information about the savannas. Müller and Nielsen (1965) provide material for the rain forest of the Ivory Coast. Some additional material for rain forests can be obtained from Odum et al. (1963) and Stocker (1935a, b). In Table 4 only mean values are given for the specific type of malacophyllous trees and shrubs; the numbers for the total of the vegetation are problematic.

For the leaf dimension quotients Table 4 demonstrates that, with increasing aridity, the surface development reduces steadily from Central Europe towards the Northern and Southern Sahara. From there on, it increases again, reaching almost Central European values in the rain forest region, for which a division into deciduous and evergreen trees could not be accomplished. The sequence in direction Central Europe → North Sahara → South Sahara → Sahel → Ivory Coast is for the malacophyllous trees and shrubs 1.10 → 0.70 → 0.47 → 0.36 → 1.03 dm^2 g^{-1}; the same arrangement is valid for the total of the vegetation. The sclerophyllous character and the degree of succulence show an inverse trend, with maxima in the southern Sahara and in the Sahel.

The decisive influence of aridity on these morphological sequences is especially manifested in the case of the salt shrub *Nitraria r.*, which, at the Dead Sea, changes leaves twice a year for the rainy and dry seasons (Shmueli, 1948). The leaf dimension quotients exhibit in this species the following changes from the rainy towards the dry season: the surface development decreases from 0.26 to 0.12, the sclerophyllous character increases from 0.94 to 1.55, as also the degree of succulence from 3.02 to 6.90. It is obvious that, from an ecological point of view, these changes are understandable as an adaptation to the increasing water stress during the dry season, which in this region is even more intensified by the high salt content of the soil (Stocker, 1960, see also this volume Parts 2 and 3).

The physiological behavior, however, manifests no constant parallelism to the degree of aridity, but there is a discrepancy between the Northern and the Southern Sahara, in transpiration as well as in photosynthesis.

The relative transpiration (Table 4) decreases from Central Europe (27%) towards the North Sahara (22%), then, however, in spite of the further increasing

Table 4. Constitution types and floristic regions

Constitution type	Floristic region	Dimension quotients				Transpiration related to			Photosynthesis related to		Efficiency of the Transpiration
		Surface developm. $\frac{dm^2}{g}$	Scleroph. character $\frac{g}{dm^2}$	Degree of succulence $\frac{g}{dm^2}$	Relative %	Surface $\frac{g}{dm^2\,d}$		dr. wt. $\frac{g}{g\,d}$	Surface $\frac{mg\,CO_2}{dm^2\,d}$	dr. wt. $\frac{mg\,CO_2}{g\,d}$	$\frac{mg\,CO_2}{g\,H_2O}$
Malacophyllous trees and shrubs	Central Europe	1.10	0.39	0.56	27	8.0		23.3	44.6	124.7	5.6
	Northern Sahara	0.70	0.50	1.05	22	19.5		40.5	10.5	21.4	0.5
	Southern Sahara	0.47	0.69	1.96	29	28.8		46.1	24.5	45.0	0.9
	Sahel	0.36	0.52	2.69	54	21.4		40.5	26.3	48.7	1.2
	Ivory Coast (Rain forest)	1.03	0.35	0.69	26	7.1		17.2	33.1	94.6	—
Total vegetation	Central Europe	0.77	0.59	0.95	32	7.9		20.2	33.9	67.3	4.3
	Northern Sahara	0.34	1.00	2.45	25	18.9		29.8	6.8	8.9	0.4
	Southern Sahara	0.31	1.46	2.34	27	35.7		32.3	22.6	21.2	0.6

aridity, it obtains in the Southern Sahara even higher values than in Central Europe (29%). The absolute rate of transpiration per leaf surface increases from Central Europe (8.0 g dm^{-2} d^{-1}) towards the North Sahara (19.5 g dm^{-2} d^{-1}), caused by the fact that the rapid increase of the saturation deficit of the air can not be counterbalanced by a decrease of relative transpiration. The discrepancy between the North and the South Sahara is manifested by a further strong rise of the transpiration rate (28.8 g dm^{-2} d^{-1}), enabled by a better water supply. The photosynthesis also produces conspicuous changes. After decrease in the North Sahara to one quarter of that of the Central European plants (10.5 versus 44.6 mg CO$_2$ dm^{-2} d^{-1}), net photosynthesis reaches 2.5 times higher values in the South Sahara (24.5 mg CO$_2$ dm^{-2} h^{-1}). The series South Sahara → Sahel → Ivory Coast (Table 4) shows, on the whole, the same relations, but of course in the inverse direction. In this aspect the data of Central Europe and of Ivory Coast become similar. This agrees with the presence of two great forest centres in the temperate and in the tropic climates.

The values in Table 4 which are based on dry weight can serve as indicators of profitableness. For the water relations these values have mainly negative consequences because in the calculation of the profitableness the water consumption is predominantly on the side of the expenses. Decisive for the calculation of profitableness are the values of the net photosynthesis per dry weight (124.7 → 21.4 → 45.0 → 48.7 → 94.6 mg CO$_2$ g^{-1} d^{-1}). These exhibit a rapid decrease from the Central European region towards the North Sahara before they rise again significantly in the South Sahara. This leads to the surprising result that the desert plants have a much more unprofitable production than the Central European deciduous forest and also the tropical rain forest plants. Again, the discrepancy is evident between the holarctic vegetation of the Northern Sahara and the palaeotropic vegetation of the Southern Sahara, which works twice as profitably.

A further aspect illustrated in Table 4 is the so-called 'efficiency of transpiration' (Maximov, 1929), expressed here as the quotient of the daily sums of net photosynthesis and transpiration (daily P/T ratio, see also Bierhuizen, this volume Part 6:C). It is again surprising that the desert plants utilize their water very ineffectively. Compared with the Central European deciduous forest vegetation the efficiency of water use of the desert vegetation is 11 times lower in the Northern (0.5 versus 5.6), and 6 times lower in the Southern Sahara (0.9 versus 5.6). This illustrates the critical situation of the desert plants in consequence of the bad water potential and the better situation of the palaeotropic South Saharan vegetation because of its higher drought resistance.

2. Constitution Types and Vegetation Boundaries

There is a 'balance of power' in which, by competition, holarctic and palaeotropic constitution types determine the boundaries of the Saharan vegetation. In this 'struggle for life', drought resistance and freezing tolerance play a decisive role. The palaeotropic African savanna plants, from which the South Saharan vegetation is descended, are adapted to the aridity of the native dry periods, but they are never confronted with cold winters. Thus the achievement of a cold tolerance was dispensable. On the other hand, the holarctic steppe plants, which are the ancestors of the North Saharan vegetation, need a freezing tolerance as well as a drought resistance. Considering that the incorporation of a new stress resistance into the complicated mechanism of the protoplasm may be delicate, it is easily conceivable that the simultaneous adjustment of two different resistances is complicated to such a degree that only one of the two devices can be realized in a perfect manner. This might be a possible reason for the more efficient drought resistance of savanna and South Saharan plant species; perhaps the progress of the C_4-research may bring more exact conceptions.

On the basis of this hypothesis the Saharan vegetation boundaries may be interpreted in such a manner: the South Sahara is a region without winter cold, so that only the higher drought resistance of the palaeotropic vegetation is decisive, whereas in the North Sahara the additional posession of a cold tolerance is indispensable. The combat zones are especially conspicuous in the Sinai, the Negev, and in the depression of the Dead Sea, where a strip of South Saharo-Sudanian vegetation advances far into the north (Zohary, 1973).

Regarding the different physiognomy of the North and the South Saharan vegetation, the following explanation seems probable: in the North Sahara, with its lower aridity, the drought resistance of the holarctic vegetation is sufficient to allow a diffusely spread vegetation, but not to render possible tree growth. In the South Sahara, the higher aridity compels the palaeotropic vegetation, in spite of its higher drought resistance, to contract in the wadis, where tree growth is possible because of the special constitution types supplying this vegetation.

V. Conclusions

The results of this broad survey confirm the principle, postulated by Goebel (1928, p. 43): 'The variety of organic structures is greater than the variety of environmental conditions', which holds true also for functional behavior (Stocker, 1931). From an ecological point of view, a 'balance of power' among the plant species is accomplished under the specific conditions of each particular environment, including the biotic factor of competition. This global problem of 'balance of power' has been examined here for the special case of a desert vegetation with special regard to the water-photosynthesis syndrome.

For the plant, the genetically fixed and limited constitution type is decisive. In the Sahara two types, the holarctic and the palaeotropic, are in competition. Drought resistance is a property of both, although in the palaeotropic type to a much higher degree than in the holarctic one, whereas freezing tolerance is developed only in the holarctic type, and is lacking in the palaeotropic one. The higher aridity in the South Sahara, and the winter cold in the North Sahara, determine the distribution areas of the two constitution types in a 'balance of power'.

In regard to the environmental conditions, the superficially monotonous looking desert landscape is composed of numerous specific biotopes, ranging from widespread landscape forms such as hamada, erg, reg, and wadi, down to a multiplicity of microhabitats (Kraus, 1911), so-called 'niches', to which great importance is also attached in the field of the zoological ecology (Fittkau, 1973). Examples for such niche plants (Stocker, 1974) are, for instance, the succulent salt-excreting *Limoniastrum feei*, and the hemispherical cushion shrub *Anabasis aretioides*, endemic in the Algerian-Maroccan desert (Killian, 1939).

In the Saharan deserts the multiplicity of environmental conditions is matched by the multiplicity of structural and functional constitutions.

References

Acocks, J. P. H.: Veld types of South Africa. Botan. Surv. South Africa, Memoir **28** (1953).
Barry, J.-P., Celles, J.-C.: Le problème des divisions bioclimatiques et floristiques au Sahara Algérien. Naturalia monspeliensia, sér. Botan. **23–24**, 5–48 (1972–73).
Evenari, M., Shanan, L., Tadmor, N.: The Negev, the challenge of a desert. Cambridge, Mass.: Harvard Univ. Press 1971.
Fittkau, E. J.: Artmannigfaltigkeit amazonischer Lebensräume aus ökologischer Sicht. Amazonia **4**, 321–340 (1973).
Goebel, K.: Organographie der Pflanzen. Bd. I. Jena: G. Fischer 1928.
Keay, R. W. J. (ed.): Vegetation map of Africa. Oxford: Univ. Press 1959.
Killian, Ch.: *Anabasis aretioides* Coss. et Moq., endémic du Sud Oranais. Bull. Soc. Hist. Nat. Afr. Nord (Alger) **30**, 422–436 (1939).
Knapp, R.: Die Vegetation von Nord- und Mittelamerika. Jena: G. Fischer 1965.
Kraus, G.: Boden und Klima auf kleinstem Raum. Jena: G. Fischer 1911.
Lange, O. L.: Untersuchungen über Wärmehaushalt und Hitzeresistenz mauretanischer Wüsten- und Savannenpflanzen. Flora **147**, 595–651 (1959).
Lange, O. L.: Über die Beziehungen zwischen Wasser- und Wärmehaushalt von Wüstenpflanzen. Veröff. Geobot. Inst. Rübel (Zürich) **37**, 155–168 (1962).
Maximov, N. A.: The plant in relation to water. London: Allen and Unwin 1929.
McGinnies, W. G.: The United States and Canada. In: Plant Ecology. Arid Zone Research (Paris UNESCO) **VI**, 250–301 (1955).
Monod, Th.: Modes "contracté" et "diffus" de la végétation Saharienne. In: Biology of deserts (ed. J. L. Cloudsley-Thompson), pp. 35–44. London: Inst. Biol. 1954.
Müller, D., Nielsen, J.: Production brute, pertes par respiration et production nette dans la forêt ombrophile tropicale. Forstl. Forsøgsvaesen Danmark **29**, 69–160 (1965).
Odum, H., Copeland, B., Brown, R.: Direct and optical assay of leaf mass of the lower montane rain forest of Puerto Rico. Proc. Nat. Acad. Sci. **49**, 429–434 (1963).
Quézel, P.: La végétation du Sahara. Stuttgart: G. Fischer 1965.
Schulze, E.-D.: Der CO_2-Gaswechsel der Buche (*Fagus silvatica* L.) in Abhängigkeit von den Klimafaktoren im Freiland. Flora **159**, 177–232 (1970).
Shmueli, E.: The water balance of some plants of the Dead Sea salins. Palestine J. Botany Jerusalem Ser. **4**, 117–143 (1948).
Stocker, O.: Der Wasserhaushalt ägyptischer Wüsten- und Salzpflanzen. Botan. Abhandl. (ed. K. Goebel) **13.** Jena: G. Fischer 1928.
Stocker, O.: Eine Feldmethode zur Bestimmung der momentanen Transpirations- und Evaporationsgröße. I und II. Ber. Deut. Botan. Ges. **47**, 126–136 (1929).
Stocker, O.: Transpiration und Wasserhaushalt in verschiedenen Klimazonen. I. Untersuchungen an der arktischen Baumgrenze in Schwedisch-Lappland. Jahrb. wiss. Botanik **75**, 494–549 (1931).
Stocker, O.: Transpiration und Wasserhaushalt in verschiedenen Klimazonen. III. Ein Beitrag zur Transpirationsgröße im javanischen Regenwald. Jahrb. wiss. Botanik **81**, 464–496 (1935a).
Stocker, O.: Assimilation und Atmung westjavanischer Tropenbäume. Planta (Berl.) **24**, 402–445 (1935b).

Stocker, O.: Meßmethoden der Transpiration. – Die Abhängigkeit der Transpiration von den Umweltfaktoren. In: Handbuch der Pflanzenphysiologie (ed. W. Ruhland), Vol. 3, S. 293–311 und 436–488. Berlin-Göttingen-Heidelberg: Springer 1956a.

Stocker, O.: Motorisierte Botanik. Umschau (Frankfurt) **1956**, 71–74 (1956b).

Stocker, O.: Die Dürreresistenz. In: Handbuch der Pflanzenphysiologie (ed. W. Ruhland), Vol. 3, S. 696–741. Berlin-Göttingen-Heidelberg: Springer 1956c.

Stocker, O.: Physiological and morphological changes in plants due to water deficiency. In: Plant-water relationships in arid and semi-arid conditions. Arid Zone Research **XV**, 63–104. Paris (UNESCO) 1960.

Stocker, O.: Der Wasser- und Photosynthesehaushalt von Wüstenpflanzen der mauretanischen Sahara. I–III. Flora **159**, 539–572 (1970); **160**, 445–494 (1971); **161**, 46–110 (1972).

Stocker, O.: Der Wasser- und Photosynthesehaushalt von Wüstenpflanzen der südalgerischen Sahara. I–III. Flora **163**, 46–88, 89–142, 480–529 (1974).

Stocker, O., Vieweg, G. H.: Die Darmstädter Apparatur zur Momentanmessung der Photosynthese unter ökologischen Bedingungen. Ber. Deut. Botan. Ges. **73**, 198–208 (1960).

UNESCO-FAO: Carte bioclimatique de la zone méditerranéenne. Recherches sur la zone aride. XXI. Paris 1963.

UNESCO-FAO: Carte de la végétation de la région méditeranéenne. Recherches sur la zone aride. XXX. Paris 1969.

Zohary, M.: On the geobotanical structure of Iran. Bull. Res. Council Israel, Sect. D, Bot. **11 D**, Suppl. (1963).

Zohary, M.: Geobotanical foundations of the Middle East. Stuttgart-Amsterdam: G. Fischer, Swets and Zeitlinger, 1973.

Index of Plant Species

Abies 370
— lasiocarpa 489
Abutilon muticum 515
Acacia 376, 508
— cyanophylla 433
— harpophylla 47, 370, 375
— raddiana 511–515
Acer 107
— monspessulanum 47
Achillea lanulosa 146
Adenium 508
Adenostoma fasciculatum 493, 498, 501
Aegialitis annulata 48
Aegiceras majus 350
Aeluropus litoralis 348
Agrostis spicatum 66
alfalfa, see Medicago sativa
algae 249, 309, 347
—, marine 356
Allium cepa 232, 373
Alnus 146
Amaranthus cruentus 427
— edulis 342, 344
Ambrosia trifida 49
Amelanchier ovalis 47
Anabasis aretioides 515, 520
Anacystis 356
— nidulans 347
Andropogon 337
Anthemis melaleuca 414
apricot, see Prunus armeniaca
Aristida 337
Artemisia 376
— herba-alba 47, 52, 441, 442, 444–447, 449
— sieberi 441, 442
Arundinella hirta 341
ash, see Fraxinus
Asphodelus microcarpus 446
Asplenium ruta-muraria 249
Astrebla 371, 380
— elymoides 378
— lappacea 47, 369, 370, 375
Atriplex 308, 354, 364, 370
— confertifolia 70
— hastata 370, 375, 378

— hymenelytra 134
— patula 379
— polycarpa 47, 52
— rosea 379
— spongiosa 378
Avicennia marina 48

Balanites 508
barley, see Hordeum
basswood, see Tilia
bean, see Phaseolus
beech, see Fagus
Beta vulgaris 49, 236, 370, 373, 497
Betula 375, 376
— pendula 48
Boscia senegalensis 515
Bouteloua 337
— gracilis 377, 435
Brassica napus 376
broad bean, see Vicia faba
Bromeliaceae 309, 310
Bromus tectorum 66
brown algae 350
Buchloe 337
Buxus sempervirens 47

Calluna vulgaris 474
Calopogonium mucunoides 368
Calotropis procera 515
Caltha infraloba 346
Camellia sinensis 48
Capparis decidua 515
— spinosa 513, 515
Capsicum 376
Caralluma negevensis 320
Cardaria draba 49
Carex pachystylis 443
Carya 107
Cassia aschrek 509–511, 515
cedar 194
Celosia argentea 427
cembran pine, see Pinus cembra
Cenchrus 337
— ciliaris 368, 375, 376
Centaurea iberica 414
— pungens 513

Ceratocystis ulmi 54
Cercidium microphyllum 47, 52
Ceriops tagal 48
Ceterach officinarum 249
Cetraria nivalis 453
— sepincola 453
Chamaegigas intrepidus 249
Chilipsis 370
Chlorella 356
— pyrenoidosa 347
— vulgaris 348
Chloris 337, 348, 367
Chrozophora brocchiana 515
Citrullus colocynthis 143–145, 513, 515
Citrus 50, 54, 175, 427
— aurantium 150–157
— sinensis 178, 179, 231
Cladonia 454, 463
— alpestris 454
— ecmocyna 453, 458
— mitis 453
— rangiferina 453
— subtenuis 453
Clostridium 11
Coffea arabica 226, 237, 427
coffee, see Coffea
Coleus 317
Colliguaya odorifera 493, 497–501
Cornus 376
— mas 47
cotton, see Gossypium
cucumber, see Cucumis sativus
Cucumis melo 515
— sativus 429
Cyclotella meneghiniana 247
Cynodon dactylon 367

Dactylis glomerata 93
Danthonia caespitosa 379
Desmodium uncinatum 375
Digitaria 337
— sanguinalis 367
Distichlis spicata 348, 436
Douglas fir, see Pseudotsuga menziesii
Dunaliella parva 247

Echinochloa 337
— crus-galli 367
Echinops spinosus 513
Elaeis guineensis 220
Eleusine 337
elm, see Ulmus
Encelia 370
Enteromorpha tubulosa 351
Eragrostis 337
Eriogonum fasciculatum 47
Erucaria boveana 414
Eucalyptus 376

Euchlaena mexicana 337
Euphorbia 508
Eurotia 370
Evernia prunastri 453
Excoecaria agallocha 48

Fagus 370
— sylvatica 101–129, 511–513
Fouquieria splendens 47, 52
Frankenia thymoides 515
Franseria deltoides 47, 52
Fraxinus 107
Froelichia 354
fungi 218, 232

Glycine max 370, 373, 375, 435
Gossypium 70, 74, 244, 370, 373, 375, 376, 427
— anomalum 208–221
— australe 209
— hirsutum 208–221
— raimondii 209
— thurberi 209, 212
Gymnocarpos fruticosum 446, 447

Hammada scoparia 47, 176, 441–445, 447–449
Helianthemum kahiricum 446
— ventosum 446
Helianthus 226, 244, 370, 373, 376
— annuus 176, 497
Hemizonia virgata 146
hemlock, see Tsuga
Heteromeles arbutifolia 493, 497, 498, 501
hickory, see Carya
Hippocrepis comosa 47
Hordeum 86–99
— murinum 414
— vulgare 49, 354, 375, 376
Hypogymnia physodes 453, 458

Impatiens noli-tangere 149
Iva frutescens 435

Juniper, see Juniperus
Juniperus 196
— californica 47
— oxycedrus 47
— phoenicea 47

Kageneckia oblonga 493, 497–501
Kalanchoë blossfeldiana 324
— daigremontiana 317, 331
Krameria grayi 47

Lactuca 146
— sativa 237
larch, see Larix
Larix 110
— decidua 474

Larrea 370
— divaricata 47, 52
Launaea arborescens 513
Laurus nobilis 149
Lavandula latifolia 47
Lecanora melanophthalma 453
Lemna minor 349, 352, 353
Leptadenia pyrotechnica 515
Leptochloa 337
lettuce, see Lactuca sativa
lichens 249, 390, 452–468
Ligustrum 376
Limoniastrum feei 515, 520
Liriodendron tulipifera 80, 433
Lithraea caustica 493, 497–501
loblolly pine, see Pinus taeda
Lolium 376
— multiflorum 127
— perenne 49, 69, 93, 373, 375
— temulentum 370
Lotus 376
Lumnizera littorea 48
Lupinus 376
Lycopersicon 233, 375
— esculentum 92, 153, 370, 376, 429

Macroptilium atropurpureum 371, 375, 376
— purpureum 375
maize, see Zea mays
Malus 205
— domestica 48
maple, see Acer
Medicago polymorpha 414
— sativa 127, 134, 427, 428
Mesembryanthemum crystallinum 309, 324–332
millet, see Panicum miliaceum
Mimosa pudica 30
Mimulus cardinalis 145, 146
— lewisii 145, 146
Mollugo verticillata 354
mosses 217
mountain pine, see Pinus mugo
Muhlenbergia 337
Myrothamnus flabellifolia 249

Nephroma arcticum 453
Neuropogon acromelanus 453
Nicotiana 226, 227, 233, 375, 376
— tabacum 49
Nitraria retusa 515, 516
Noaea mucronata 446
Notholaena maranthae 249

Ochromonas malhamensis 247
Olea 370
— europaea 497

Opuntia 52
— basilaris 47, 319
Osbornia octodonta 48

Panicum 337
— maximum 369–373, 375, 376
— miliaceum 377, 427
Parmelia 456, 463
— borreri 453
— caperata 453, 454, 458–462, 465
— fuliginosa 453
— olivacea 453
— sulcata 453, 459–461
Paspalum 337, 367
Peganum harmala 443, 513
Peltigera polydactyla 453, 457, 458
Pennisetum 337, 347, 354
— typhoides 342, 348, 349, 351, 352, 375, 376
Phalaris minor 414
Phaseolus 227, 228, 231, 373, 376
— vulgaris 49, 236, 370, 375
Phleum pratense 93
Phoenix sylvestris 48
Phragmites communis 49
Phytophthora infestans 213
Picea 375, 376
— abies 48, 101–129, 474–489
— engelmannii 489
— sitchensis 48, 196
pine, see Pinus
Pinus 107, 193, 194, 196, 370, 375, 376
— aristata 489
— cembra 474–488
— contorta 48
— — var. murrayana 198
— echinata 80
— flexilis 489
— leucodermis 489
— mugo 474, 488
— peuce 489
— ponderosa 196
— resinosa 48, 198
— sylvestris 48, 149
— taeda 371
Pisum sativum 93, 272–277
Platymonas subcordiformis 247
Poa pratensis 146
ponderosa pine, see Pinus ponderosa
Populus 375, 376
Portulaca oleracea 346, 347, 354
Prosopis glandulosa 47
— juliflora 47, 52
Prunus armeniaca 174–184
— ilicifolia 493, 497, 498, 501
— serotina 48
Pseudoperonospora cubensis 213
Pseudotsuga menziesii 48, 110, 189–200

Pyrus 370
— amygdaliformis 47
— communis 48

Quercus 370
— agrifolia 47
— coccifera 47
— douglasii 47
— ilex 497
— prinus 48
— robur 434
— rubra 48
— wislizenii 146
Quillaya saponaria 493, 497–501

Ramalina farinacea 453
— maciformis 453, 455, 458
Ramonda nathaliae 249
Reaumuria hirtella 370
— negevensis 47, 447
Reboudia pinnata 414
red clover, see Trifolium pratense
redwood, see Sequoia sempervirens
Retama retam 513
Rhanterium adpressum 513
Rhus ovata 493, 497, 498, 501
— tripartita 513, 515
Robinia 375
Rosmarinus officinalis 47
Rumex pulcher 229

Saccharomyces cerevisiae 12
Saccharum 370
— officinarum 336, 337
Salvadora 508
— persica 515
Saxifraga 146
Scirpus olneyi 435, 436
Sempervivum 310
Sequoia sempervirens 194
Sequoiadendron giganteum 48
Sesamum indicum 176–181
Setaria 337, 367
— anceps 371, 375, 376
— italica 377
Simmondsia chinensis 47, 52
sitka spruce, see Picea sitchensis
Solanum lycopersicum, see Lycopersicon esculentum
— tuberosum 49, 373
Sonneratia alba 48
Sorghastrum 337
Sorghum 206, 213, 294–302, 337, 347, 354, 371, 374, 377, 427, 428
— bicolor 49, 369, 370, 375, 376, 380, 427, 428
— halepense 367
— sudanense 377

soybean, see Glycine max
Spartina patens 436
Spiraea 146
Sporobolus 337
spruce, see Picea
Stipa capensis 414
Streptococcus faecalis 10, 11
Stylosanthes humilis 375
sugar beet, see Beta vulgaris
sugarcane, see Saccharum officinarum
sunflower, see Helianthus

Taeniatherum asperum 66
Taxus baccata 48
Themeda australis 367
Thyridolepis mitchelliana 378
Tidestromia oblongifolia 346, 355
Tilia 107
— cordata 434
tobacco, see Nicotiana tabacum
tomato, see Lycopersicon esculentum
Tortula ruralis 248
Trifolium pratense 373
Trigonella arabica 414
Triodia 355, 374
— basedowii 376
Triticum 69, 71, 72, 86–99, 349, 375, 405, 406, 427, 435
— aestivum 49
— durum 49
Tsuga 107
— canadensis 48

Ulmus 54, 107
Ulva lactuca 350
Urochloa 337

Veratrum 146
Vicia faba 49, 71, 92
Vigna luteola 176
— sinensis 375
Vitis 375
— vinifera 48, 233

Welwitschia mirabilis 309, 310
wheat, see Triticum

Xanthium strumarium 434
Xanthoria aureola 458

Zea mays 12, 13, 49, 73, 79, 86–99, 212, 230, 236, 237, 292–301, 336–339, 343, 369–380, 427
Zilla macroptera 513
Ziziphus lotus 509–513
Zoysia 337
Zygophyllum album 515
— dumosum 47, 52, 441, 442, 444–449
— fontanesii 142–145

Subject Index

ABA, C_4 pathway 356
— —, regulation 351
—, CAM regulation 330
—, chlorophyll degradation 235
—, effect on CK level 227
—, guard cells 234
—, ion fluxes 233
— level, increase 230
— —, kinetin 231
— —, salinity 230
— —, water stress 237
— —, wilting 231
—, malate dehydrogenase, C_4 plants 350
—, PEP carboxylase, C_4 plants 349
—, RuDP carboxylase, C_4 plants 349
—, stomata 179, 181, 233–235
—, transpiration 233, 234
abscisic acid, see ABA
acclimation 390
—, stomatal response to humidity 175
— to dry soil, leaf resistance 173
acid hydrolases, drought 208
— phosphatase 208, 209
— —, solubilization 220
— ribonuclease 208
amino acids, semipolar, toxic effect 256
amylases 208
annual grasses, root extension 66
arid regions, productivity 392, 418, 419, 440, 498, 509
— — —, available water 419
aridity 509
—, leaf morphology 516
arido-active plants 389
arido-passive plants 389
aspartic acid, C_4 pathway 326
assimilate partition, water stress 287
auxin, stomatal response 167

Biomass, desert, annual fluctuation 441
— —, annuals 443, 444
— —, perennials 443
— —, root 444
—, deserts 440, 441
—, vegetation formations 440

bladder cells, malate fluctuations 330
boundary layer, air 138
— —, aqueous 6
— — —, viscosity 8
— — resistance, T/P ratio 424
bulk water, structure 6, 16
bundle sheath cells 326, 337
— — —, drought resistance 213

C_3-C_4 intermediates 354
C_4 grasses 355, 364
— —, geographical distribution 310
— metabolism 308
— —, ecological evidence 310
— pathway, algae 347
— —, brown algae 350
— —, characteristics 335
— plants, RuDP, water stress 245, 246
calcium-bridge hypothesis 277
CAM 308
—, characteristics 313
—, ecological evidence 310
—, guard cells 160–167
—, induction 328
— plants, climate relation 310
— —, functioning of stomata 166
—, shift from C_3 metabolism 324
canopy drip, forest 101, 105–110
capillary conductivity in soil 65
— forces 120
carbohydrates, freezing tolerance 254
carbon balance, sclerophylls 501
— isotope discrimination 328
— — —, definition 345
carboxylating enzymes, water stress 244
carboxylation relations, C_4 plants 338
Casparian strip 87
cavitation in vessels 51
cell elongation 27
— walls, pressure relations 26
chemical equilibrium, effect of solutes 28
— —, pressure effect 28
— potential of water 21
— reaction rates 29
chilling tolerance, C_4 plants 367

chlorophyll degradation 235
chloroplasts, acid phosphatase activity 212
—, ultrastructure, C_4 plants 341–343
— —, osmotic stress 216
CK, CAM regulation 330
— concentration, ABA treatment 228
—, drought 227
— levels, mannitol stress 229
—, osmotica 226
—, salinity 226
—, stomata 235
—, xylem exudates 226
cluster model, water structure 7
CO_2 assimilation, water stress 284
— compensation point, C_4 plants 344
— exchange, CAM plants 316, 317
— internal, stomatal response, see stom. resp.
cohesion theory 32
cold resistance, C_4 grasses 365
colligative protection, definition 259
compartmentation, hydrolytic enzymes 211
conductance, leaf, predicted 199
— —, water potential 196
—, root, radial pathway 90
— to CO_2, mesophyll 174
conductivity, see also resistance, permeability
—, soil 112
convergent evolution, mediterranean climate 492
cooperative photosynthesis, C_4 plants 341
Crassulacean acid metabolism, see CAM
crop yield, C_4 syndrome 378
cryoprotective agents 259
cryoscopy 45
cuticle, pore membrane model 149
cuticular layer, altitude 482, 483
— —, length of growing season 484
— —, needle age 483
— —, photosynthetic production 486
— —, thickness, water loss 483
— membrane, diffusion of non-electrolytes 156
— —, permeability 148–158
— —, pore area 152
— — — number 157
— — — radius 154–157
— —, swelling 153
— —, water content 153
— resistance, see resistance
— transpiration 149, 160, 175, 481, 482
— —, affected by humidity 149
— —, affected by pH 149
— —, affected by swelling 150
— —, needles 480
— —, shoot maturation 485
— —, wax abrasion by snow 487
cytochrome oxidase, water stress 218
cytokinin, see CK

cytoplasma, effect of pressure 27
—, ultrastructure, osmotic stress 216
cytorrhysis 26, 27

Darcy equation 37, 112, 119, 122
dark respiration, water stress 218
deep seepage, forest 112
desert communities, annuals 444, 445
— —, biomass production 445–447
— — —, fluctuations 443
— — —, precipitation 445
— — —, runoff 445
— —, diversity 450
— —, reestablishment 449
— —, seed numbers 444
— —, stability 450
— plants, arido-active 441
— —, arido-passive 441
— —, C_4 pathway 355
— —, CAM 318
— —, constitution types 513, 516
— —, drought injury 27
— —, functional adaptation 519
— —, irrigation effect 448
— —, leaf surface 513, 515
— — — temperature 142–145
— —, life forms 513
— — —, net photosynthesis 515
— — — —, transpiration 513, 515
— —, maximal production 448
— —, morphological characteristics 510, 513
— —, osmotic potential 52
— —, partial death 447
— —, permanent biomass changes 448
— — photosynthesis 518
— — —, daily course 510–512
— — —, daily sum 513
— — —, seasonal course 511
— —, sclerophylly 513, 515
— —, stem growth 447
— —, succulence 513, 515 .
— —, transpiration, daily course 510–512
— —, water potential 46
— vegetation 507
— —, constitution types 509
— —, contracted type 508
— —, diffuse type 507
— — zonation 506
deserts, hot, C-metabolism of plants 310
—, rainfall, Sahara 508
desiccation resistance, conifers 478
— —, cormophytes 249
—, winter, needles 477
development, crop, water stress 299, 301
developmental stage, stress response 283
dewpoint hygrometry 43

Subject Index

dextrose-sucrose density gradient, enzyme distribution 211
diffusion resistance, see resistance
disjoining pressure 7, 15
— —, aqueous electrolytes 14
— —, temperature dependence 8
drought avoidance 50
— —, acid phosphatase 209
—, mechanisms of injury 27
— resistance, ABA level 230
— —, desert plants 518, 519
— tolerance 50
— —, C_4 grasses 374
— —, hormones 238
— —, values 376

Ectodesmata 148
efficiency of transpiration, desert plants 518
electric charge, cells, CAM 329
electrolyte layers, structure 13
— solutions, structure 9
endodermis 73, 89, 91
—, development 87
—, disruptions 74
energy balance, lichens 457
— —, reflectance of leaves 432
— —, scattering of radiation 432
— budget equation, leaf 138
— fixation, vegetation formations 394
enzyme activity, CAM, regulation 316
enzymes, hydrolytic, water stress 207, 208, 211
—, poikilohydric plants 249
—, structure-bound, solubilization 208
epidermis, degree of hydration 175
—, water potential 175
erosion, forest 111
ethylene production, drought 231
eutectic freezing, injury 254
evaporation, forest 113
—, winter, timberline 480
evaporative cooling 21, 141, 179
evapotranspiration, forest, simulation 123–128
—, world 397

Fick's first law 34, 152
fine structure, root cells 212
flooding tolerance 61
flowering hormones, water stress 237
fog, forest water balance 107
forest ecosystems, water exchange 101–129, 189
— limit, alpine 488
freezing injury, membrane inactivation 254
— point, water layers 7
— tolerance, desert plants 519
frictional potential 49
— —, definition 42
— —, determination 44

— —, trees 49
— resistance 196
frost drought, timberline 478, 488
— injury, sensitive sites of cells 253
β-fructofuranosidase 208
fruit yield, water stress during plant development 288
fusicoccin, stomatal response 167

α-Galactosyl glycides, osmoregulation 247
germination, kinetin, water stress 237
—, models 411
gibberellic acid, C_4 pathway, regulation 352
glycerol, cryoprotection 260
glycolate metabolism, C_4 plants 344
grain yield, irrigation 297
— —, water stress 300
gravitational potential 89
— —, soil water 112, 122
— —, definition 42
— —, determination 43
growing season, critical length, timberline 485, 486
growth, altitude 485
—, diurnal, water status 294
—, length of growing season 486
—, lichens 454, 463
—, model 412
—, rhythmic in trees 200
—, shoot maturation, altitude 485
—, temperature, C_4 pathway 347, 348
— — effect 486
—, water stress 284
guard cells, hydraulic isolation 175
— —, see also stomata
guttation 46, 90

Halophytes, chloroplasts, freezing tolerance 258
—, proline accumulation 248
Hartlaubcharakter 496
heartwood, water storage 194
heat-pulse measurement, water flow 195
heat resistance, water stress 179
— tolerance, C_4 plants 368
Hill reaction, inhibition 244
homoiohydric plants 60
hormones, C_4 pathway 351
—, water stress 225
Huber-Gradmann-van den Honert theory 34, 50, 89
humidity, lichen water content 455
—, stomatal response, see stomatal resp.
hydration cluster 15
—, cytoplasm 23
—, enzymatic activity 29
—, protein 22, 23, 25
—, protoplasm 52

hydrature 19, 23
hydraulic potential, soil water 112
hydrolabile plants 53
hydrostable plants 53
hypertonic stress, membranes 257
hysteresis, stomatal response to light 182

IAA oxidase activity, water stress 231
ice crystals, mechanical influence 254
— formation, tree stem, timberline 480
indoleacetic acid, see IAA
interception, forest 101, 104, 110
interchloroplastic bodies 217
— —, water stress 213
International Biological Program 388
irrigation threshold 429

Kinetin, ion fluxes 233
— metabolism, ABA treatment 227
— —, desiccation effects 228
— —, drought 227
—, protein synthesis 236
Kirchhoff's laws 35
"Kranz"-syndrome, definition 341
"Kranz"-type, leaf anatomy 308, 341
— — —, arid conditions 355
Krummholz 474, 488

Leaf area development, water stress 284
— elongation, diurnal course 296
— energy exchange 137–147
— expansion, water stress 372
— growth, water stress 289
— ontogeny, C_3 to C_4 shift 346
— temperature, affected by transpiration 141
— —, over-air 141–146
— —, under-air 141–146
— —, winter, timberline 480
— water status, regulation 286
lenticels 73, 80
lichens 390, 452–468
—, CO_2 exchange during rewetting 457, 458
—, general ecology 464
—, growth 454
light, effect on CAM 332
— response, C_4 grasses 366
lipases, stress activation 218
—, water stress 220
lipid membranes, permeability 150
— —, polar pathway 150
— —, pore size 151
lipophilic nonelectrolytes, permeation 276
liquid phase resistance 64
litter-interception, forest 114
lysimeter 113
lysosomal hypothesis 208
lysosomes, hydrolytic activity 212

Malate dehydrogenase 314
— fluctuations, salinity influences 327
malic acid, C_4 pathway 336
— —, CAM 314
— —, regulation 315
— enzyme, guard cells 164
mammals, body temperature 13, 14
mangrove plants, C_4 pathway 350
— —, frictional potential 49
— —, water potential 48
mannitol, osmoregulation 247
matric potential 45
— —, range 51–53
— —, soil water 112, 122
mechanical equilibria within the leaf 30
mediterranean climate, areas 494
— —, evapotranspiration 493, 495
— —, geography 492
— —, rainfall 493, 495
membrane equilibria 29
— proteins, denaturation 257
— structure, change 263
— structure, poikilohydric plants 249
membranes, conformational alteration 277
—, freezing tolerance 254
—, inactivation 256
—, leaky 38
—, permeability 26
—, protection 257
—, semipermeable 38, 39
—, toxic solutes 257
mesophyll cells, drought resistance 213
— resistance 174, 427
— —, T/P ratio 424, 428
mineral uptake, ABA level 232
— —, kinetin effect 232
mitochondria, membranes 218
—, pressure relations 26
—, ultrastructural response 213
—, ultrastructure, osmotic stress 216, 217
morphology model 413
movements of cells and organs 30

NaCl, interaction with succinate 261, 262
—, NAD-malate dehydrogenase, C_4 plants 350
—, PEP carboxylase, C_4 plants 349
—, photophosphorylation 255
—, RuDP carboxylase, C_4 plants 349
—, toxic effect 259, 262
NAD-malic enzyme species 337
NADP-malic enzyme species 337
needle temperature, winter, timberline 480
net photosynthesis, see photosynthesis
— primary productivity, see productivity
nicotinamide-adenine-dinucleotide phosphate,
    see NADP
nitrogen nutrition, C_4 pathway 351

nuclear magnetic resonance 9
nucleic acid degradation 237
— acids, freezing tolerance 254

Osmometry 45
osmoregulation 19, 27, 54, 245–247
—, CAM plants 331
osmotic cell 2
— gradients, during transpiration 90
— potential, definition 22, 23
— —, determination 44
— — difference 38
— —, diurnal pattern 297
— —, minimum values 52
— —, physiological significance 23
— —, plant development 284
— —, range 51–53
— —, timberline 476
— —, tissue of conifers 196
— pressure, regulation 247
— spectra 52
— stress, cell organells 216
— —, hydrolytic enzymes 209
oxygen, influence on C_4 plants 346
— uptake, sporangia, water stress 218

PEP 314
— carboxylase 314
— — activity, guard cells 162–167
— —, feed-back regulation 316
— — species 337
— —, water stress 244
percolation, forest soil, simulation 123–128
pericycle 87
periderm, disruptions 72
— formation 71
permanent wilting point 36, 61, 65, 69, 117
permeability, see also resistance
—, ABA effect 232
—, change, freezing tolerance 205
— constant 271
—, kinetin effect 232
—, membrane 269
— —, alteration by water stress 272, 273
— —, composition 277
— —, drought injury 275
— —, freezing 256, 257
— —, measurement 270
— —, nonelectrolytes 274
— —, plant development 273, 274
— —, sugars 278
—, roots 71
peroxisomes, ultrastructure, osmotic stress 216
phosphate, inorganic, CO_2 uptake 220
phosphoenol-pyruvate, see PEP
photophosphorylation, cyclic, inhibition 244

—, inactivation 257
—, uncoupling 254
photorespiration 245, 246
—, C_4 plants 344
—, water stress 244
— — —, C_4 plants 371
photosynthesis, adaptation, lichens 453
—, ambient CO_2 426
—, assimilate accumulation 427
—, daily course 184
— — —, desert plants 512
—, internal CO_2 426
—, lichens 458
— —, diurnal course 455
— —, maximal rates 452, 453
— —, model 464–466
— —, water content 457, 459, 460
—, light 426
—, negative daily balance, sclerophylls 500
—, ontogenetic drift, C_4 grasses 373
—, response to fatty acids 220
—, seasonal course, lichens 453
— — —, sclerophylls 492–500
—, temperature 426
—, water stress 244, 283, 511
— — —, sclerophylls 500, 501
photosystem II, bundle sheath chloroplasts 326
pit, bordered 194
plasmolysis 26
—, metabolism 27
—, salinity 39
plasmolyticum, permeability measurement 271
poikilohydric plants 60, 390, 452
— —, photosynthesis 249
— —, vacuolar content 249
Poiseuille equation 93–97, 154
polar pores, cuticular membranes 152
pollutants, entrance into leaf 160
polyethylene glycol, cellular ultrastructure 213, 216
polyribosomes 217
porometer 170
precipitation, effective 191, 411
—, forest 105–110
pressure chamber technique 43, 196
— differences across membranes 25
— effect on chemical equilibrium 28
— potential, definition 22, 24
— —, determination 45
— —, range 51–53
— —, stomatal closure 173
— relations, chemical reactions 24
— —, physiological importance 24
primary production, deserts 440
— —, relative to biomass 440, 441
— —, vegetation formations 440
— productivity, see productivity

production, dry matter, irrigation 302
productivity, available water 414
—, CO$_2$ supply 426
—, correlation models 392, 396, 400
— — —, accuracy 403, 406
— — —, evapotranspiration 399
— — —, multivariable 401, 405
—, driving variables 409
—, evaporation 400
—, evapotranspiration 401
—, fertilizer effect 405
—, growth models 412
—, humidity 401
—, models, selection of variables 410
—, natural vs. introduced species 419
—, nitrogen fertilization 416
— — supply 417, 418
—, precipitation 400, 401, 403
—, radiation 401
—, rate variables 409
—, simulation models 408
—, state-variable models 408, 409
— — —, application 418
— — —, crop development 412
— — —, initialization 412, 414
— — —, plant morphology 413
— — —, rate of dying material 412
— — —, rooting system 413
— — —, verification 414
—, statistical vs. analytical models 419
—, temperature 403
—, vegetation formations 392, 394
— — period 401
—, world 392
proline, cryoprotection 259
—, osmoregulation 247, 248
protein content, drought 211
— synthesis, CK effect 236
— —, water stress 248
proteins, cryoprotection 262, 263
—, soluble, freezing tolerance 254
proton gradient, freezing of membranes 256
protoplasm, destruction 221
—, water permeability 270, 271
psychrometry 43, 45
P/T ratio 424, 518
pulse chase method, C$_4$ plants 345

Radiation, solar, C$_4$ grasses 378
rainfall, annual variability 413
—, spacial distribution, forest 107
reflectance, biomass measurement 435
—, communities, water stress 435
—, leaf energy balance 432, 433
—, water stress 433
reflection coefficient 38, 153, 155

relative water content, see also water content
— — —, acid phosphatase 209
— — —, tissue of conifers 196–198
reproduction, water stress 288
resistance, see also permeability
—, boundary layer 137
— changes, salt-induced 40
—, cuticular 481
—, diffusive 34, 170
—, flow in plants 54
— — — root 86–99
— — — the SPAC 34
— — — trees 196
— — — vessels 93
—, internal cell walls 174
—, leaf, measuring technique 170
—, potential differences 39
—, rhizospheric 67–84
—, roots, cytoplasmic streaming 74
— —, diurnal periodicity 73
— —, flux dependence 73
— —, longitudinal 72, 91
— —, radial 72
— —, suberized 76, 78
— —, temperature dependence 73, 74
— —, unsuberized 76, 78
—, soil 36, 67–84, 95–98
—, solute gradients 38
—, stomatal 137
—, water loss, lichens 456
—, xylem 72, 73
respiration, growth 412
—, hormones, water loss 237
—, lichens 458
— —, water content 457
—, model 412
resurrection plants 365, 374
rhizospheric resistance 67–84
ribonuclease, see RNAse
ribonucleic acid, see RNA
ribosomes, ratio, water stress 248
—, water stress 217
ribulose-1,5-diphosphate, see RuDP
rime, forest water balance 109
RNA, water stress 217
RNAse activity, hormones 236
root, anatomic development 71, 86
— anatomy 80, 86
—, axial pathway 89, 92–95
— biomass, conifers 194
— cap 71
— cortex 86
— diameter 69
— distribution 67, 96
— epidermis 86
— extension 63–84
—, flooding, CK levels 230

—, free space 91
— geometry 96
— growth, plant development 429
— —, seasonal course 70
— —, soil water 429
— —, water stress 289, 290
— hairs 71, 86, 88
— length density, root resistance 291
— — —, water uptake 293
—, models 413
— observation chamber 69
—, osmotically induced flow 90
— pressure 90
— production 63
—, radial pathway 73, 89–92
— resistance, water transport 290, 291
— shrinkage 193
—, suberization 71, 72, 79, 80, 87
— system, models 411
— —, perennials 69
— —, static 69
—, vascular cylinder 87
—, water entry 81, 86, 89
— — flow 89–95
— — stress, crop 298
— zone water, forest 191
—, zones of water absorption 87
rooting density 67, 69, 70, 76, 88, 96–98
— depth, conifers 193
— volume, trees 191
RuDP carboxylase, water stress 244, 245
— — activity, guard cells 162–167

Sahara, geography 506
—, northern, vegetation 513
—, plant geography, causal interpretation 519
—, rainfall 506
—, southern, vegetation 515
—, vegetation 507
— —, life forms 515
Sahel 508
salinity 39
—, C_4 pathway 348
—, CAM 324
— —, CO_2 exchange 326
— —, malate synthesis 327
— —, transpiration 326
—, cellular ultrastructure 213
—, growth reduction 27
—, mechanism of injury 27
salt injury 29
— transport, kinetin effect 232
sapwood, water storage 193–195
sclerophylls, evolutionary strategy 502
—, palaeohistory 502
—, stress adaptations 503

sclerophylly, CO_2 uptake 498
—, drought avoidance 497
— index 496
— —, seasonal change 497
senescence, kinetin 235
—, water stress response 235
short-grass prairie species 355
snow, forest water balance 109
sodium phenylpyruvate, cryoprotection 264
— —, photophosphorylation 255
— succinate, photophosphorylation 255
soil-plant-atmosphere continuum 19, 32–41, 46, 114
soil, capillary forces 116
—, conductivity 115
— drought, plant water potential 53
—, infiltrability 111
— moisture, mediterranean climate 496
— — diffusivity 65
— — gradient, forest 107
— —, recharge 79
— — transport 410
—, pore size 122
— — volume 115
— resistance 290, 291
—, salinity, CAM plants 332
— temperature, models 411
— —, seasonal change, timberline 479
— —, under snow 478
— water, availability 21, 61, 115, 411
— — conductivity 101
— —, depletion 115–119, 411
— —, evaporative loss 411
— —, field capacity 61, 115, 118
— — flow equation 123
— — holding capacity 413
— — infiltration 115–119
— — — wave 119
— —, matric potential profiles 119–121
— — movement 64–84, 89, 115–119
— — —, horizontal 122
— — —, vertical 122
— — potential 34, 42, 65
— — —, determination 43
— — —, root zone 191
— — —, transpiration rate 35
— — —, turgor pressure 35
— —, recharge 191
— —, retention 101
— — storage 115–119
— —, unsaturated flow 123
— —, yield 429
solid-liquid interfaces 6
solute gradients 38
— potential, see osmotic potential
solutes, effect on chemical equilibrium 28

SPAC, see soil-plant-atmosphere continuum
stem-flow, forest 101, 105–110
steppe shrubs, root growth 69
stomata, ABA effect 233
—, anaerobic respiration 165
—, functions 160
—, glycolysis 165
—, hydropassive movement 162–167
—, K^+ exchange 162–166
—, midday closure 50, 181, 294, 501
—, oscillations 53, 169
—, osmotic potential 162–167
—, PEP carboxylase activity 162–167
—, respiration 165
—, RuDP carboxylase activity 162–167
—, starch conversion 165
stomatal closure, during night 198
— —, threshold water deficit 172, 196
— —, water stress 286
— movement and age 54
— resistance, diurnal course 183–185, 296
— —, measuring technique 170
— —, T/P ratio 424
— —, water potential 172–177, 369–371
— — — stress, C_4 plants 373, 375
— response, after-effect of temperature 183
— — — of water stress 183
— —, air pollutants 54
— —, biochemical mechanisms 160–167
— —, CAM plants 318
— —, crop, water stress 298
— —, DCMU, FCCP 167
— —, dynamic model 169
— — to environment, scheme 171
— —, external CO_2 179
— —, feed back systems 171–181, 234–239
— —, hormones 54, 167, 171, 181
— —, humidity 50, 172–177, 196
— — —, influence of light 175
— — —, leaf water status 175
— — —, species differences 175
— —, internal CO_2 160–167, 179–182
— — — —, C_4, C_3 species 180
— — — —, influence of humidity 180
— — — O_2 160–167
— —, leaf pressure potential 173
— — — water potential 196
— —, light 182, 183, 198
— — —, leaf age 183
— — —, threshold 182, 198
— —, photoactive opening 161–163
— —, plant water status 172–177
— —, prediction of diurnal courses 183, 184
— —, scotoactive closing 163–165
— —, steady state model 169–185
— —, temperature 177–179, 196
— —, water deficit 196

subsurface flow, forest 111
succinate, effect on membranes 261, 262
succulence, CAM 329
sucrose, osmoregulation 246
sugars, protection 259, 262
supercooling 10
surface-evaporation, forest 104
surface run-off, forest 111
swamp plants 61

Temperature adaptation, C_4 plants 367
—, lichen thallus 457
— of mammals 13, 14
— optimum, photosynthesis, C_4 plants 346
—, root growth 66
—, stomatal response 177–179, 196
— tolerance, C_4 grasses 378
tensiometer 113
thermal anomalies of water 10
— conductivity in water layers 9
thinning, effect on throughfall 110
throughfall of rain, forest 101, 105–110
thylakoid membranes, inactivation 255
thylakoids, desiccation 254
—, freezing 257
— — tolerance 254
—, swelling 213
—, toxic agents 256
timberline 474
—, alpine 487
—, altitude 473
—, causal explanation 487
—, microclimate 478
—, polar 487
—, shoot maturation 483
—, soil temperature 479
—, summer temperature 475
—, tree species 473, 489
—, water relations 476
tonoplast, destruction 213
total water potential, see also water pot.
— — —, definition 42
— — —, plant development 284
— — —, soil 122
T/P ratio 377, 427
—, agronomy 428
—, C_4 plants 374
—, CO_2 effect 425
—, definition 423, 424
—, vapour pressure deficit effect 424
translocation, lichens 454
transpiration, affecting leaf temperature 141
—, altitude 481
—, changes in water potential 53
— coefficient 419
—, conifers 189–200
—, control by drought 198–200

Subject Index

—, driving force 137, 195
—, energy budget equation 139
—, forest 104, 114
—, gas diffusion process 137
— model 412
—, night 198
—, physiological importance 21
— rate, stomatal resistance 173
— ratio, see T/P ratio
—, rhizospheric resistance 78
—, soil water movement 65, 66
— stream, resistances 116
—, total plant resistance 73
—, wind speed influence 139
transport across membranes 30
—, active 38
— of solutes 21
tree limit, alpine 488
turgidity, relative 19, 20
turgor pressure 19, 26, 35
— —, cell elongation 26
— —, chemical reactions 25
— —, definition 33
— —, diurnal course 297
— —, function of environment 33
— —, growth 296
— —, negative 52
— —, plant development 284
— —, range 51–53
— —, stomatal movement 26, 36

Ultrastructure, cellular, water stress 237

Van't Hoff's law 33
vesicles, cytoplasmatic, water stress 213
viscosity of water, anomalous 7

Water and morphological convergence 471
— — plant geography 470
— absorption 63–84
— balance, conifers 189–200
— — equation, ecosystem 104
—, chemical potential 21
— content, lichens 455
— — —, death by continuous saturation 463
— — —, measurement 456
— — —, optimum 462, 463
— —, needles, altitude 476
— —, physiological importance 20
— —, relative, tissue 19, 498
— —, seasonal course, sclerophylls 499
— —, shoots, timberline 477
— —, timberline, seasonal change 476, 477
— —, transpiration, timberline 482
water deficit, leaf 196
— —, physiological importance 20
— —, tree tissue 194

— flow, branched conduits 35, 44
— —, maximum speed in conifers 195
— —, model, tree 190
— —, nonisothermal 46
— —, roots, osmotically induced 90
— —, SPAC 34, 63
— —, transient state 37
—, ice-like structure 10
— loss, lichens 454, 456
water-photosynthesis syndrome 506
water potential (generally) 3, 19, 34, 42
— —, C_4 syndrome 379
— —, chemical reactions 22
— —, components 3, 22, 42, 89
— —, conifers 196
— —, crop, photosynthetic rate 299
— —, definition 3
— —, determination 3, 43
— —, flow of water in roots 89
— —, fluctuations 53, 54
— —, frictional component, see frict. pot.
— —, gradients, rhizospheric 68
— — —, soil 65
— — —, tree 195, 196
— —, gravitational component, see grav. pot.
— —, leaf, ABA 234
— — —, C_4 grasses 365, 369–371
— — —, CO_2 exchange 234
— — —, diurnal fluctuation 294, 296, 297
— — — elongation 295
— — —, water stress 286
— —, matric component, see matr. pot.
— —, maximum values 46
— —, minimum values 46–50
— —, osmotic component, see osmot. pot.
— —, photorespiration, C_4 plants 371
— —, physiological importance 21
— —, pre-dawn values 46, 196
— —, pressure component, see press. pot.
— —, pressure-free component 23
— —, range 46–50, 497
— —, seasonal course, sclerophylls 499, 500
— —, soil, see soil water potential
— —, stomatal resistance, C_3/C_4 plants 375
— —, terminology 3
— —, tissue of conifers 196
— —, total, definition 42
— — vs. water content, sclerophylls 497, 498
— —, wilting point, values 376
— reserve, conifers, extraction 194, 195
— status in plants 19–32, 42–58
— storage, conifers 189–200
— — —, extensible tissue reserve 193
— — —, mature wood reserve 193
— — —, recharge 193, 194
— —, lichens 455
— —, roots 193

— stress, adaptation, hormones 238, 239
— —, C_4 grasses 378
— —, CAM 328
— —, carbon metabolism 244
— —, concept, plant growth 284
— —, enzymatic activity 208
— —, hydrolytic enzymes 207, 208, 211
— —, leaf reflectance 433
— —, — temperature 434
— —, mechanisms of injury 26
— —, metabolic processes 235, 237
— —, mild 289
— —, net photosynthesis 368, 498, 509
— —, ontogenetic state of leaf 373
— —, plant development 237
— —, primary lesion 19
— —, protein synthesis 236
— —, protoplasmic structure 208
— —, radiant thermometry 434
— —, stomatal response to humidity 175
— —, structure effects 207
— —, yield 429
— structure 6–18
— supply, application efficiency 421–423
— —, root extension 66
—, thermal anomalies 10
— transport 290
— — system, trees 189
— —, temperature influence 12, 194
— uptake, affected by low temperature 194
— —, conifers 189–200

— —, lichens 454, 455
— —, models 411
— —, passive 452
— —, rhizospheric resistances, model 75–84
— —, roots 95–99, 292
— — — of cereals 87
— — —, cylindrical model 82
— — —, spherical model 82
— —, suberized roots 71, 72, 74, 76, 79, 80, 88
— —, unexplored soil 64
— —, unsuberized roots 76, 79, 80
— —, winter, timberline 479
water-use efficiency 32, 134, 175, 419, 421
— —, C_4 grasses 379
— —, CAM plants 319
— —, definition 423
— —, primary productivity models 412
wilting 30

Xeromorphism 389
—, leaf reflectance 434
xerophytes, cuticular transpiration 149
xerothermic index 508
xylem sap, solute concentration 45
— —, water potential 51

Yield, crop plants, water stress 281

Zero turgor method 44
zymogens, activation 209
—, de novo synthesis 209

Ecological Studies
Analysis and Synthesis

Editors: W.D.Billings, F.Golley, O.L.Lange, J.S.Olson

Vol. 1: **Analysis of Temperate Forest Ecosystems**
Editor: D.E. Reichle
1st corrected reprint
91 figs. XII, 304 pages. 1973

Vol. 2: **Integrated Experimental Ecology.** Methods and Results of Ecosystem Research in the German Solling Project
Editor: H. Ellenberg
1971. Out of print

Vol. 3: **The Biology of the Indian Ocean**
Editor: B. Zeitzschel in cooperation with S.A. Gerlach
286 figs. XIII, 549 pages. 1973

Vol. 4: **Physical Aspects of Soil Water and Salts in Ecosystems**
Editors: A. Hadas, D. Swartzendruber, P.E. Rijtema, M. Fuchs, B. Yaron
221 figs. 61 tables. XVI, 460 pages. 1973

Vol. 5: **Arid Zone Irrigation**
Editors: B. Yaron, E. Danfors, Y. Vaadia
181 figs. X, 434 pages. 1973

Vol. 6: K. STERN, L. ROCHE **Genetics of Forest Ecosystems**
70 figs. X, 330 pages. 1974

Vol. 7: **Mediterranean Type Ecosystems**
Origin and Structure
Editors: F. di Castri, H. A. Mooney
88 figs. XII, 405 pages. 1973

Vol. 8: **Phenology and Seasonality Modeling**
Editor: H. Lieth
120 figs. XVI, 444 pages. 1974

Vol. 9: B. SLAVÍK
Methods of Studying Plant Water Relations
181 figs. XVIII, 449 pages. 1974

Vol. 10: **Coupling of Land and Water Systems**
Editor: A.D. Hasler
95 figs. XVI, 309 pages. 1975

Vol. 11: **Tropical Ecological Systems.** Trends in Terrestrial and Aquatic Research
Editors: F.B. Golley, E. Medina
131 figs. XV, 398 pages. 1975

Vol. 12: **Perspectives of Biophysical Ecology**
Editors: D.M. Gates, R.B. Schmerl
215 figs. XIII, 609 pages. 1975

Vol. 13: **Epidemics of Plant Diseases.** Mathematical Analysis and Modeling
Editor: J. Kranz
46 figs. X, 170 pages. 1974

Vol. 14: **Primary Productivity of the Biosphere**
Editors: H. Lieth, R.H. Whittaker
67 figs. 46 tables. VIII, 339 pages. 1975

Vol. 15: **Plants in Saline Environments**
Editors: A. Poljakoff-Mayber, J. Gale.
54 figs. VII, 213 pages. 1975

Vol. 16: **Fennoscandian Tundra Ecosystems**
Part 1: **Plants and Microorganisms**
Editor: F.E. Wielgolaski
Editorial Board: P. Kallio, T. Rosswall
90 figs. 96 tables. XV, 366 pages. 1975

Vol. 17: **Fennoscandian Tundra Ecosystems**
Part 2: **Animals and Systems Analysis**
Editor: F.E. Wielgolaski
Editorial Board: P. Kallio, H. Kauri, E. Øsbye, T. Rosswall
81 figs. 97 tables. 1975

Vol. 18: **Remote Sensing for Environmental Sciences**
Editor: E. Schanda
178 figs. (7 color plates). 31 tables. XIII, 367 pages. 1976

Vol. 19: **Water and Plant Life**
Problems and Modern Approaches
Editors: O.L. Lange, L. Kappen, E.D. Schulze
178 figs. 64 tables. XX, 536 pages. 1976

Vol. 20: F.B. CHRISTIANSEN, T.M. FENCHEL
Theories of Populations in Biological Communities
68 figs. 5 tables. Approx. 160 pages. 1977

Vol. 21: V.Y. ALEXANDROV
Cells, Macromolecules and Temperature
Conformational Flexibility of Macromolecules and Ecological Adaptation. Translated from the Russian by V.A. Bernstam
74 figs. 30 tables. Approx. 420 pages. 1977

Distribution rights for the whole series for U.K., Commonwealth and the Traditional British Markets (excluding Canada):
Chapman & Hall Ltd., London

**Springer-Verlag
Berlin
Heidelberg
New York**

Residue Reviews
Residues of Pesticides and Other Contaminants in the Total Environment

Editor: Francis A. Gunther
Assistant Editor: Jane Davies Gunther
Advisory Board: F. Bar, F. Bro-Rasmussen, D.G. Crosby, S. Dormal-van den Bruel,
C.L. Dunn, H. Egan, H. Frehse, K. Fukunaga, H. Geissbühler, G.K. Kohn, H.F. Linskens,
N.N. Melnikov, R. Mestres, P. de Pietri-Tonelli, I.S. Taylor, R. Truhaut, I. Ziegler

Residue Reviews
A series providing concise, critical reviews of timely advances, philosophy, and significant areas of accomplished or needed endeavor in the total field of residues of pesticides and other foreign chemicals.

Residue Reviews
is continuing and expanding its fourteen-year concern with ecology and the total environment in the form of authoritative reviews about the residues of contaminants in water, air, and soil; the significance of their occurrence in plants, animals, and men; and their metabolism by plants, animals, and microorganisms.
Summaries in German and French.

Archives of Environmental Contamination and Toxicology

Editor: William E. Westlake, Twain Harte, California
Coordinating Board of Editors: John W. Hylin, Department of Agricultural Biochemistry, University of Hawaii, Honolulu; Francis A. Gunther, Department of Entomology, University of California, Riverside.

Archives of Environmental and Toxicology is a unified repository of important full-length articles in English describing original experimental or theoretical research work pertaining to the scientific aspects of contaminants in the environment. It provides a place for the archival publication of detailed, definitive reports of significant advances and discoveries in the fields of air-, water-, and soil-contamination and pollution, and in disciplines concerned with the introduction, presence, and effects of waste and deleterious substances in the total environment.

Bulletin of Environmental Contamination and Toxicology

Editor-in-Chief: John W. Hylin, Department of Agricultural Biochemistry, University of Hawaii, Honolulu
Editorial Coordinating Committee: Francis A. Gunther, Editor-Residue Reviews; W. Ellis Westlake, Editor-Archives of Environmental Contamination and Toxicology
Associate Editors: Francis A. Gunther, Department of Entomology, University of California, Riverside; Professor Frederick Sargent II, School of Public Health, The University of Texas at Houston, Houston, Texas; Arthur C. Stern, Department of Environmental Sciences and Engineering, School of Public Health, University of North Carolina, Chapel Hill, North Carolina

The Bulletin of Environmental Contamination and Toxicology covers one of today's most pressing problems – air, soil, water, and food contamination. All disciplines which in any way shed light on the introduction, presence, and effects of toxicants on the total environment are included. Descriptions of new methods, procedures, and techniques are sufficiently detailed so that other researchers can readily adopt them. The journal's use of the rapid photo-offset method for its publication enables the reader to have the latest research results without delay.
Subscription information and sample copies upon request.

Springer-Verlag Berlin Heidelberg New York